Elementary Statistics
Methods and Applications

Second Edition

Khairul Islam
Texas A&M University-Kingsville

Tanweer Shapla
Eastern Michigan University

CENGAGE Learning

Elementary Statistics: Methods and Applications, Second Edition

Elementary Statistics: Methods and Applications
Khairul Islam I Tanweer Shapla

ISBN: 978-1-337-04879-8

WCN: 01-100-101

Cengage Learning
20 Channel Center Street
Boston, MA 02210
USA

Cengage Learning is a leading provider of customized learning solutions with office locations around the globe, including Singapore, the United Kingdom, Australia, Mexico, Brazil, and Japan. Locate your local office at: **www.international.cengage.com/region.**

Cengage Learning products are represented in Canada by Nelson Education, Ltd.

For your lifelong learning solutions, visit **www.cengage.com/custom.**

Visit our corporate website at **www.cengage.com.**

Contents

Preface

Elementary statistics: Methods and Applications is a product of many years of dedication in teaching and learning statistics. This book aims at explaining essential concepts of statistics by simple examples. All examples are very object oriented. The focus is given to all necessary steps to make the understanding of the methods involved clear. In most cases, technology implementation of an underlying method is demonstrated using a graphing calculator and Minitab.

The main feature of the text involves the followings:

- Important key words in definitions or paragraphs are typed in **bold** and *italic*.
- Every definition is concise and demonstrated with examples.
- Examples are simple, precise and aimed at referring to main ideas and concepts.
- Too much description often obscures the main idea, so unnecessary elaboration is avoided.
- Attempts are made to implement technology for any method discussed.
- About 200 examples are incorporated throughout the text to illustrate the statistical techniques and their applications.
- Graphs, charts, and output windows for different inference methods are presented for better understanding of the underlying methods.
- For technology implementation, important steps and option selections in Minitab and TI-84 Plus calculator are provided in detail in reference to specific examples so that users can reproduce the results of their own.

It is of great expectation that students and instructors will find the book indispensable for understanding the practice of statistics they need at elementary level.

Khairul Islam

Tanweer Shapla

Fall 2015

CHAPTER 1
DATA AND VARIABLES

Introduction

In this modern world, data are everywhere now. We need to know what the data are about. How to analyze the data? What inference can be made from the analytic results of the data? Answers to such questions involve statistics. Indeed, ***Statistics*** is the science of decision making or inference using data. The process of inference generally involves collecting, analyzing and interpreting data or testing hypotheses about some predetermined questions in a study.

It is very clear that data are of central attraction to us in the process of inference. Therefore, in this chapter, we will mainly learn about data and data related phenomena before making any transition to the learning of statistical methods.

It is worth mentioning that data are plural, but often data is used in a singular sense as a mass noun, like information. We do not want to get involved in a "data is" or "data are" debate. Instead, let us adapt both "data is" and "data are" terminologies, as we often do in the conversation.

Section 1.1: Data and variables

Data consist of information and facts about the participants of a study. The participants of the study are referred to as ***subjects*** or ***individuals***. A ***subject*** or an ***individual***, used interchangeably, may refer to a person, an animal or a thing that we are actually interested in the study.

For example, if a study is about the unemployment rate by the level of the educational attainment of people in the United States, then the people of the United States who are part of the study are individuals or subjects.

We have collected some unidentifiable and insensitive characteristics of student volunteers using a Google document questionnaire. The goal was not only to collect data but also to familiarize students with the questionnaire that may generate data. A partial list of students' characteristics together with the information relating to the characteristics is presented as Example 1.1.

Example 1.1

The table below is a partial list of information about basic characteristics of several students in a class. The characteristics are Gender, Age (in years), Weight (in pounds), Smoking status (Smoking), Credit hour taking (Hour), Class standing (Standing) and Fast food restaurant preference (Restaurant).

Gender	Age	Weight	Smoking	Hour	Standing	Restaurant
Female	22	150	No	12	Senior	Burger
Female	21	120	Yes	15	Junior	Arby's
Female	19	125	No	12	Sophomore	None
Male	25	175	No	12	Senior	McDonalds
Male	21	190	No	13	Junior	Taco bell
Female	18	150	Yes	16	Freshman	Sonic

Answer the following questions:

(a) Who are the subjects in the table?
(b) What forms the data in the table?
(c) What students' characteristics are collected in the table?
(d) How many subjects are listed in the table?

Solution

(a) Students are the subjects in the table.
(b) Students' information forms the data.
(c) Students' characteristics collected are Gender, Age (in years), Weight (in pounds), Smoking status (Smoking), Class standing (Standing) and Fast food restaurant preference (Restaurant).
(d) Six subjects are listed in the table.

Variable

*A **variable*** is any characteristic of individuals in a study. A variable may take different values for different individuals.

Example 1.2

Let us consider the following characteristics of individuals:

- Monthly income
- Monthly cell phone bill
- Monthly expenditure
- Age
- BMI
- Yearly medical cost
- Blood type

These characteristics may take different values for different individuals, and hence they all are examples of variables.

Q: Can you list some variables not mentioned in above list?

Section 1.2: Types of variables

Broadly speaking, a variable can be of two basic types- a ***categorical*** or ***qualitative variable*** and a ***quantitative variable***.

Qualitative variable

A variable whose different values represent different groups is called a ***categorical*** or ***qualitative variable***. The different values of a categorical variable are generally known as ***labels***, ***classes*** or ***categories***.

Example 1.3

Followings are some examples of categorical/qualitative variables:

- **Gender** has two classes: Male and Female.
- **Color** may have three classes: Blue, Red and White.
- **Region** may have four classes: North, South, East and West.
- **Size** may have four classes: Small, Medium, Large and Extra large.

Q: What happens if we assign numerical values to represent different classes of a categorical variable?

It is okay to assign numerical values to represent different classes of a categorical variable. However, an arithmetic operation such as addition, subtraction, division and multiplication on these numeric values is meaningless.

For example, if we represent two classes of gender by 1 for male and 2 for female, then a total of 3 (1+2=3) or an average of 1.5 [(1+2)/2=1.5] of these two numbers means nothing and cannot be interpreted.

Q: How can you summarize a categorical variable?

We can summarize a categorical variable by frequency, relative frequency or percent frequency for each class of the variable using tables, graphs and charts, which we will discuss in detail in chapter 2.

Q: What are different types of a categorical variable?

A categorical variable can be of two types- a ***nominal*** variable and an ***ordinal*** variable. Different values of a ***nominal variable*** have no order or rank values associated with them. For example, region is a nominal variable since the four classes of region, namely, north, south, east and west have no order or rank values.

On the other hand, the different values of ***an ordinal*** variable have inherent order or rank values associated with them. For example, self rated health status of a patient with five classes, say, very poor (1), poor (2), good (3), very good (4) and excellent (5) is an ordinal variable. Clearly, each value is higher in order or rank than the previous one. However, the extent of difference of two successive values of an ordinal variable may be different and unknown. For example, the difference between "very good" and "good" may not be same as the difference between "excellent" and "very good".

Example 1.4

Some examples of a nominal variable are:

- **Employment status**: Employed, Unemployed.
- **Hair color**: Blonde, Brunette, Red, Black.
- **State of birth**: Indiana, Ohio, Michigan, etc.
- **Smoking status**: Yes, No.

Example 1.5

A few examples of an ordinal variable are:

- Class standing: Freshman, Sophomore, Junior, Senior.
- Rank: First, Second, Third, etc.
- Health condition: Very poor, Good, Very good, Excellent.
- Time of day: Night, Dawn, Noon, Afternoon, Evening.
- Stage of cancer: I, II, III, IV.
- Level of agreement: Strongly disagree, Disagree, Neutral, Agree, Strongly agree.
- Letter grade in a course: A, B, C, D, F.

Quantitative variable

A variable is quantitative if it takes ***numeric*** values and allows arithmetic operations among its numeric values with meaningful interpretations. The values of a quantitative variable are measured using a ***scale*** with a fixed unit of measurement between its possible values. For example, if height of a group of students is measured in inches, then one inch is the fixed unit of measurement for height.

Example 1.6

Some examples of a quantitative variable are:

- Income
- Weight
- BMI
- Amount of snowfall recorded in different cities in a given year
- Number of accidents in a major highway in a given year
- Number of donors with blood type B at a blood drive center at a given day

Types of quantitative variables

A quantitative variable can be of two different types- a ***discrete variable*** and a ***continuous variable***.

Discrete variable

A ***discrete variable*** is a count taking on integer values. All possible values of a discrete variable form a finite set. For example, the number of cars sold in a dealership at any given day is a discrete variable which may take values 0, 1, 2, etc and form a finite set of integers.

Example 1.7

Some examples of a discrete variable are:

- Number of car accidents on a given day at an intersection
- Number of credit hours each senior student is taking in a given semester
- Number of cities a student has ever visited
- Number of dependents in a family
- Number of earning members in your family
- Year at which a student was born

Continuous variable

A quantitative variable that is not discrete is called a ***continuous variable***. A continuous variable make take values on an interval of numbers.

Example 1.8

Some examples of continuous variables are:

- Body temperature
- Height
- Weight
- Body mass index

Whatever values these variables take may be observed in some interval of numbers.

Types of a continuous variable

A continuous variable can be of two different types- a ***ratio variable*** and an i***nterval variable***.

Ratio variable

A ***ratio variable*** has an inherently defined zero, and the ratio of any two values is meaningful.

Weight, time, height, distance etc. are examples of a ratio variable.

Note that a weight of zero pounds is "no weight at all" and a weight of 90 pounds is three times a weight of 30 pounds. Note that a zero value is inherently defined for weight, and a ratio of two different values of weight is meaningful.

Similarly, a job finished by 10 minutes is 2 times faster than the job finished by 20 minutes, and zero minutes mean "no time at all". Similar property applies to height and distance variables as well.

Interval variable

An ***interval variable*** has an arbitrary zero value and the difference (not the ratio) of any two values is meaningful.

Temperature is an example of an interval variable. Note that 0°F is an arbitrary value (it does not mean "the total absence of temperature"; it just indicates a very cold situation). It makes no sense to say that 90°F is three times as warm as 30 °F (i.e., the ratio is not meaningful). However, we can say that 90°F is 60°F warmer than that of 30°F (i.e., the difference is meaningful).

Section 1.3: Types of data

Depending on what variables the data are collected from, data could be

(a) Quantitative - discrete or continuous (either ratio or interval)

(b) Qualitative or categorical (either nominal or ordinal)

Quantitative data are observed values of a quantitative variable, while ***qualitative data*** are observed values of a qualitative or categorical variable.

Depending on the nature of time of data collection, data can also be divided as cross-sectional and time-series data.

Cross-sectional data

Cross-sectional data are collected at the same time point on multiple subjects.

Example 1.9

Some examples of cross-sectional data are:

- The maximum temperature from each of the 50 states of the US on December 25 in a given year
- The monthly expenditure of a sample of freshman students at a college in a given month
- The number of cars sold by each of 15 randomly selected dealers in given year
- The number of auto accidents reported in each of the 50 states of the US in a given year
- Annual income for each of 20 randomly selected households of a city in a given year
- Unemployment rates of 50 states of the US in a given year

Q: Give some examples of cross-sectional data not listed above.

Time series data

The ***time series data***, also called the ***longitudinal data***, refers to data collected on the same subject at multiple time points.

Example 1.10

Some examples of time series data are:

- The maximum temperature of a city for each month for the last three years
- Yearly unemployment rate in the USA over the last twenty years
- The number of employees of an industry by year for 10 years
- Yearly new motor vehicle sales of a given dealer from 2000 to 2014
- The number of federal hospitals in the US from 1975 to 2013
- Average monthly household income of people of a city for the last two years

Definition: Time series plot

A ***time series plot***, also called a ***runs*** **plot**, is of great importance to business men to study the trend, cyclical or seasonal variation and other changes over time of their business phenomena.

Panel data

The ***panel data***, also called the ***time series cross-sectional data,*** refers to the collection of measurements on multiple subjects over time. Therefore, the panel data represents the combination of both time series and cross-sectional data. An analysis of panel data examines changes in variables over time and differences in variables between subjects.

The crime rates of 50 states in the US from 2000 to 2014 is an example of a panel data.

Example 1.11

The table below represents uninsured people (in millions) in the USA from 2000 to 2010.

Year	Uninsured (in millions)
2000	36.6
2001	38
2002	39.8
2003	41.9
2004	41.8
2005	43
2006	45.2
2007	44.1
2008	44.8
2009	49
2010	49.9

Source: *Income, Poverty and Health Insurance Coverage in the United States: 2010 (Washington, D.C.: U.S. Census Bureau, Sept. 2011)*

What kind of data is presented in above table?

Of course, it is a time series data.

Section 1.4: Population and sample

A ***population*** is a collection of all objects or individuals for which an inference is sought for.

Example 1.12

What is the average annual expenditure of all students attending four year colleges in the USA?
In order to answer this question, we might consider annual expenditure from all students in four year colleges in the USA. Here, the collection of expenditures from all students attending four year colleges is the population.

Example 1.13

Below is a list of populations that one may be interested in and collect information to answer some predetermined unknown phenomena.

- Students who have ever attended a U.S. college
- All female students attending a university
- All faculties in the US colleges
- Patients who visited a doctor's office
- All diabetic patients treated with an insulin

Definition: Sample

Often, getting data from the entire population is not an efficient idea. For example, obtaining information from all students attending four year colleges in the USA may be expensive, time consuming or even impossible at the time point the information is sought for. How about collecting annual expenditure information from a representative part of all students attending four year colleges?

A ***sample*** is a representative part of the population from which researchers actually collect information for making an inference about the population which otherwise is unknown.

Example 1.14

Suppose a researcher is interested in determining the average speed of all cars passing through a given neighborhood on a given day. To this end, the researcher measured the speed of all cars passing through a particular point of the neighborhood on a given day. The speed of all cars collected is the population.
The speed of 40 cars selected at random from the speed of all cars passing through the neighborhood on a given day is a sample.

Example 1.15

President Obama's bailout of the auto industry in 2009 was a source of great controversy in 2012 presidential election. Do you support the government bailout plan of the auto industry? With this question in mind, suppose a congressman mailed a questionnaire on this issue to a randomly selected 1000 voters out of all voters in his district. It turned out that only 18% of questionnaires mailed were returned with information sought for.
Under this scenario, the population is
(a) the 1000 voters received the questionnaire.
(b) the 68% questionnaire returned.
(c) those voters supported the bailout plan.
(d) all voters in his district.

Types of statistics

There are two types of statistics, namely, descriptive statistics and inferential statistics.

Definition: Descriptive statistics

Descriptive statistics is any form of summary of data regarding a sample or a population in terms of tables, graphs and charts or numeric methods such as mean, median, variance, standard deviation, etc.

Example 1.16

The table below refers to the percent of health insurance coverage status by type of coverage and nativity in the U.S. in 2013:

Nativity	Not covered	some insurance	private insurance	Government health plan	Medicaid	Medicare
Native	11.2	88.8	66.0	35.3	17.5	16.0
Foreign-born	27.7	72.3	52.1	27.6	16.0	13.0
Naturalized citizen	15.9	84.1	60.3	33.8	15.0	20.9
Not a citizen	38.1	61.9	44.9	22.2	16.8	6.0

Source: *U.S. Census Bureau, Current Population Survey, 2014 Annual Social and Economic Supplement. (Numbers in thousands)*

Q: What kind of statistics is this problem about?

This problem is about descriptive statistics, because it provides a summary of percent distribution of the US people by nativity and type of insurance coverage in a tabular form. There is no indication of the use of a sample, and no inference is being made about health insurance status by nativity.

Definition: Inferential statistics

Inferential statistics makes inferences via estimation and/ or test of hypothesis about a population characteristic on the basis of a sample from the population. A sample and making of an inference are essential parts of an inferential statistics.

Example 1.17

On the basis of a ***sample*** of 43 students from a large class, it was concluded that about 40% students in the class is freshman students.
What kind of statistics is this problem about?

This problem is about inferential statistics, because it uses ***sample estimate*** to make inference of percentage of freshman students (beyond 43 students) in the class.

Section 1.5: Data sources

The main sources of data are

- census
- sample survey
- designed experiment
- observational study

Census is the complete count of the population. In other words, we use census to collect data from all individuals in the population. Generally, a census is time consuming, expensive and may be messy. Still census is often a useful method of collecting data on nationally important variables. For example, the US census takes place every ten years and collects data on many nationally important variables.

Sample survey is a very popular and convenient way of collecting data from a representative part of the population, called ***sample***.

Designed experiment

In a designed experiment, researchers use two random groups, one called the ***treatment group*** who receives a treatment (a new drug or procedure) and the other called the ***control group*** who receives nothing (a placebo or traditional treatment). Researchers then follow up two groups for an outcome or a desired response. This way of collecting data, by allowing a manipulation of a factor, is called a designed experiment.

Example 1.18

In order to determine if smoking really causes a lung problem, researchers select treatment and control groups at random from nonsmokers with similar health conditions, without a lung problem. The treatment group is allowed smoking for a certain period of time, say, 5 years, while the control group remained free of smoking for 5 years. At the end of 5 years of study period, the researchers collect data from two groups to investigate the presence of any lung problem. Note that this design allows the manipulation of the factor and hence is an example of a designed experiment.

If the incidence of lung problem is higher in treatment group than the control group, the researcher may conclude that smoking is a significant factor for the development of the lung problem.

Observational study

In an ***observational study***, researchers collect data by observing characteristics of individuals in the sample without any manipulation of factor.

Example 1.19

Suppose that a researcher is interested to know if smoking is a factor for the development of lung problem. To this end, the researcher selects a random sample and collects data to investigate the status of smoking and lung problem in the sample. Note that the researcher has done nothing to manipulate the factor in this example. Thus, this is an example of an observational study.
If it turns out that the incidence of lung problem among smokers is higher than the non-smokers, then the researcher may conclude that smoking is a significant factor for developing a lung problem.

Chapter 1: Practice problems

(1.1) A researcher, as a part of his survey, collects information about type of treatment the participants have received during the past six months along with other information such as age, gender, marital status, weight, change of medical history, etc. Which of the followings is true?

(a) age and weight are quantitative variables.
(b) gender and marital status are qualitative variables.
(c) age and marital status are categorical variable.
(d) both (a) and (b).

(1.2) The variable "change of medical history" in problem (1.1) is collected as 1 if medical history has changed and 2 if medical history remains unchanged. The change of medical history in this case is an example of a

(a) categorical variable.
(b) quantitative variable.
(c) continuous variable.
(d) discrete variable.

(1.3) A housing company uses a leaflet to provide information such as school district, distance of house from the school, size of the house (in square-feet), number of bedrooms, year the house was built, etc. for customers interested in purchasing any house. The variable school district is
(a) categorical.
(b) continuous.
(c) discrete.
(d) quantitative.

(1.4) The variable "number of bedrooms" as described in problem (1.3) is
(a) quantitative.
(b) discrete.
(c) both (a) and (b).
(d) none.

(1.5) The variable "size of the house" (in square-feet) as described in problem (1.3) is
(a) categorical.
(b) discrete.
(c) quantitative and continuous.
(d) all of the above.

(1.6) Write some examples of a quantitative variable.

(1.7) Give some examples of a discrete quantitative variable.

(1.8) Give some examples of a continuous variable.

(1.9) A survey is conducted on students taking a statistics class. Several variables are measured in the survey. Which of these variables listed below is **not** quantitative?
(a) the number of credit hours that a student takes in the semester
(b) the gender
(c) the monthly average expenditure
(d) the High School GPA

(1.10) A description of different houses on the market includes the following variables. Which of the variables is quantitative?
(a) the size of the house (in square-feet)
(b) the monthly gas bill
(c) the monthly electric bill
(d) all of the above

(1.11) A couple is looking to purchase a house. Their primary concern is on the following information: school district, if public school bus is available or not, type of heating: electric or gas, number of bedrooms, number of bathrooms etc. Which of the variables is qualitative?

(a) school district
(b) type of heating
(c) availability of school bus
(d) all of the above

(1.12) While initial sorting for mails to be sent takes place, the following information are of the concern: Zip code, State, City, Street Address, etc. Which of the following variables is categorical?

(a) State
(b) City
(c) Name of the street
(d) all of the above

(1.13) The data below represents percentage of adults diagnosed with diabetes receiving treatment in the USA, during 2007-2009.

Treatment Type	Percentage
Insulin only	12
Insulin and oral medication	14
Oral medication only	58
No medication	16

Source: *2007–2009 National Health Interview Survey*

(a) What type of variable and data are in column 1?
(b) What type of variable and data are in column 2?
(c) What kind of statistics is this problem about?
(d) Does this study relate to a sample or census? Justify your answer.

(1.14) The data below is the percentage distribution of parent-reported severity of autism spectrum disorder (ASD) by age of children in the United States in 2011–2012.

Severity of autism	Age of children	Percentage ASD
Mild	6-9	56.1
Moderate		34.1
Severe		9.7
Mild	10-13	47.8
Moderate		36.7
Severe		15.5
Mild	14-17	55.1
Moderate		29.6
Severe		15.3

Source: *National Health Statistics Reports, Number 65, March 20, 2013.*

(a) What type of variables and data are in columns 1, 2 and 3?
(b) What kind of statistics is this problem about? Justify your answer.

(1.15) The data below is the percentage distribution of persons aged 19–25 and 26–35 who were uninsured in the United States by quarters of 2011.

Age groups	Quarters	Percentage Uninsured
19-25	Q1	30.4
	Q2	27.3
	Q3	28.4
	Q4	25.2
26-35	Q1	26
	Q2	28.3
	Q3	28.4
	Q4	28.2

Source*: Early Release of Health Insurance Estimates Based on Data from the National Health Interview Survey, January–September 2012.*

(a) What type of variables and data are in column 1, 2, and 3?
(b) What kind of statistics is this problem about? Justify your answer.

(1.16) A medical doctor is asking question about the "Type of Allergy" his patients are suffering from. Three types of allergy information are obtained: Food Allergy, Skin Allergy and Respiratory Allergy.
The variable "Type of Allergy" is an example of a
(a) discrete variable
(b) continuous variable
(c) qualitative variable
(d) quantitative variable

Different values of variable "Types of Allergy" in the example are termed as:
(a) discrete data
(b) continuous data
(c) labels
(d) all of the above

(1.17) Every three months, a doctor sends survey questionnaires to a randomly selected 100 patients out of all eligible patients treated by the doctor to understand their treatment outcome. In one occasion, 78 patients sent back the survey questionnaire to the doctor with the information sought for.
The population in this problem is:
(a) 100 patients who received survey questionnaire.
(b) 78 patients who returned the survey questionnaire.
(c) all eligible patients treated by the doctor.
(d) all visited the doctor's office.

The sample in this problem is:

(a) 100 patients who received survey questionnaire.
(b) 78 patients who returned the survey questionnaire.
(c) all eligible patients treated by the doctor.
(d) all visited the doctor's office.

(1.18) How much trust and confidence do you have in our government handling of the immigration reform system? In response to a question by a political scientist, the answer obtained may be either of the following types: great deal, fair amount, not very much, not at all, no opinion. How many values can the variable take? What do you call the different values of this variable?

(1.19) What different sizes of pizza do you sell? The salesman replies "We have small, medium, large and extra large". What type is the variable "size of pizza"?

(1.20) One can order a coffee at McDonald's in the following sizes: small, medium, or large. What do we call the values for the variable "size of coffee"? The variable "size of coffee" is an example of _________________ variable and the data represented by the values of "size of coffee" is an example of _________________ data.

(1.21) Give an example of nominal data, ordinal data and ratio data.

(1.22) What kind of data does the variable "household income" of a group of people at a given month take?

(1.23) What kind of data is the monthly expenditure from all students in a given month?

(1.24) Give examples and distinguish between
(a) qualitative and quantitative variables
(b) discrete and continuous variables
(c) nominal and ordinal variables
(d) interval and ratio variables

(1.25) Identify each of the following examples as time series data (TSD) and cross-sectional data (CSD):
(a) the maximum temperature recorded on January 1st, 2013 in 50 states of the United States
(b) the daily closing price of a certain stock for the last three months
(c) the average closing price of each of 10 stocks in 2012
(d) the yearly unemployment rate of the USA over the last twenty years
(e) the average monthly expenditure for each of a sample of students in a university in a given year

(1.26) Which of the following variables are discrete and which are continuous?
(a) Number of car accidents in a highway on a given day
(b) Body temperature for a group of patients
(c) Total credit hours taking by students of a large class in the Fall of 2013
(d) The number of speeding tickets received by employees of a firm since last summer
(e) Speeds at which tickets are issued on a given day
(f) Number of printing mistakes in a manuscript
(g) January 2013 electric bills for each residence of a neighborhood
(h) Number of cities a large company has branch offices

(i) Number of dependents in a family

(1.27) The data below refer to unemployment rates and earnings for persons aged 25 and over by educational attainment in the US in 2012.

Educational attainment	Unemployment rate (Percent)	Median weekly earnings (in $)	Rank by earnings
Doctoral degree	2.5	1,624	2
Professional degree	2.1	1,735	1
Master's degree	3.5	1,300	3
Bachelor's degree	4.5	1,066	4
Associate's degree	6.2	7,85	5
Some college, no degree	7.7	7,27	6
High school diploma	8.3	6,52	7
Less than a high school diploma	12.4	4,71	8

Source: *U.S. Bureau of Labor Statistics, Current Population Survey.*

Write out the type of variables and data in each column.

(1.28) Identify nominal and ordinal variables from the list below:

(a) Status of health: very poor, poor, good, very good and excellent
(b) Status of smoking: Yes, No
(c) Size of pizza: small, medium, large, extra large
(d) Level of agreement: strongly disagree, disagree, neutral, agree, strongly agree
(e) Meal preference: breakfast, lunch, dinner
(f) Ethnicity: Caucasian, African American, Asian, Pacific Islander, etc
(g) Job satisfaction level: very dissatisfied, dissatisfied, neutral, satisfied, very satisfied
(h) State of birth: Indiana, Ohio, Michigan, etc.

(1.29) Which of the following variables is **not** an ordinal variable?

(a) Taste of food: strongly dislike, dislike, neutral, like, strongly like, etc.
(b) Age group in years: 0-5, 6-10, 11-15, etc.
(c) Time of day: night, dawn, noon, afternoon, evening
(d) Level of education: elementary, secondary, college
(e) Hair color: blonde, brunette, red, black

(1.30) Which of the following variables is **not** a nominal variable?

(a) Stage of cancer: stages I, II, III, IV
(b) Gender: Male, Female
(c) Race: White, Black, Others
(d) Region: North, South, East, West

(1.31) Which of the followings is a not time-series data?

(a) The average monthly household income of people in a residential area for the last two years
(b) New motor vehicle sales by each of 10 dealers of a state in a given month
(c) New motor vehicle sales of a dealer by year from 2000 to 2013

(d) The number of staffs in an industry by year for 5 years

(1.32) Which of the following examples correspond to time series data (TSD) and cross-sectional data (CSD)?

(a) the average monthly expenditure by month for a sample of students at a university for the last 12 months
(b) the number of cars sold by a particular dealer by year from 2000 to 2012
(c) the data on household income of people in a residential area in 2012
(d) the total number of cars sold in 2012 by each of 15 randomly selected dealers
(e) the maximum temperature recorded by each month for the last three years

(1.33) Which of the followings are examples of a discrete or continuous variable?

(a) Number of earning members in a family
(b) BMI for group of patients in a hospital taking treatment on a given day
(c) Household income of all families in a large city
(d) Number of voters by each family of a neighborhood
(e) Time it takes to go to work by each employee of a firm

(1.34) Which of the followings is an example of a nominal variable?

(a) Rank: first, second, third, etc.
(b) Pain level: no pain, mild, moderate, severe, worst, etc.
(c) Political affiliation: democratic, republican
(d) Income: \$0-\$19,999, \$20,000-\$39,999, \$40,000-\$59,999, etc.

(1.35) As explained in Example 1.1, below is a list of basic characteristics of several students in a class.

Gender	Age	Weight	Smoking	Hour	Standing	Restaurant
Female	22	150	No	12	Senior	Burger
Female	21	120	Yes	15	Junior	Arby's
Female	19	125	No	12	Sophomore	None
Male	25	175	No	12	Senior	McDonalds
Male	21	190	No	13	Junior	Taco bell
Female	18	150	Yes	16	Freshman	Sonic

(a) How many variables are listed in the above table?
(b) Name those variables.
(c) How many variables are nominal variables?
(d) Is there any ordinal variable? If so, which one?

Chapter 1: Answer to selected problems

(1.1) (d), 1.2) (a), 1.3) (a), 1.4) (c), 1.5) (c), 1.9) (b), 1.10) (d), 1.11) (d), 1.12) (d)

(1.13) (a) nominal, (b) continuous, (c) descriptive statistics, (d) census; it includes complete count in the USA.

(1.14) (a) Column 1- ordinal, Column 2- ordinal, Column 3- Continuous (b) descriptive statistics

(1.15) (a) Column 1- ordinal, Column 2- ordinal, Column 3- Continuous (b) descriptive statistics

(1.16) (c), (c); (1.17) Population: (c), sample: (b), an effective sample size.

(1.18) 5 values, labels; (1.19) ordinal; (1.20) ordinal, ordinal; (1.22) Quantitative continuous

(1.25) (a) CSD, (b) TSD, (c) CSD, (d) TSD, (e) CSD

(1.26) (a) discrete, (b) continuous, (c) discrete, (d) discrete, (e) continuous, (f) discrete, (g) continuous, (h) discrete, (i) discrete

(1.27) Column 1- ordinal, Column 2- continuous, Column 3- continuous, Column 4- ordinal

(1.28) (a) Ordinal, (b) Nominal (c) Ordinal (d) Ordinal, (e) Nominal, (f) Nominal, (g) Ordinal, (h) Nominal

(1.29) (e); 1.30) (a); 1.31) (b)

(1.32) (a) TSD, (b) TSD, (c) CSD, (d) CSD, (e) TSD

(1.33) (a) discrete, (b) continuous, (c) continuous, (d) discrete, (e) continuous

(1.34) (c); (1.35) (a) 7 (c) 3 (d) Yes, Standing.

CHAPTER 2
GRAPHS AND CHARTS

Introduction

Tables, graphs and charts are pictorial ways of presenting descriptive statistics of a sample or a population. The idea is to investigate data by looking at the summary presented in an easy-to-follow table, graph or chart, which is often called the ***exploratory data analysis***. In this chapter, we will learn how to present descriptive statistics for qualitative and quantitative variables using tables, graphs and charts.

Section 2.1: Descriptive statistics with tables

Descriptive statistics of a qualitative or categorical variable include frequency, relative frequency or percent frequency for each class of the variable. Often, these descriptive statistics are well presented in a tabular form. Let us start with some definitions to make our discussion easy.

Definitions: Frequency, relative frequency and percent frequency

The ***frequency*** of a class is the count of individuals in that class.
The ***relative frequency*** of a class is the proportion of individuals in that class.
The ***percent frequency*** of a class is the relative frequency of that class multiplied by 100.
If a dataset contains n individuals, the ***relative frequency*** of a class is computed by

$$Relative\ frequency\ of\ a\ class = \frac{frequency\ of\ the\ \ class}{n}$$

Definition: Distribution of a categorical variable

Generally, the ***distribution of a categorical variable*** refers to a table with frequency for each class of the variable. Specifically, a table with ***frequency***, ***relative frequency*** or ***percent frequency*** for each class of the variable is referred to as a ***frequency distribution***, ***relative frequency distribution*** or ***percent frequency distribution***, respectively.

Example 2.1

Job Satisfaction: On the whole, how satisfied are you with the work you do? The question was asked to 53,043 survey people. Table 2.1(a) summarizes the responses from the survey.

The frequency of each class presented in Table 2.1(a) is computed from the reported percentage and the total number of subjects surveyed (53,043).

Table 2.1(a): Frequency distribution of job satisfaction for a total of 53,043 survey people.

Job satisfaction	Frequency
Very Satisfied	25249
Moderately Satisfied	20262
A Little Dissatisfied	5304
Very Dissatisfied	2228
Total	53043

***Source**: http://www.statisticbrain.com/u-s-job-satisfaction-statistics/ due to the University of California Berkeley (Date Verified: 1.1.2014).*

(a) Compute relative frequency and percent frequency for each class of job satisfaction.
(b) Explain how you computed relative and percent frequency in part (a).
(c) Interpret results of relative frequency or percent frequency for the class "Very Satisfied".

Solution

(a) The columns of relative frequency and percent frequency are reported in Table 2.1(b) below.
(b) The relative frequency of the class "Very Satisfied" is 25,249/53,043=0.476 and the percent frequency (%) of this class is $0.476 \times 100 = 47.6$.
Similar approaches are applied to other computations.
(c) The relative frequency of 0.476 or equivalently, the percent frequency of 47.6 means that 47.6% of the surveyed people are "Very Satisfied" with whatever job they are doing.

Table 2.1(b): Frequency, relative frequency and percent frequency distributions of job satisfaction labels of 53,043 survey people.

Label of job satisfaction	Frequency	Relative Frequency	Percent Frequency (%)
Very Satisfied	25249	0.476	47.6
Moderately Satisfied	20262	0.382	38.2
A Little Dissatisfied	5304	0.100	10.0
Very Dissatisfied	2228	0.042	4.2
Total	53043	1.000	100.0

Example 2.2

Table 2.2(a) provides the cost of Americans 65 years and older with Alzheimer's disease and other dementias by payer type for the year 2013. Compute relative and percent frequency for each class.

Table 2.2 (a): Aggregate costs of care, of 203 billion dollars, by payer type for Americans 65 years and older with Alzheimer's disease and other dementias in 2013.

Payer Type	Costs (in billions of dollars)
Medicare	107
Medicaid	35
Out-of-pocket	34
Other	27

***Source**:* http://www.alz.org/downloads/facts_figures_2013.pdf

Solution
We have reported the relative and percent frequency for each class in Table 2.2(b).

Table 2.2(b): Relative and percent frequency for each class by payer.

Payer Type	Costs	Relative Frequency	Percent
Medicare	107	0.53	52.71
Medicaid	35	0.17	17.24
Out-of-pocket	34	0.17	16.75
Other	27	0.13	13.30
Total	203	1.00	100.00

Section 2.2: Graphs for qualitative variables

The popular graphs for summarizing qualitative variables are ***bar diagram*** and ***pie chart***.
A ***bar diagram***, also called ***a bar chart*** or ***bar graph***, consists of bars with heights as frequencies, relative frequencies or percent frequencies for different categories of the variable. All bars in a bar diagram are of equal width, with a uniform gap in between them, to mean that they are all distinct categories. Often, bars are used to compare quantitative measures by different classes of a qualitative variable. For example, bar diagram in Example 2.3 below compares unemployment rate (percent) for persons aged 25 years and over by their educational attainment in the USA in 2012.

Example 2.3

Table 2.3 provides the unemployment rate (percent) for persons aged 25 years and over in the USA in 2012 by the level of education attained. The information of the table is translated into a bar diagram in Figure 2.1 drawn using the Minitab software.

Table 2.3: Distribution of unemployment rate (percent) for persons aged 25 years and over by educational attainment in the USA in 2012

Education attained	Unemployment rate (%)
Doctoral	2.5
Professional	2.1
Master	3.5
Bachelor	4.5
Associate	6.2
Some college	7.7
HS diploma	8.3
<HS diploma	12.4

Source: *Bureau of Labor Statistics, Current Population Survey; available at* *http://www.bls.gov/emp/ep_chart_001.htm*

Figure 2.1: Bar diagram of unemployment rate by the educational attainment

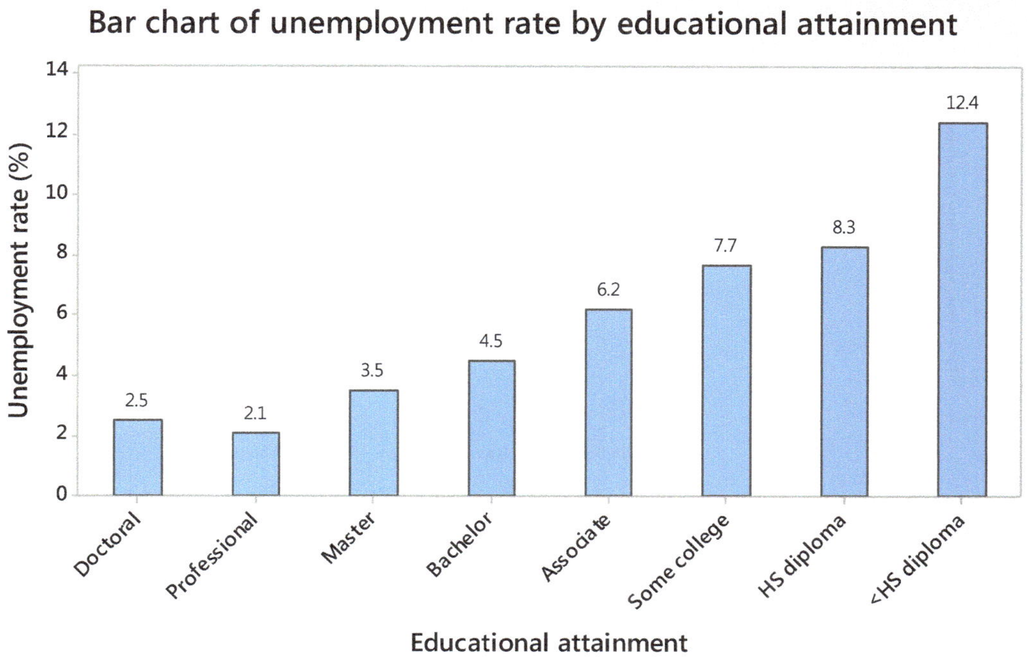

Example 2.4

Data in Table 2.4 below represents the number of people uninsured (in thousands) in 2012 in the USA by nativity.

Table 2.4: Distribution of people without any health insurance by the nativity in the USA in 2012.

Nativity	Uninsured
Native	35,436
Foreign-born	13,177
Naturalized citizen	3,431
Not a citizen	9,746
Total	61,790

Source: *U.S. Census Bureau, Current Population Survey, 2012 Annual Social and Economic Supplement.*

(a) Draw a well labeled bar chart.
(b) What percent of people not covered by any health insurance is not a citizen?

Solution

(a) Figure 2.2 contains the bar diagram drawn using the Minitab software.
(b) Percent not covered by any insurance who is not a citizen $= \frac{9{,}746}{61{,}790} \times 100 = 15.8\%$

Figure 2.2: Bar diagram of people without any health insurance by nativity in the USA in 2012.

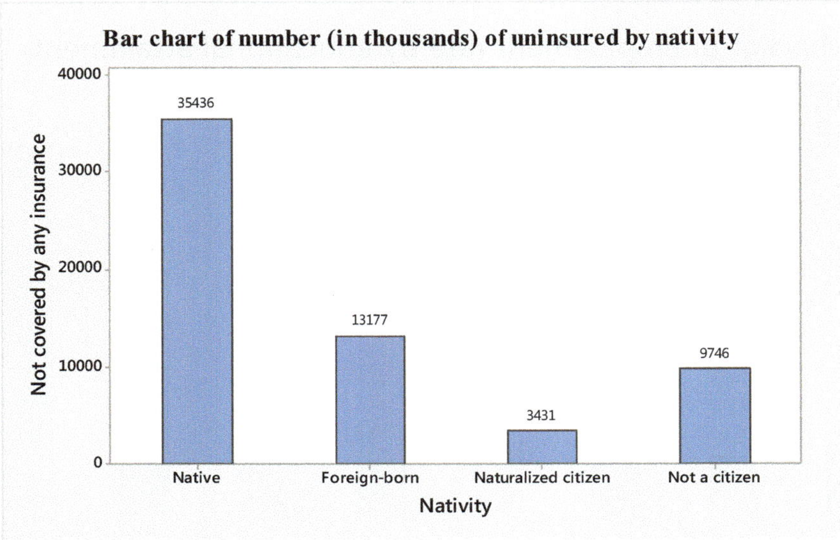

Pie chart

A pie chart consists of slices of a pie; with each slice representing the frequency or percent frequency of a class.

In a pie chart, the total pie accounts for the 100% data counts by sharing 360 degrees proportionately among different classes. If a dataset contains n individuals, then a class with frequency f has $\left(\frac{f}{n} \times 360\right)$ degrees and $\left(\frac{f}{n} \times 100\right)$ percent in the pie occupied by the class. It is very easy to make a pie chart in Excel, Minitab and other statistical software.

Caution:
While making a pie chart, make sure to justify that all distinct categories add up to 100% of data points.

Example 2.5

Revisit the job satisfaction data presented in Example 2.1.

Label of job satisfaction	Frequency
Very Satisfied	25249
Moderately Satisfied	20262
A Little Dissatisfied	5304
Very Dissatisfied	2228

(a) Can you make a pie chart for the above data? Justify your answer with an explanation.
(b) Make a pie chart if your answer in (a) is yes.

Solution

(a) Yes, we can make a pie chart for the above data, because the total percentage adds up to 100%.
(b) Figure 2.3 contains the pie chart for the job satisfaction data drawn using the Minitab software.

Figure 2.3: Pie chart for the job satisfaction data of Example 2.1

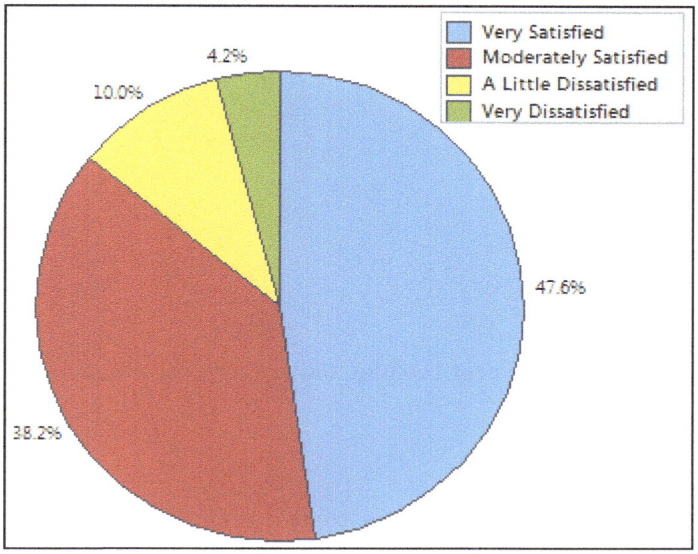

Example 2.6

Table 2.5 contains the percentage distribution of persons aged 19–25 who were uninsured in the United States by quarters of 2011.

Table 2.5: Distribution of uninsured people aged 19-25 years in the USA by quarters of 2011

Quarters	Percentage Uninsured
Q1	30.4
Q2	27.3
Q3	28.4
Q4	25.2

Source: *Early Release of Health Insurance Estimates Based on Data from the National Health Interview Survey, January–September 2012.*

(a) Can you make a pie chart for the above data? Justify your answer with an explanation.
(b) If a pie chart is not appropriate, what graph would you recommend to represent above data?

Solution

(1) We cannot draw a pie chart for the above data because the total percentage is more than 100% (111.3%). This may be due to the fact that the percentage of uninsured people by quarters must be overlapping in different quarters, making total percentage above 100%.
(2) We would recommend a bar diagram to represent the above data.

Section 2.3: Graphs for quantitative variables

For a quantitative variable, the commonly used graphs for presenting descriptive statistics are the followings:

- Histogram
- Stemplot
- Timeplot
- Dotplot
- Boxplot
- Scatterplot

This section discusses all graphs except boxplot and scatterplot, which are discussed in chapters 3 and 4, respectively.

Histogram

A ***histogram*** consists of rectangles with bases representing distinct classes or intervals formed from a continuous variable and heights as frequencies of classes. If heights of rectangles represent relative frequencies or percent frequencies, we call it a ***relative frequency histogram*** or ***a percent frequency histogram***.

For example, suppose we consider a variable 'time', which measures the amount of time a student spends to finish a 30-minute quiz. Then, time is a continuous variable. We may form some intervals of time such as 0-5, 5-10, 10-15, 15-20, 20-25 and 25-30, and then count how many students finish the quiz in each interval of time.

Note that, we have constructed six classes or intervals of time. The lower value of a class is called the ***lower class limit***, and the upper value is the ***upper class limit***. The difference between the upper and the lower class limits is called the ***length*** or ***width*** of the class. The class ***midpoint*** of a class is the average of the lower and upper class limits. For example, the time interval 15-20 has the lower class limit 15 and upper class limit 20. The width of this class is $20 - 15 = 5$. The midpoint of this class is $\frac{15+20}{2} = 17.5$.

It is important that we pay attention while counting how many students finish the quiz at a specified interval. Two methods could be considered in this counting process. One of the two methods, called the ***upper limit inclusive*** method, counts a student in a class if the finishing time of the student is less than or equal to the upper limit and greater than the lower limit of the class. The other method, called the ***lower limit inclusive*** method, on the other hand, counts a student in a class if the finishing time of the student is greater than or equal to the lower limit and less than the upper limit of the class.

Example 2.7

The table in Figure 2.4(a) gives summary of the time (in minutes) to finish a quiz by 28 students from an elementary statistics class arranged by 6 classes, each of length 5.

We draw the frequency histogram in Figure 2.4(b) using the Minitab software (see Section 2.5 for details about drawing a histogram). Note that there is no space between two adjacent rectangles of the histogram. It means that all the values of the continuous variable are covered by the non-overlapping classes.

Figure 2.4 a-b: Frequency distribution and related histogram for the finishing time of a quiz

Classes	Frequency
0-5	1
5-10	5
10-15	6
15-20	8
20-25	5
25-30	3
Total	28

(a)

The frequency count for each class is done by considering the ***lower limit inclusive*** method.

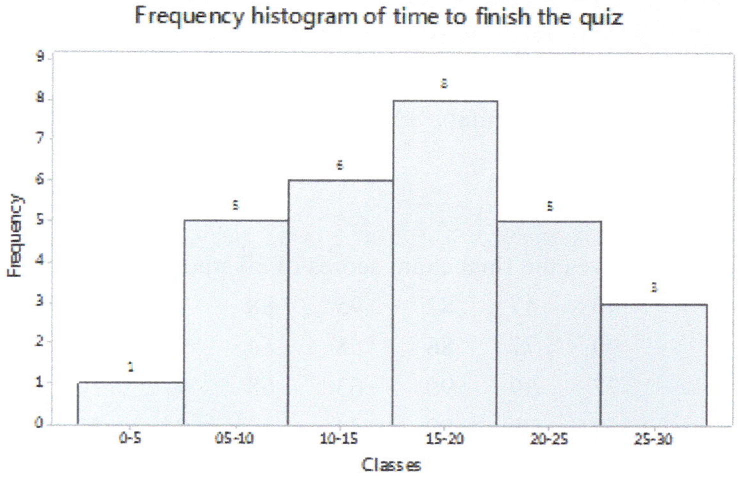

(b)

Histograms help us understand the shape of distribution of a variable and investigate the presence of outliers (any value showing unusual pattern) in the dataset.

How to decide the number of classes of a histogram?

There is no universal rule to decide the number of classes for a histogram. Generally, the number of classes k of a histogram for a dataset with N individuals is determined by the following rules of thumb:

Rule (1): k is the smallest whole number such that $2^k > N$.

Rule (2): k is equal to $\sqrt{N}$, rounded to the nearest integer so that all values of the variable are covered.

Example 2.8

If the total number of data points is $N = 68$, then determine the number of classes k to be used for constructing a histogram.

Solution

By the rule (1), k=7 becuase $2^7 = 128 > 68$.

By the rule (2), $k = 8$ because $\sqrt{68} = 8.25$.

Example 2.9

If the total number of data points is $N = 28$.

By the rule (1), k=5 because $2^5 = 32 > 28$.

By the rule (2), $k = 5$ because $\sqrt{28} = 5.3$.

Generally speaking, it is the researcher who decides the number of classes to meaningfully construct a histogram.

In general, any number between 5 and 10, inclusive, is a good choice for the number of classes.

How to determine the width or length of a class?

Once the number of classes (k) is determined, the length or width (l) of each class is the quantity

$$l = \frac{range}{k}, \text{ rounded up to the nearest integer}$$

where, $range = highest\ data\ value - lowest\ data\ value$

Statistical software such as Minitab, SAS, SPSS, etc. automatically determines the classes and length for users.

Example 2.10

The following table gives the final exam scores of 50 students in a statistics class.

60	47	82	95	88	72	67	66	68	**98**
90	77	86	58	64	95	74	72	88	74
77	**39**	90	63	68	97	70	64	70	70
58	78	89	44	55	85	82	83	72	77
72	86	50	94	92	80	91	75	76	78

Draw a histogram with any suitable number of classes and width, with reasonable justification of your choice. Find class midpoints for classes.

Solution

We have $range = 98 - 39 = 59$.

Note that using rule (1), $k = 6$ and $l = 59/6 = 9.8$

Using rule (2), $k = 7$ and $l = 59/7 = 8.5$.

For the sake of the convenience, we take $k = 7$ and $l = 10$ and consider classes 35-45, 45-55, ..., 95-105. Then, we count the frequency of each class using the **lower limit inclusive** method so as to cover all data points.

The frequency histogram drawn using the Minitab software appears in Figure 2.5 (b).

Figure 2.5 a-b: Frequency distribution and related histogram of exam score of 50 students.

Classes	Frequency
35-45	2
45-55	2
55-65	7
65-75	13
75-85	11
85-95	11
95-105	4

(a)

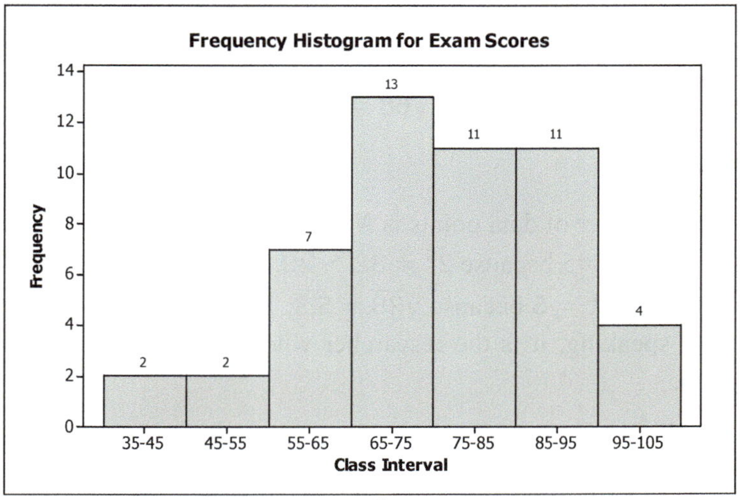

(b)

Frequency polygon

A ***frequency polygon*** is drawn by plotting frequency of each class against the midpoint of the class and then connecting the points for all classes by line segments. By plotting relative frequency or percent frequency against the midpoint, we get the ***relative frequency polygon*** or ***percent frequency polygon***.

Ogive

An **ogive** is a smooth curve through the points obtained by plotting the cumulative frequency of each class against the upper limit of the corresponding class. We can have a **relative frequency ogive** or **percent frequency ogive** by plotting cumulative relative frequency or cumulative percent frequency of each class against the upper limit of the corresponding class.

Example 2.11

In reference to Example 2.10, we have computed the cumulative frequency, cumulative percent frequency, midpoint and upper limit for each class in Table 2.6.
Now, draw the following graphs with an explanation of how you draw it:

(a) Ogive with cumulative frequency
(b) Frequency polygon

Solution
The Table 2.6 contains cumulative frequency, midpoint and upper limit for each of the classes.
Table 2.6: Frequency, midpoint and upper limit of each class

Classes	Frequency	Cumulative frequency	Cumulative percent frequency	Midpoint	Upper class limits
35-45	2	2	4	40	45
45-55	2	4	8	50	55
55-65	7	11	20	60	65
65-75	13	24	48	70	75
75-85	11	35	70	80	85
85-95	11	46	92	90	95
95-105	4	50	100	100	105

(a) The ogive is shown in Figure 2.6(a). We draw the ogive by plotting the cumulative frequency against the upper limit of each class and then connecting points for all classes by a smooth curve. The curve is extended to the lower limit of the lowest class at the height of 0 to mean that nobody scored below 35. Note that 100% of the score is less than 105. We use Minitab for drawing the ogive (see the drawing instruction in section 2.5).

(b) The frequency polygon is shown in Figure 2.6(b). The frequency polygon is drawn by plotting the frequency of each class against the midpoint, and then connecting all the points with line segments. Generally, the curve is extended to the x-axis at both ends.

Figure 2.6: Ogive (a) and frequency polygon (b) for the exam score data.

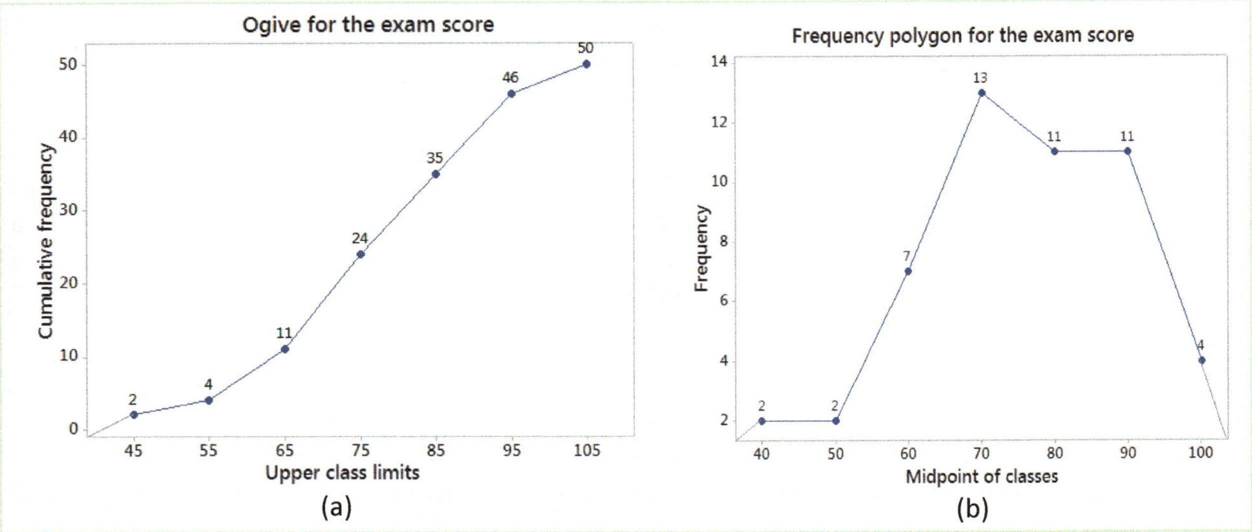

Timeplot

A timeplot consists of points of time-series measurements over time. It reveals trends, cyclical or seasonal variation and other changes over time which may be of great importance to the researchers.

Example 2.12

The data below is the number of uninsured (in millions) in the USA from 2000 to 2010.

Year	2000	2001	2002	2003	2004	2005	2006	2007	2008	2009	2010
Uninsured	36.6	38	39.8	41.9	41.8	43	45.2	44.1	44.8	49	49.9

Source*: Income, Poverty and Health Insurance Coverage in the USA: 2010, Washington, D.C.: U.S. Census Bureau, Sept. 2011.*

As we see from the timeplot of uninsured people in Figure 2.7, there is mostly an increased trend in the number of uninsured in the USA over time.

Dotplot

A ***dotplot***, also called a ***dotchart***, consists of stacks of dots for each occurrence of data points. It shows how each value is distributed in the dataset and depicts any extreme high or low data value.

Example 2.13

The data below represents the total number of credit hours (sorted in order of magnitude) taken by 25 students of a class in the fall of 2012:

5	5	5	6	6	6	6	8	8	8	10	10	10
12	12	12	12	12	12	14	15	16	16	18	18	

(a) Draw a dotplot of the above dataset.
(b) Use the information in the dotplot to determine the percent of students taking more than 12 credit hours.

Solution

(a) The dotplot is drawn using Minitab (see Minitab instruction in section 2.5).
(b) From the dotplot we can say that 20% of the students is taking more than 12 credit hours.

Figure 2.7: Timeplot of uninsured (in millions) in the USA from 2000 to 2010

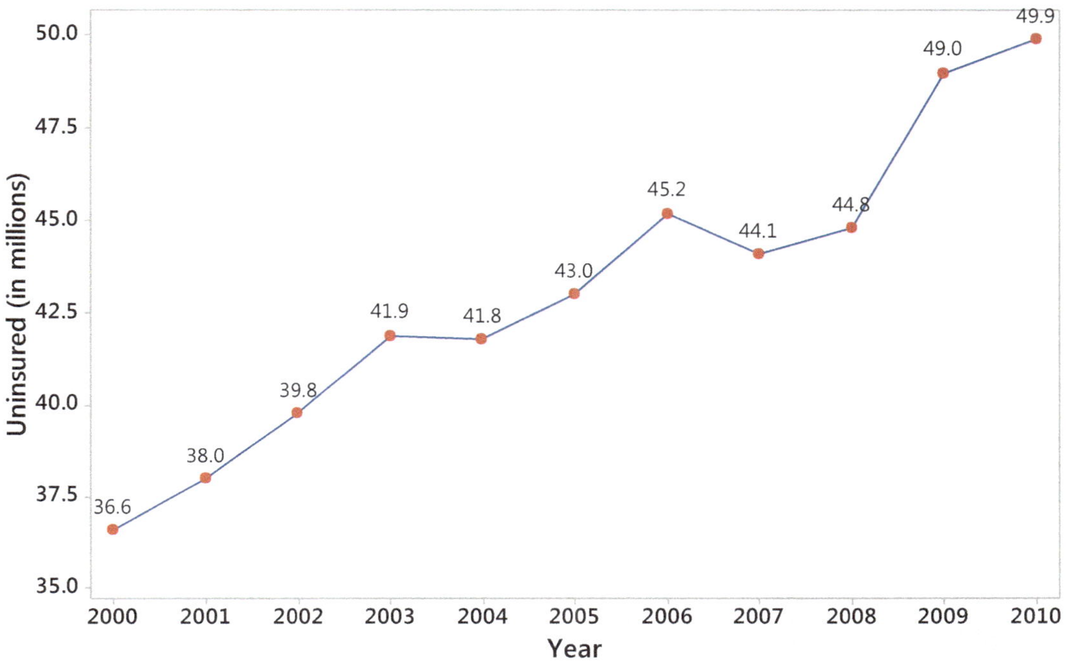

Figure 2.8: Dotplot of credit hours of 25 students

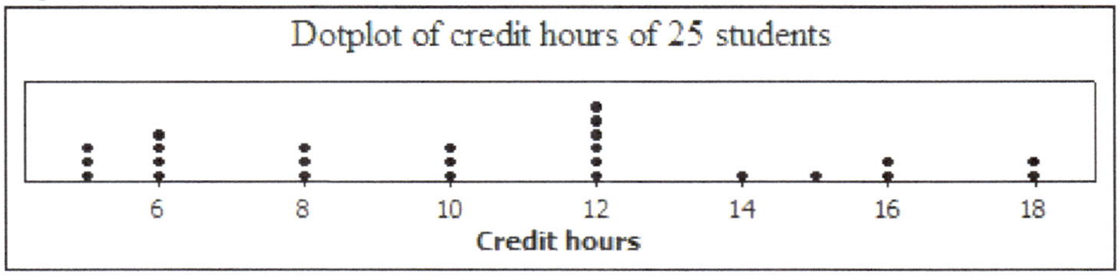

Stem-and-leaf plot

Given a number, a ***leaf*** is the final or rightmost digit of the number, and a ***stem*** is the string of digits without the leaf. A ***stem-and-leaf plot*** (or simply ***stemplot***) consists of a vertical list of stems with their leaves written in respective rows, separated from the stems by a vertical line.

Below is an algorithm to construct a stemplot manually:

- Write out all the stems in increasing order from the lowest to the highest, inclusive, in a vertical column and draw a vertical line to the right of this list.

- Read each number and write its leaf next to its stem.

A stem without a leaf refers to the nonexistence of any number having the stem in the dataset, which might explain important data structure.
By ordering leaves from the lowest to the highest, we get an ***ordered stem-and-leaf diagram***. A stemplot may be used to summarize the data because it preserves the data values and may provide useful information about the shape of the distribution.

Extended stemplot

An ***extended stem-and-leaf diagram*** is the one created by writing each stem twice in separate lines. In an extended stemplot, leaves 0-4 go with the upper stem and leaves 5-9 go with the lower stem. It is useful when dataset contains only few stems with small range of values.

Example 2.14

Time (in minutes) to finish a quiz by 12 students in an Elementary Statistics Class is given below:

10	15	20	13	10	5	7	30	20	25	25	30

(a) Draw a stemplot for the above dataset and answer questions (b)-(e) using the stemplot.
(b) What percentage of students finished before 15 minutes?
(c) What percentage of students finished after 25 minutes?
(d) What percentage of students finished between 15 and 30 minutes, inclusive?
(e) Do you think an extended stemplot with two lines per stem will be more appropriate? Justify your answer.

Solution

(a) In the above dataset, we have one and two-digit numbers between 5 and 30. A one digit number has a stem of 0 and leaf is the number itself. Thus, we have stems of 0, 1, 2, 3 and the stemplot looks like:

0	5 7
1	0 0 3 5
2	0 0 5 5
3	0 0

(b) $\frac{5}{12} \times 100 = 41.67\%$
(c) 16.7%
(d) 58.3%
(e) Not necessarily, because we have only a few leaves per stem.

Below is the Minitab output for the stemplot of the above dataset, with two lines per stem, where appropriate. Minitab created an ordered stem-and-leaf plot. The second column represents stems from 0 to 3 and third column representing leaves. There are two rows per stem for stems 1 and 2, with the first row containing leaves 0-4 and the second row containing the leaves 5-9.
Minitab tells us that $N = 12$ and the leaf unit is 1.0, so that a stem of 0 and a leaf of 5 in the first row, for example, represent the number 05.

```
Stem-and-leaf of Time to finish a quiz N = 12
Leaf Unit = 1.0

 2  0  57
 5  1  003
 6  1  5
 6  2  00
 4  2  55
 2  3  00
```

The first column accumulates frequencies from the top and the bottom until it converges in the middle or reaches the middle value or median. The first 6 in the first column means that there are 6 values at or below **15** and second 6 means that there are 6 values at or above **20**. The median is $\frac{15+20}{2} = 17.5$.

Example 2.15

The data below represents the number (in millions) of persons aged 19-25 years who lacked health insurance coverage in the USA, during 1997–2011.

1997	1998	1999	2000	2001	2002	2003	2004	2005	2006	2007	2008	2009	2010	2011
7.7	7.4	7.3	8.1	8.1	7.9	8.9	8.9	8.8	9.3	8.8	8.9	9.5	10	8.4

Draw a stem-and-leaf diagram using Minitab or manually.

Solution

Below we provide a stem-and-leaf diagram drawn in Minitab.

```
Stem-and-leaf of Number uninsured N = 15
Leaf Unit = 0.10

 2   7   34
 4   7   79
 7   8   114
(5)  8   88999
 3   9   3
 2   9   5
 1   10  0
```

Minitab has created an ordered extended-stem-and-leaf plot. The second column contains stems from 7 to 10 and third column contains leaves. For each of the stems 7, 8 and 9, there are two lines per stem with leaves 0-4 appearing with the upper stem and leaves 5-9 appearing with the lower stem. Minitab uses one or two rows for a given stem depending on the data values.
Minitab tells us that $N = 15$ and the leaf unit is 0.10, so that a stem of 7 and a leaf of 3 in the first row, for example, together represent the number 7.3.

The first column accumulates frequencies from the top and the bottom until it reaches the row containing the middle value or median. The number 5 (in parenthesis) tells there are 5 data points between 8.8 and 8.9. The number 7 above 5 (in parenthesis) means there are 7 data points at or below 8.4. The number 3 below 5 (in parenthesis) means there are 3 data points at or above 9.3.

The median belongs to the row with number 5 (in parenthesis) in the first column and the median is 8.8.

Section 2.4: Shape of a distribution

How the shape of a distribution looks like is of the greatest interest to the statisticians and decision makers. The shape of the distribution of a variable can be symmetric or asymmetric (skewed).

Definition: Symmetric shape

The shape of a distribution is ***symmetric*** if the right tail and left tail of the histogram appear to be mirror image of each other, i.e., the two halves of the distribution coincide each other. A bell shaped or normal curve and a uniform or rectangular curve are examples of a symmetric shape. A bell shaped distribution is also called a ***mound shaped*** distribution because most of the observations pile up in the middle.

Definition: Asymmetric shape

The shape of a distribution is ***asymmetric*** if it is not symmetric. An asymmetric shape can be negative skewed and positive skewed.

An asymmetric shape is ***negatively skewed*** or ***left skewed*** if it has a longer left tail. An asymmetric shape is ***positively skewed*** or ***right skewed*** if it has a longer right tail. An existence of a longer left or right tail means that there are some extreme data values in the tail.

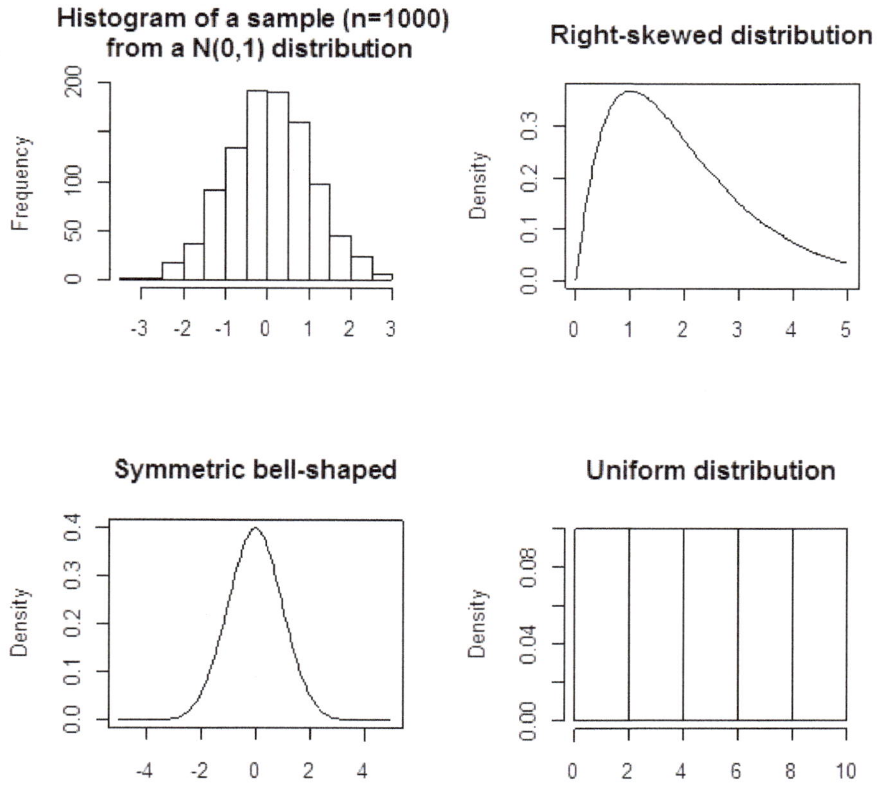

Section 2.5: Drawing graphs using Minitab

In this section, we discuss how we can draw some common graphs such as bar graph, pie chart, histogram, dotplot, timeplot, etc. in Minitab.

Bar chart for summarized data

Ref: Example 2.4

(1) Enter the categorical variable "Nativity" in C1 and summary variable "Not covered" in C2.
(2) Select **Graph>Bar Chart**.
(3) In "**Bars represent**" pull-down menu, select "**Values from a table**" and highlight "Simple." Press OK.
(4) For "Graph variables" box, select C2 and "Categorical variable" box, select C1.
(5) Use Labels option to add "Title/footnote" and "Data Labels" to the graph and Click OK to obtain the bar graph.

Pie chart for summarized data

Ref: Example 2.5

(1) Enter "Label of job satisfaction" in C1 and "Frequency" in C2.
(2) Select **Graph** and highlight **Pie Chart**.
(3) Click the "Chart values from a table" radio button.
(4) Enter C1 in the "Categorical variable" box and C2 in the "Summary variable" box.
(5) Use **Labels** option to add title/footnote and then **slice Labels** to add category name, frequency and percent in the graph. Click OK to obtain the pie chart.

Histogram for grouped data

Ref: Example 2.7

Suppose C1 represents Classes (0-5, 05-10, 10-15, etc.) and C2 represents Frequency for Classes in C1. Use the following step to draw histogram for this grouped data

(1) Click **Graph > Bar Chart**.
(2) In "**Bars represent**" pull-down menu, select "**Values from a table**" and highlight "**Simple**." Press OK.
(3) For "**Graph variables**" box, select C2 and "**Categorical variable**" box, select C1.
(4) Use **Labels** option to add "Title/footnote" and "**Data Labels**" to the graph and Click OK.
(5) Double-click the x-axis or any x-axis value to Edit Scale: Under **Space Between Scale Categories** <u>uncheck</u> **Gap between clusters** and enter 0. Click **OK**.

Time series plot

Ref: Example 2.12

We plot **Uninsured** versus **Year** using the following steps:

(1) Enter Year in C1 and Uninsured in C2.
(2) Select **Graph>Time Series Plot...>Simple** and press **[OK]**.

(3) For **Series**, select C2.
(4) Click on **Time/Scale**; for **Stamp** select C1.
(5) Press OK.
(6) Use **Labels...**>*Data labels>Use y-value labels* option to labels your plot; or after graph is done, right click on the graph>Add> *Data Labels>Use y-value labels* to get data label in the plot. Add title…etc. as shown in the graph or as you wish.

Dotplot

Ref: Example 2.13

We have used the following steps to draw a dotplot of credit hours data:

(1) Enter credit hour in C1
(2) Select **Graph>Dotplot...>Simple** and press **[OK]**.
(3) For graph variables, select C1 and use **Labels...**for subtitle, etc. and press **[OK]**

Ogive

Ref: Example 2.11

Minitab does not have a default method to draw an ogive. So, we will basically plot cumulative frequency or cumulative relative frequency against upper the class limit. It requires us to have necessary calculations beforehand.

(1) Type the Upper Class Boundaries in column C1 and the Cumulative relative frequency in column C2.
(2) Use **Graph -> Scatterplot -> With Connect Line** and **OK**.
(3) Select C2 for the **Y-variable** and C1 for the **X-variable**.
(4) Use **Data View...** option, select both **Symbols** and **Connect line**. Optionally, use **Labels...** options for **Titles/Footnotes** and **Data label**s for **y-value labels**. By clicking on **OK**, you get an ogive.
(5) After the graph is created, Right-click on the X-axis of the graph and use **Edit X scale** and **Position of ticks** 45:100/10 to tick marks from 45 to 105 in steps of 10 to show upper class limits.

Frequency polygon

Ref: Example 2.11

Minitab does not have a default method to draw a frequency polygon. Therefore, we will use similar technique we employ for drawing ogive with the exception that we plot frequency of each class against midpoint of the class.

(1) Type the Midpoints of classes in column C1 and the frequency in column C2.
(2) Use **Graph -> Scatterplot -> With Connect Line** and **OK**.
(3) Select C2 for the **Y-variable** and C1 for the **X-variable**.
(4) Use **Data View...** option, select both **Symbols** and **Connect line**. Optionally, use **Labels...** options for **Titles/Footnotes** and **Data label**s for **y-value labels**. By clicking on **OK**, you get a frequency polygon.

(5) After the graph is created, Right-click on the X-axis of the graph and use **Edit X scale** and **Position of ticks** 40:100/10 to tick marks from 40 to 100 in steps of 10 to show midpoints of classes.

Practice problems on Chapter 2

2.1) The data in table below refers to the number and percentage of death of people aged 25-44 years by five leading causes of death.

Cause of death	Number of death	Percent of total death
Accident	29492	26
Cancer	14746	13
Heart disease	13612	12
Suicide	12478	11
Homicide	6806	6
Others	36298	32
Total	113432	100

***Source**: National Vital Statistics System, Mortality in the USA in 2011 (preliminary).*

(a) Is it possible to make pie chart for showing various causes of death for the above data? Justify your answer.
(b) If your answer in (a) is Yes, make a pie chart for the cause of death.
(c) What percent of people die due to either cancer or heart disease?
(d) Will it make sense to draw a bar diagram showing the cause of death? Why or why not?

2.2) The data below represents the unemployment rate for persons aged 25 years and over by their educational attainment in the USA in 2013.

Educational degree attainment	Unemployment rate in 2013
Doctoral	2.2
Professional	2.3
Master	3.4
Bachelor	4.0
Associate	5.4
Some college	7.0
HS diploma	7.5
<HS diploma	11.0

***Source**: Bureau of Labor Statistics, Current Population Survey*
http://www.bls.gov/emp/ep_chart_001.htm

(a) Present these data in a well-labeled bar graph.
(b) Would it also be correct to make a pie chart? Justify your answer with appropriate argument.

2.3) The data below is the fast food preference of a number of students in several sections of Elementary Statistics class.

Fast Food Preference	Count
Taco Bell	12
McDonalds	19
Burger King	17
Subway	14
Wendy	10
Pizza Hut	15
Others	10
Total	97

(a) Make a pie chart for the above data.
(b) Make a bar chart for the above data.
(c) Does it matter whether you present the above data using a bar or a pie chart? Justify your answer.

2.4) The data below corresponds to the number of credit hours taken by a sample of 20 students in an introductory statistics class.

13	16	14	17	16	13	15	12	15	14
16	18	9	17	16	12	12	17	12	12

(a) Draw a histogram for the above dataset using the class intervals 8-10, 10-12, 12-14, 14-16 and 16-18 using the upper limit inclusive method.
(b) Can you draw a histogram for the above dataset using the intervals specified in (a) with the lower limit inclusive method? If not, how can you modify the class interval to draw a histogram using the lower limit inclusive method?
(c) If you wish to draw a stem-and-leaf diagram for the data set, how should you proceed?

2.5) The data below represents a frequency distribution of the serum HDL (mmHg) for a sample of 40 people.

Class Interval	*Frequency*
20-30	1
30-40	6
40-50	11
50-60	15
60-70	6
70-80	1
Total	40

Suppose the upper limit inclusive is used to construct above frequency distribution. For example, an individual with HDL of 30 mmHg is considered in group 20-30, not in group 30-40.

(a) What percent of people has Serum HDL above 50 mmHg?
(b) What percent of people has Serum HDL at or below 60 mmHg?
(c) Draw a histogram for the above dataset.

2.6) The data below is the percentage distribution of type of insurance coverage for people aged 19–25 years at the time of interview by quarters in the United States during January 2008–September 2012.

Type of Coverage	Q1 (%)	Q2(%)	Q3(%)	Q4(%)
Uninsured	27.9	30.4	27.3	28.4
Private Health Insurance coverage	56.3	53.2	58.4	54.7
Public Plan Coverage	16.8	16.9	15.2	17.8

(a) Can you make a pie chart for quarter 1?
(b) What percent of people are covered by any plan in quarter 1?
(c) What percent of people do not have any coverage in quarter 1?

2.7) The following list of numbers corresponds to how many credit cards each of 25 students owns.

3	1	0	2	4	1	3	0	2	1	2	3	2
2	0	1	2	1	2	2	3	1	3	1	2	

(a) Construct a dot-plot for the above dataset.
(b) What is your comment about the shape of the distribution of the number of credit cards of the students?
(c) Compute percent of students having 3 or 4 credit cards.
(d) Can you draw a stem-plot for the above dataset? Why or why not?

2.8) The data below refers to the median weekly earnings (in dollars) for persons aged 25 years and over in the USA in 2012 by levels of education.

Education attained	Median weekly earnings
Doctoral	1624
Professional	1735
Master	1300
Bachelor	1066
Associate	785
Some college	727
HS diploma	652
<HS diploma	471

Source: *Bureau of Labor Statistics, Current Population Survey.*

(a) Can you represent the above data by a bar diagram? If so, draw a well labeled bar diagram.
(b) What label of education corresponds to the highest median weekly earnings?

2.9) Facebook users differ greatly in the number of friends they have in their networks. Recent Pew Research Center survey findings found the following distribution of the number of friends among the facebook users in their networks.

Number of friends	Percent
1-100	39
101-250	23
251-500	20
500+	15

Source: http://www.pewresearch.org/fact-tank/2014/02/03/6-new-facts-about-facebook/

(a) Can you represent the above table in a pie chart? Why or why not?
(b) What graph or chart do you think is appropriate to represent the data?
(c) What adjustment of the table would you recommend to be able to represent the data in a pie chart?

2.10) In 2011, the estimated number of diagnoses of HIV infection in the United States, by age at diagnosis, was as follows:

Age groups	Estimated number of HIV infection
≤ 24	10539
25-39	18978
40-54	15268
55+	4489
Total	49274

Source: CDC, http://www.cdc.gov/hiv/statistics/basics/index.html

(a) Find the relative frequency and percentage frequency distribution of HIV infection by age groups.
(b) Can you represent the percent frequency distribution in a pie chart?
(c) Can you represent the percent frequency distribution in a bar chart?

2.11) In 2011, the estimated number of AIDS diagnosis in the United States, by age at diagnosis, was as follows:

Age groups	Estimated number of AIDS diagnosis
≤ 24	2999
25-39	11505
40-54	13344
55+	4204
Total	32052

Source: CDC, http://www.cdc.gov/hiv/statistics/basics/index.html

(a) What graph would be appropriate to represent data in the above table? Why?

(b) Find the relative frequency and percent frequency distribution of AIDS diagnosis by age groups.
(c) What percent of estimated diagnosis of AIDS is before the age of 25 or after the age of 54?

2.12) Below is a graph for the salary of different categories of employees of non-exempt titles in a university.

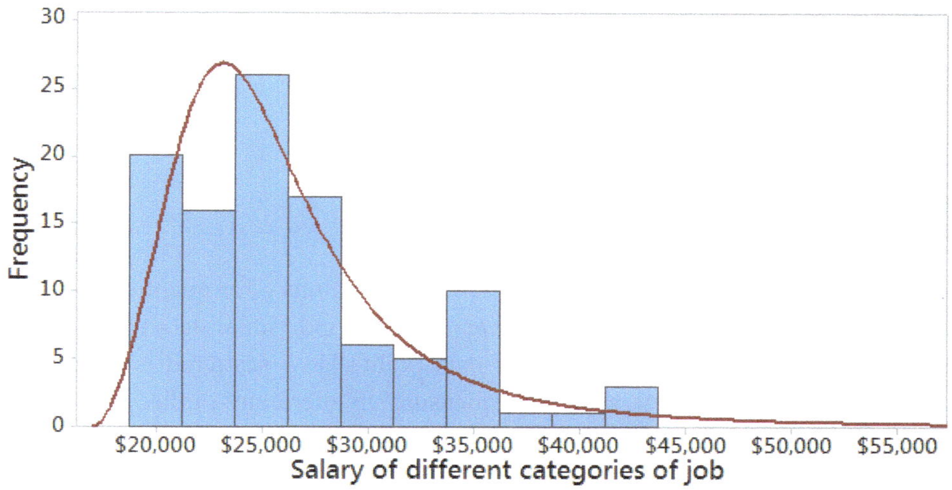

a) What is the name of the graph?
b) What is your comment about the shape of the graph?

CHAPTER 3
NUMERIC DESCRIPTIVE STATISTICS

Introduction

This chapter devotes to descriptive statistics using numerical summary for quantitative variables. Given a dataset, what is the average of a variable? How spread are the individual data points from the average? How far is the highest data point from the lowest data point? How scattered is the middle 50% of the data? Is the descriptive statistic we use a resistant measure? Is there any outlier in the dataset? We will address answers to these types of questions using various numeric descriptive statistics measure such as

- measures of central tendency
- measures of variation
- measures of percentiles and quartiles
- detection of outlier and boxplot

Section 3.1: Measures of center

The ***measures of center*** refer to the center of a data around which most of the values tend to cluster. Three popular measures of center are:

- Mean
- Median
- Mode

These measures may apply to a dataset referring to a sample or a population. Suppose, we want to know the mean age of sophomore students in a college. Imagine two scenarios:

Scenario (1): We have a sample of sophomore students from the college with their ages. We compute mean age of these students.

Scenario (2): Suppose we are lucky enough to know the age of all sophomore students from the college database. We compute mean age of these students.

How does the mean age in scenario (1) differ from the mean age in scenario (2)?

Of course, the mean in scenario (1) is based on a sample and thus we have a sample mean. If we take a different sample instead of the one we have already used, the mean age may change. In other words, different samples will yield different values of mean. Thus, a sample mean is a variable.

On the other hand, the mean in scenario (2) is computed from a population and thus we have a population mean. Since we have used ages of all sophomore students of the college, the mean age is always a constant or fixed value. In other words, the population mean is constant for any population. Any such descriptive measure obtained from a population is called a ***parameter*** and a parameter is always a constant.

Definition: Statistic and parameter

Any descriptive measure obtained from a sample is called a ***statistic***. For example, mean, median, mode or any other descriptive characteristic of a sample will be a statistic. Value of a statistic may vary from sample to sample, and therefore, a statistic is a variable. Given a sample, the value of the statistic is always known.

Any descriptive measure obtained from a population is called a ***parameter***. For example, mean, median, mode or any other descriptive characteristic of a population will be a parameter. The value of a parameter is a constant and is generally unknown. We wish to estimate the unknown parameter using the value of the corresponding statistic from a sample.

In general, there is a notational difference between a parameter and a statistic. For example, a population mean is denoted by μ (pronounced "**mew**"), whereas a sample mean is denoted by $\bar{x}$ (pronounced "**x bar**").

Definition: Mean

The ***mean*** of a set of data points is their sum or total divided by the number of data points in the set. The mean of a sample data is called the ***sample mean*** and the mean of a population data is called the ***population mean***.

Mean of a sample of n data points $x_1, x_2, \ldots, x_n$, denoted by $\bar{x}$, is computed using the formula

$$\bar{x} = \frac{x_1 + x_2 + \cdots + x_n}{n} = \frac{\sum x}{n}$$

Mean of a population of N data points $X_1, X_2, \ldots, X_N$, denoted by by μ, is computed using the formula

$$\mu = \frac{X_1 + X_2 + \cdots + X_N}{N} = \frac{\sum X}{N}$$

The expressions $\sum x$ and $\sum X$ represent the sum of sample and population data points, respectively.

Example 3.1

Time (in minutes) to finish a quiz by 12 students in a class is given below:

10	15	20	13	10	5	7	29	20	25	26	30

Compute the sample mean $\bar{x}$ and interpret your result.

Solution

$$\bar{x} = \frac{\sum x}{n} = \frac{10 + 15 + \cdots + 30}{12} = \frac{210}{12} = 17.5$$

This means that, on average, each student spends 17.5 minutes to finish the quiz.

Definition: Median

The median, denoted by $\tilde{x}$ (pronounced "**x tilde**"), of a sample is the middle most value of the ordered sample. Median divides the entire dataset into two equal halves; one half of the ordered values are less than or equal to the median and the other half of the ordered values are greater than or equal to the median.

How to find median?

Given a sample of n data points,

(1) Arrange n values in ascending or descending order.
(2) If n is **odd**, the median is the middle value.
(3) If n is **even**, the median is the average of two middle values.

Example 3.2

Find the median time to finish the quiz by 12 students reported in Example 3.1.

Solution

The sorted numbers in ascending order look like:

5 7 10 10 13 **15** **20** 20 25 26 29 30

Then, the median is the mean of middle two values 15 and 20:

$$\tilde{x} = (15 + 20)/2 = 35/2 = 17.5$$

The median in this case is the same as the mean computed in example 3.1.

Example 3.3

Replace the maximum time of 30 minutes by 60 minutes in Example 3.1, and then compute median and mean. Make a comment about the change of results.

Solution

Replacing 30 by 60, the sorted list looks like:

5 7 10 10 13 **15** **20** 20 25 26 29 60

The median and mean are respectively: (the mean of middle two values 15 and 20) is

$$\tilde{x} = (15 + 20)/2 = 35/2 = 17.5$$

$$\bar{x} = \frac{\sum x}{n} = \frac{240}{12} = 20$$

Comment: As we see, the median is still the same but mean is changed from 17.5 to 20. A single extreme value may cause the mean to change to a greater extent.

This is why we say that the ***mean is not a resistant measure***, but the ***median is a resistant measure***.

A statistic is a ***resistant measure*** if it is less affected or not affected by an extreme large or small sample value.

Useful remarks

(1) In general, the population mean μ is unknown.
(2) A point estimate of μ is the sample mean $\bar{x}$.
(3) A point estimate is a single value calculated from the sample.
(4) $\sum_{i=1}^{n}(x_i - \bar{x}) = 0$, and is left as an exercise to curious students.
(5) Mean is not a resistant measure of center, but median is a resistant measure.
(6) Unless otherwise is specified, the data points used in this chapter is assumed to be a sample.

Definition: Mode

The ***mode*** of a sample refers to the most frequently occurred value in the sample.

Example 3.4

The data below represents the total number of credit hours (sorted in order of magnitude) taken by a sample of 25 students of a class in the fall of 2012:

5 5 5 6 6 6 6 8 8 8 10 10 10 12 12 12 12 12 12 14 15 16 16 18 18

Find the mode of the number of credit hours.

Solution

As appears from the data, the most frequently occurred number is 12. Then, the mode is 12 credit hours.

Are mean, median and mode the same?

(1) For a symmetric distribution, mean, median and mode are the same.
(2) For a skewed distribution, mean, median and mode are different.
(3) Of course, if a dataset contains identical values, mean, median and mode are the same.

Indeed, for a left skewed data, the mean is less than the median, and the median is less than the mode. For a right skewed data, the mean is greater than the median and the median is greater than the mode. This situation is explained by the fact that the mean is pulled more towards the tail containing extreme values than the median, as has been shown in Figure 3.1.

Figure 3.1: Distributions showing location of varying measures of center

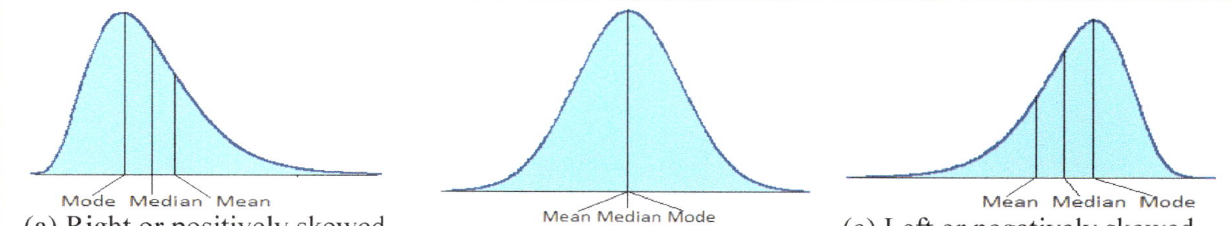

(a) Right or positively skewed (b) Symmetric or bell shaped (c) Left or negatively skewed

Section 3.2: Measures of variation

The ***measure of variation*** or ***spread***, also called ***the dispersion***, of a set of data points refers to the amount of variation, scatterness or variability in the dataset.

If all the values of a dataset are the same, then there is no variation, spread or dispersion in the dataset.

Example 3.5

Consider two simple data sets:

Dataset 1: 1, 3, 2

Dataset 2: 2, 2, 2

Which of the two data sets above has more spread or variability?

Solution

Of course, the dataset 1 has some sort of spread or variability. The dataset 2 has no spread or variability at all, because its values are identical.

There are different ways to measure the variability of a dataset. However, some widely used measures of variation include:

- range
- variance
- standard deviation
- coefficient of variation

Why measures of variation?

Measures of center cannot determine the existing variation in the dataset. Several datasets with the same measures of center may vary internally. For example, below we have summary measures from two samples with the same measures of center but varying measures of variation and shape (as appeared in the box-plot for two samples).

Sample	Count	Mean	Median	Mode	Range	IQR	SD	Variance
1	16	70	70	70	18	8.25	5.51	30.4
2	16	70	70	70	19	13	6.72	45.2

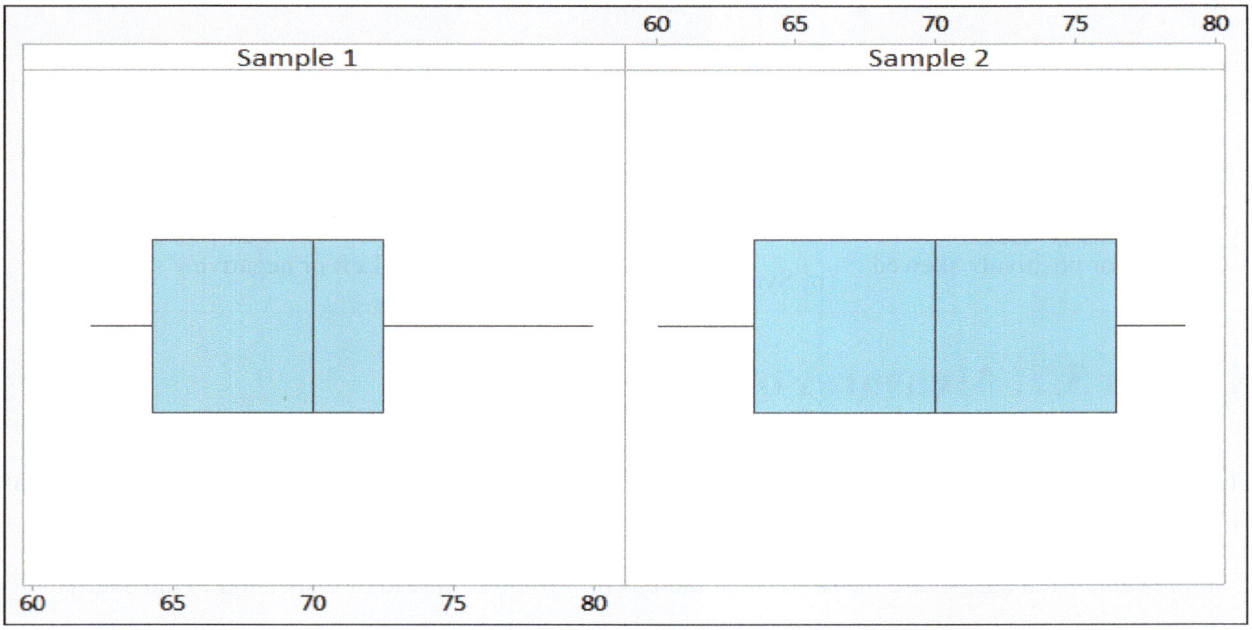

Also, look at the three distributions with the same measures of center but varying standard deviations in Figure 3.2. It appears from the figure that measures of center do not address variability of a dataset adequately. Therefore, we need measures of variability to adequately describe a dataset, and compare several datasets with the same center but varying variability.

Figure 3.2: Distributions with the same measures of center but varying standard deviations

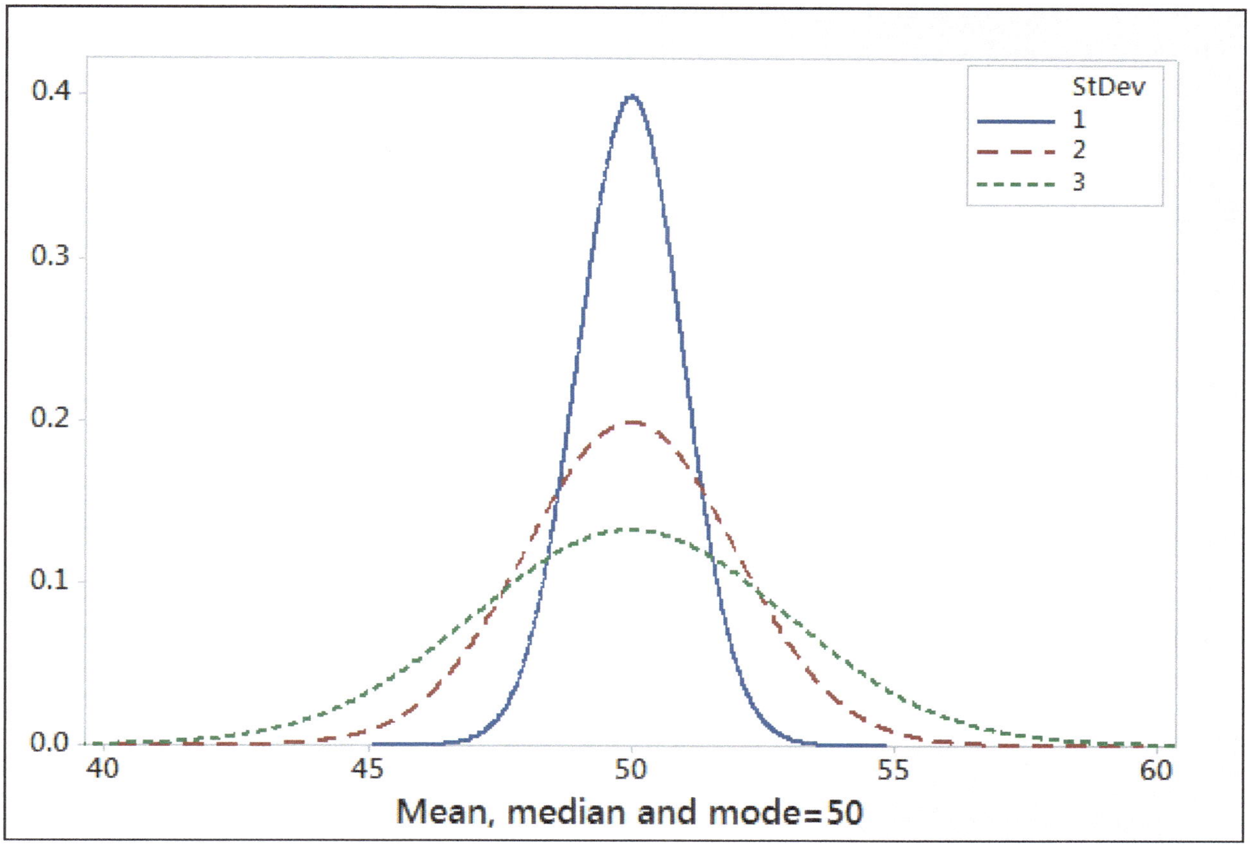

Definition: Range

The ***range***, the simplest of all measures of variation, is the maximum value minus the minimum value of a dataset.

Range=maximum value–minimum value.

Range is not a resistant measure (see Example 3.7).

Example 3.6

The sorted data for time to finish the quiz by 12 students of Example 3.2 is

5 7 10 10 13 15 20 20 25 26 29 30

Find the range of the data.

Solution

Here, the maximum=30 and minimum=5.

So, the range is $(30 - 5) = 25$.

Example 3.7

Replace 30 by 60 in Example 3.6, compute the range and comment on the result.

Solution

5	7	10	10	13	15	20	20	25	26	29	60

Here, the maximum=60 and minimum=5.

The range will be 60-5=55.

Comment

The ***range is not a resistant measure*** because it is severely affected by extreme large and small values.

Definition: Population variance and population standard deviation

The population variance and population standard deviation both measure the amount of variation in the population in two different ways.

The ***population variance***, denoted by σ^2, is the average of squared differences or deviations of all data values from their mean μ.

If a population contains N data points $X_1, X_2, \ldots, X_N$, then the population variance is

$$\sigma^2 = \frac{\sum(X-\mu)^2}{N}$$

Where,

μ is the population mean $\mu = \frac{\sum X}{N}$,

$\sum(X-\mu)^2$ is the sum of the squared deviations of all the data points from their mean μ.

The ***population standard deviation***, denoted by σ, is the positive square root of the population variance σ^2: $\sigma = +\sqrt{\sigma^2}$

The major step in finding the variance σ^2 is to find the sum of the squared deviations of the data points from their mean. Overall, the computation of σ^2 involves the following five steps:

(1) Compute μ.
(2) Compute the deviation of each data point from μ: $X - \mu$.
(3) Square each deviation in Step 2: $(X-\mu)^2$.
(4) Sum all squared deviations found in Step 3: $\sum(X-\mu)^2$.
(5) Divide the sum found in Step 4 by N to get σ^2: $\sigma^2 = \frac{\sum(X-\mu)^2}{N}$.

Example 3.8

The data below is the percentage (rounded to the nearest integer) of persons aged 19–25 years who were uninsured in the United States by four quarters of 2011.

(a) Considering the percentage uninsured in four quarters a population, compute the population variance by following all steps.

(b) Find the population standard deviation.

Quarters	Percentage Uninsured
Q1	30
Q2	27
Q3	28
Q4	25

Source: *Early Release of Health Insurance Estimates Based on Data from the National Health Interview Survey, January–September 2012.*

Solution

(a) We just follow the 5 steps described earlier:

(1) The population mean is

$$\mu = \frac{\sum X}{N} = \frac{30 + 27 + 28 + 25}{4} = 27.5$$

(2) Compute deviations $X - \mu$

$$30 - 27.5 = 2.5, \quad 27 - 27.5 = -0.5, \quad 28 - 27.5 = 0.5, \quad 25 - 27.5 = -2.5$$

(3) Compute squared deviations $(X - \mu)^2$

$$(2.5)^2 = 6.25, \qquad (-0.5)^2 = 0.25, \qquad (0.5)^2 = 0.25, \qquad (-2.5)^2 = 6.25,$$

(4) Find the total squared deviations $\sum(X - \mu)^2$

$$6.25 + 0.25 + 0.25 + 6.25 = 13$$

(5) Divide the total squared deviation by N to get variance $\sigma^2 = \frac{\sum(X-\mu)^2}{N}$

$$Variance, \sigma^2 = 13/4 = 3.25$$

Often, making a computational table below is helpful:

X	$X - \mu$	$(X - \mu)^2$
30	2.5	6.25
27	-0.5	0.25
28	0.5	0.25
25	-2.5	6.25
Sum	0	13

$$\sum(X - \mu)^2 = 13, \quad \sigma^2 = \frac{13}{4} = 3.25$$

(b) The population standard deviation, $\sigma = +\sqrt{\sigma^2} = +\sqrt{3.25} \approx 1.803$

Note: The second column total is the numeric proof that $\sum(X - \mu) = 0$, true for any population dataset.

Example 3.9

By rolling a fair die, one can get a population of six numbers: $\{1, 2, 3, 4, 5, 6\}$.

Apply all 5 steps to compute the population variance using a computational table and then compute the standard deviation.

Solution

The mean is μ =3.5

X	$X-\mu$	$(X-\mu)^2$
1	-2.5	6.25
2	-1.5	2.25
3	-0.5	0.25
4	0.5	0.25
5	1.5	2.25
6	2.5	6.25
Sum	0	17.5

$$\sum(X-\mu)^2 = 17.5, \quad \sigma^2 = \frac{17.5}{6} \approx 2.92$$

The population standard deviation is $\sigma = \sqrt{\sigma^2} = \sqrt{2.92} \approx 1.71$

Example 3.10

The blood pressure of an individual is expressed in two ways - systolic and diastolic blood pressures. The systolic blood pressure for most healthy adults ranges between 90 and 120 millimeters of mercury (mm Hg). Normal diastolic blood pressure ranges between 60 and 80 mm Hg. High blood pressure is a significant risk factor of heart attack, stroke, kidney disease, eye damage and hardening of arteries. Low blood pressure may cause nausea, dizziness, loss of balance, headaches, shortness of breath and many other life-threatening symptoms. The data below is the measurements of systolic blood pressure (SBP) of a patient on 5 consecutive visits:

120 115 118 117 140

Find the population variance and standard deviation for the above measurements of the SBP.

Solution

The mean is μ =122.

X	$X-\mu$	$(X-\mu)^2$
120	-2	4
115	-7	49
118	-4	16
117	-5	25
140	18	324
Sum	0	418

$$\sum (X - \mu)^2 = 418, \quad \sigma^2 = \frac{418}{5} = 83.6$$

The population standard deviation is

$$\sigma = \sqrt{\sigma^2} = \sqrt{83.6} \approx 9.143$$

Definitions: Sample variance and sample standard deviation

For a sample of size n, the ***sample variance*** is the total squared deviation of all sample values from their mean divided by $(n-1)$.

For a sample of n values $x_1, x_2, \ldots, x_n$, the sample variance, denoted by s^2, is defined by

$$s^2 = \frac{\Sigma(x - \bar{x})^2}{n - 1}$$

The ***sample standard deviation***, denoted by s, is the square root of the sample variance s^2:

$$s = \sqrt{s^2}$$

The computational steps for a sample variance are the same as the population variance except that we divide the total squared deviation of sample data points from their mean by $(n-1)$.

Some of the useful facts about variance and standard deviation

(1) The unit of σ and s is same as the unit of the measurement of x.
(2) The unit of σ^2 and s^2 is the square of the unit of the measurement of x.
(3) The values of σ and s are always positive and are equal to 0 when all observations are the same.
(4) σ^2, s^2, σ and s are all extremely affected by outliers.

Example 3.11

The table gives the yearly average high and low temperatures (in °F) from 1981 to 2010 for selected places in the Thumb and Southeast Michigan.

No	Place	High °F	Low °F
1	Ann Arbor	59	40
2	Dearborn	60	40
3	Detroit	59	42
4	Flint	57	38
5	Harbor Beach	52	38
6	Monroe	58	41
7	Port Huron	56	41
8	Saginaw	57	38

Source: *National Climatic Data Center* ***NOAA's 1981-2010 Climate Normals.***

(a) Considering the above dataset as a sample, find the sample standard deviation of average high and low temperatures separately.
(b) Which temperature has higher variability, high or low?

Solution

(a) For high temperature:
The mean is $\bar{x} = 57.25$.

x	$x - \bar{x}$	$(x - \bar{x})^2$
59	1.75	3.0625
60	2.75	7.5625
59	1.75	3.0625
57	-0.25	0.0625
52	-5.25	27.5625
58	0.75	0.5625
56	-1.25	1.5625
57	-0.25	0.0625
Mean=57.25	Sum=0	Sum=43.5

$$\sum (x - \bar{x})^2 = 43.5$$

Then, the sample variance is $s^2 = \frac{1}{n-1}\sum(x - \bar{x})^2 = \frac{43.5}{7} = 6.21$ sq °F.

The sample standard deviation is $s = \sqrt{6.21} = 2.49$ °F.

In a similar manner, one can verify that the standard deviation of low temperature is

$$s = 1.58°F$$

(c) The average high temperature has a higher variability than the average low temperature.

Why $(n - 1)$ in the denominator of s^2 formula?

Recall that the formula to compute sample variance is

$$s^2 = \frac{\sum(x_i - \bar{x})^2}{n - 1}$$

(1) Using $(n - 1)$ in the denominator makes s^2 an unbiased estimate of σ^2.
(2) For large sample size, using n or $(n - 1)$ does not make a significant difference in the result. However, for small sample size, using n in the denominator underestimates the population variance σ^2.
(3) n deviations $x_i - \bar{x}; i = 1,2, \ldots, n$, used to compute the sample variance or sample standard deviation, always add up to 0, i.e., $\sum(x - \bar{x}) = 0$. Therefore, knowing $(n - 1)$ deviations determines the last one automatically. In other words, since only $(n - 1)$ squared deviations can vary freely, we average the total squared deviation by dividing $(n - 1)$. The quantity $(n - 1)$ is called the degrees of freedom for the sample variance.

Example 3.12

Waiting time (in minutes) of a customer in four consecutive visits of oil change are 57, 46, 42, 55.

a) Consider these four numbers as a sample data and compute sample variance and sample standard deviation.
b) Again, consider these four numbers as a population data and compute population variance and population standard deviation.

Solution

We use the following computational table:

x	$\bar{x}$	$x - \bar{x}$	$(x - \bar{x})^2$
57	50	7	49
46	50	-4	16
42	50	-8	64
55	50	5	25
Total		$\sum(x - \bar{x}) = 0$	$\sum(x - \bar{x})^2 = 154$

(a) Considering a sample: $\bar{x} = 50$.
The sample variance is

$$s^2 = \frac{\sum(x_i - \bar{x})^2}{n - 1} = \frac{154}{3} = 51.3$$

The sample standard deviation is

$$s = \sqrt{s^2} = \sqrt{51.3} = 7.2$$

(b) Considering a population: $\mu = 50$.
The population variance is

$$\sigma^2 = \frac{\sum(X - \mu)^2}{N} = \frac{154}{4} = 38.5$$

The population standard deviation is

$$\sigma = \sqrt{\sigma^2} = \sqrt{38.5} = 6.2$$

Example 3.13

Suppose we have two samples, one of size 5 ($n_1 = 5$) and the other of size 7 ($n_2 = 7$), of systolic blood pressure (SBP), measured in millimeters of mercury (mmHg).

Suppose sample 1 is taken from patients who visited a doctor's office between 8:30am and 11:00am on a given week day, and sample 2 is taken from patients who visited a doctor's office between 8:30am and 11:00am on a given weekend. Suppose two sample measurements are as follows:

Sample 1:	120	115	159	117	140		
Sample 2:	137	121	158	121	169	145	120

Verify the following computations:

Sample 1	$\bar{x}_1 = 130.2\ mmHg$	$s_1^2 = 358.7\ sq\ mmHg$	$s_1 = 18.94\ mmHg$
Sample 2	$\bar{x}_2 = 138.7\ mmHg$	$s_2^2 = 384.9\ sq\ mmHg$	$s_2 = 19.62\ mmHg$

Make a conclusion about the variability of the two samples using sample variance and sample standard deviation.

Conclusion

Looking at the variances and standard deviations, we can conclude that sample 2 has more variability than sample 1.

Note that the two samples in Example 3.13 vary in mean, variance and standard deviation. How can we make a better comparison of these two samples, given that there are variability in measures of center and variation? Well, we can use a new measure, called the coefficient of variation.

Definition: Coefficient of variation

The coefficient of variation (cv) measures the ***percent of the standard deviation relative to the mean*** and is defined as follows:

$$cv = \frac{standard\ deviation}{mean} \times 100$$

For a population, the cv is defined as

$$cv = \frac{\sigma}{\mu} \times 100$$

The sample version of the cv is

$$cv = \frac{s}{\bar{x}} \times 100$$

Note that the cv is a unit free measure because mean and standard deviation have the same unit of measurement.

Use of cv

(1) cv is very useful to compare variability of two or more datasets with varying means, standard deviations and units of measurement.

(2) When comparing the variability of several groups using cv, the one with the lowest value of cv refers to the best group incurring lower risk, statistically.

Example 3.14

Compare the two samples using coefficient of variation for Example 3.13.

Solution

For sample 1, coefficient of variation is

$$cv_1 = \frac{s_1}{\bar{x}_1} \times 100 = \frac{18.94}{130.2} \times 100 = 14.55$$

For sample 2, coefficient of variation is

$$cv_2 = \frac{s_2}{\bar{x}_2} \times 100 = \frac{19.62}{138.7} \times 100 = 14.15$$

Looking at the values of cv, sample 2 seems to be better than sample 1 incurring lower variability relative to mean.

Example 3.15

Suppose for two investment options, an annual report reveals the followings:
Option 1: Mean return μ =9% and standard deviation σ =8 %.
Option 2: Mean return μ =9.4% and standard deviation σ =11.5 %.
Note that means and standard deviations of two investment options are different.

Which investment option do you think is riskier than the other?

Solution

$$cv(option\ 1) = \frac{\sigma}{\mu} \times 100 = \frac{8}{9} \times 100 = 88.9$$

That is the standard deviation is 88.9% of the mean return.

$$cv(option\ 2) = \frac{\sigma}{\mu} \times 100 = \frac{11.5}{9.4} \times 100 = 122.3$$

Conclusion

Relative to the mean, the variability in mean return is higher for option 2 than option 1. Therefore, investment option 2 seems riskier than option 1.

Summary points about measures of variability

Measures of Variability (Descriptive statistics)	**Population (Parameter)**	**Sample (Statistic)**
Range *Range is sensitive to outliers or extreme values, why?*	$Range = Max - Min$	$Range = Max - Min$
Interquartile Range (IQR) *IQR is better than range*, why?	$IQR = Q_3 - Q_1$ It's the range of the middle 50% data.	$IQR = Q_3 - Q_1$ IQR is useful in determining the outliers in the data.
Variance Unit: Square unit *Variance is sensitive to outliers or extreme values, why?*	Population variance $\sigma^2 = \frac{\sum(X_i - \mu)^2}{N}$ Average of squared deviations of all data points in the population from their mean.	Sample variance $s^2 = \frac{\sum(x_i - \bar{x})^2}{n-1}$ Average of squared deviations of all data points in the sample from their mean.

Standard deviation Unit: Linear unit *Standard deviation is sensitive to outliers or extreme values, why?*	$\sigma = +\sqrt{\sigma^2}$ Standard deviation is the positive square root of variance.	$s = +\sqrt{s^2}$
Coefficient of Variation (cv) Unit: None *cv is sensitive to outliers or extreme values, why?*	$cv = \frac{\sigma}{\mu} \times 100$ Percent of the standard deviation relative to mean.	$cv = \frac{s}{\bar{x}} \times 100$

Section 3.3: Measures of position

Given a quantitative variable, summary measures such as **Z-score**, **percentiles**, **quartiles** and **five-number summary** describe relative position of a data point in the dataset. For examples, Z score of a raw score X indicates how many standard deviations above or below the score X is from the mean score. The pth percentile of a dataset has $p\%$ of the data points at or below it. In other words, $(100 - p)\%$ of the data points is at or above the pth percentile. This section devotes to these positional measures.

Definition: Z-score

The **Z-score**, also called the **standardized score**, of a variable X with mean μ and standard deviation σ is defined as follows:

$$Z = \frac{X - \mu}{\sigma}$$

The sample version of Z-score is defined by

$$z = \frac{x - \bar{x}}{s}$$

Z-score measures how many standard deviations above or below a given score is from the mean score. For example, a Z-score of 1.25 means that a given score X is 1.25 standard deviations above the mean score. A Z-score of -1.25 means that a given score X is 1.25 standard deviations below the mean score.

Does Z-score have any unit?

No, Z- score does not have any unit of measurement. Why?

Example 3.16

The mean height of students in a class is 65 inches and the standard deviation is 3.5 inches.

(a) What is the z-score of height of a student who is 60 inches tall?
(b) Interpret your result in (a).

Solution

(a) The value of Z score for a height of 60 inches will be

$$z = \frac{60 - 65}{3.5} = -1.43$$

(b) The $z = -1.43$ means that the student with the height of 65 inches is 1.43 standard deviations below the mean height of the class.

Example 3.17

Intelligence quotient (IQ) measures a person's cognitive ability (intelligence) in relation to the age. Roughly, 68% of the population has an IQ between 85 and 115. Suppose, the mean IQ of people is 100 and standard deviation is 15.

Comment on IQ of 85 and 115.

Solution

For an IQ of 85, the z-score is

$$z = \frac{85 - 100}{15} = -1.0$$

Therefore, an IQ of 85 is 1 standard deviation below the mean IQ.

For an IQ of 115, the z-score is

$$z = \frac{115 - 100}{15} = 1.0$$

Therefore, an IQ of 115 is 1 standard deviation above the mean IQ.

Definition: Percentiles

Height, weight or test score are often expressed as percentile to describe the relative position of individual scores in the entire dataset. Percentiles divide the entire dataset into 100 equal parts. Given an ordered dataset, the ***pth percentile***, denoted by P_p, is a value with approximately lowest $\boldsymbol{p}\%$ data points at or below it. Also, the pth percentile has highest $(\mathbf{100} - \boldsymbol{p})\%$ of the data points at or above it.

How to compute percentile P_p of a dataset?

There are several algorithms for computing percentiles. Results of percentiles with different algorithms may often vary depending on the dataset or the number of data points. Below is one of the algorithms for computing the pth percentile of a dataset:

(1) Sort the data points in ascending order.

(2) Compute the location l of pth percentile using the formula:

$$l = \frac{p}{100} \times n; \quad p = 1, 2, \dots, 99$$

where,

p is the desired percentile, e.g., for 10th percentile, $p = 10$.

n is the number of data points.

l is the position or location for desired percentile.

(3)

(a) When l is fraction, round it up to the next integer for the location of the pth percentile.
(b) When l is an integer, the average of the two measurements at locations l and $(l+1)$ is the pth percentile.

Caution!

Remember that neither l nor $(l+1)$ is the value of the percentile; l and $(l+1)$ are used to locate the value of a desired percentile.

Example 3.18

The sorted data of time (in minutes) it takes to finish a quiz of 12 students in a class is

5 7 10 10 13 **15** **20** 20 25 25 30 30

Find 10th, 25th, 30th, 50th, 75th and 90th percentiles of the finishing time.

Solution

10th percentile:
The location for the 10th percentile is

$$l = \left(\frac{10}{100}\right) 12 = 1.2$$

Rounding up to the next integer we have, $l = 2$.

Then, the 10th percentile is the measurement at location 2, which is 7. That is, $P_{10} = 7$.

How to interpret this result?

This means, 10% of the students finish the quiz at or before 7 minutes. In other words, 90% of the students finish the quiz at or after 7 minutes.

25th percentile:
The location for the 25th percentile is

$$l = \left(\frac{25}{100}\right) 12 = 3$$

Then, the 25th percentile is the average of measurements at positions 3 and 4, which is $\frac{10+10}{2} = 10$. That is, $P_{25} = 10$.

30th percentile:
The location for the 25th percentile is

$$l = \left(\frac{30}{100}\right) 12 = 3.6$$

Rounding up to the nearest integer greater, we have $l = 4$. Then, the 30th percentile is at the location 4, which is 10. That is, $P_{30} = 10$.

50th percentile:
The location for the 50th percentile is

$$l = \left(\frac{50}{100}\right) 12 = 6$$

Then, the 50th percentile is the average of measurements at positions 6 and 7, which is $\frac{15+20}{2} = 17.5$. That is, $P_{50} = 17.5$.

75th percentile:

The location for the 75th percentile is

$$l = \left(\frac{75}{100}\right) 12 = 9$$

Then, the 75th percentile is the average of measurements at positions 9 and 10, which is $\frac{25+25}{2} = 25$. That is, $P_{75} = 25$.

90th percentile:

The location for the 90th percentile is

$$l = \left(\frac{90}{100}\right) 12 = 10.8$$

Rounding up to the integer greater than 10.8 is 11.

Then, the 90th percentile is the measurement at location 11, which is 30. That is, $P_{90} = 30$.

How to compute percentile of a given value in the dataset?

Very often we are interested to determine the percentile of a given value x in the dataset. For example, referring to Example 3.18, we might be interested to know the percentile of the student who finishes the quiz in 15 minutes.

Given a dataset with n values, the percentile corresponding to a given value x is computed using the following formula:

$$p[x] = \frac{(number\ of\ values < x) + 0.5}{n} \times 100$$

where,

$p[x]$ is the percentile given to the value x,

n is the number of values in the dataset.

The general practice is to round $p[x]$ up to the next integer.

Example 3.19

In reference to Example 3.18, find the percentile of the student who finishes the quiz in 15 minutes.

Solution

The sorted data of time to finish the quiz for 12 students is

5	7	10	10	13	**15**	20	20	25	25	30	30

We see that the number of students who finish the quiz in less than 15 minutes is 5. Then, by using the formula:

$$p[x] = \frac{(number\ of\ values < x) + 0.5}{n} \times 100$$

$$p[15] = \frac{(5) + 0.5}{12} \times 100 = 45.83$$

By rounding up, the value of 15 minutes corresponds to 46th percentile in the dataset.

Now, let us consider computing the 46th percentile of the dataset:

46th percentile:

Location for the 46th percentile is

$$l = \left(\frac{46}{100}\right) 12 = 5.52$$

Rounding up to the nearest integer greater, we have $l = 6$. Then, the 46th percentile is at the location 6, which is 15. That is, $P_{46} = 15$.

Definition: Quartiles

Three quartiles divide the entire dataset into four equal parts.

The first quartile, denoted by Q_1, is the 25th percentile which divides the lower half of the entire dataset into two equal halves.

The second quartile, denoted by Q_2, is the 50th percentile which divides the entire dataset into two equal halves.

The third quartile, denoted by Q_3, is the 75th percentile which divides the upper half of the entire dataset into two equal halves.

It follows that 25% of the data points are at or below Q_1 and 75% of the data points are at or below Q_3. Indeed, it follows that

$$\boldsymbol{Median = Q_2 = P_{50}}$$
$$\boldsymbol{Q_1 = P_{25}}$$
$$\boldsymbol{Q_3 = P_{75}}$$

Interquartile range

In the middle 50% of the data points, the minimum is the first quartile ($\boldsymbol{Q_1}$), which divides the lower half of the entire dataset into two equal halves and maximum is the third quartile ($\boldsymbol{Q_3}$), which divides the upper half of the entire dataset into two equal halves. Therefore, the ***interquartile range*** or the ***range of the middle 50% data***, denoted by IQR, is defined as

$$\boldsymbol{IQR = Q_3 - Q_1}$$

If data contains outliers, then IQR is preferable to range because IQR is less affected by extreme values than does the range.

An IQR is important in detecting outliers. The detection of outliers will be discussed shortly.

Example 3.20

Find three quartiles Q_1, Q_2 and Q_3 and hence the value of IQR using quiz scores for 13 students in an Elementary Statistics Class given by:

20	15	10	15	5	20	10	7	11	9	20	15	20

Solution

Sorting these numbers we get:

5 7 9 **10** 10 11 **15** 15 15 20 20 20 20

Here, $n = 13$.

Note that $Q_1 = P_{25}$, $Q_2 = P_{50}$ and $Q_3 = P_{75}$.

Computation of Q_1:

The location for Q_1, i.e., 25th percentile is

$$l = \left(\frac{25}{100}\right) 13 = 3.25$$

Rounding up to the integer greater than 3.25 is 4.

Then, the 25th percentile or Q_1 is measurement at position 4, which is 10.

That is, $Q_1 = 10$.

Computation of Q_2:

The location for Q_2, i.e., 50th percentile is

$$l = \left(\frac{50}{100}\right) 13 = 6.5$$

Rounding up to the integer greater than 6.5 is 7.

Then, the 50th percentile or Q_2 is measurement at position 7, which is 15.

That is, $Q_2 = 15$.

Computation of Q_3:

The location for Q_3, i.e., 75th percentile is

$$l = \left(\frac{75}{100}\right) 13 = 9.75$$

Rounding up to the integer greater than 9.75 is 10.

Then, the 75th percentile or Q_3 is measurement at position 10, which is 20. That is, $Q_3 = 20$.

Computation of IQR:

The interquartile range, $IQR = Q_3 - Q_1 = 20 - 10 = 10$.

Five number summary and boxplot

The five-number summary is a comprehensive way of describing a quantitative variable and its distribution. Roughly speaking, before doing any sophisticated analysis of a variable, it is a good practice to inspect the five-number summary of the dataset for the variable.

Definition: Five-number summary

Descriptive statistics by means of five summary numbers, namely, the minimum, the first quartile, the median, the third quartile, and the maximum, is collectively called the five-number-summary.

A graphical display of five-number summary is the boxplot, also called the box-and-whisker diagram.

Below is an algorithm to draw the boxplot:

(1) Find five-number summary.
(2) Draw three vertical lines locating Q_1, Q_2 and Q_3 on a number line and complete a box connecting these three quartiles.
(3) Mark the minimum and the maximum by two vertical line segments and connect them with the box by horizontal lines.

Example 3.21

Calculate five-number summary and draw a boxplot for time (in minutes) it takes to finish a quiz of 12 students, using the 12 ordered values below:

5 7 10 10 13 **15** **20** 20 **25** **25** 30 30

Here, Min=5, Q_1= 10, Q_2= 17.5 (Median), Q_3= 25 and Max=30.

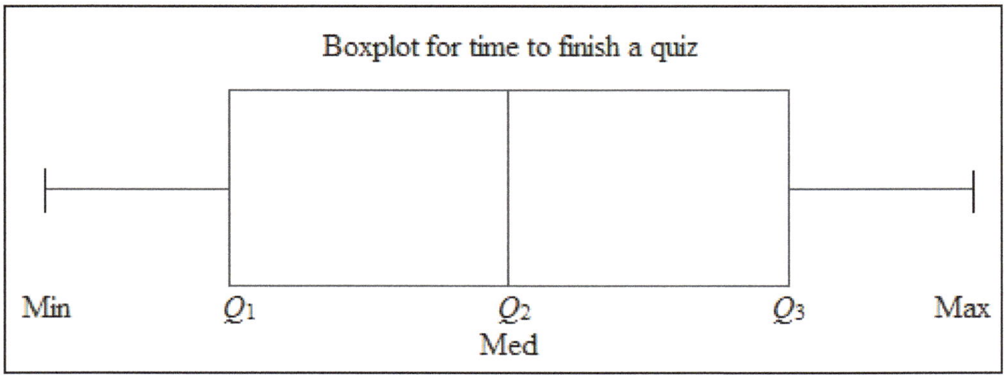

Detection of outliers

Outliers are individual data values that stand apart from the rest of the data values showing an unusual pattern. Some graphs such as histogram, stem-plot and boxplot provide information about the existence of possible outliers in the dataset.

How to detect the existence of outliers in a dataset in a quantitative way?

The answer to the above question is due to Professor John Tukey (1977). He considered the values of "$Q_1 - 1.5 \times IQR$" and "$Q_3 + 1.5 \times IQR$" as the "***lower fence***" and the "***upper fence***", respectively, to mark off the "reasonable" values from the "outliers". Any values outside the fences $Q_1 - 1.5 \times IQR$ and $Q_3 + 1.5 \times IQR$ are **outliers**.

The two quantities $Q_1 - 1.5 \times IQR$ and $Q_3 + 1.5 \times IQR$ are also known as the ***lower*** and ***upper limits***:

$$Lower\ Limit = LL = Q_1 - 1.5 \times IQR$$
$$Upper\ Limit = UL = Q_3 + 1.5 \times IQR$$

Then, an observation is outlier if it is below LL or above UL.

Example 3.22

Reference to Example 3.21, investigate if there is any outlier in the time to finish the quiz.

Solution

By sorting, we have the following arrangement of the data points:

5	7	**10**	**10**	13	**15**	**20**	20	**25**	**25**	30	30

We have,

$$Q_1 = 10, \quad Q_2 = 17.5, \quad Q_3 = 25\,, \quad IQR = Q_3 - Q_1 = 15$$

$$IQR \times 1.5 = 15 \times 1.5 = 22.5$$

$$LL = Q_1 - 1.5 \times IQR = 10 - 22.5 = -12.5$$

$$UL = Q_3 + 1.5 \times IQR = 25 + 22.5 = 47.5$$

Since there is no observation below -12.5 and above 47.5, there is no outlier in the dataset.

Example 3.23

The blood pressure of an individual is expressed in two ways - systolic and diastolic blood pressure. The systolic blood pressure for most healthy adults ranges between 90 and 120 millimeters of mercury (mm Hg). Normal diastolic blood pressure ranges between 60 and 80 mm Hg. High blood pressure is a significant risk factor of heart attack, stroke, kidney disease, eye damage and hardening of arteries. Low blood pressure may cause nausea, dizziness, loss of balance, headaches, shortness of breath and many other life-threatening symptoms.

The data below is the measurements of systolic blood pressure of a patient on 5 consecutive visits

115	**117**	**118**	**120**	140

Find if there is any outlier in the data.

Solution

We find that

$$Q_1 = 117, \quad Q_3 = 120\,, \quad IQR = 3$$

$$IQR \times 1.5 = 3 \times 1.5 = 4.5$$

$$LL = Q_1 - 1.5 \times IQR = 117 - 4.5 = 112.5$$

$$UL = Q_3 + 1.5 \times IQR = 120 + 4.5 = 124.5$$

As we see, there is no value in the dataset below 112.5, however, 140 is above the UL of 124.5. Thus, 140 is an outlier.

Mean and variance for grouped data

For grouped data where data are presented as class interval and frequency corresponding to each class interval, we need to modify the definition of the mean and variance formula described earlier. Consider the grouped data of Example 2.7, where we have time (in minutes) to finish a quiz by 28 students, arranged in 6 classes, each of length 5. See the arrangement in table below:

Classes	Mid-values (X)	Frequency (f)
0-5	2.5	1
5-10	7.5	5
10-15	12.5	6
15-20	17.5	8
20-25	22.5	5
25-30	27.5	3

In the table, X is the mid-value of a class, and f is the frequency of that class. For this sort of arrangement of a population, the mean (μ) and variance (σ^2) are computed using the formulas given below:

$$\mu = \frac{\Sigma fX}{\Sigma f}$$

$$\sigma^2 = \frac{\Sigma f(X-\mu)^2}{\Sigma f}$$

Example 3.24

Compute population mean, variance and standard deviation of the time to finish a quiz of 28 students using the grouped data arrangement below

Classes	Mid-values (X)	Frequency (f)
0-5	2.5	1
5-10	7.5	5
10-15	12.5	6
15-20	17.5	8
20-25	22.5	5
25-30	27.5	3

Solution

The formulas for computing mean and variance for the grouped data are

$$\mu = \frac{\Sigma fX}{\Sigma f} \text{ and } \sigma^2 = \frac{\Sigma f(X-\mu)^2}{\Sigma f}$$

Here, X is the mid-value of a class, and f is the frequency of that class.

From table below, we have $\sum fX = 450$ and $\sum f = 28$. Therefore, the population mean is

$$\mu = \frac{\sum fX}{\sum f} = \frac{450}{28} = 16.07$$

Classes	X	f	fX	$f(X-\mu)^2$
0-5	2.5	1	2.5	184.1449
5-10	7.5	5	37.5	367.2245
10-15	12.5	6	75	76.4694
15-20	17.5	8	140	16.3592
20-25	22.5	5	112.5	206.7245
25-30	27.5	3	82.5	391.9347
Sum	90	$\sum f$ =28	$\sum fX$ =450	$\sum f(X-\mu)^2$ =1242.8572

The population variance is

$$\sigma^2 = \frac{\sum f(X-\mu)^2}{\sum f} = \frac{1242.8572}{28} = 44.388$$

The population standard deviation is

$$\sigma = +\sqrt{\sigma^2} = \sqrt{44.388} = 6.662$$

Section 3.4: Implementation of Technology

In this section, we will provide Minitab and TI-84 Plus Instruction for computing numeric descriptive statistics. We will apply technology to do the following example:

Example 3.25

The data below represents online price in dollars (X) of 15 Xbox 360 games on a given day.

26.89	20.17	35.99	46.67	20.02	39.99	15.49	35
58.99	8.59	13.49	59.99	50.73	28.99	19.99	

We will compute descriptive statistics of X using Minitab and TI-84 Plus calculator with necessary instructions.

Minitab Instruction for numeric descriptive statistics

Ref: Example 3.25

Below are steps that can be followed to get desired descriptive statistics in Minitab

(1) Enter the price of Xbox 360 games into C1 column of Minitab Worksheet and label as X.
(2) From Menu Bar option, use: **Stat>Basic Statistics>Display Descriptive Statistics...**
(3) In Display Descirptive Statistics dialog box, select C1 for Variables: and Click **[Statistics...]** and **check** boxes for desired statistics and **uncheck** other boxes.
(4) Click **[OK]** twice.

See the dialog boxes below with option selections for selected statistics.

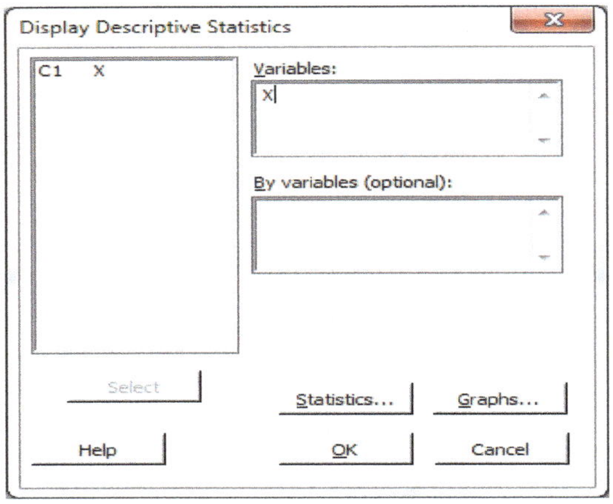

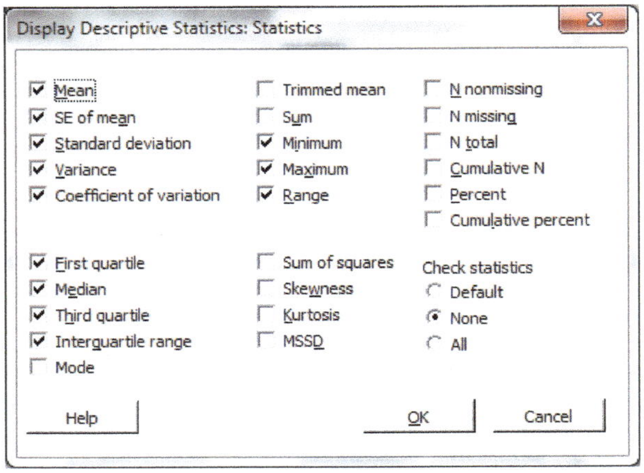

After all these steps, we get the following results in the Minitab session window for data in Example 3.25.

Descriptive Statistics: X

Variable	Mean	SE Mean	StDev	Variance	CoefVar	Minimum	Q1	Median	Q3	Maximum
X	32.07	4.25	16.46	270.91	51.33	8.59	19.99	28.99	46.67	59.99

Variable	Range	IQR
X	51.40	26.68

Boxplot in Minitab

For Ref: Example 3.25, below is the necessary steps to draw a boxplot in Minitab:

(1) Enter the price of Xbox 360 games into C1 column of Minitab Worksheet and label as X.
(2) From Menu Bar option, use: **Graph>Boxplot…**
(3) For one Y, click on box **Simple** and click [**OK**].
(4) For **Graph variables:** Select C1, use **Labels**… option to add Title, Footnote, etc. and check **Transpose value and category scales** from **Scale**… option to get a horizontal boxplot.
(5) Click **[OK]** twice.

See the dialog boxes below with option selections and the boxplot produced.

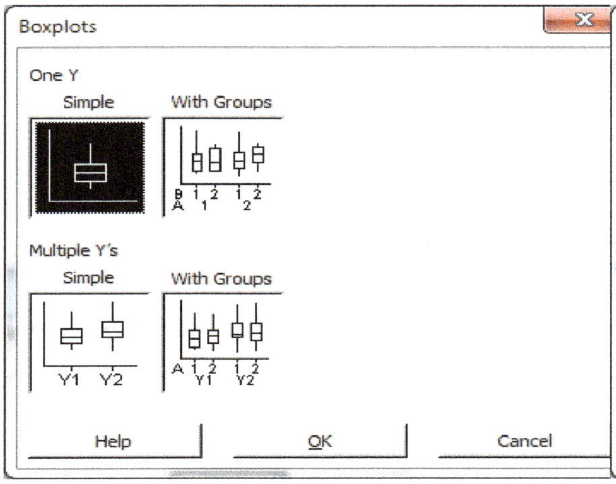

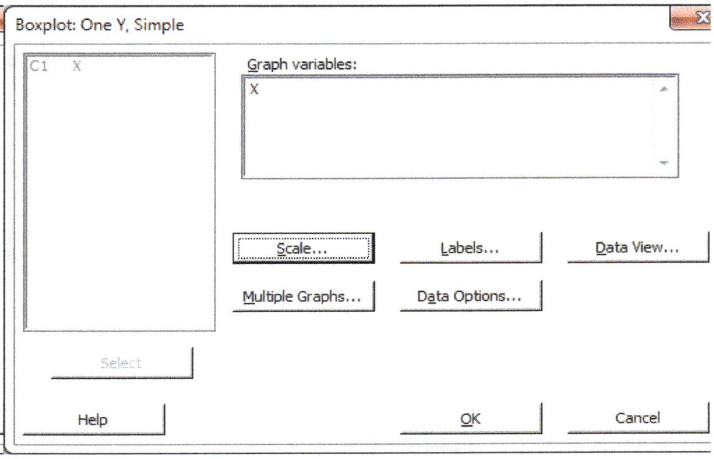

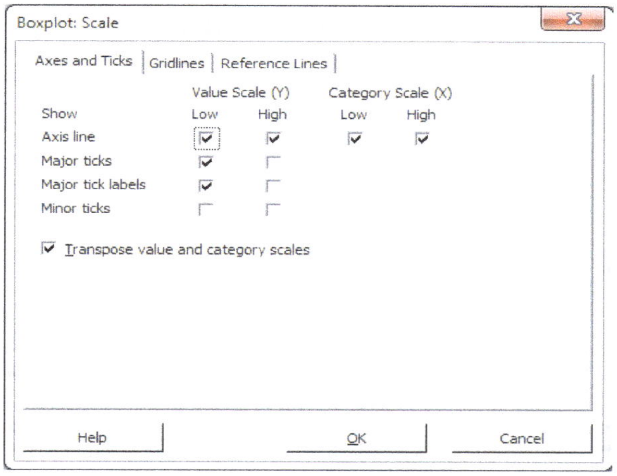

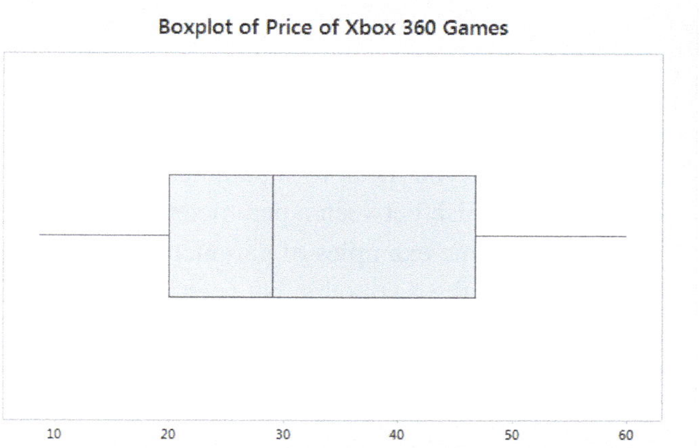

TI-84 Plus instruction for descriptive statistics

Data Entry

(1) STAT>EDIT>1:Edit… (Enter data in any of L1,L2,…L6), we enter price of Xbox 360 data in L1.

(2) 2nd >MODE[:Quit] (to get back the empty screen after the data entry)

Getting Descriptive Statistics

(1) STAT> CALC>1: 1-Var Stats>ENTER

(2) 2nd 1 [:L1] or 2 [L2]… to tell where the data is

(3) Hit ENTER

(4) Use Arrow-key to go down for all statistics you need

The result of the steps as described can be demonstrated in figures below:

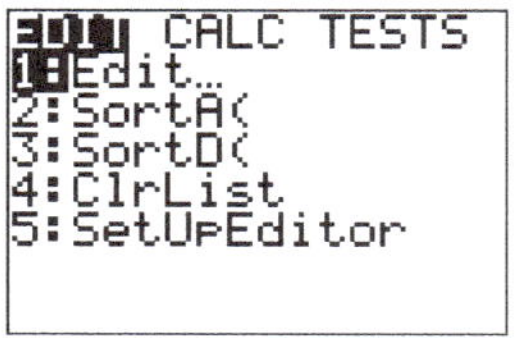

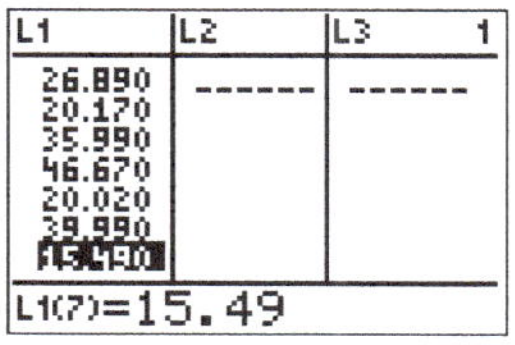

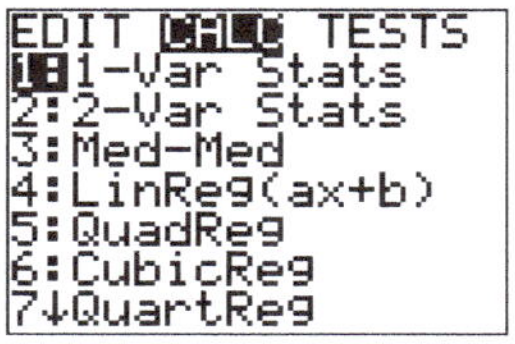

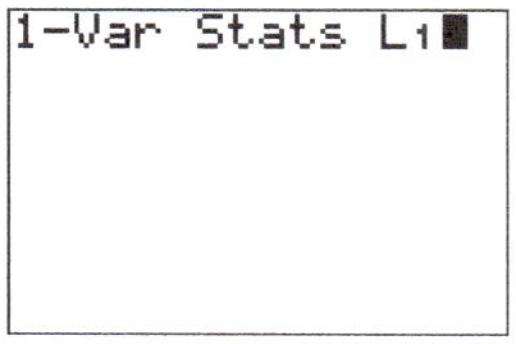

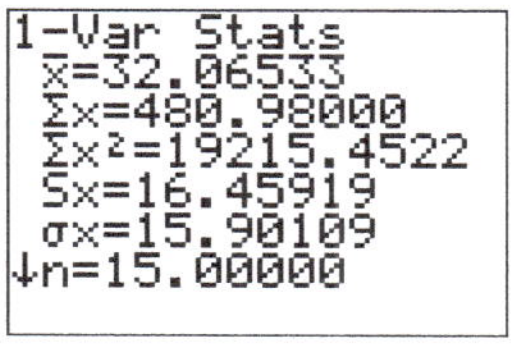

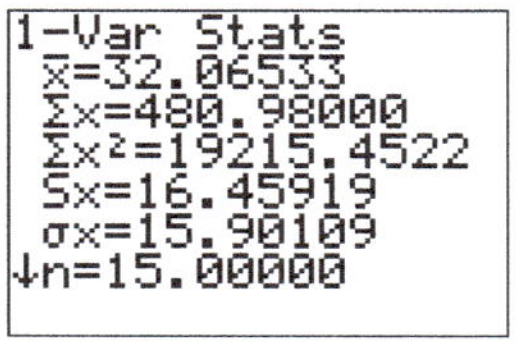

Chapter 3: Practice problems

(3.1) Briefly, answer the following questions:

(a) What do you mean by a parameter and a statistic?
(b) Distinguish between a parameter and a statistic.
(c) Give some examples of parameter and statistic.
(d) What is the distinction between μ and $\bar{x}$?

(3.2) Briefly, answer the following questions:

(a) Define mean, median and mode.
(b) How would you compute the median from a dataset with an even number of observations?
(c) How would you compute the median from a dataset with an odd number of data points?

(3.3) Compare the mean, median and mode of a population with a shape that is

(a) symmetric or bell shaped.
(b) left or negatively skewed.
(c) right or positively skewed.

(3.4) Comment on the shape of a population for which

(a) mean=median=mode.
(b) mean=20, median=15 and mode=10.
(c) mean=12, median=17 and mode=21.

(3.5) Data below refers to the amount of carbohydrate (in grams) from a sample of 6 General Mills' cereal bars.

23	28	41	22	20	23

(a) Compute the mean carbohydrate in the sample.
(b) What is the median amount of carbohydrate in the sample?
(c) What is the mode amount of carbohydrate in the sample?
(d) Looking at the values of mean, median and mode, what can you say about the shape of the distribution of the amount of carbohydrate in the population of the General Mills' cereal bars?

(3.6) The following numbers correspond to the price (in dollars) of seven varieties of golf bags searched on the internet by a golfer.

83	105	149	159	190	100	380

(a) Find the mean price of golf bags.
(b) What is the median price of golf bags?
(c) Considering above price as a sample, compute sample variance and sample standard deviation of the price.
(d) Compute Q_1, Q_3 and IQR of the price.
(e) Verify if there is any outlier in the price.
(f) What is the 50th percentile of the price?

(3.7) The data below refers to the price (in dollars) of top 10 books as searched on the internet on a given day:

17	29	10	40	13	17	26	29	6	26

(a) Find the 25th, 50th and 75th percentiles of price of top 10 books.
(b) What are the five-number summary to the list of price?
(c) What percentile price does $17 correspond to the list?
(d) Investigate if there is any outlier in the listed price.

(3.8) The mean price of golf bags is $165 and the standard deviation is $101.
(a) Find the Z-score for the price of a bag of $320.
(b) How would you interpret the z-score obtained in (a)?

(3.9) Which of the following statements are true?
(a) Mean is highly affected by extreme values.
(b) Median is less affected by extreme values.
(c) The five number summary of a dataset consists of the minimum, the first quartile, the mode, the third quartile and the maximum.
(d) The third quartile is the same as the 75th percentile.
(e) If the shape of a variable is left skewed, then the mean is less than the median.
(f) If the distribution of a variable is symmetric, then the mean, median and mode are the same.

(3.10) Fill in the blanks with appropriate word(s) for the following sentences.
(a) Mean is not a ________________________ measure, but median is a __________________ measure.
(b) Median has ________% data points at or below it and ____________% data points at or above it.
(c) Three _____________ divide the entire data sets into four equal parts.
(d) Z-score measures how many standard deviations above or below a given data point is from the ___________ value.
(e) The ______________________________________ measures the size of the standard deviation relative to the size of the mean.
(f) The range of the middle 50% data points is called the ___________________________________.
(g) The five number summary of a data set consists of the _______________, __________________, _______________, ________________________, and ____________________________.

(3.11) Suppose that an employee with a BBA (Honors), on average, earns $63,200 with a standard deviation of $9,620.
(a) What is the Z-score of a person whose salary is $70,500?
(b) Interpret your result of Z-score in (a).

(3.12) As searched in http://www.cheapoair.com, on November 4, 2013, the data below refers to the price (in dollars) of an adult 1+ stops one way flight trip from Detroit Wayne (DTW) airport to San Diego (SAN) California airport on November 20, 2013:

187 192 243 258 261 304 314 340 384 787

(a) Find the mean and median price of the trip.
(b) Find the sample standard deviation of the price of the trip.
(c) Find the 25th percentile and 50th percentile of price.
(d) What are the range and IQR of price? How do you compare the result of IQR from that of the range?
(e) Is there any outlier in the price?

(3.13) A Z-score of $Z = 1.3$ means that a given score X is
(a) 1.3 standard deviations above the mean score.
(b) 1.3 standard deviations below the mean score
(c) 1.3 standard deviations above or below the mean score.
(d) all of the above.

(3.14) A Z-score of $Z = -1.3$ means that a given score X is
(a) 1.3 standard deviations above the mean score.
(b) 1.3 standard deviations below the mean score
(c) 1.3 standard deviations above or below the mean score.
(d) all of the above.

(3.15) The data below represents online price (in $) of 10 Xbox 360 games on a given day.

26.89	20.17	35.99	46.67	20.02
58.99	8.59	13.49	59.99	50.73

(a) Compute range, Q_1, Q_3 and IQR of above price.
(b) Compute the sample variance of the price.
(c) Compute the sample standard deviation of the price.
(d) How is the standard deviation different from the variance?
(e) Find 75th percentile of the price.
(f) What percentile is the price value 8.59 in the dataset?

Chapter 3: Answer to selected problems

(3.4) (a) Symmetric (b) Positive or right skewed (c) Negative or left skewed

(3.5) (a) 26.17 grams (b) 23 grams (c) 23 (d) Negative or left skewed

(3.6) (a) \$166.57 (b) \$149.00 (c) Sample variance $s^2 = 10278.9$ and sample standard deviation $s = 101.4$ (d) $Q_1 = 100$, $Q_3 = 190$, IQR=190 (e) $LL = -35$ and $UL = 325$. Therefore, \$380 is an outlier. (f) \$149.00.

(3.7) (a) $Q_1 = \$13$, $Q_2 = \$21.5$, $Q_3 = \$\,29$ (b) 6, 13, 21.5, 29, 40 (c) 35th percentile (d) None

(3.8) (a) 1.53 (b) Try yourself

(3.9) (a) T (b) T (c) F (d) T (e) T (f) T

(3.10) (a) resistant, resistant (b) 50% (c) Quartile (d) mean (f) IQR (g) min, Q1, Q2, Q3 and max.

(3.12) (a) \$327 (b) Sample standard deviation $s = \$173.11$ (c) \$243, \$282.5 (d) $Range = 600$, $IQR = 97$ (e) $LL = 97.5$ and $UL = 485.5$; therefore, 787 is an outlier.

(3.13) (a) **(3.14)** (b)

CHAPTER 4
CORRELATION AND REGRESSION

Introduction

In this chapter, we explore the relationship between two variables for each individual in the dataset via **scatterplot**, **linear correlation** and **linear regression**. Values of two variables are said to form a **bivariate data** if they come from the same individual. For example, values of serving size (in grams) and calorie count of various fast food, current age and expected life of males in the US or the cost of airline fuel consumption of US domestic and international carriers form a bivariate data. The questions that most frequently arise about a bivariate data are:

- How are two variables related?
- Is the relationship linear?
- If so, what is the strength of the linear relationship?
- Can we fit a line to approximate the relationship?
- Can the fitted line be used for the prediction?
- How good is the prediction?

The main objective of this chapter is to address answers to these questions and provide some applications.

Section 4.1: Scatterplot and correlation

Let us start the section with some definitions and related examples.

Definition: Bivariate data

Values of two variables coming from each of all individuals in a study are said to form ***bivariate data***. For example, measurements of systolic and diastolic blood pressures, or height and weight for a group of people form a bivariate data.

A bivariate data is denoted by an ***ordered pair*** $(\boldsymbol{x}, \boldsymbol{y})$, where x and y are measurements of two variables from the same subject. Similarly, given a group of individuals, the followings are some examples of a bivariate data (x, y):

- (Age, body-mass-index)

- (Hours of study, GPA)
- (Age at current, expected life)

Given a bivariate data (x, y), we want to know if the two variables are linearly related and if they do, find the strength and direction of the relationship between x and y.

Definition: Scatterplot

A ***scatterplot*** is a graph of a bivariate data (x, y), where x is placed along X-axis and y along Y-axis. A scatterplot is the first step to understand the existence, and /or pattern of relationship between two variables.

Why a scatterplot?

A scatterplot is a useful tool because

(1) it gives an initial clue of whatever model best represents the data.
(2) it helps to understand the ***direction*** (positive-negative), ***form*** (linear-nonlinear) and ***strength*** of linear relationship existing between two variables.
(3) it helps detecting outliers.

Definition: Correlation

Two variables are said to have a ***correlation*** if a change in one variable causes a change in the other variable. Correlation can be positive or negative depending on the direction of the change.

(1) Two variables have a ***positive correlation*** if they change in the same direction. A positive correlation corresponds to a ***positive linear relationship*** between two variables in that the scatterplot seems to form a straight line with an upward direction from left to right. For example, the correlation between the cost of airline fuel and airline ticket or hours of study and GPA for a group of students is generally positive.
(2) Two variables have a ***negative correlation*** if they change in the opposite direction. A negative correlation corresponds to a ***negative linear relationship*** between two variables in that the scatterplot seems to form a straight line with a downward direction from left to right. For example, the correlation between current age and expected life or price and demand for a commodity for a group of people is negative.
(3) The correlation is ***perfect positive*** or ***perfect negative*** if the scatterplot form a straight line with a positive or negative slope, respectively. A perfect positive or negative correlation corresponds to a perfect positive or negative linear relationship, respectively.

In Figures 4.1 (a)-(f), we have demonstrated relationship and correlation between x and y using scatterplots.

Figure 4.1: Scatterplots showing relationship and correlation between x and y.

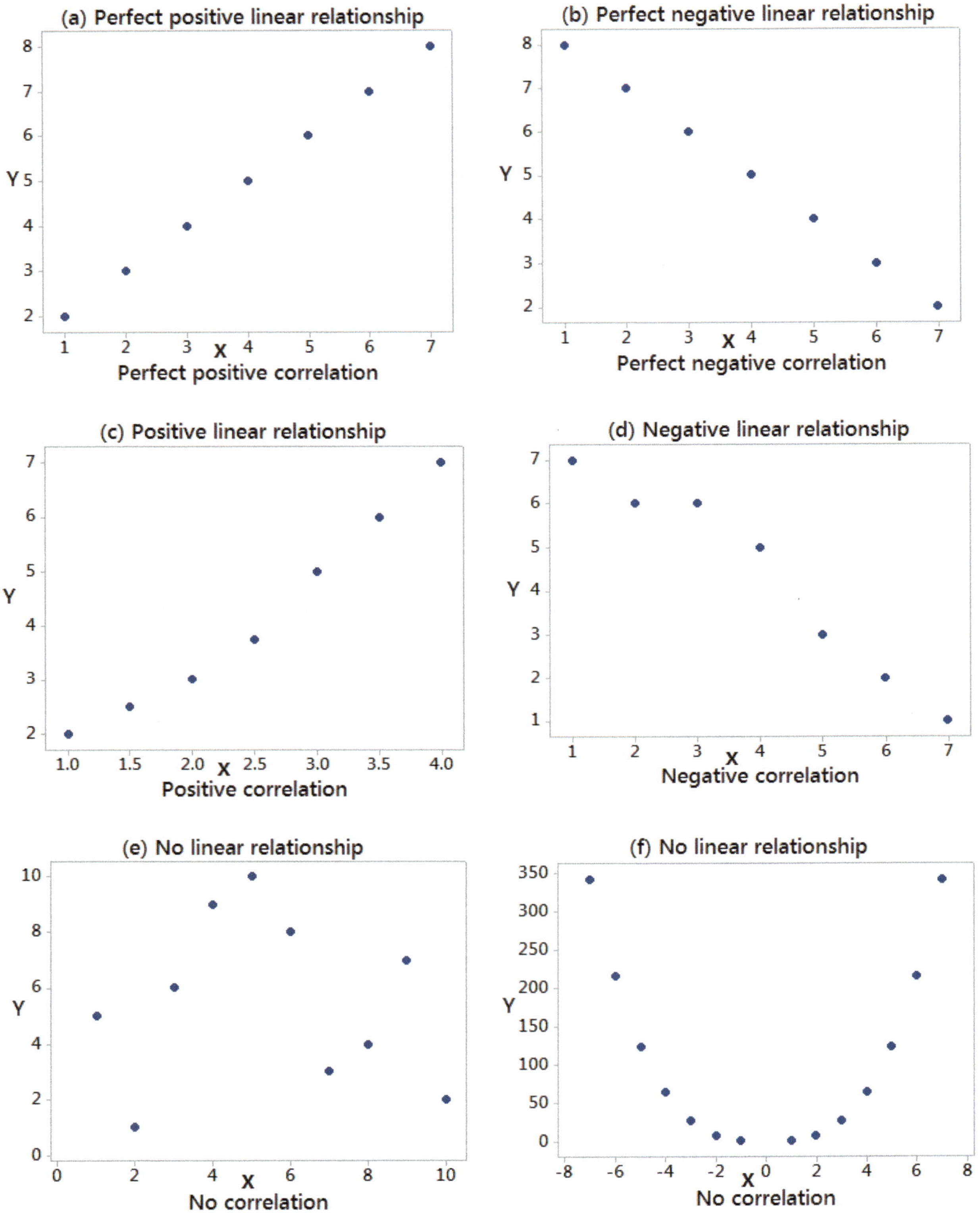

Example 4.1

We have a bivariate data (x, y), where x measures calorie count and y measures fat content (in grams) from a serving size of 28 grams of various nuts and seeds.

Types of nuts and seeds	x	y
Almonds	171	15.3
Brazil nuts	193	19.1
Cashew nuts	164	13.7
Hazelnuts	183	17.8
Sesame seeds	171	15.8
Sunflower seeds	164	13.6
Walnuts	194	19.2

Source: *Calorie Counter* **http://www.weightlossresources.co.uk/r/calories/calorie_counter.htm**

(a) Draw a scatterplot of y versus x.

(b) Does the scatterplot suggest a relationship between x and y? If it does, what form of relationship is there?

(c) Does the plot provide any indication of outlier? Why do you think so?

Solution

Figure 4.2: Scatterplot of fat content (in grams) against calorie count

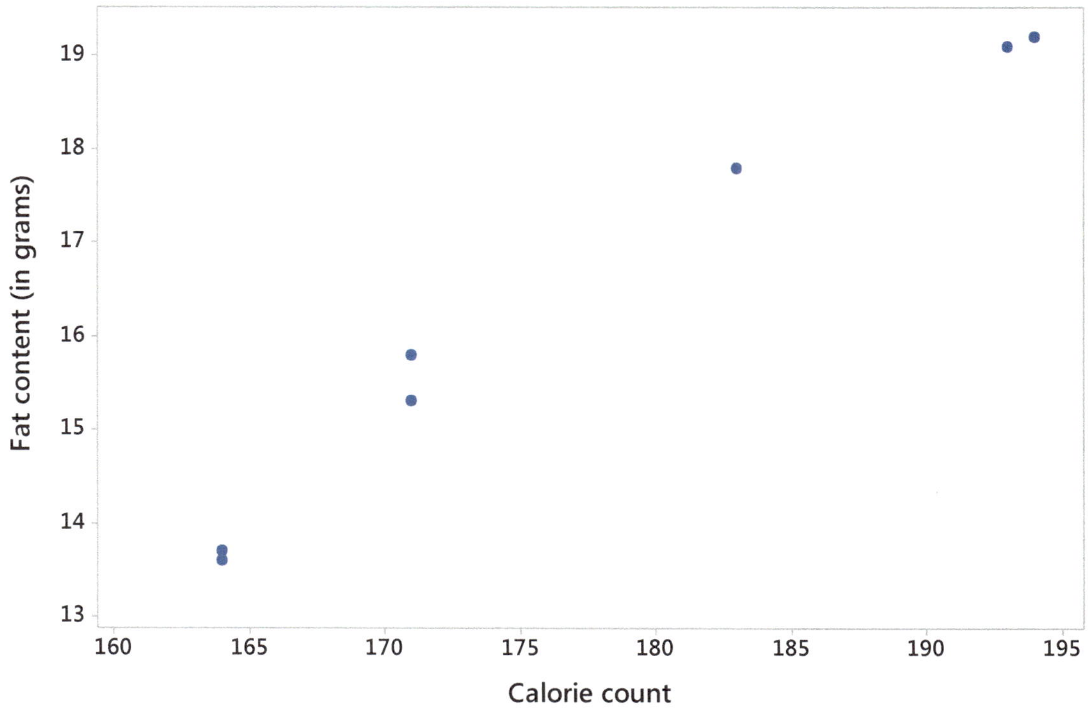

(a) The scatterplot of y versus x appears in above Figure 4.2.
(b) Yes, the scatterplot suggest a linear relationship between x and y, and the relationship is positive.
(c) No, the scatterplot does not provide any indication of the presence of outliers.

Definition: Coefficient of correlation

Let us have a sample of n ordered pairs $(x_1, y_1), (x_2, y_2), \dots, (x_n, y_n)$ of a bivariate data (x, y). The ***coefficient of correlation***, or simply ***correlation*** is a measure of the strength and direction of linear relationship between x and y. Correlation is denoted by r, and can be computed using the following formula:

$$r = \frac{\sum(x - \bar{x})(y - \bar{y})}{\sqrt{\sum(x - \bar{x})^2} \times \sqrt{\sum(y - \bar{y})^2}}$$

where,
$\sum(x - \bar{x})(y - \bar{y})$ is the sum of the product of deviations of x and y values from their respective means, $\bar{x}$ and $\bar{y}$,
$\sum(x - \bar{x})^2$ is the sum of the squared deviations of x from its mean $\bar{x}$,
$\sum(y - \bar{y})^2$ is the sum of the squared deviations of y from its mean $\bar{y}$.

For computational convenience and simplicity, it is a good practice to compute columns of $(x - \bar{x})(y - \bar{y})$, $(x - \bar{x})^2$ ***and*** $(y - \bar{y})^2$ ***for their sum to be used in the computing formula of*** r.
There are several other alternative formulas for computing r. One of them is as follows:

$$r = \frac{n\sum xy - (\sum x) \times (\sum y)}{\sqrt{n\sum x^2 - (\sum x)^2} \times \sqrt{n\sum y^2 - (\sum y)^2}}$$

where,
n is the number of ordered pairs (x, y) in the sample,
$\sum xy$ is the sum of the product of x and y values,
$\sum x$ is the sum of x values,
$\sum y$ is the sum of y values,
$\sum x^2$ is the sum of squared x values,
$\sum y^2$ is the sum of squared y values.
One can use any of the formulas for computing r. Once we understand the process involved in the computation of r, we mostly prefer to use TI graphing calculator, Minitab or other statistical software to compute it.

Sign of r for the direction of the linear relationship

The **sign** of r provides the direction of the linear relationship between two variables.

(1) If the sign of r is positive, then there is a positive linear relation between two variables.
(2) If the sign of r is negative, then the relation between two variables is negative.

Some useful facts about correlation coefficient r

(1) The value of r lies between -1 *and* $+1$, *inclusive,* i.e., $-1 \leq r \leq 1$.
(2) If $r = +1$, then there is a perfect positive linear relationship between x and y.
(3) If $r = -1$, then there is a perfect negative linear relationship between x and y.
(4) If $r = 0$, then there is ***no linear*** relationship between x and y.
(5) r is symmetric with respect to x and y, that is, the correlation between x and y is the same as the correlation between y and x.
(6) r is a **unit** free measure. Therefore, the change of the unit of either variable will not affect the value of r.
(7) r is very sensitive to outliers i.e. a few outlying observations may affect r severely.

Example 4.2

Consider Example 4.1 with calorie count (x) and fat content in grams (y) from a serving size of 28 grams of various nuts and seeds.

Types of nuts and seeds	x	y
Almonds	171	15.3
Brazil nuts	193	19.1
Cashew nuts	164	13.7
Hazelnuts	183	17.8
Sesame seeds	171	15.8
Sunflower seeds	164	13.6
Walnuts	194	19.2

(a) Can you justify the computation of r, given the evidence of relationship between x and y in Figure 4.2?
(b) Compute the value of r and comment on the direction of relationship on the basis of the computed value of r.
(c) How will the computed value of r be affected if the fat content is measured in kilograms instead of grams?

Solution

(a) From the scatterplot (Figure 4.2), it follows that x and y are linearly related. Therefore, we can compute the correlation r to evaluate the strength of the linear relationship.
(b) For computational convenience, we will complete five columns: x, y, $(x-\bar{x})(y-\bar{y})$, $(x-\bar{x})^2$ and $(y-\bar{y})^2$. We have $\sum x = 1240$ and $\sum y = 114.5$. Therefore, $\bar{x} = \frac{\sum x}{n} = \frac{1240}{7} = 177.1$ and $\bar{y} = \frac{\sum y}{n} = \frac{114.5}{7} = 16.4$.
The first row of the last three columns of the computational table are obtained as follows:

$$(x-\bar{x})(y-\bar{y}) = (171 - 177.1) \times (15.3 - 16.4) = 6.71$$

$$(x-\bar{x})^2 = (171 - 177.1)^2 = 37.21$$

$$(y - y)^2 = (15.3 - 16.4)^2 = 1.21$$

Other rows are computed in a similar way to complete the table.

x	y	$(x - \bar{x})(y - \bar{y})$	$(x - \bar{x})^2$	$(y - y)^2$
171	15.3	6.71	37.21	1.21
193	19.1	42.93	252.81	7.29
164	13.7	35.37	171.61	7.29
183	17.8	8.26	34.81	1.96
171	15.8	3.66	37.21	0.36
164	13.6	36.68	171.61	7.84
194	19.2	47.32	285.61	7.84
Sum: 1240	114.5	180.93	990.87	33.79

We have $\sum(x - \bar{x})(y - \bar{y}) = 180.93$, $\sum(x - \bar{x})^2 = 990.87$ and $\sum(y - \bar{y})^2 = 33.79$. Therefore,

$$r = \frac{\sum(x - \bar{x})\,(y - \bar{y})}{\sqrt{\sum(x - \bar{x})^2} \times \sqrt{\sum(y - \bar{y})^2}} = \frac{180.93}{\sqrt{990.87} \times \sqrt{33.79}} = 0.989$$

Thus, we have a positive correlation between x and y.

(c) The value of r will be unaffected by changing unit from grams to kilograms.

Example 4.3

Consider the table below of bivariate data (x, y), where x (in °F) is the yearly average maximum temperature and y (in °F) is the minimum temperature from 1981 to 2010 for selected places in the Thumb and Southeast Michigan.

No	Place	High °F (x)	Low °F (y)
1	Ann Arbor	59	40
2	Dearborn	60	40
3	Detroit	59	42
4	Flint	57	38
5	Harbor Beach	52	38
6	Monroe	58	41
7	Port Huron	56	41
8	Saginaw	57	38

Source: National Climatic Data Center

(a) Draw a scatterplot of x and y.
(b) Do you suspect any outlier from the scatterplot?
(c) Does the plot suggest any linear relationship between x and y?
(d) Guess what would be the strength of the linear relationship, if there is any.
(e) Compute the correlation r between x and y.
(f) Comment on the value of r.
(g) If the unit of temperature is converted from °F to °C, will the value of r change? Justify your answer.

Solution

Figure 4.3: Scatterplot of minimum low °F against maximum high °F temperature

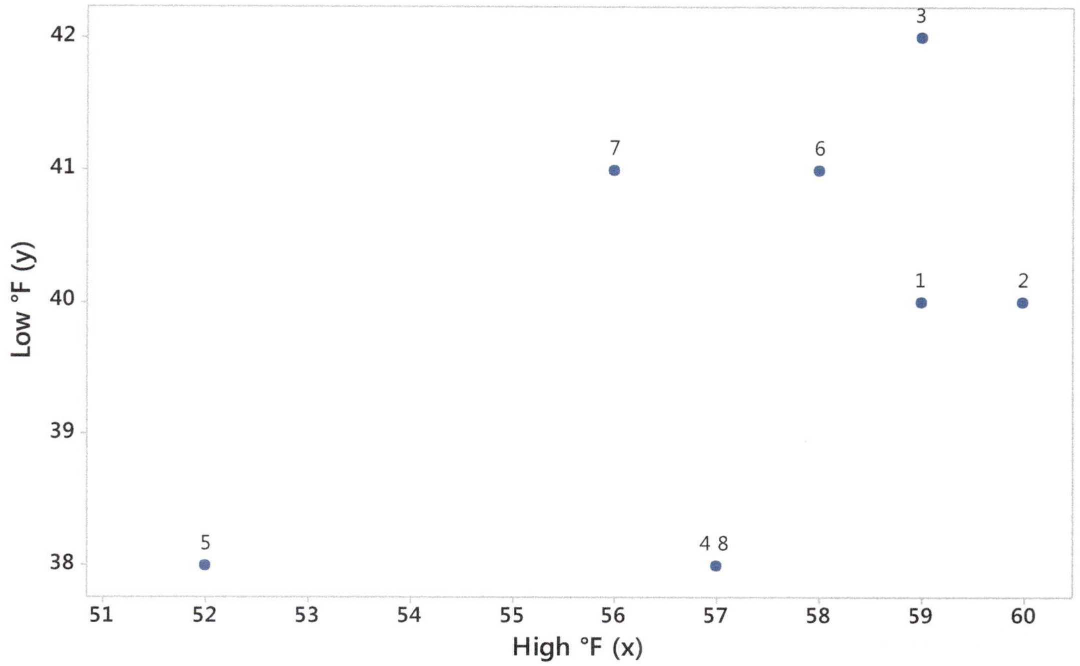

(a) The scatterplot of low temperature versus high temperature appears in Figure 4.3.
(b) Looking at the scatterplot, the 4th and 8th observation points (57,38), appear to be suspect of outliers.
(c) Yes, the scatterplot suggests a weaker form of a positive linear relationship between high and low temperatures.
(d) Make your own guess of a value of r, looking at the scatterplot.
(e) Below is a computational table of r, using the formula:

$$r = \frac{n\sum xy - (\sum x) \times (\sum y)}{\sqrt{n\sum x^2 - (\sum x)^2} \times \sqrt{n\sum y^2 - (\sum y)^2}}$$

x	y	x^2	y^2	xy
59	40	3481	1600	2360
60	40	3600	1600	2400
59	42	3481	1764	2478
57	38	3249	1444	2166
52	38	2704	1444	1976
58	41	3364	1681	2378
56	41	3136	1681	2296
57	38	3249	1444	2166
Sum: 458	318	26264	12658	18220

We have: $n = 8$, $\sum x = 458$, $\sum y = 318$, $\sum x^2 = 26264$, $\sum y^2 = 12658$ and $\sum xy = 18220$. Substituting these values in the formula we have

$$r = \frac{8 \times 18220 - 458 \times 318}{\sqrt{8 \times 26264 - 458^2} \times \sqrt{8 \times 12658 - 318^2}} = 0.53$$

(f) The value of r suggests that there is a moderate positive linear relationship between high and low temperature.

(g) If the unit of temperature is converted from °F to °C, the value of r will not change, because the change of units of variables does not affect the value of r.

Example 4.4

The data in table below represent airline fuel cost per gallon (in dollars) for the US domestic and international carriers with scheduled service by month from July 2014 to December 2014.

Month	Domestic (x)	International (y)
July	3.00	2.84
August	2.98	2.84
September	2.89	2.81
October	2.69	2.67
November	2.58	2.55
December	2.30	2.30

Source: Bureau of transportation statistics at http://www.transtats.bts.gov/fuel.asp

(a) Draw a scatterplot to see if there is any linear relationship.

(b) If the scatterplot in (a) seems linear, compute the value of r and comment on your result.

Solution

Figure 4.4: Scatterplot of fuel cost per gallon (in $) of the US international airline against the domestic airline for six months from July 2014 to December 2014

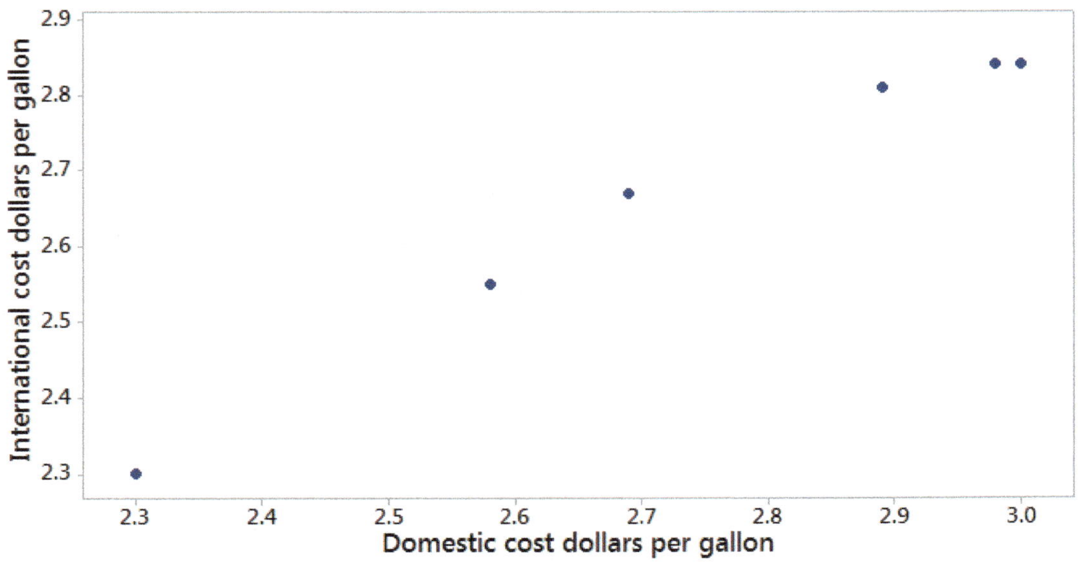

(a) The scatterplot of airline fuel cost per gallon (in dollars) of the US international against the domestic airline appears in Figure 4.4.
(b) The scatterplot in (a) seems linear and hence we can compute the value of r to determine the strength of the linear relationship quantitatively.

x	y	$(x - \bar{x})(y - \bar{y})$	$(x - \bar{x})^2$	$(y - y)^2$
3.00	2.84	0.044	0.068	0.029
2.98	2.84	0.041	0.058	0.029
2.89	2.81	0.021	0.023	0.020
2.69	2.67	0.000	0.003	0.000
2.58	2.55	0.019	0.026	0.014
2.30	2.30	0.163	0.194	0.137
Sum:16.44	16.01	0.288	0.372	0.229

From computation table, we have $\sum(x - \bar{x})(y - \bar{y}) = 0.288$, $\sum(x - \bar{x})^2 = 0.372$ and $\sum(y - \bar{y})^2 = 0.229$. Then,

$$r = \frac{\sum(x - \bar{x})(y - \bar{y})}{\sqrt{\sum(x - \bar{x})^2} \times \sqrt{\sum(y - \bar{y})^2}} = \frac{0.288}{\sqrt{0.372} \times \sqrt{0.229}} = 0.99$$

The computed value of r implies that there is a very strong and positive linear relationship between cost of domestic and international airlines.

Example 4.5

We have a sample of 10 ordered pairs (x, y), where x denotes the home runs for American League (AL) and y denotes the home runs for National League (NL). The scatter plot of bivariate data (x, y) is given below:

x	y
2499	2708
2546	2840
1649	1470
1086	1042
1552	1550
2742	2220
1551	1211
688	656
2290	1523
160	225

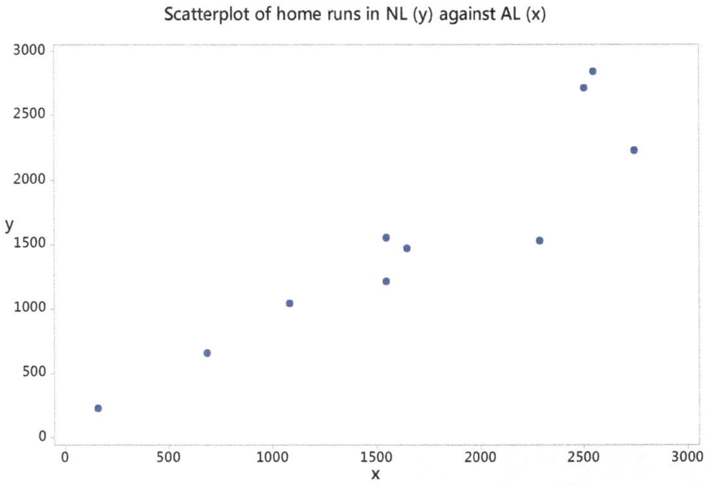

Source: Samples are generated by Minitab using data from 1901 to 2013 available at http://www.baseball-almanac.com/hitting/hihr6.shtml

(a) Does the scatterplot demonstrate any relationship?
(b) Is there any indication of outlier in the plot?

(c) Guess a value of the correlation coefficient r looking at the scatter plot.
(d) Compute the correlation between x and y using the following information under the usual notations:

$$\bar{x} = 1676, \bar{y} = 1545, \sum(x-\bar{x})(y-\bar{y}) = 6009220,$$

$$\sum(x-\bar{x})^2 = 6601810 \text{ and } \sum(y-\bar{y})^2 = 6388637$$

Solution

(a) The scatterplot exhibits a positive linear relationship.
(b) There is no strong indication of any outlier from the plot.
(c) Make a guess of your own!
(d) The correlation between x and y is given by

$$r = \frac{\sum(x-\bar{x})(y-\bar{y})}{\sqrt{\sum(x-\bar{x})^2} \times \sqrt{\sum(y-\bar{y})^2}} = \frac{6009220}{\sqrt{6601810} \times \sqrt{6388637}} = 0.925$$

Section 4.2: Simple linear regression model

In this section, we study a simple linear regression model. A ***simple linear regression model*** is an equation relating a predictor x, a response y and an error term ε. In other words, it is a true linear equation of how x and y are related. The main objective is to estimate a simple linear regression model to predict a response using an explanatory variable. Let us start with some definitions:

Definition: Response and predictor variables

A ***response*** variable (y) is an outcome, also called the dependent or an ***explained*** variable, whose value depends on another variable, called the ***predictor*** or ***explanatory*** variable (x). For example, fat content (in grams) is a response, whose value depends on the serving size (in grams), called the predictor or explanatory variable.

Definition: Simple linear regression model

Statistically, a simple linear regression model is an equation of true relationship between y and x given by

$$y = \beta_0 + \beta_1 x + \varepsilon$$

In this equation, $\mu_y = \beta_0 + \beta_1 x$ is called the mean value of the response y when the value of the explanatory variable is x,

β_0 is the y-intercept of the mean value of y when x=0,

β_1 is the slope of the equation of mean value of y. It measures the **change** in the mean value of y for one unit increase in x,

ε is a random error term that accounts for the variability in y, not explained by x.

Predicted mean of y

The equation $\hat{y} = b_0 + b_1 x$ is called the ***predicting equation*** of y on x. The quantity $\hat{y}$ is called the ***predicted mean*** of y at x.

The b_0 and b_1 in the predicting equation are the point estimates of β_0 and β_1, respectively obtained by the least squares method.

Least squares method

The estimates b_0 and b_1 are obtained by minimizing the quantity $\sum(y-\hat{y})^2$. The method of minimizing $\sum(y-\hat{y})^2$ is called the ***least squares method***. By the principle of optimization of calculus, $\sum(y-\hat{y})^2$ is minimized if

$$b_1 = \frac{\sum(x-\bar{x})(y-\bar{y})}{\sum(x-\bar{x})^2}$$

and

$$b_0 = \bar{y} - b_1\bar{x}$$

The derivations for the expressions b_0 and b_1 are beyond the scope of this book. It is better off we present the least square estimates by the following definition.

Definition: Least squares predicting equation

Given a sample of n ordered pairs $(x_1, y_1), (x_2, y_2), \ldots, (x_n, y_n)$ for a bivariate data (x, y), the equation

$$\hat{y} = b_0 + b_1 x$$

is called the ***least squares predicting equation*** for predicting y using x if

$$b_1 = \frac{\sum(x-\bar{x})(y-\bar{y})}{\sum(x-\bar{x})^2}$$

and

$$b_0 = \bar{y} - b_1\bar{x}$$

The following terminologies are used interchangeably for the ***least squares predicting equation*** $\hat{y} = b_0 + b_1 x$:

(1) the regression equation
(2) the equation of the best fit
(3) the least square regression equation
(4) the estimated regression line

Example 4.6

The current age (x) and expected life remaining (y) for the US male population for selected ages are given below:

x	30	35	40	45	50	55	60
y	52	47.2	42.5	37.8	33.2	28.8	24.5

We want to find a predicting equation of y on x. To this end,

(a) What is the response and predictor for this problem?
(b) Find slope and intercept of the predicting equation.
(c) Write out the predicting equation of y on x
(d) Find the predicted value of the expected life if current age is 45 years.
(e) Find the predicted value of the expected life if current age is 70 years.

Solution

(a) The response is expected life remaining (y) and predictor is current age (x).

(b) Note that $\sum x = 315$ and $\sum y = 266$. Then, $\bar{x} = \frac{\sum x}{n} = \frac{315}{7} = 45$ and $\bar{y} = \frac{\sum y}{n} = \frac{266}{7} = 38$.

Therefore, we can complete the table below for the computational need.

x	y	$(x-\bar{x})(y-\bar{y})$	$(x-\bar{x})^2$
30	52	-210	225
35	47.2	-92	100
40	42.5	-22.5	25
45	37.8	0	0
50	33.2	-24	25
55	28.8	-92	100
60	24.5	-202.5	225
Sum: 315	266	-643	700

From the table, $\sum(x-\bar{x})(y-\bar{y}) = -643$ and $\sum(x-\bar{x})^2 = 700$.
Therefore, the slope is

$$b_1 = \frac{\sum(x-\bar{x})(y-\bar{y})}{\sum(x-\bar{x})^2} = -\frac{643}{700} = -0.92$$

The intercept is

$$b_0 = \bar{y} - b_1\bar{x} = 38 - (-0.92)(45) = 79.4$$

(c) The predicting equation of y on x is

$$\hat{y} = b_0 + b_1 x = 79.4 - 0.92(x)$$

(d) The predicted value of the expected life at a current age of 45 years is

$$\hat{y} = 79.4 - 0.92(45) = 38$$

(e) The predicted value of the expected life at a current age of 70 years is

$$\hat{y} = 79.4 - 0.92(70) = 15$$

Definition: Residual or prediction error

The residual or prediction error, denoted by e, of a predicted value $\hat{y}$ at x is defined by

$$e = \textit{observed } y - \textit{predicted } y = y - \hat{y}$$

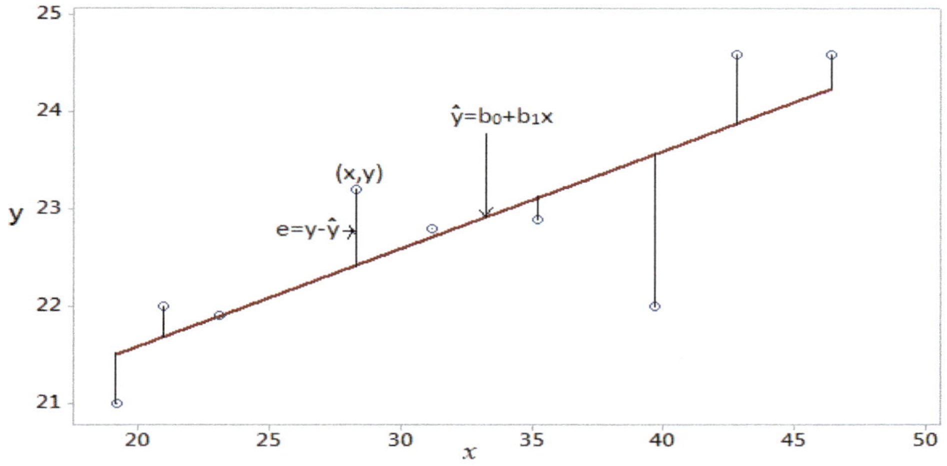

Definitions: Interpolation and extrapolation

An interpolation is a prediction of the response at a predictor's value within its given range. An extrapolation is the prediction of the response at a predictor's value beyond its given range.

An extrapolation at a predictor's value far beyond its given range is dangerous and should always be done carefully or avoided.

Example 4.7

The data in table below refers to the calorie count (y) for different grams of serving sizes (x) of various fast foods. We wish to predict calorie count for given values of serving size of the fast food using the equation of y on x.

Type of fast food	x (Size in grams)	y (Calorie)
Big Mac	215	492
Cheeseburger	141	379
Hamburger	108	254
Quarter Pounder with cheese	206	515
Kentucky Fried Chicken	67	195
KFC fries	100	294
McDonalds fries	78	207
Pizza Deluxe	66	171
Pizza	135	263
Potato Wedges	135	279

For above data, we have $n = 10$, $\bar{x} = 125.1$, $\bar{y} = 304.9$, $\sum(x-\bar{x})(y-\bar{y}) = 54378.1$, $\sum(x-\bar{x})^2 = 25084.9$ and $\sum(x-\bar{x})^2 = 129366.9$.

(a) Identify the predictor and response variables.
(b) Find the intercept and slope for predicting equation of y on x.
(c) Write out the least squares predicting equation for predicting calorie (y) using serving size (x).
(d) Predict the value of calorie when the serving size is 230 grams. What do you call this prediction?
(e) Predict the value of calorie when the serving size is 70 grams. What do you call this prediction?
(f) Find the residual for estimating calorie by the serving size of 100 grams.

Solution

(a) The predictor is serving size and response is calorie count.
(b)

$$b_1 = \frac{\sum(x-\bar{x})(y-\bar{y})}{\sum(x-\bar{x})^2} = \frac{54378.1}{25084.9} = 2.17$$

$$b_0 = \bar{y} - b_1\bar{x} = 304.9 - 2.17 \times 125.1 = 33.43$$

The intercept of the predicting equation is b_0=33.43 and slope is b_1=2.17.
The slope of 2.17 means that for every one gram increase of the serving size, the mean calorie increases by 2.17.

(c) $Calorie = 33.43 + 2.17 \times Serving\ size$

(d) The predicted calorie with a serving size of 230 grams is:

$$Calorie\ (for\ x = 230) = 33.43 + 2.17 \times 230 = 532.53$$

This prediction is called **extrapolation**, because the value of $x = 230$ is outside the given range of x being used to estimate the model.

(e) The predicted calorie with a serving size of 70 grams is:

$$Calorie\ (for\ x = 70) = 33.43 + 2.17 \times 70 = 185.33$$

This prediction is called **interpolation**, because the value of $x = 70$ is within the given range of x being used to estimate the model.

(f) $Calorie\ (for\ x = 100) = 33.43 + 2.17 \times 100 = 250.43$

$$residual = observed\ y - predicted\ y = y - \hat{y} = 294 - 250.43 = 43.57$$

Section 4.3 Coefficient of determination

The coefficient of determination is a measure of the percent of variation in the response variable explained by the estimated regression line.

In the regression terminology, the ***total variation*** in the response is the total sum of squares (SST) defined by

$$SST = \sum (y - \bar{y})^2$$

The ***explained variation*** is the sum of squares due to regression (SSR) and is defined by

$$SSR = \sum (\hat{y} - \bar{y})^2$$

The ***coefficient of determination***, denoted by R^2, is defined by

$$R^2 = \frac{Explained\ variation}{Total\ variation} = \frac{SSR}{SST} = \frac{\sum(\hat{y} - \bar{y})^2}{\sum(y - \bar{y})^2}$$

R^2 is interpreted as the percent of variation in the response explained by the estimated regression line.

For a simple linear regression model, the quantity $\frac{\sum(\hat{y}-\bar{y})^2}{\sum(y-\bar{y})^2}$ turns out to be the square of the correlation coefficient, r^2. Therefore, for a simple linear regression model, R^2 is simply r^2. In other words, to compute the coefficient of determination for a simple linear regression model, we first compute the correlation coefficient r and then square it to obtain r^2.

Note that $0 \leq r^2 \leq 1$. Therefore, the coefficient of determination lies between 0 and 1, inclusive. The higher the value of r^2, the greater is the predictability of the model. For example, if $r^2 = 0.95$, then 95% of the total variation in the response is explained by the estimated regression equation.

Example 4.8

Reference to the Example 4.7, find the value of the coefficient of determination and interpret the result.

Solution

Given $n = 10$, $\bar{x} = 125.1$, $\bar{y} = 304.9$, $\sum(x - \bar{x})(y - \bar{y}) = 54378.1$, $\sum(x - \bar{x})^2 = 25084.9$ and $\sum(x - \bar{x})^2 = 129366.9$.

$$r = \frac{\sum(x-\bar{x})(y-\bar{y})}{\sqrt{\sum(x-\bar{x})^2} \times \sqrt{\sum(y-\bar{y})^2}} = \frac{54378.1}{\sqrt{25084.9} \times \sqrt{129366.9}} = 0.955$$

Then, $r^2 = 0.955^2 = 0.91$.

Therefore, the coefficient of determination is 0.91. This means that 91% of the total variation in calorie is explained by the serving size using the estimated regression model, which also implies that the predicted model is reasonably good.

Example 4.9

We have a bivariate data (x, y), where y refers the top five Metropolitan Area Exporters export values (in billions of dollars) in 2011 and x refers to their respective value in 2010. We want to predict the export value in 2011 (y) using the export value in 2010 (x).

Export in 2010 (x)	Export in 2011 (y)
85.1	105.1
80.6	104.5
62.2	72.7
44	49.4
35.9	43.1

Given the dataset we have, $n = 5$, $\bar{x} = 61.6$, $\bar{y} = 75.0$, $\sum(x-\bar{x})(y-\bar{y})$ =2536.9, $\sum(x-\bar{x})^2$ =1883.9 and $\sum(y-\bar{y})^2$ =3454.5.

Using above information, answer questions (a)-(f) below:

(a) Compute the coefficient of correlation between x and y.
(b) Compute intercept and slope for predicting y using x.
(c) Write out the predicting equation.
(d) Compute the coefficient of determination and interpret the result.
(e) On the basis of the result in (d), comment on the predictability by the regression equation.
(f) Predict the value of export in 2011 if the export in 2010 is 44 and find the prediction error.

Solution

(a) The coefficient of correlation is

$$r = \frac{\sum(x-\bar{x})(y-\bar{y})}{\sqrt{\sum(x-\bar{x})^2} \times \sqrt{\sum(y-\bar{y})^2}} = \frac{2536.9}{\sqrt{1883.9} \times \sqrt{3454.5}} = 0.99$$

(b) The slope is

$$b_1 = \frac{\sum(x-\bar{x})(y-\bar{y})}{\sum(x-\bar{x})^2} = \frac{2536.9}{1883.9} = 1.35$$

The intercept is

$$b_0 = \bar{y} - b_1\bar{x} = 75 - 1.35(61.6) = -8.2$$

(c) The predicting equation for y is $\hat{y} = -8.2 + 1.35(x)$

(d) The coefficient of determination is $r^2 = .99^2 = 0.98$. It means that 98% of the variation in the response y (i.e., export in 2011) is explained by the estimated regression line.
(e) Since the coefficient of determination is 0.98, the predictability of the model is very good.
(f) The predicted export in 2011 at export value 44 in 2010 is $\hat{y} = -8.2 + 1.35(44) = 51.2$ billions. The prediction error is $e = y - \hat{y} = 49.4 - 51.2 = -1.8$ billion dollars.

Section 4.4: Technology implementation for correlation and regression

Example 4.10

Suppose we have the following sample dataset about the measurements of x and y:

x	2	3	4	5	5	6
y	4	5.2	6.9	10	9	11.9

We will provide TI-84 Plus calculator to compute correlation, regression line and prediction.

Drawing a scatterplot in TI-84 Plus calculator

Ref: Example 4.10

(1) Press **STAT**, Select **1:Edit** and press ENTER.
(2) Type x-values in L1 column and y-values in L2 column.
(3) Press **2nd Y=** and select **1:Plot** and press ENTER. Press ENTER to select **On**.
(4) Move cursor to **Type:** scatter plot and prress ENTER.
(5) Press **2nd, 1** to specify Xlist: L1 and **2nd 2** to specify Ylist:L2.
(6) Go down to choose a symbol from Mark and press ENTER.
(7) Press ZOOM, select **9:ZoomStat**. Press ENTER.

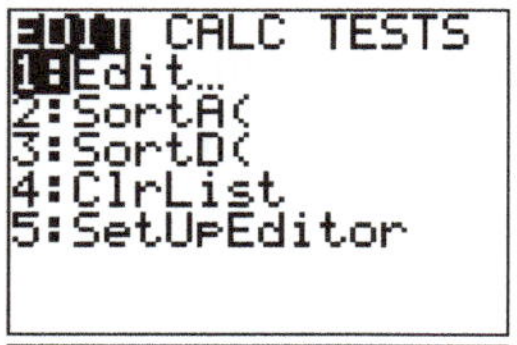

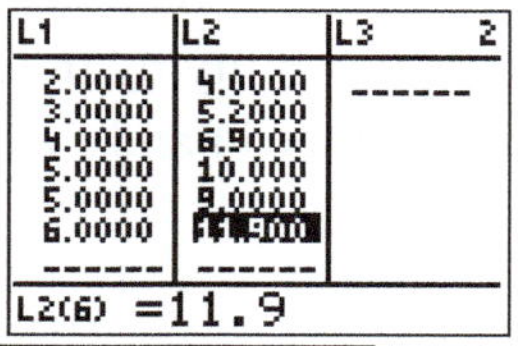

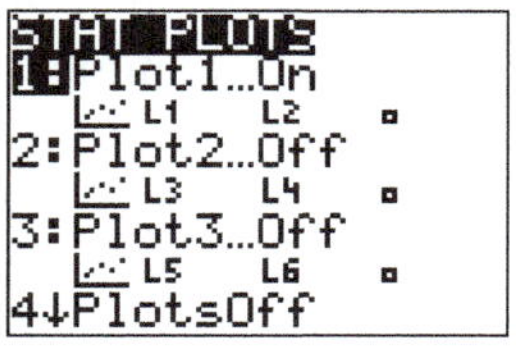

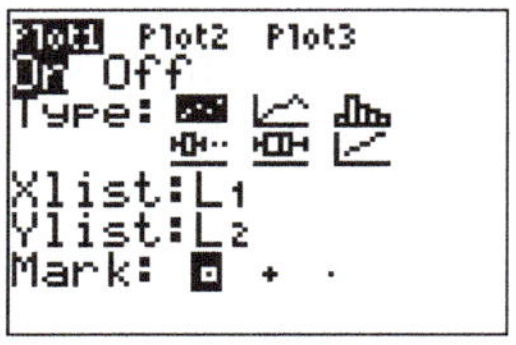

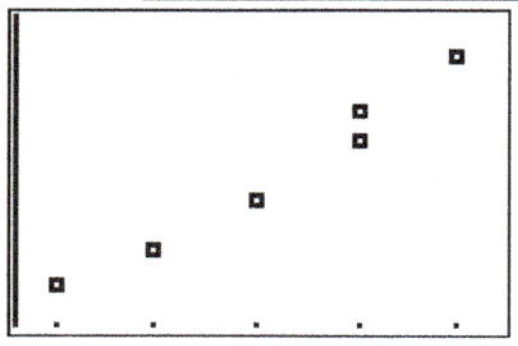

Correlation coefficient r

1. Press 2nd 0, scroll down to Diagonostic On, press ENTER twice to display “DiagnosticOn Done”.
2. Press STAT, select CALC, select **4:LinReg**. Hit ENTER.
3. LinReg will be pasted in the home screen. Now type the column name where you have x variable stored (in this case, L1 column), type the column name where you have y variable stored (in this case, L2 column), and press ENTER.

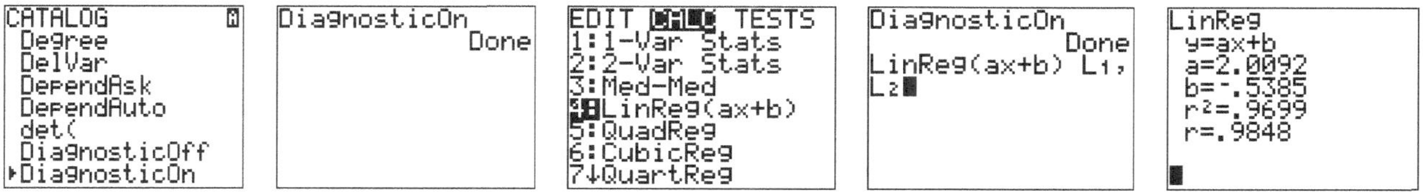

Regression and prediction in TI-84 Plus calculator

(1) Press **STAT**, Select **1:Edit** and press ENTER
(2) Type x-values in L1 column and y-values in L2 column.
(3) Press **STAT**, **CALC** select **4:LinReg(ax+b).**
(4) Type **L1, L2, press VARS**, choose **Y-VARS 1:Function**, hit ENTER, select **Y1,** hit ENTER. The home screen shows: **LinReg(ax+b) L1, L2, Y1**; (**Y1** is optional if you want to save the predicted equation in Y1 and use it for prediction at any X value)
(5) Hit **ENTER**.
(6) Y1(7) will predict Y at $x = 7$; Choose this option by: **press VARS**, choose **Y-VARS 1:Function**, hit ENTER, select **Y1,** hit ENTER and then type **(7)** and ENTER.

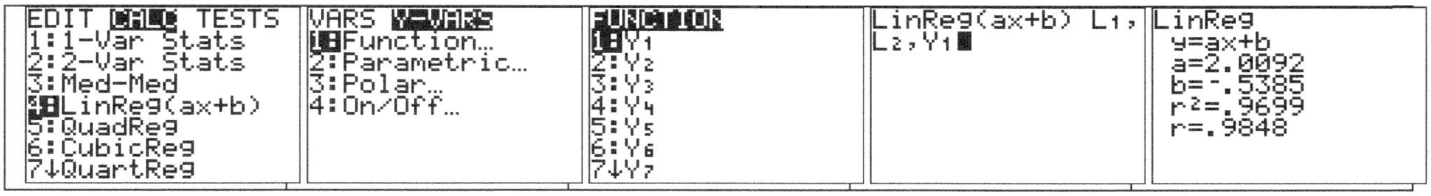

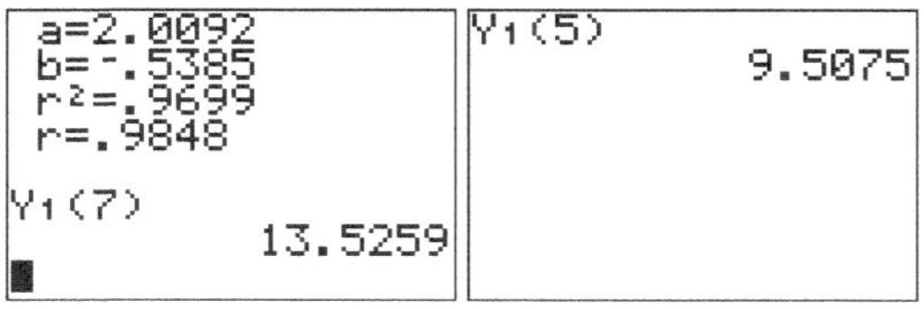

Again, option Y1 in above is required if you wish to save results from the linear regression line in Y1 Variable, which can be used for prediction very easily:

For example, to predict the value of y when $x = 5$ follow the procedure below:

(1) Press **VARS**, choose **Y-VARS**, press **1**, select **Y1**. Y1 will be appeared in the home screen.
(2) Type (5) to enter the value of the independent variable.
(3) Press **ENTER**.

Set Diagnostic on

If the regression model does not provide r and r^2, we need to set diagnostic on:

(1) Press 2nd 0, scroll down to Diagnostic On, press ENTER twice.
(2) Then re-do the regression model.

Scatterplot, correlation and regression in Minitab

Ref: Example 4.10

Scatterplot steps

(1) Type x values in C1 and y values in C2 columns

(2) Click **Graph** > **Scatterplot…** and highlight "**Simple**." Press OK.

(3) For Y variable select C2 and X variable select C1

(4) Use Data View… to select symbols and **Labels** option to add "Title/footnote" and Click OK.

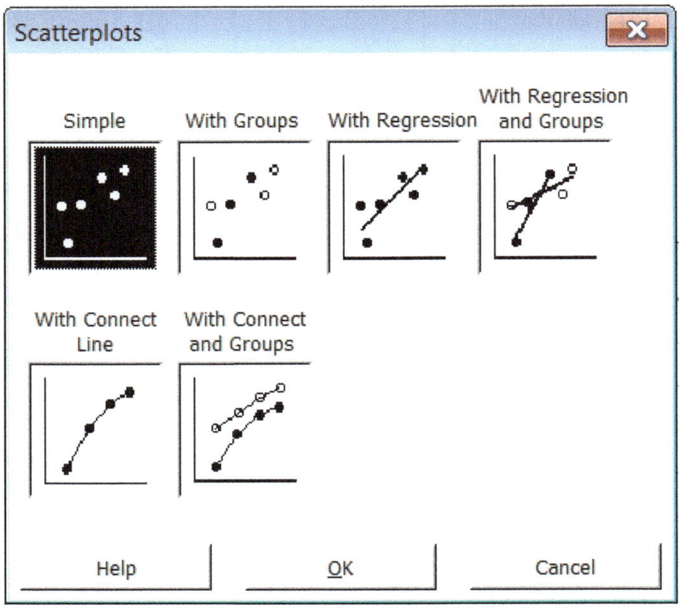

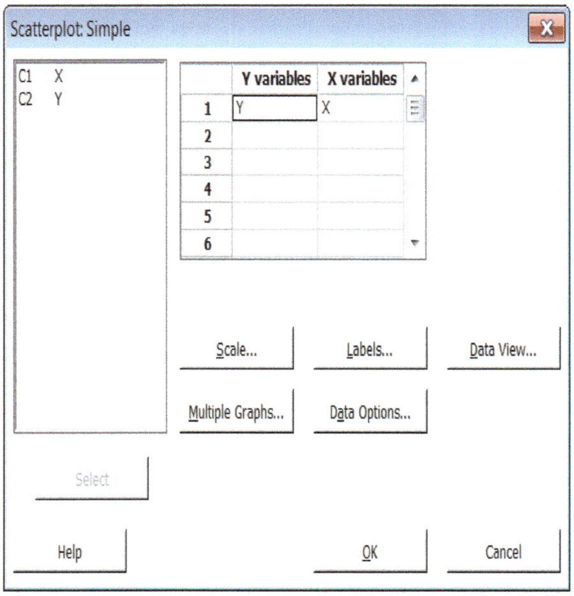

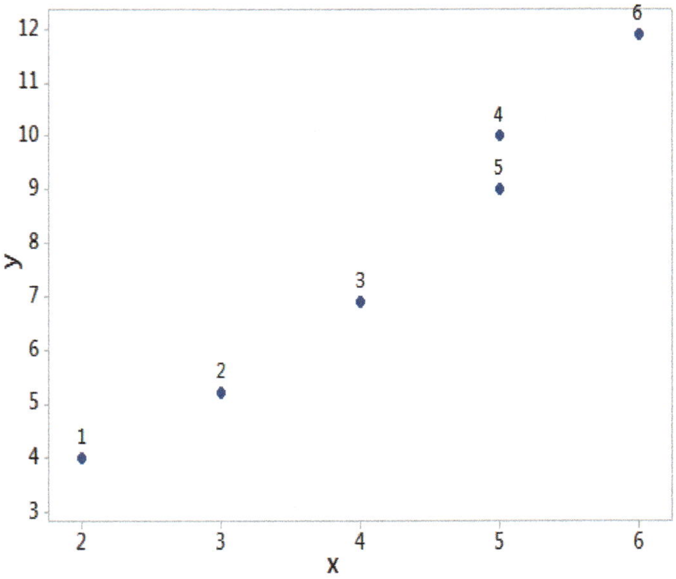

Correlation steps

(1) Type x values in C1 column and y values in C2 column

(2) Select **Stat** > **Basic Statistics** > **Correlation**

(3) For **Variables**: select x and y by double clicking on C1 and C2 and click OK.

Results are displayed as follows:

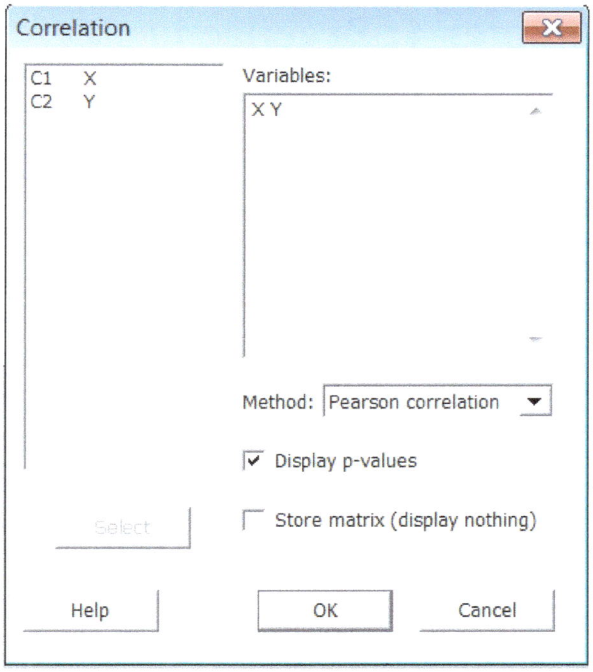

Correlation: X, Y

Pearson correlation of X and Y = 0.985

Regression steps

(1) Type x values in C1 column and y values in C2 column

(2) Select **Stat** > **Regression** > **Regression** > **Fit Regression Model**

(3) For **Responses**: select Y and for Continuous predictors: select x

(4) Optionally use **Graphs…** and **Results…** to select desired graphs and results.

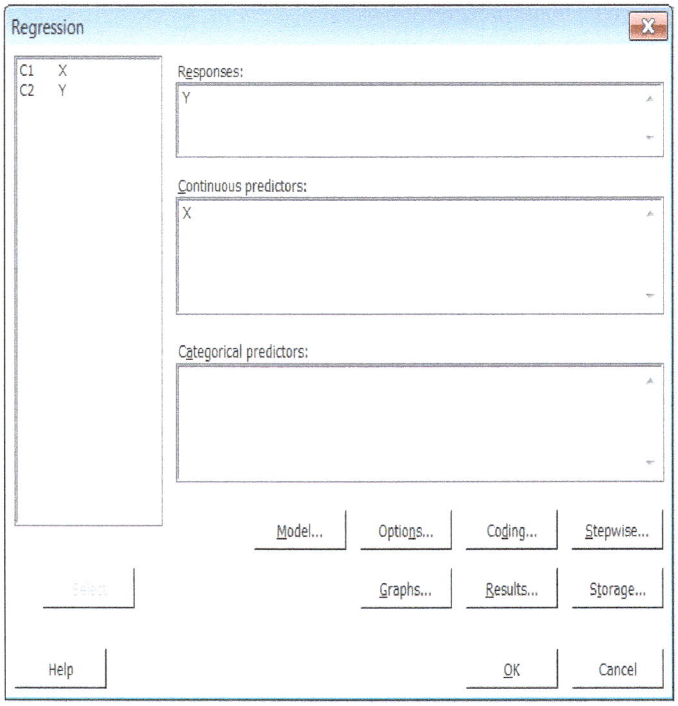

Results displayed

Regression Analysis: Y versus X

Regression Equation

Y = -0.538 + 2.009 X

Chapter 4: Practice problems

(4.1) Height (y) and age (x) for a group of students are measured to predict height using age. The coefficient of correlation between the height and age for the group is $r = 0.75$. Identify each of the following statements being True (T) or False (F).

(a) The scatterplot of (x, y) may suggest a linear relationship between x and y.
(b) There is a linear relationship between height and age.
(c) There is a non-linear relationship between height and age.
(d) The relationship between height and age is linear and positive.
(e) The relationship between height and age is linear and negative.
(f) The slope of the estimated regression equation will be positive.
(g) The slope of the estimated regression equation will be negative.
(h) If a linear regression model is attempted to predict height using age, about 56% of variation in the height will be explained by the estimated regression model.

(4.2) A regression model is intended to predict the fiber content in nuts and seeds using their protein content. The response for the model is:

(a) Fiber content
(b) Protein content
(c) Both (a) and (b)
(d) None of the above

(4.3) The slope of the regression model is interpreted as

(a) The change in the response per unit increase in the predictor
(b) The change in the predictor per unit increase in the response
(c) Correlation between the response and predictor
(d) All of the above

(4.4) For a regression model to predict the fiber content using protein content, the coefficient of determination is obtained to be r^2=.55.
The value of r^2=0.55 means that

(a) 55% of the variation in protein content is explained by the estimated regression line.
(b) 55% of the variation in fiber content is explained by the estimated regression line.
(c) the correlation between protein and fiber is 0.55.
(d) all of the above.

(4.5) The coefficient of determination for any regression model determines

(a) the proportion of variation in response explained by the estimated regression line.
(b) the proportion of variation in predictor explained by the estimated regression line.
(c) the correlation between the response and the predictor.
(d) the change in response variable per unit increase in the predictor.

(4.6) The coefficient of determination of a model to predict a response variable using a predictor is r^2= 0.81 and the estimated regression equation is

$$Response = 2.5 - 0.23(predictor)$$

The value of the correlation between the response and the predictor is

(a) 0.81
(b) -0.81
(c) 0.90
(d) -0.90

(4.7) The co-efficient of determination of a model to predict a response variable using a predictor is r^2= 0.81 and the estimated regression equation is

$$Response = 2.5 + 0.23(predictor)$$

The value of the correlation between the response and the predictor variable is

(a) 0.90
(b) -0.90
(c) 0.81
(d) -0.81

(4.8) Suppose that the least-squares regression line for a dataset has a slope of 20.5. Suppose that after removing a data point from the set, the new least-squares regression line has a slope of 10.5. The point removed may be a suspect of

(a) an outlier
(b) an influential observation
(c) (a) or (b) or both
(d) none of the above

(4.9) By removing one data point from the computation, the coefficient of correlation between two variables has changed from 0.85 to 0.25. The point removed may be a suspect of

(a) an outlier
(b) an influential observation
(c) (a) or (b) or both
(d) none of the above

(4.10) Which of the followings is true:

(a) Correlation r takes a value between -1 and +1, inclusive.
(b) If r=0, then there is no linear relationship between two variables.
(c) r is a unit-free measure.
(d) All of the above.

(4.11) Which of the followings is not true:

(a) r is severely affected by outliers.
(b) If r=0, then there is no relationship between two variables.
(c) If r=1, there is a perfect positive linear relationship between two variables.
(d) The correlation between x and y is the same as the correlation between y and x.

Answer questions 4.12-4.16 using the following information:
A study is intended to predict calorie count (y) in fast-food menus using their fat content (x, measured in grams). Suppose the estimated regression line is:

$$Calories = 200 + 10.5 \times (Fat)$$

The correlation between Calories and Fat is $r = .90$. The average number of calories in fast-food menus is 662 (i.e., $\bar{y} = 662$), and the average fat content in menu items is 44 grams (i.e., $\bar{x} = 44$).

(4.12) The predicted value of calorie count for a menu with 44 grams of fat is
(a) 262 calories
(b) 462 calories
(c) 662 calories
(d) 862 calories

(4.13) Fat content in the study is
(a) a predictor
(b) a response
(c) an outlier
(d) all of the above

(4.14) Which of the following statements is true?
(a) About 90% of the variation in calories for menu items is explained by the estimated regression line.
(b) About 90% of the variation in fat content for menu items is explained by the estimated regression line.
(c) About 81% of the variation in calories for menu items is explained by the estimated regression line.
(d) About 81% of the variation in fat content for menu items is explained by the estimated regression line.

(4.15) Which of the following statements is true?
(a) According to the model, by adding one gram of fat to a menu would have an increase of 10.5 calories in the menu.
(b) According to the model, any non-fat menu would have 200 calories in it.
(c) By adding 2 grams of fat to a menu would increase the number of calories by 21.
(d) All of the above.

(4.16) The model also imply that
(a) 90% of the data points fall on the estimated regression line.
(b) The estimated regression line passes through the point (44,662).
(c) 10% of the data points are outliers.
(d) All of the above

(4.17) Generally speaking, an estimated regression line is
(a) the line that passes through most of the data points.
(b) the line that has about 50% of the data points falling above it and 50% falling below it.

(c) the best line with sum of the squares of the vertical distances of the data points from the line is the minimum.
(d) all of the above.

(4.18) Generally speaking, the coefficient of determination of an estimated regression line is
(a) the correlation between the underlying two variables
(b) the slope of the line
(c) the square of the correlation between the underlying two variables
(d) all of the above

(4.19) We want to determine the correlation between serving size (in grams) and fat content (in grams) of KFC's fast food menu. We have serving size and fat content of two fast food menus.

Fast food type	Serving Size (in grams)	Fat (in grams)
Kentucky Fried Chicken	67	12
KFC fries	100	14.8

(a) Find the correlation between serving size and fat content on the basis of above two data points.
(b) If a regression line is fitted to predict fat content using serving size, what will be the value for the coefficient of determination of the model?
(c) Is the slope of the model positive or negative?

(4.20) We have two data points of current age and expected life, both measured in years, of U.S. female population as given below:

Current age (in years)	Expected life (in years)
50	33.1
70	16.5

(a) Find the correlation between current age and expected life for above two data points.
(b) If a regression line is fitted to predict expected life using current age, what will be the value for the coefficient of determination of the model?
(c) Is the slope of the model positive or negative?

(4.21) A researcher is interested to determine the effect of speed (x) on the fuel consumption (y) for a certain model of car on the basis of a sample of data points (x, y). The speed has been measured in kilometers per hour and fuel consumption measured in liters of gasoline used per 100 kilometers traveled. The estimated regression line is

$$Fuel\ consumption = -0.01 \times (\square peed\ limit) + 12.5$$

The correlation between fuel consumption and speed is found to be $r^2 = 0.03$.
(a) Identify the response and predictor variables for this problem.
(b) Should you recommend a linear regression model for predicting the fuel consumption?
(c) What is your stand on the predictability of the model?
(d) What is the value of the correlation r?
(e) What is the unit for r?

Answer questions (4.22)-(4.25) using the scatterplot below:
A scatterplot, with an estimated fitted line, of fiber versus protein for selected types of nuts and seeds, each with serving size of 100 grams is given below.

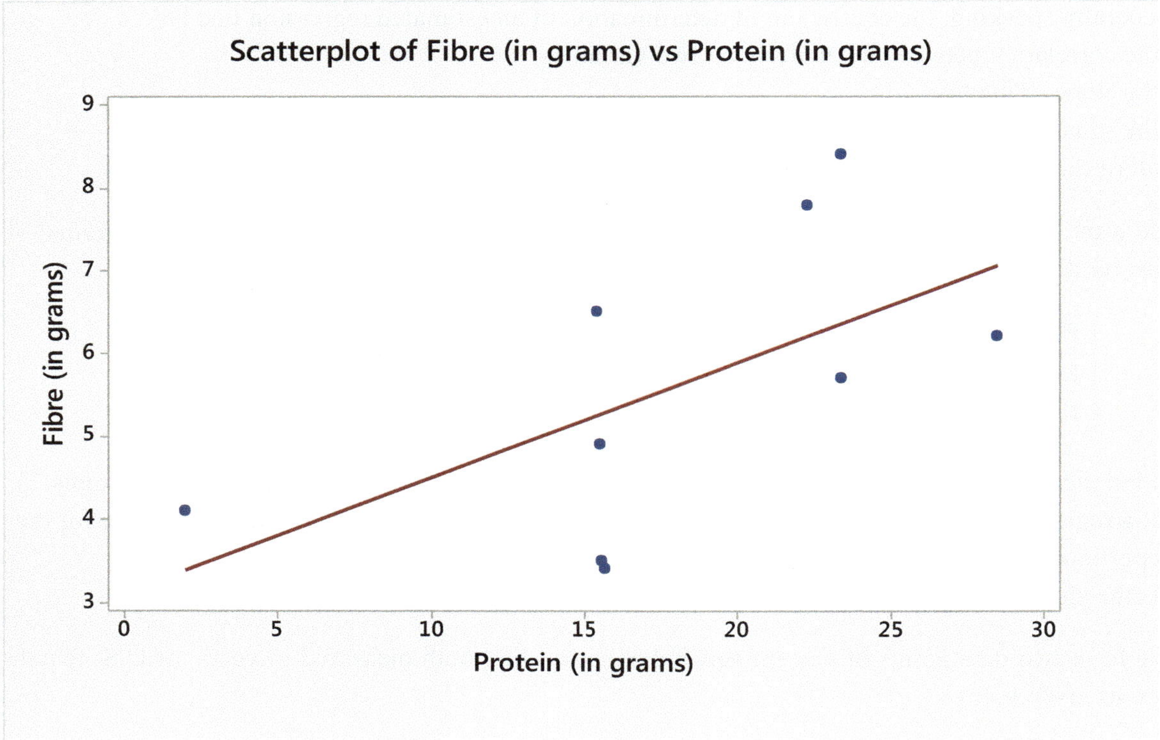

(4.22) Above scatterplot suggests that the correlation between fiber and protein is
(a) positive
(b) negative
(c) zero
(d) cannot be decided.

(4.23) The possible value of r could be:
(a) +1
(b) -1
(c) +0.60
(d) -0.60

(4.24) The slope of the estimated line is
(a) Positive
(b) Negative
(c) Indeterminate
(d) All of the above

(4.25) We can make a conclusion that
(a) the estimated line will have the smallest possible sum of the squares of the vertical distances of the data points from the line.

(b) a perfect linear relationship is evident between fiber and protein content.
(c) there is certainly some outliers in the scatterplot.
(d) all of the above

(4.26) We have a bivariate data (x, y), where y refers the top 10 Metropolitan Area Exporters export values (in billions of dollars) in 2011 and x refers to their respective value in 2010. We want to predict the export value in 2011 (y) using the export value in 2010 (x).

Export in 2010 (x)	Export in 2011 (y)
85.1	105.1
80.6	104.5
62.2	72.7
44	49.4
35.9	43.1
35.4	41.1
33.7	39.5
26.3	26.7
22.5	26.6
23.2	26.2

(a) Identify response and explanatory variables for this problem.
(b) Find the correlation between x and y and comment on your result.
(c) Will you recommend a linear regression model for predicting y using x?
(d) Write out the predicting equation.
(e) Find the predicted value of export value in 2011 when the export value in 2010 is 22.5 and find the residual for this prediction.
(f) Overall, how would you assess the predictability of the estimated model? (Hints: use the coefficient of determination of the model).

(4.27) The data below refers to the calorie and fat contents from the serving size of 100 grams of various nuts and seeds. We wish to predict fat using calorie for nuts and seeds.

Types of nuts and seeds	Calories (kcal)	Fat (gram)
Whole almond	611.6	54.8
Brazil nuts	687.4	68.3
Plain cashew nuts	584.5	48.9
Chestnuts	170	2.7
Hazelnuts	655	63.5
Salted peanuts	622	53
Sesame seeds	610.5	56.4
Sunflower seeds	585.3	48.7
Walnuts	691.3	68.5

(a) Identify response and explanatory variables for this problem.
(b) Fit a regression model of fat on calorie.
(c) Find the correlation and the coefficient of determination for the model and comment on the results.
(d) Predict fat content when calorie content in a serving size of 100 grams is 600 kcal.

(4.28) As reported in *National Vital Statistics Report* (Volume 60, Number 4, January 11, 2012), the current age (x) and life expectancy (y) for selected ages of the US male is given below:

Age (x)	Life expectancy (y)
25	56.9
30	52
35	47.2
40	42.5
45	37.8
50	33.2
55	28.8
60	24.5
65	20.3
70	16.5
75	12.9
80	9.7
85	7

Source: http://www.cdc.gov/nchs/data/nvsr/nvsr60/nvsr60_04.pdf.

(a) Draw a scatterplot of y versus x for male.
(b) What form of relationship exists between current age and expected life for male data?
(c) Find the correlation between current age and the expected life for male data.
(d) What model will you recommend for predicting expected life using current age for male data?

(4.29) Given data points are measurements on x and y:

x	-4	-3	-2	-1	0	1	2	3	4
y	4	3	2	1	0	1	2	3	4

(a) Make a scatterplot of these data.
(b) What form of relationship between x and y is evident from the scatter plot?
(c) Based on the scatterplot, what is your guess about the correlation r?
(d) Find the value of r. Does your guess in (c) agree with this value?
(e) If you fit a linear regression equation, what would be the value of the coefficient of determination?

(4.30) The Table below contains the US census population data (rounded to the nearest million) from 1910-2010. We are interested to find a regression model of the U.S. population on year.

Year (x)	U.S. Population (y)
1910	92
1920	106
1930	123
1940	132
1950	151
1960	179
1970	203
1980	227
1990	249
2000	281
2010	309

Source: http://www.msnbc.msn.com/id/40764172/ns/us_news-life/t/population-growth-slowest-census-shows/

(a) Identify the response and explanatory variables in this problem.
(b) Fit a linear model of y on x.
(c) How do you assess the "goodness of the fit" of the model?
(d) Use your model in (b) to predict the population in 2005.
(e) Extrapolate the population in 2020.

(4.31) Given data points are measurements on x and y:

x	-4	-3	-2	-1	0	1	2	3	4
y	16	9	4	1	0	1	4	9	16

(a) Make a scatterplot of these data.
(b) What form of relationship between x and y is evident from the scatter plot?
(c) Based on the scatterplot, what is your guess about the correlation r?
(d) Find the value of r and comment on your result.

(4.32) Suppose from a sample of Kelly Blue Book information on age (in years) and price (in hundreds of dollars) of 20 randomly selected cars of a certain model between the ages of 1 and 7 years, the following estimated regression equation is obtained:

$$Price = 275.99 - 17.99 \times Age$$

(a) What are the predictor and response variables for this model?
(b) How would you comment on the value of the slope for this problem?
(c) Predict the price of the car when it is 4 years old.

(d) Find the residual at observed value of (age, price) = (4, 220).
(e) Predict the price of a car when age=5 years. What do you call this prediction?
(f) Predict the price of a car when age=10 years. What do you call this prediction? Does this prediction make sense to you?

(4.33) The data in table below refer to the percent E-commerce of total manufacturing shipments (x) and the percent E-commerce of merchant wholesale trade sales (y) by year from 2002-2010.

Year	x	y
2010	46.4	24.6
2009	42.8	24.6
2008	39.7	22
2007	35.2	22.9
2006	31.2	22.8
2005	28.3	23.2
2004	23.1	21.9
2003	21	22
2002	19.2	21

Source: Summary of U.S. Shipments, Sales, Revenues, and E-commerce, available at

http://www.census.gov/econ/estats/2010/all2010tables.html

(a) Draw a scatterplot of y versus x.
(b) What trend or relationship does the scatterplot show between x and y?
(c) What is the direction of the relationship, if there is any?
(d) Is there any presence of outliers in the data? Why do you think so?
(e) Compute the value of r.
(f) Fit a regression or predicting equation for predicting the percent E-commerce of merchant wholesale trade sales (y) using the percent E-commerce of total manufacturing shipments (x).
(g) Predict a value of y when $x = 21$. What do you call this prediction? Find the error of this prediction.
(h) Predict a value of y when $x = 18$. What do you call this prediction?

(4.34) Data below refers to the U.S. percentage E-commerce of total manufacturing shipments (y) by Years (x), 2000-2010. Answer the following questions:

(a) Make a scatterplot of y versus x.
(b) Does the plot suggest any linear relationship? If so, what is the direction of the relationship?
(c) Guess the strength of the linear relationship and assign a value for the linear correlation between x and y.
(d) Compute the value of *r* and comment on your result.
(e) Fit the regression or predicting equation of y on x.
(f) Predict the value of y when x=2000. What do you call this prediction? Find the error of this prediction.

x	y
2010	46.4
2009	42.8
2008	39.7
2007	35.2
2006	31.2
2005	28.3
2004	23.1
2003	21
2002	19.2
2001	18.2
2000	18

(4.35) The data below refers to the calorie and fat contents from the serving size of 100 grams of various nuts and seeds. We wish to predict fat using calorie for nuts and seeds.

Types of nuts and seeds	Calories (kcal)	Fat (gram)
Whole almond	611.6	54.8
Brazil nuts	687.4	68.3
Plain cashew nuts	584.5	48.9
Chestnuts	170	2.7
Hazelnuts	655	63.5
Salted peanuts	622	53
Sesame seeds	610.5	56.4
Sunflower seeds	585.3	48.7
Walnuts	691.3	68.5

(a) Identify response and explanatory variables for this problem.
(b) Find the correlation between fat content and serving calories.
(c) Fit a regression model of fat on calorie.
(d) Predict fat content when calorie content in a serving size of 100 grams is 600 kcal. What do you call this prediction? Can you estimate the residual of your predicted value?
(e) Predict fat content when calorie content in a serving size of 100 grams is 655 kcal. What do you call this prediction? Can you estimate the residual of your predicted value? If so, find the residual of your prediction.
(f) What is the coefficient of determination of the predicted model?
(g) Make an interpretation of the coefficient of determination of your predicted model.

(4.36) The following table presents the percent distributions of domestic (x) and international passengers (y) traveling by Delta Air Lines for each month from January through December 2013.

x	7.1	6.8	8.7	8.3	8.7	9.1	9.4	9.3	8.0	8.6	7.8	8.2
y	7.0	6.3	8.3	7.4	8.3	10.0	10.7	10.7	8.5	8.0	6.8	8.0

Source: Bureau of Transportation Statistics T-100 Market data from http://www.transtats.bts.gov/Data_Elements.aspx?Data=1.

(a) Find the correlation between x and y.
(b) Find the estimated regression line to predict y using x.
(c) Find a predicted value of y at $x = 9.1$ and find the residual of the predicted model. What do you call this prediction?
(d) Find a predicted value of y at $x = 10$. Can you find the residual of the predicted model? What do you call this prediction?

Chapter 4: Answer to selected problems

(4.1) (a) T (b) T (c) F (d) T (e) F (f) T (g) F (h) T
(4.2) (a) Fiber content
(4.3) (a) The change in the response per unit increase in the predictor
(4.4) (b) 55% of the variation in fiber content is explained by the estimated regression line.
(4.5) (a) the proportion of variation in response explained by the estimated regression line.
(4.6) (d) -0.90
(4.7) (a) 0.90
(4.8) (c) (a) or (b) or both
(4.9) (c) (a) or (b) or both
(4.10) (d) All of the above.
(4.11) (b) If r=0, then there is no relationship between two variables.
(4.12) (c) 662 calories
(4.13) (a) a predictor
(4.14) (c) About 81% of the variation in calories for menu items is explained by the estimated regression line.
(4.15) (d) All of the above.
(4.16) (b)The estimated regression line passes through the point (44,662).
(4.17) (c) the best line with sum of the squares of the vertical distances of the data points from the line is the minimum.
(4.18) (c) the square of the correlation between the underlying two variables
(4.19 (a) 1 (b) 1 (c) positive
(4.20) (a) -1 (b) 1 (c) negative
(4.21) (a) response: Fuel consumption, Predictor: Speed limit (b) No (c) Very bad model (d) 0.1732 (e) No unit, r is a unit free measure.
(4.22) (a) positive
(4.23) (c) +0.60
(4.24) (a) positive
(4.25) (a) the estimated line will have the smallest possible sum of the squares of the vertical distances of the data points from the line.

(4.26)

(a) Response is export in 2011 and predictor is export in 2010.
(b) r=0.997, almost a perfect positive linear relationship.
(c) Yes, I would recommend a linear model for predicting y using x.
(d) The predicting equation is $\hat{y} = -5.3 + 1.31 \times x$.
(e) The predicted export in 2011 when 2010 export is 22.5 is $\hat{y} = -5.3 + 1.31 \times 22.5 = 24.2$. The residual for this prediction is $e = 26.6 - 24.2 = 2.4$.
(f) The coefficient of determination is $r^2 = 0.99$. This means 99% of the response variation is explained by the predicted regression line.

(4.27)

(a) The response is fat content and explanatory variable calorie.

(b) $\hat{y} = -20.02 + 0.124 \times x$.
(c) r=0.99 and r^2=0.98. Therefore, the coefficient of determination is 0.98, which means that 98% of the response variation is explained by the predicted regression line.
(d) The predicted fat content when the calorie content is 600 kcal is $\hat{y} = -20.02 + 0.124 \times 600 = 54.4$.

(4.28)

(a) Try yourself.
(b) Negative linear relationship.
(c) r=-0.998.
(d) Linear regression model.

(4.29)

(a) Try yourself
(b) A non-linear relationship. More specifically, the scatterplot looks a curve of an absolute function.
(c) Guess $r = 0$, because the scatterplot definitely suggest a non-linear relationship.
(d) The computed value of $r = 0$, which confirms the guess.

(4.30)

(a) Response is the US population and explanatory variable is year.
(b) $\hat{y} = -4099 + 2.2(x)$.
(c) $r = 0.99$ and $r^2 = 0.98$. Therefore, 98% of variation of the response is explained by the model predictor. Therefore, the model is vey good.
(d) $\hat{y} = -4098.7 + 2.19(2005) = 292$.
(e) $\hat{y} = -4098.7 + 2.19(2020) = 325$.

(4.33)

(a) Try yourself
(b) A linear relationship.
(c) Positive.
(d) The point (39.7,22) seems to be an outlier.
(e) $r = 0.81$.
(f) $\hat{y} = 19.56 + 0.1(x)$.
(g) $\hat{y} = 19.56 + 0.1(21) = 21.7$. Interpolation. $e = 22 - 21.7 = 0.3$.
(h) $\hat{y} = 19.56 + 0.1(18) = 21.4$. Extrapolation.

(4.34)

(a) Try yourself
(b) Yes, a positive linear relationship.
(c) Try yourself.
(d) $r = 0.98$.
(e) $\hat{y} = -6138.7 + 3.08(x)$.
(f) $\hat{y} = -6138.7 + 3.08(2010) = 52.1$. Interpolation. $e = 46.4 - 52.1 = -5.7$.

CHAPTER 5
ELEMENTS OF PROBABILITY

Introduction

Very often we hear statements like

- there is a 85% chance of having a heavy rain tonight
- I am 90% confident that I will win the race
- I may get a ticket if I park in the faculty area
- Our team will win

How do we make statements like these?
Well, they may be based on:

- *scientific reasoning*
- *previous experience*
- *personal guess*
- *belief or wish*

Whatever the reason, there are always some degrees of uncertainty associated with these statements.
The ***probability*** is the measure of the degree of ***uncertainty*** of events.
In this chapter, we will learn some basic elements which will help us understand the concept of probability and compute probability using some basic rules of probability.

In section 5.1 we will define terms and terminologies necessary to introduce the concept of probability, define probability of an event and go over the simple properties of probability. In section 5.2, we will discuss addition rules of probability. In section 5.3, we will introduce conditional probability and independence of events.

Section 5.1 Concept of probability basics

In this two section, we will learn about

- experiment
- outcome
- equally likely outcomes
- sample space
- events

- define probability of an event

Let us start with a simple example:

Example 5.1

A coin with one side labeled as H (head) and the other side as T (tail) is tossed one time. What do we observe? Of course, we will observe either an H or a T.
The tossing of the coin is an ***experiment***.
Observing an H or a T is an ***outcome***.
Outcomes H and T are ***equally likely***, because either of them has the same chance of being appeared.
The set of two possible outcomes $\{H, T\}$ is the ***sample space***.
Each outcome, H or T, is a ***sample point***.
There are two sample points associated with this experiment.
We use the notation S for sample space and $n(S)$ for the number of sample points or outcomes associated with the experiment.
Then, we might write $S = \{H, T\}$ and $n(S) = 2$.

Some useful definitions

An ***experiment*** is any act whose results are uncertain or unknown until the act is performed.
Any result of the performed experiment is called an ***outcome***.
Outcomes of an experiment are ***equally likely*** if each of them has the same chance of being observed.
The set of all possible outcomes of an experiment is called the ***sample space***.
Every individual outcome is a ***sample point***. The sample space is usually denoted by the letter $\boldsymbol{S}$ and $n(\boldsymbol{S})$ is the count of sample points in the sample space.

Example 5.2

An experiment consists of tossing a fair coin three times, each of which results in either an H or a T.

(a) Write out the sample space.
(b) Argue if the outcomes of the sample space are equally likely or not.
(c) Find the value of $n(S)$.

Solution

(a) $S = \{HHH, HHT, HTH, THH, HTT, THT, TTH, TTT\}$
(b) All eight outcomes in S are equally likely, because any of these outcomes would have been observed.
(c) $n(S) = 8$.

Often, a tree diagram helps better understand the all possible outcomes for a repetitive trial of an experiment. Note how all eight possible outcomes result from the tossing of a coin three times.

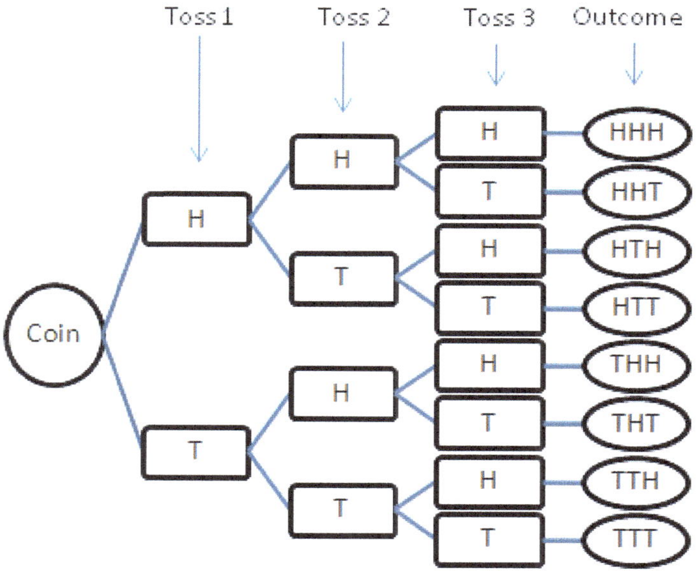

Definition: Event

An ***event*** is any collection or subset of outcomes of the sample space S.
Events are represented by any letter other than S.
If A represents an event, then $n(A)$ represents the number of sample points in event A.
An event with no outcome of the sample space is called an ***impossible event***, usually denoted by $\emptyset$ or $\{\}$.
An event with all possible outcomes is called a ***sure*** or ***certain event***. The sample space itself is a sure event.

Example 5.3

Think of the coin toss experiment of Example 5.1. We have, $S = \{H, T\}$ and $n(S) = 2$.
Four events of this experiment may be:
$A = \{H\}, B = \{T\}, C = \{\}, D = \{H, T\}$.

(a) Find the number of sample points for each event.
(b) Identify the sure event. Does the sure event differ from the sample space?

Solution

(a) $n(A) = 1, n(B) = 1, n(C) = 0$ and $n(D) = 2$.
(b) The sure event is D. The sure event and the sample space are the same. Note that in this example, $n(S) = n(D) = 2$.

Notation for the probability of an event A

The notation $P(A)$ denotes the probability of an event A.

Classical definition of $P(A)$

If an experiment consists of $n(S)$ equally likely outcomes and if an event A consists of $n(A)$ of outcomes, then the probability of the event A is

$$P(A) = \frac{n(A)}{n(S)}$$

Example 5.4

Revisit Example 5.3 and compute probability of each of the four events.

Solution

We have, $n(A) = 1$, $n(B) = 1$, $n(C) = 0$ and $n(D) = 2$.

Then,

$$P(A) = n(A)/\,n(S) = 1/2$$
$$P(B) = n(B)/\,n(S) = 1/2$$
$$P(C) = n(C)/\,n(S) = 0/2 = 0$$
$$P(D) = n(D)/\,n(S) = 2/2 = 1$$

Properties of probability

(1) The probability of an impossible event is 0, i.e., $P(\{\}) = 0$.
(2) The probability of a sure event is 1, i.e., $P(S) = 1$.
(3) The probability of any other event A satisfies $0 \leq P(A) \leq 1$.

The verification of above properties follows easily from Example 5.4.

Other definition of probability may be useful for events that do not consist of equally likely outcomes.

Definition: Empirical or experimental probability of an event

Empirical or experimental probability of an event applies to an experiment whose outcomes are not equally likely. The probability of an event associated with such an experiment is computed from the proportion of time the event is observed in many repeated trails of the experiment.

If n_{A} is the number of occurrences of the event A in n repeated trials of the experiment, then

$$P(A) = \frac{\text{number of occurrences of the event A}}{\text{number of trials of the experiment}} = \frac{n_{\mathrm{A}}}{n}$$

The larger the number of trials of the experiment, the closer is the empirical probability to the classical probability.

Example 5.5

Suppose we toss a coin 50 times and observe 27 heads (Successes) and 23 tails (Failures).

(a) Compute $P(H)$ and $P(T)$ by using the empirical definition.
(b) Compare $P(H)$ and $P(T)$ obtained using the empirical definition from that of the classical definition.

Solution

(a) Using the empirical definition

$$P(H) = 27/50 = .54$$
$$P(T) = 23/50 = .46$$

(b) With classical approach where H and T are equally likely events. Therefore, we will have $P(H) = \frac{1}{2} = 0.50 = P(T)$.

Subjective probability

If outcomes of an experiment are not equally likely and empirical probabilities cannot be used, then probability of an event is assigned by personal judgment, and is called the ***subjective probability***.

Example 5.6

All students in an introductory statistics class were asked the following two questions at the beginning of the semester:

(a) What do you mean by probability?

(b) What is the probability that you will use the knowledge of probability in your major field of study?

The responses from a randomly chosen student were:

(a) Probability is about gambling.

(b) 50%

The probability in (b) is just the intuitive personal judgment of the student. Such kind of assignment of probability is a subjective probability.

If same question is asked at the end of the semester, the response could be different.

Example 5.7

The table below is the distribution of people without health insurance coverage by selected income groups of people in the United States of America in 2009.

Income (in $)	Uninsured (in thousands)
<25,000	15,320
25,000-49,999	14,999
50,000-74,999	8,997
75,000+	9,669
Total	48,985

A person is selected at random from the uninsured people in 2009. Find the probability of the following events:

(a) The person selected has income between 25,000 and $74,999.

(b) The person selected has income below 25,000.

(c) The person has income either below 25,000 or above 75,000.

(d) The person has income between 25,000 and 49,999.

(e) The person does not have income between 25,000 and 49,999.

Solution

Define the following events:

A: The person selected has income between 25,000 and $74,999.

B: The person selected has income below 25,000.

C: The person has income either below 25,000 or above 75,000.

D: The person has income between 25,000 and 49,999.

E: The person does not have income between 25,000 and 49,999.

Note that each family member is equally likely and total count, $n(S) = 48{,}985$ thousand.

(a) $P(A) = \frac{n(A)}{n(S)} = \frac{14{,}999+8.997}{48{,}985} = \frac{23{,}996}{48{,}985} = 0.490.$

(b) $P(B) = \frac{n(B)}{n(S)} = \frac{15{,}320}{48{,}985} = 0.313$

(c) $P(C) = \frac{n(C)}{n(S)} = \frac{15{,}320+9{,}669}{48{,}985} = \frac{24{,}989}{48{,}985} = 0.51$

(d) $P(D) = \frac{n(D)}{n(S)} = \frac{14{,}999}{48{,}985} = 0.306$

(e) $P(E) = 1 - P(D) = 1 - \frac{14{,}999}{48{,}985} = 0.694$

Section 5.2: Addition rules of probability

In this section, we will introduce Venn diagram for representing events in the sample space and states additive rules of probability for two or more events.

Definition: Venn diagram

A Venn diagram is a pictorial representation of events in a sample space. The sample space S is represented by a rectangle and the events are displayed within the sample space by circles or ovals. For example, two disjoint or mutually exclusive events A and B are displayed using the Venn diagram in Figure 5.1.

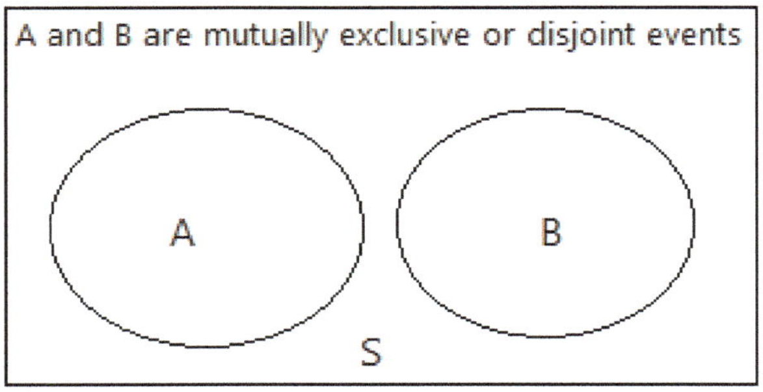

Figure 5.1

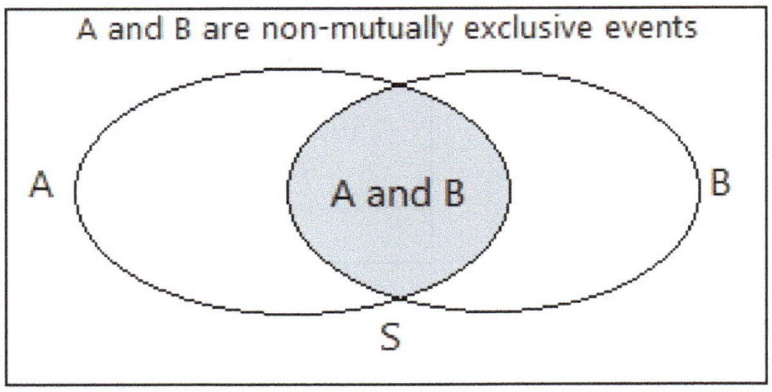

Figure 5.2

Definition: Disjoint events

Two events A and B are disjoint or mutually exclusive if they cannot occur together. This means that two events A and B have no outcome in common.
Events A and B in Figure 5.1 are disjoint events.

Definition: Non-disjoint events

Two events A and B are non-mutually exclusive or non-disjoint if they are not disjoint. Events A and B in Figure 5.2 are non-mutually exclusive or non-disjoint events.

Definition: Complement event

For any event A, the complement of A, denoted by A^c, is the event that A does not occur.

Below we demonstrate various events of common interest using Venn diagrams.

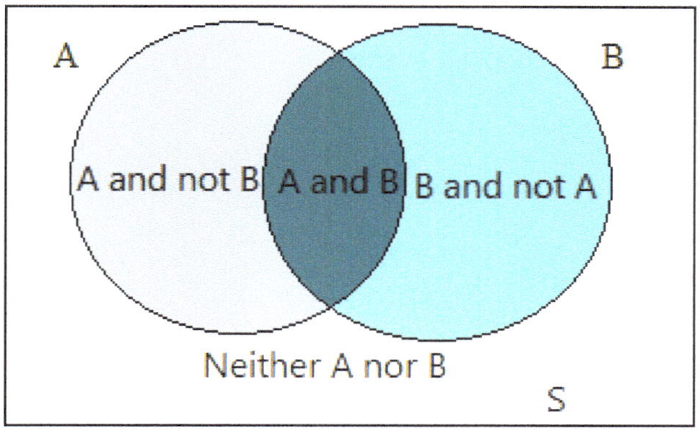

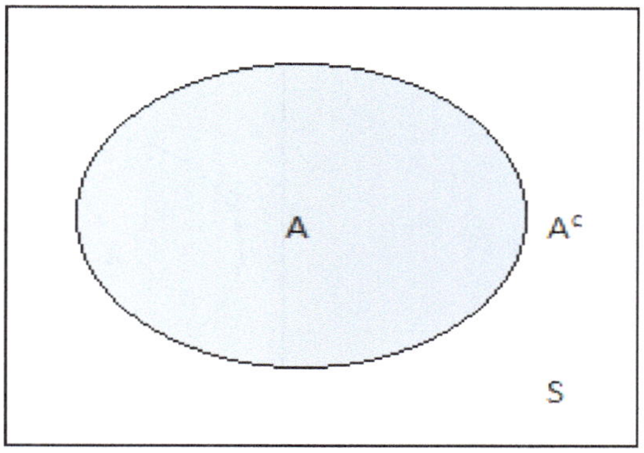

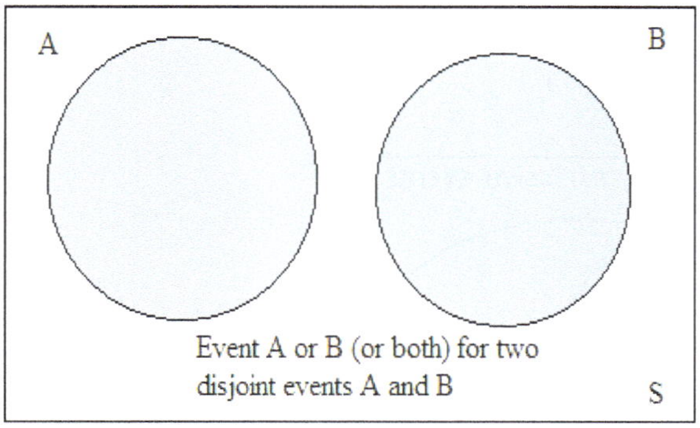

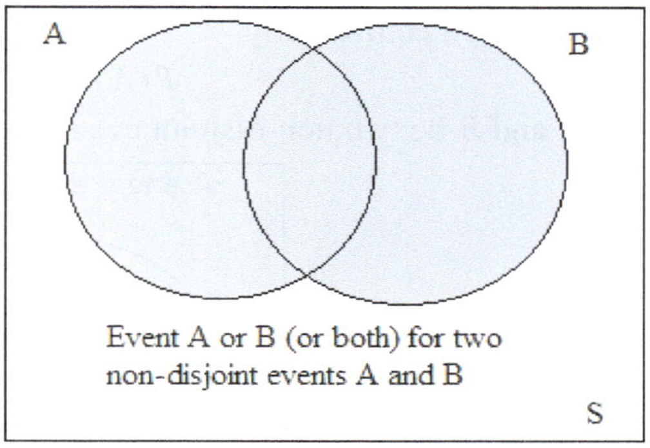

Notations

(1) $\boldsymbol{A \cap B} \equiv (\boldsymbol{A\ and\ B})$: refers to the intersection of events A and B; the part they both share together.

(2) $\boldsymbol{A \cup B} \equiv (\boldsymbol{A\ or\ B}) \equiv \boldsymbol{at\ least\ one\ of\ events\ A\ and\ B}$: refers to the event A or B (or both) will occur.

Probability of event A^c

For any event A,

$$P(A^c) = 1 - P(A)$$

Or, expressed equivalently,

$$P(A\ does\ not\ occur) = 1 - P(A)$$

Or,

$$P(A) + P(A^c) = 1$$

Example 5.8

Let R be the event that it will rain tomorrow.

Then, $P(R) + P(R^c) = 1$

If $(R) = \frac{1}{3}$, then $P(R^c) = 1 - P(R) = 1 - \frac{1}{3} = \frac{2}{3}$.

Addition rules of probability for two events

(a) Let A and B be two disjoint or mutually exclusive events.

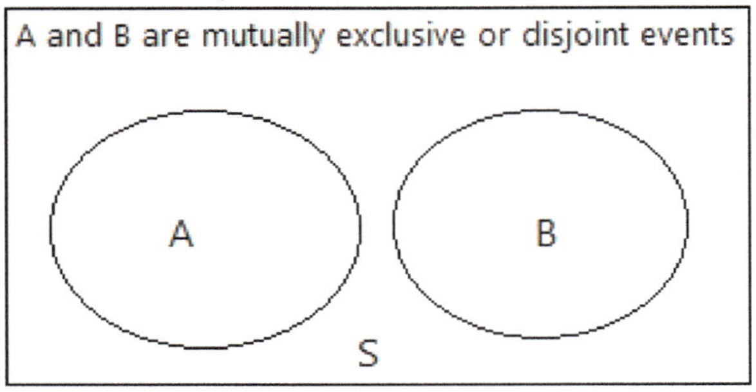

Then,

$$P(A \cup B) = P(A) + P(B)$$

Or, expressed equivalently,

$$P(A \text{ or } B) = P(A) + P(B)$$

(b) Let A and B be two non-disjoint events.

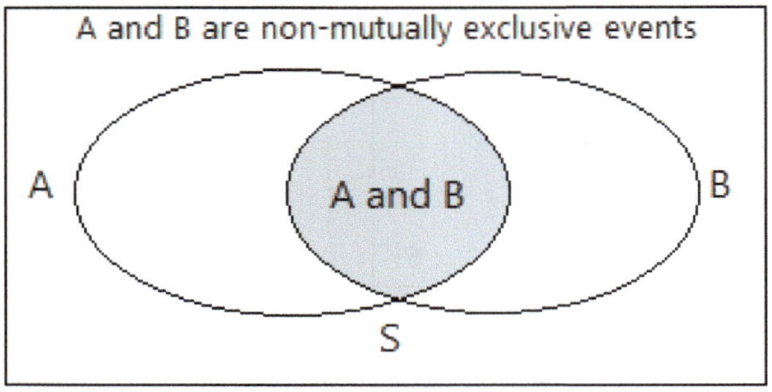

Then,

$$P(A \cup B) = P(A) + P(B) - P(A \cap B)$$

Or, expressed equivalently,

$$P(A \text{ or } B) = P(A) + P(B) - P(A \text{ and } B)$$

Example 5.9

Table below provides the percentage distribution of blood type in the USA by different ethnic groups.

	Ethnic Group			
Blood Type	Caucasians	African-American	Hispanic	Asian
O	45%	51%	57%	40%
A	40%	26%	31%	28%
B	11%	19%	10%	25%
AB	4%	4%	2%	7%
Total	100%	100%	100%	100%

Source: 2012 American Red Cross, Blood Types and the Populations
http://www.redcrossblood.org/learn-about-blood/blood-types;

An African-American is selected at random.

(a) Find the probability that the randomly chosen African-American has either blood type O or B, i.e., find $P(O \cup B)$.
(b) A person with blood type A can receive blood transfusions from people with blood types O and A. What is the probability that a randomly chosen African-American can donate blood to persons with blood type A?
(c) What is the probability that a randomly chosen African-American cannot donate blood to persons with blood type A?
(d) What is the probability that a randomly chosen African-American's blood type is not AB type?

Solution

(a) $P(O \cup B) = P(O) + P(B) = 0.51 + 0.19 = 0.70$
(b) $P(O\ or\ A) = 0.51 + 0.26 = 0.77$
(c) $1 - P(O\ or\ A) = 0.23$
(d) $P(not\ AB) = 1 - P(AB) = 0.96$

Example 5.10

Among those diagnosed with diabetes in the United States in the year 2007-2009, 26% is treated by insulin, 72% is treated by oral medication and 14% is treated by both insulin and oral medication.
A person is randomly selected from those diagnosed with diabetes in the year 2007-2009.

(a) Find the probability that the person is treated by insulin or oral medication.
(b) Find the probability that the person is treated by neither insulin nor oral medication.

Solution

Let us define

I: the person is treated by insulin
O: the person is treated by oral medication

Then

(a) $P(I\ or\ O) = P(I) + P(O) - P(I\ and\ O) = 0.26 + 0.72 - 0.14 = 0.84$
(b) Treated by neither insulin nor oral medication is the event (not "$I\ or\ O$") and hence we have $P(not\ \text{"}I\ or\ O\text{"}) = 1 - 0.84 = 0.16$.

Section 5.3: Conditional probability multiplicative rule and independence

Consider two events:
A: a randomly selected person is a lung cancer patient.
B: a randomly selected person is a smoker.
Then, the event "$A|B$" stands for the event A given that B has occurred. We want to compute the probability of the event $A|B$, denoted by $P(A|B)$.
In other words, we are interested in the probability of those who have lung cancer among smokers.

Definition: $P(A|B)$

The conditional probability of A given B is

$$P(A|B) = \frac{P(A \cap B)}{P(B)}, P(B) > 0$$

In a similar manner, the conditional probability of B given A is given by

$$P(B|A) = \frac{P(A \cap B)}{P(A)}, P(A) > 0$$

Remark: It follows that the conditional probability $P(A|B)$ can be computed using the formula $P(A|B) = \frac{n(A \cap B)}{n(B)}$ directly using the reduced sample space B.

Multiplicative rule for any two events

From the definition of conditional probability $P(A|B)$, it follows that

$$P(A \cap B) = P(A|B)P(B)$$

and this rule is called the multiplicative rule of probability.
Similarly, from the definition of $P(B|A)$, it follows that

$$P(A \cap B) = P(B|A)P(A)$$

Example 5.11

Among people aged 25 years and above in the USA in 2012, 20% earned a Bachelor degree and 83% were native born. Given a Bachelor degree, 85% are native born. (**Source**: U.S. Census Bureau, Current Population Survey, 2012 Annual Social and Economic Supplement).
Let us define two events:
B: The selected person has a Bachelor's degree
N: The selected person is a Native born

(a) What do 20%, 83% and 85% represent in terms of the probability of events?
(b) A person is selected at random among those who were 25 years and above in the USA in 2012. What is the probability that the person has a Bachelor degree and is a native born?
(c) Given native born, what is the probability that the person has a Bachelor degree?

Solution

(a) $P(B) = 0.20,\ P(N) = 0.83$ and $P(N|B) = 0.85$
(b) $P(B\ and\ N) = P(N|B) \times P(B) = 0.85 \times 0.20 = 0.17\ or\ 17\%$.
(c) $P(B|N) = P(B\ and\ N)/P(N) = 0.17/0.83 = 0.20\ or\ 20\%$.

Example 5.12

Table below is the distribution of cigarette smoker by gender among the employees of a firm:

	Male (M)	Female (M^c)	Total
Smoker (C)	6	7	13
Non-Smoker (C^c)	6	5	11
Total	12	12	24

An employee is randomly selected from the firm.
Let us define the following events:

C: selected employee is a smoker
C^c: selected employee is a non-smoker

M: the selected employee is a male

M^c: the selected employee is a female

Evaluate the following probabilities:

(a) $P(M \cap C)$, $P(C)$ and $\frac{P(M \cap C)}{P(C)}$

(b) $P(M|C)$

(c) $P(M \text{ or } C)$

Solution

(a) $P(M \cap C) = \frac{6}{24}$, $P(C) = \frac{13}{24}$ and $\frac{P(M \cap C)}{P(C)} = \frac{\frac{6}{24}}{\frac{13}{24}} = \frac{6}{13}$.

(b) $P(M|C) = \frac{P(M \cap C)}{P(C)} = \frac{6}{13}$

(c) $P(M \text{ or } C) = P(M) + P(C) - P(M \text{ and } C) = \frac{12}{24} + \frac{13}{24} - \frac{6}{24} = \frac{19}{24} = 0.7917$

Definition: Independence of two events

Two events A and B are independent if and only if

$$P(A|B) = P(A)$$

Or equivalently,

$$P(B|A) = P(B)$$

This means that the occurrence of one event does not change or affect the probability of the occurrence of the other event.

If two events A and B are independent, then

$$P(A \cap B) = P(A)P(B).$$

The above condition is a consequence of the definition of conditional probability and independence together. The above condition is also called the multiplication rule for two independent events A and B.

If k events $A_1, A_2, \ldots, A_k$ are all independent then

$$P(A_1 \cap A_2 \cap \ldots \cap A_k) = P(A_1)P(A_2)P(A_3)\ldots P(A_k)$$

Example 5.13

Revisit Example 5.10. Suppose four persons are randomly chosen among those diagnosed with the diabetes in the year 2007-2009. What is the probability that all are treated with insulin?

Solution

Let us define

I_i: the randomly selected ith person is treated by insulin, $i = 1, 2, 3, 4$

Then, $P(I_1 \cap I_2 \cap I_3 \cap I_4) = P(I_1)P(I_2)P(I_3)P(I_4) = (0.26)^4 = 0.0046$

Chapter 5: Practice problems

(5.1) A fair six-sided die is rolled. The set of possible outcomes is $\{1, 2, 3, 4, 5, 6\}$. This set is called

(a) an event.
(b) the experiment.
(c) the sample space.
(d) the probability.

(5.2) The following table summarizes the probability distribution of uninsured people by different age groups (in years) in the USA in 2011.

Age (in years)	Probability
<18	0.14
18-24	0.16
25-34	0.24
35-44	0.17
45-54	0.16
55-64	0.11
65 +	0.01

Source: https://www.census.gov/hhes/www/cpstables/032012/health/h01_000.htm, the U.S. Census Bureau.

A person was selected at random among those who were uninsured in 2011.

(a) Compute the probability that the person was 55 years or older.
(b) Compute the probability that the person was in the age group 18-24 or 45-54.
(c) Can you compute the probability that the person was exactly 25 years old? Justify your answer.

(5.3) A couple plans to have three children. Let B denote a boy and G a girl. Then, one of the possible outcomes will be, for example, BBG, which means that the first two children are boys and the third one is a girl.

(a) Write the sample space of all possible arrangement of boys and girls a couple might have.
(b) Argue why all outcomes are equally likely. What is the probability of each of these arrangements?
(c) What is the probability that the couple would have at least one boy?
(d) What is the probability that the couple would have no boy?

(5.4) Table below contains the distribution of uninsured people (in thousands) by region in the USA in 2010.

Region	Uninsured
Northeast	6,779
Midwest	8,605
South	21,665
West	12,855
Total	49,904

Source: http://www.census.gov/hhes/www/hlthins/data/incpovhlth/2010/table8.pdf.
U.S. Census Bureau, Current Population Survey, 2010 and 2011 Annual Social and Economic Supplements

A person is selected at random from those uninsured in the USA in 2010. Find the probability of the followings:

(a) The person selected is an uninsured from the Northeast region.
(b) The person selected is either from the Midwest or West.
(c) The person selected is not from the South.

(5.5) Assume that A and B are two disjoint events with respective probabilities .31 and .30.

(a) Find the probability that either event A or B occurs.
(b) Find the probability that A and B both occurs.
(c) Which of the followings must be true?
   i. The probability that (A or B) does not occur is 0.39.
   ii. A and B must include all events in the sample space.
   iii. The probability that A does not occur is 0.61

(5.6) Assume that A and B are two disjoint events with respective probabilities .49 and .51

(a) Find the probability that either event A or B occurs.
(b) Find the probability that A and B both occurs.
(c) Which of the followings must be true?
   i. The probability that (A or B) does not occur is 0.
   ii. A and B must include all events in the sample space.
   iii. The probability that A does not occur is 0.51
   iv. All of the above.

(5.7) Two events A and B are independent with P(A)=0.31 and P(B)=0.30. Determine the following probabilities:

(a) P(A and B)=?
(b) P(A|B)=?
(c) P(B|A)=?

(5.8) Two events A and B occur with probabilities P(A)=0.4 and P(B)=0.5, respectively. It also follows that P(A and B)=0.20. Then events A and B must be

(a) disjoint
(b) independent
(c) non-disjoint
(d) none of the above

(5.9) Two events A and B occur with probabilities P(A)=0.4 and P(B)=0.5, respectively. It also follows that P(A and B)=0.20. Then, which of the following must be true?

(a) P(A|B)=0.4.
(b) P(B|A)=0.5.

(c) Events A and B are independent.
(d) All of the above.

(5.10) Two events A and B occur with probabilities P(A)=0.7 and P(B)=0.3, respectively. It also follows that P(A and B)=0.21. Then, determine the followings:
(a) P(A|B)=?
(b) P(B|A)=?

(5.11) Among those diagnosed with diabetes in the United States in the year 2007-2009, 26% is treated by insulin, 72% is treated by oral medication and 14% is treated by both insulin and oral medication. A person is randomly selected from those diagnosed with diabetes in the year 2007-2009.
(a) Find the probability that the person is treated by insulin or oral medication.
(b) Find the probability that the person is treated only by oral medication.
(c) If three persons are randomly chosen among those diagnosed with the diabetes in the year 2007-2009, what is the probability that all are treated with oral medication only?

(5.12) A card is drawn at random from a deck of 52 playing cards.
Let us define the following events:
A: the selected card is a red card.
B: the selected card is a black card.
C: the card is an ace.
Compute the following probabilities:
(a) P(A)
(b) P(A or B)
(c) P(C)
(d) P(C|A)
(e) P(not C)

(5.13) American Red Cross estimates that among Caucasian American, 45% has Type O blood, 40% Type A, 11% Type B, and 4% Type AB.
If an US Caucasian blood donor is chosen at random, what is the probability that the donor has blood type
(a) A or B
(b) Not AB
(c) Not O
If four potential donors are chosen at random, what is the probability that
(d) all are Type O
(e) no one is Type AB

(5.14) In a small neighborhood, it appears that 68% of the residents own a native car, 59% own a foreign car and 33% own both a native car and a foreign car. A resident is selected at random. What is the probability that the randomly selected resident owns
(a) a native or a foreign car
(b) a native car only
(c) a foreign car only

(d) If two residents are selected at random, what is the probability that both will own only native cars?

(5.15) An instructor wants to know the media of social network in his class. It reveals that 64% students use Facebook for social network, 48% students use Skype and 30% students use both Facebook and Skype. A student is randomly selected from the class.

(a) What is the probability that the student uses either a Face book or a Skype?
(b) What is the probability that the student only uses Skype?
(c) What is the probability that the student only uses Facebook?
(d) If three students are selected at random, what is the probability all three students use only Face book?

(5.16) Among all women aged 25 years and above in the USA in 2012, 55% were employed and 30% were High School graduates. Given a High School graduate, 46% were employed.
Let us define following two events:

E: women aged 25 years and above were employed in the USA
H: women aged 25 years and above were HS graduates in the USA

(a) Express 55%, 30% and 46% specified in terms of probability of associated events.
(b) If a woman is selected at random among those 25 years and above, what is the probability that she is employed and HS graduate?
(c) If a woman is selected at random among those 25 years and above, what is the probability that she is High School graduate given that she is employed?

Chapter 5: Answer to selected problems

(5.1) (c)
(5.2) (a) 0.12, (b) 0.32, (c) No, because we do not have any information on exact age
(5.3)

(a) $S = \{BBB, BBG, BGB, GBB, BGG, GBG, GGB, GGG\}$
(b) All eight outcomes are equally likely because each outcome has the same chance of being observed. So each arrangement has the probability of 1/8.
(c) 7/8, d) 1/8

(5.4)

(a) $\frac{6,779}{49,904}$ or 0.136
(b) $\frac{8,605+12,855}{49,904} = \frac{21460}{49904}$ or 0.43
(c) $1 - \frac{21,665}{49,904} = \frac{28239}{49904}$ or 0.57

(5.5) (a) 0.61, (b) 0, (c) i
(5.6) (a) 1, (b) 0, (c) iv
(5.7) (a) 0.093, (b) 0.31, (c) 0.30
(5.8) (b); **(5.9** (d); **(5.10** (a) 0.7 (b) 0.3
(5.11)
Let us define

I: the person is treated by insulin
O: the person is treated by oral medicine

Then

(a) $P(I\ or\ O) = P(I) + P(O) - P(I\ and\ O) = 0.26 + 0.72 - 0.14 = 0.84$
(b) $P(O, not\ I) = P(O) - P(I\ and\ O) = 0.72 - 0.14 = 0.58$
(c) $0.58^3 = 0.1951$

(5.12)

(a) $P(A) = 26/52 = 1/2$
(b) $P(A\ or\ B) = P(A) + P(B) = 26/52 + 26/52 = 1$
(c) $P(C) = 4/52 = 1/13$
(d) $P(C|A) = 2/26 = 1/13$
(e) $P(not\ C) = 12/13$

(5.13)

(a) $P(A\ or\ B) = P(A) + P(B) = 0.40 + 0.11 = 0.51$
(b) $P(not\ AB) = 1 - 0.04 = 0.96$
(c) $P(not\ O) = 1 - P(O) = 1 - 0.45 = 0.55$
(d) $(0.45)^4 = 0.041$
(e) $(0.96)^4 = 0.849$

(5.14)
Define

N: Native car
F: Foreign car

Given $P(N) = 0.68,\ P(F) = 0.59,\ P(N\ and\ F) = 0.33.$

(a) $P(N\ or\ F) = P(N) + P(F) - P(N\ and\ F) = 0.68 + 0.59 - 0.33 = 0.94.$
(b) $P(N, not\ F) = P(N) - P(N\ and\ F) = 0.68 - 0.33 = 0.35.$
(c) $P(F, not\ N) = P(F) - P(N\ and\ F) = 0.59 - 0.33 = 0.26.$
(d) $[P(N, not\ F)]^2 = 0.35 * 0.35 = 0.1225$

(5.15)

Let us define

F: Facebook
S: Skype

Given $P(F) = 0.64,\ P(S) = 0.48,\ P(F\ and\ S) = 0.30.$

(a) $P(F\ or\ S) = P(F) + P(S) - P(F\ and\ S) = 0.64 + 0.48 - 0.30 = 0.82.$
(b) $P(S, not\ F) = P(S) - P(F\ and\ S) = 0.48 - 0.30 = 0.18.$
(c) $P(F, not\ S) = P(F) - P(F\ and\ S) = 0.64 - 0.30 = 0.34.$
(d) $[P(F, not\ S)]^3 = 0.34 * 0.34 * 0.34 = 0.04$

(5.16)

(a) $P(E) = 0.55,\ P(H) = 0.30,\ P(E|H) = 0.46.$
(b) $P(E\ and\ H) = P(E|H) \times P(H) = 0.46 \times 0.30 = 0.138\ or\ 13.8\%.$
(c) $P(H|E) = P(E\ and\ H)/P(E) = 0.138/0.55 = 0.2509\ or\ 25\%.$

CHAPTER 6
PROBABILITY OF DISCRETE RANDOM VARIABLES

Introduction

In this chapter, we will introduce the concept of discrete random variables and probability distribution of discrete random variables. The specific learning objectives will involve

- Discrete random variables
- Probability distribution of discrete random variables
- Computing expected value and variance for discrete random variables
- Binomial probability distribution
- Poisson distribution
- Applications

Section 6.1: Random variables

A random variable represents each outcome of an experiment numerically. Let us give an example and then proceed with the formal definition.

Example 6.1

A couple plans to have two children. Then, the sample space is $S = \{BB, BG, GB, GG\}$ and $n(S) = 4$. Note that each outcome is equally likely and has probability of 1/4.

Now, let us assign each outcome a numeric value by counting the number of girls associated with it. Let X represents the number of girls associated with the outcomes of the experiment. Then,

$$X(BB) = 0$$
$$X(BG) = 1 = X(GB)$$
$$X(GG) = 2$$

We call X a discrete random variable, which takes three values 0, 1 and 2.

Example 6.2

Let us consider an experiment of surveying 3 random persons of their usage of any dietary supplement containing calcium. Each person is equally likely to be a user or non-user of dietary supplement. Let us assign a Y to a person who uses dietary supplement and an N to a person who does not use it. Then, the sample space would be

$$S = \{NNN, NNY, NYN, YNN, NYY, YNY, YYN, YYY\}$$

Let X be the number of persons with any usage of dietary supplement. Then,

$$X(NNN) = 0$$
$$X(NNY) = X(NYN) = X(YNN) = 1$$
$$X(NYY) = X(YNY) = X(YYN) = 2$$
$$X(YYY) = 3.$$

Thus, the discrete random variable X takes four values 0, 1, 2 and 3.

Definition: Random variable

A ***random variable*** is a numeric valued variable defined on the sample space of an experiment. A random variable can be of two types, namely, a discrete random variable and a continuous random variable. A random variable is called a ***discrete random variable*** if it represents count. A random variable is called a ***continuous random variable*** if it is not discrete. A continuous random variable takes values from an interval of real numbers.

Of course, the random variables X in examples 6.1 and 6.2 are discrete random variables.

Example 6.3

Let X be the number of speeding tickets given in a highway on a given day. Then, X could take values from a countable finite list of numbers $\{0, 1, 2, \cdots\}$. Thus, X is a discrete random variable.

Example 6.4

Consider the following random variables:

- intelligence ratio (IQ) score of a group of people,
- gas mileage of light trucks,
- body temperature or body mass index of patients waiting for a surgery,
- time it takes for a light bulb to fail,

etc.

These are all examples of a continuous random variable because whatever values these variables take may be observed in some interval of real numbers.

Section 6.2 Discrete probability distribution

A ***discrete probability distribution*** of a discrete random variable is any formula, table or graph that assigns probability to each value of the random variable.

Let X be a discrete random variable and x be any specific value of X.

Notation

$p(x) = P(X = x)$ refers to the probability assigned to a particular value x of the random variable X. For example, $p(2) = P(X = 2)$.

Example 6.5

Going back to the Example 6.1, the random variable X, the number of girls a couple with two children may have, takes values 0, 1 and 2 with respective probabilities 1/4, 2/4 and 1/4. Expressing in tabular form using $p(x)$ notation, we will have a discrete probability distribution given by

x	$p(x)$
0	1/4
1	2/4
2	1/4

From above table, we have $p(2) = P(X = 2) = 1/4$, which means that a couple with two children will have 2 girls with probability $\frac{1}{4}$.

Example 6.6

Reference to Example 6.2, the random variable X, the number of persons with any usage of dietary supplement in survey of three random persons, takes values 0, 1, 2 and 3. Table below is a discrete probability distribution of X.

x	$p(x)$
0	1/8
1	3/8
2	3/8
3	1/8

Properties of a discrete probability distribution $p(x)$

The function $p(x)$ represents a discrete probability distribution if

1) $0 \le p(x) \le 1$
2) $\sum_x p(x) = 1$

In other words, the probability of a given value x lies between 0 and 1 inclusive and the total probability over all possible values of the random variable is 1.

Definition: Cumulative distribution function $F(x)$

The function $F(x) = P(X \le x)$ is called the cumulative distribution function. It refers to the total probability assigned to all values less than or equal to x. For example, $F(3) = P(X \le 3) = p(0) + p(1) + p(2) + p(3)$.

If X is a discrete random variable, and a and b are two positive integers, $a < b$, then it follows that

$$P(a \le X \le b) = F(b) - F(a-1)$$
$$P(a < X < b) = F(b-1) - F(a)$$

For example, $P(1 \le X \le 5) = p(1) + p(2) + p(3) + p(4) + p(5) = F(5) - F(0)$

$$P(1 < X < 5) = p(2) + p(3) + p(4) = F(4) - F(1)$$

Definition: Expected value of a random variable X

The ***expected value*** or ***mean*** of a discrete random variable X with probability distribution function $p(x)$ is defined as

$$\mu = E(X) = \sum_{All\ x} x \times p(x)$$

Example 6.7

Reference to Example 6.6, find the expected value of X and interpret the result.

Solution

Note that X denotes the number of persons with any usage of dietary supplement in a survey of 3 random persons. The probability distribution of X is given by

x	$p(x)$
0	1/8
1	3/8
2	3/8
3	1/8

Then, the expected value or mean of X is

$$\mu = \sum_{x} x \times p(x) = 0 \times p(0) + 1 \times p(1) + 2 \times p(2) + 3 \times p(3)$$

$$= 0 \times \frac{1}{8} + 1 \times \frac{3}{8} + 2 \times \frac{3}{8} + 3 \times \frac{1}{8} = \frac{12}{8} = 1.5$$

Interpretation

The expected value of 1.5 means that if three persons are surveyed, on average, there will be 1.5 persons with any usage of dietary supplement.

Q: Does it make sense to have 1.5 persons with any usage of dietary supplement in 3 surveyed persons?

Realistically, it is not possible to have 1.5 persons with the usage of dietary supplement in 3 surveyed persons. However, probabilistically, it is possible. Indeed, it means that if we keep surveying three persons over and over, on average, we will have 1.5 persons with any usage of dietary supplement per survey. Interpreting linearly, if we make 10 surveys, on average, $10 \times 1.5 = 15$ persons will have any usage of dietary supplement.

Example 6.8

As appears in National Vital Statistics Reports (Volume 62, January 6, 2014), the probability of a 35 years old person dying between ages 35-36 year is 0.00126. An insurance company is decided to sell a \$25,000 whole life insurance policy to a person aged 35 years for an annual premium of \$350.

(a) Find the expected profit per policy per year.

(b) What premium would you recommend if the company wants to make an expected profit of \$300 per policy per year?

Solution

(a) Let X be the profit from selling a whole life insurance to a person aged 35 years who dies in the interval 35-36 with probability 0.00126. Then the probability distribution of X is given by:

x	$p(x)$
$\$350$, if policy holder lives	0.99874
$\$350 - \$25000$, if policy holder dies	0.00126

Then the expected profit is

$$\mu = \sum_x xp(x) = 350(0.99874) + (350 - 25000)(0.00126)$$
$$= 350(0.99874) + 350(0.00126) - 25000(0.00126)$$
$$= 350 \times (1) - 31.5 = 318.5$$

Therefore, the expected profit per policy per year is \$318.50.

(b) Let the premium be $\$y$ per policy per year. Then, we will have the following distribution of the profit X.

x	$p(x)$
y, if policy holder lives	0.99874
$y - \$25000$, if policy holder dies	0.00126

We need to solve for y using the equation

$$300 = \sum_x xp(x)$$
$$300 = y \times 0.99874 + (y - 25000) \times 0.00126$$
$$300 = y \times 0.99874 + y \times 0.00126 - 25000 \times 0.00126$$
$$300 = y \times (0.99874 + 0.00126) - 25000 \times 0.00126$$
$$300 = y \times (1) - 31.5$$
$$y = 331.5$$

Therefore, we would recommend the annual price per policy to be \$331.50.

Definition: Variance and standard deviation of a discrete random variable X

The ***variance*** of a discrete random variable X with probability distribution function $p(x)$ is defined as

$$\sigma^2 = V(X) = \sum_{All\ x} (x - \mu)^2 \times p(x)$$

The ***standard deviation*** of X is the positive square root of the variance of X, and is given by

$$\sigma = +\sqrt{\sigma^2}$$

Example 6.9

Reference to Example 6.6, compute variance of X.

Solution

Note that in Example 6.7, we computed the expected value of X, which is $\mu = 1.5$.

The variance of X is

$\sigma^2 = \sum_x (x-\mu)^2 \times p(x)$
$= (0-1.5)^2 \times p(0) + (1-1.5)^2 \times p(1) + (2-1.5)^2 \times p(2) + (3-1.5)^2 \times p(3)$
$= 2.25 \times \frac{1}{8} + 0.25 \times \frac{3}{8} + 0.25 \times \frac{3}{8} + 2.25 \times \frac{1}{8}$
$= 0.75$

The standard deviation of X is

$\sigma = +\sqrt{\sigma^2} = +\sqrt{0.75} = 0.87$

Example 6.10

A gambler rolls a die in gambling. In each roll, he wins \$50 if it shows up an even number, and loses \$50 if it shows up an odd number. The gambler is using a die that has probability of 0.6 of showing up an even number and 0.4 of showing up an odd number. What is the expected profit of the gambler from a roll?

Solution

Let X be the profit of the gambler from a roll. Here, X takes two values: \$50 with probability 0.6 in case of a win and -\$50 with probability 0.4 in case of a loss.

x	$p(x)$
\$50, if wins	0.60
-\$50, if loses	0.40

Then, the expected value of X is

$$\mu = \sum_x x \times p(x) = 50 \times 0.6 + (-50) \times 0.4 = 10$$

On average, the gambler will make \$10 from each roll.

The variance of X is

$$\sigma^2 = \sum_{All\ x} (x-\mu)^2 \times p(x) = (50-10)^2 \times 0.6 + (-50-10)^2 \times 0.4 = \$2400$$

Therefore, the standard deviation of X is

$$\sigma = +\sqrt{\sigma^2} = +\sqrt{2400} = \$48.99$$

Section 6.3: Binomial distribution

In this section, we will briefly discuss one of the most useful discrete probability distributions, called the Binomial distribution.

Let us consider the following four assumptions:

(1) An experiment consists of n trials, where n is a fixed number,
(2) The n trials are independent,
(3) Each trial consists of two outcomes- success or failure,
(4) The probability of a success, denoted by p, remains constant from trial to trial.

Let X denote the number of successes of an experiment of n trials. Under assumptions (1)-(4), the probability that X takes the value x is computed using the following formula:

$$p(x) = P(X = x) = \binom{n}{x} p^x (1-p)^{n-x}; \quad x = 0, 1, 2, \dots, n.$$

The above formula is referred to as the ***binomial distribution***, ***binomial probability distribution***, or ***binomial probability function***. In above formula,

n = number of trails in the experiment

x = number of successes in n trials; $x = 0, 1, 2, \dots, n$

$\binom{n}{x}$ = number of ways x successes may result from n trials

$p(x)$=probability of x successes in n trials; $x = 0, 1, 2, \dots, n$

p=probability of a success which remains fixed from trial to trial

In general, a binomial distribution with two parameters n and p is often denoted by $b(x;\ n, p)$. The two parameters n and p denote the number of trials and the success probability of a binomial distribution, respectively.

The quantity $\binom{n}{x}$, read as "n choose x", is computed using the formula

$$\binom{n}{x} = \frac{n!}{x!\,(n-x)!}$$

where $n!$ is pronounced as "n factorial" and is calculated as follows:

$$n! = n \times (n-1) \times \dots \times 1$$
$$0! = 1$$

For example,

$$4! = 4 \times 3 \times 2 \times 1 = 24$$

$$\binom{10}{4} = \frac{10!}{4!\,(10-4)!} = \frac{10 \times 9 \times 8 \times 7 \times 6 \times 5 \times 4 \times 3 \times 2 \times 1}{(4 \times 3 \times 2 \times 1)(6 \times 5 \times 4 \times 3 \times 2 \times 1)} = 210$$

$$\binom{10}{8} = \frac{10!}{8!\,(10-8)!} = \frac{10 \times 9 \times 8 \times 7 \times 6 \times 5 \times 4 \times 3 \times 2 \times 1}{(8 \times 7 \times 6 \times 5 \times 4 \times 3 \times 2 \times 1)(2 \times 1)} = 45$$

We can have them computed in TI-84 Plus calculator as follows:

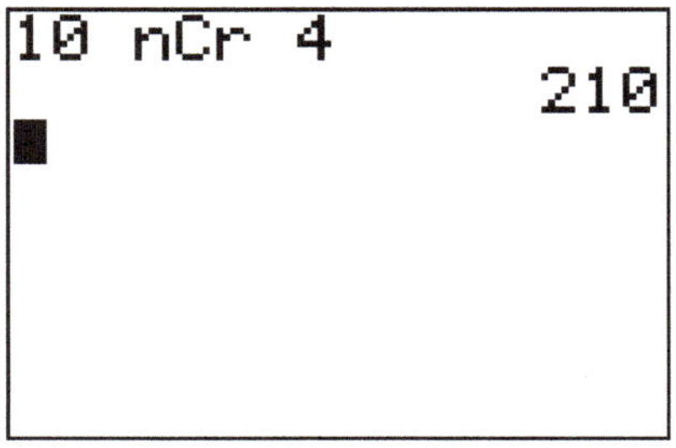

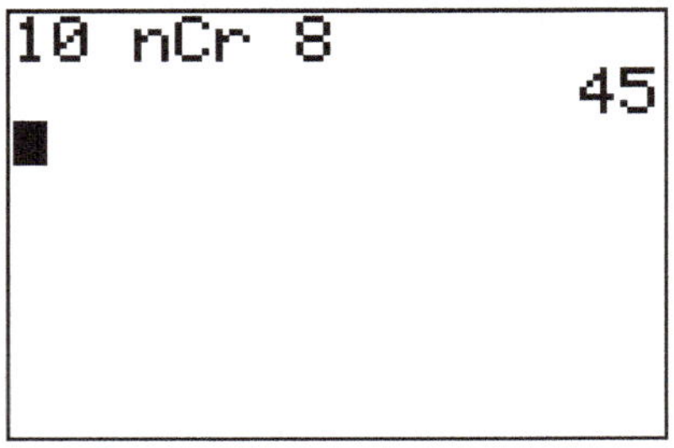

Example 6.11

Suppose that a basketball player shoots 10 free throws. Each time he shoots a throw, the probability is 0.8 that he makes a basket. We wish to compute the probability that he will make exactly 4 baskets. Answer the following questions:

(a) In regard to the formula $p(x) = \binom{n}{x} p^x (1-p)^{n-x}$, find values of n, p and x.

(b) How is the use of the formula in (a) justifiable?

(c) Compute $p(4)$.

(d) Interpret the result in (c).

Solution

(a) $n = 10, p = 0.8$ and $x = 4$.

(b) Four assumptions of the binomial probability are justified in this problems:

(1) We have 10 trials,

(2) 10 trials are independent,

(3) There are two outcomes for each trial: making the basket or not

(4) Probability of making a basket is 0.8 and it remains constant from trial to trial

(c) $p(4) = \binom{10}{4}0.8^4(1-0.8)^{10-4} = 210 \times 0.8^4 \times 0.2^6 = 0.0055$

We also get $p(4) = 0.0055$, using **Table T1** of Binomial Probability in Appendix with $n = 10, p = 0.8$ and $x = 4$.

TI-84 Plus calculator can be used in two different ways to compute $p(4)$:

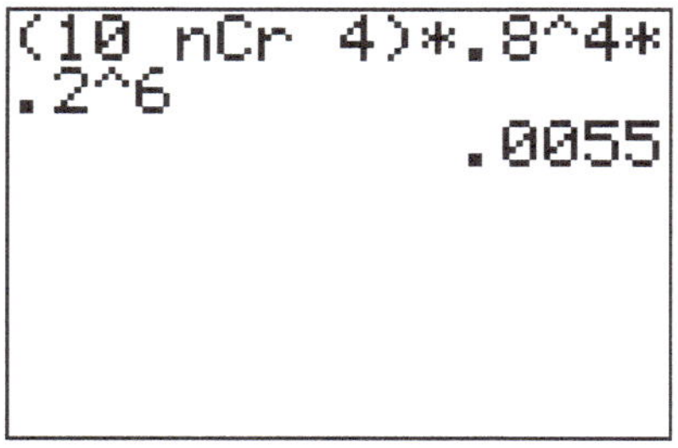

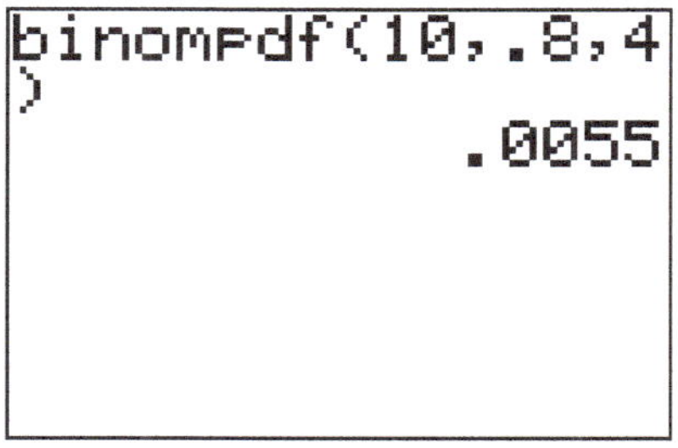

TI-84 Plus calculator has three arguments n, p, x in the $\boldsymbol{binompdf}()$ function to compute probability $p(x)$: $\boldsymbol{binompdf(n, p, x)}$.

(d) The probability is 0.0055 that the player will make exactly four baskets if he shoots 10 free throws, each having the probability 0.8 of making a basket.

Useful results for a binomial distribution

The mean, variance and standard deviation of a binomial random variable X with parameters n and p are as follows:

$$\mu = np$$
$$\sigma^2 = np(1-p)$$
$$\sigma = \sqrt{np(1-p)}$$

Example 6.12

A poultry farm wants to test the efficiency of a new incubator for hatching. Suppose that with the new incubator, the probability is 0.90 that each incubated egg will be hatched. Initially, 10 eggs are considered for incubation.

Find the followings:

(a) Write out the formula to obtain the probability that x eggs will be hatched out of 10 eggs.

(b) Justify how four assumptions are verified for the applicability of formula in (a).

(c) What is the probability that exactly 8 eggs will be hatched?

(d) What is the probability that at most 8 eggs will be hatched?

(e) Compute the mean and variance of the number of eggs that will be hatched.

Solution

(a) The probability that x eggs will be hatched out of 10 eggs is

$$p(x) = \binom{10}{x}(0.90)^x(0.10)^{10-x};\ x = 0,1,\ldots,10.$$

(b) The four assumptions of binomial probability required for the validity of answer in (a) are:

(1) There are n trials (here 10 eggs),
(2) Hatching or not hatching of 10 eggs is independent,
(3) There are two outcomes, hatching or not hatching,
(4) Probability of hatching of each egg is 0.90.

(c) The probability that 8 eggs will be hatched is

$$p(8) = \binom{10}{8}(0.90)^8(0.10)^{10-8} = (45)(0.90)^8(0.10)^2 = 0.1937$$

We also get $p(8) = 0.1937$, using **Table T1** of Binomial Probability in the Appendix with $n = 10$, $p = 0.9$ and $x = 8$, or using TI-84 Plus calculator as follows:

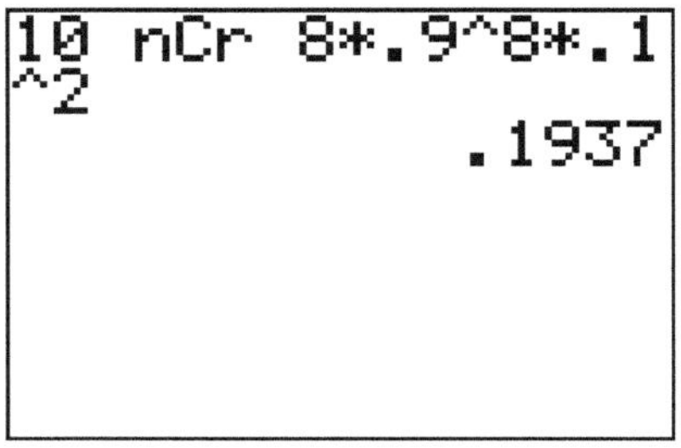

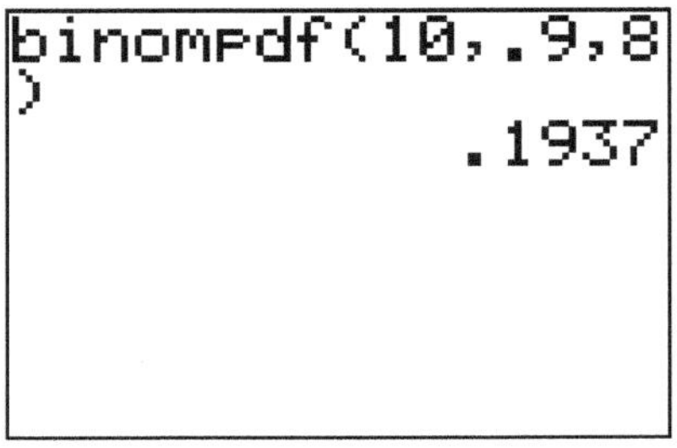

(d) The probability that at most 8 eggs will be hatched is given by

$p(x \le 8) = 1 - p(x > 8) = 1 - [p(9) + p(10)] = 1 - 0.3874 - 0.3487 = 0.2639$, using Table T1.

Recall that $F(x) = P(X \le x)$ is the cumulative distribution function.

One can use TI-84 Plus calculator function $\boldsymbol{binomcdf}()$ with three arguments n, p, x to compute $F(x)$. See the use of the $\boldsymbol{binomcdf(n,p,x)}$ for computing $F(8) = p(x \le 8)$ (see section 6.5 for detail instruction):

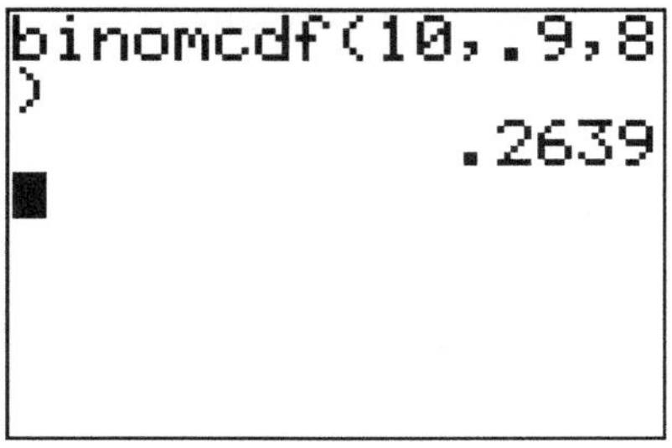

(e) The mean and variance of the number of eggs that will be hatched is

$$Mean = np = 10(0.90) = 9$$
$$Variance = np(1-p) = 10(0.90)(0.10) = 0.9$$

Example 6.13

A quiz consists of 10 true-false questions. A student makes a random guess to answer each question. The probability that the student will make a correct guess to answer each of the questions is 0.5. Let X be the random variable that denotes the number of correct guesses.

(a) Set up a probability function to determine the probability that the student will answer x questions correctly. What are the possible values of x?

(b) Find the probability that the student will answer at least five questions correctly.

Solution

(a) The probability that the student will answer x questions correctly is given by

$$p(x) = \binom{10}{x}(0.5)^x(0.5)^{10-x};\ x = 0,1,\ldots,10$$

The possible values of x are $0, 1, \ldots, 10$.

(b) The probability that the student will answer at least five questions correctly is

$$P(x \geq 5) = p(5) + \cdots + p(10) = .2461 + .2051 + .1172 + .0439 + .0098 + .0010$$
$$= 0.6231, \text{ using Table T1.}$$

Note that $P(x \geq 5) = 1 - p(x \leq 4) = 1 - F(4)$. Then, one can also use the TI-84 Plus calculator to compute the above probability via the usage of $binomcdf$ function:

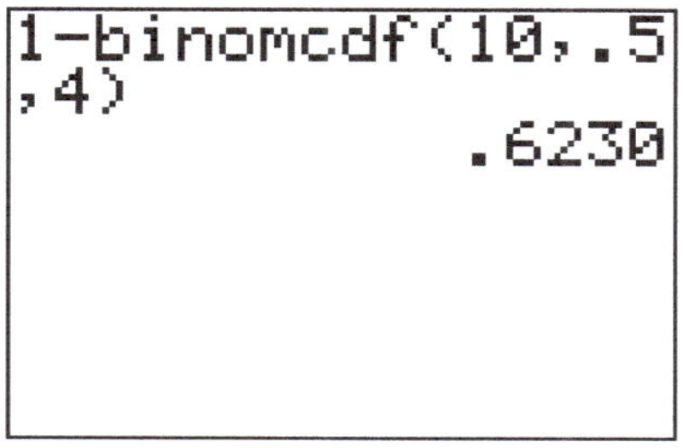

Section 6.4: Poisson distribution

Poisson distribution is another discrete probability distribution that is used to model the occurrence of rare events, e.g., the number of patients diagnosed with Ebola at a specified region, the number of accidents at an intersection per week, the number of cars a salesman sell per day, the number of days a school is closed per year due to an emergency, etc. The events may take place in any unit of time, space or region.

A discrete random variable X is said to have a ***Poisson distribution*** with parameter θ, read as "theta", if its probability distribution is given by the following function

$$p(x) = \frac{e^{-\theta}\theta^x}{x!};\ x = 0, 1, 2, \ldots$$

The parameter θ is called the mean of the distribution.

Useful results for a Poisson distribution

The mean μ and variance σ^2 of a Poisson distribution with parameter θ are as follows:

$$\mu = \theta$$

$$\sigma^2 = \theta$$

Note that the mean and variance of a Poisson distribution are the same.

Example 6.14

Suppose, on average, an auto salesman sells 5 cars per day. Let X be number of cars he sells in a given day of the week. Answer the following questions:

(a) What is the appropriate probability function for computing the probability of the number of cars he sells on a given day of the week?
(b) What is the probability that he will sell no car at all on a given day?
(c) What is the probability that he will sell exactly 5 cars on a given day?
(d) What is the probability that he will sell not more than 5 cars on a given day?
(e) What is the probability that he will sell at least 6 cars on a given day?

Solution

(a) We can use a Poisson probability distribution with parameter $\theta = 5$ for computing the probability of the number of cars the salesman sells on a given day. The probability function to be used is given by

$$p(x) = \frac{e^{-5}5^x}{x!};\ x = 0, 1, 2, \ldots$$

(b) The probability that he will sell no car at all on a given day is

$$p(0) = \frac{e^{-5}5^0}{0!} = 0.0067$$

We also get $p(0) = 0.0067$ from **Table T2** in the Appendix of Poisson probability with mean $\mu = 5$.

(c) The probability that he will sell exactly 5 cars on a given day is

$$p(5) = \frac{e^{-5}5^5}{5!} = 0.1755$$

Using **Table T2**, we also get $p(5) = 0.1755$.

(d) The probability that he will sell not more than 5 cars on a given day is

$$p(x \le 5) = p(0) + p(1) + p(2) + p(3) + p(4) + p(5)$$
$$= 0.0067 + 0.0337 + 0.0842 + 0.1404 + 0.1755 + 0.1755 = 0.616$$

We used **Table T2** for getting values of $p(0)$, $p(1)$, …, $p(5)$.

(e) The probability that he will sell at least 6 cars on a given day is

$$p(x \ge 6) = 1 - p(x \le 5) = 1 - 0.616 = 0.384$$

Even though we can use **Table T2** for computing Poisson probability, but we prefer to use technology such as Minitab or TI-84 Plus calculator for computing any Poisson probability.

Example 6.15

As per CDC, on average, about 15 out of 1000 children get identified with autism spectrum disorder (ASD) (i.e., about 1 in every 68 children). Let X be the number of children identified with ASD in the next 1000 children inspected. Answer the following questions:

(a) Write an appropriate probability model for the distribution of X.
(b) What is the probability that there will be exactly 5 ASD children?
(c) What is the probability that there will be at most 5 ASD diagnosis?
(d) What is the probability that there will be 6 or more ASD diagnosis?

Solution

(a) We can use a Poisson probability distribution with parameter θ =15 given by

$$p(x) = \frac{e^{-15}15^x}{x!}; \; x = 0, 1, 2, \ldots$$

to model the distribution of X.

(b) The probability that there will be exactly 5 ASD diagnosis is given by

$$p(5) = \frac{e^{-15}15^5}{5!} = 0.0019$$

The TI-84 Plus calculator function ***poissonpdf***() with arguments θ and x can be used to compute $p(x)$ (see section 6.5 for detail instruction):

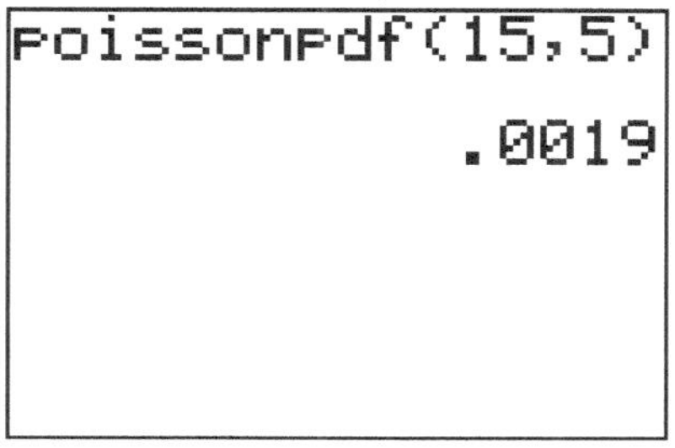

(c) The probability that there will be at most 5 ASD diagnosis is given by

$$F(5) = p(x \leq 5) = p(0) + p(1) + p(2) + p(3) + p(4) + p(5) = 0.0028$$

TI-84 Plus calculator function ***poissoncdf***() with arguments θ and x can be used to compute $F(x)$ (see section 6.5 for detail instruction):

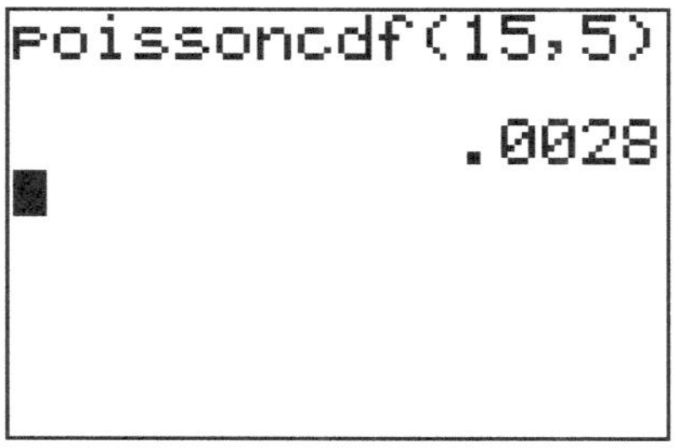

(d) The probability that there will be 6 or more ASD diagnosis is

$$p(x \geq 6) = 1 - p(x \leq 5) = 1 - 0.0028 = 0.9972$$

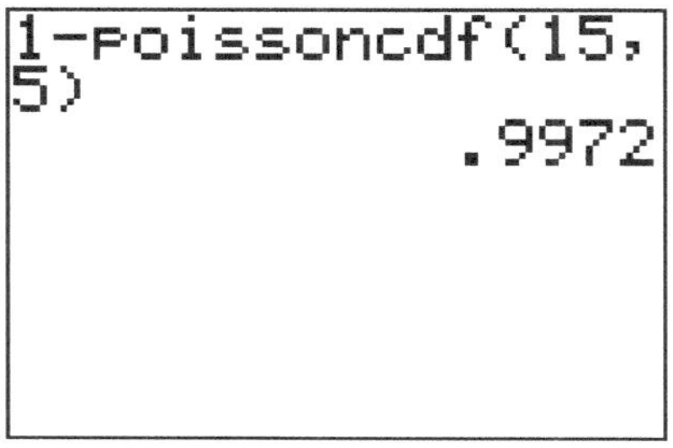

Section 6.5: Computing Binomial and Poisson probability using Technology

In this section, we demonstrate how to use Minitab and TI-84 calculator for computing binomial and poisson probability.

Example 6.16

Suppose X has a binomial distribution with $n = 5$ and $p = 0.45$. We wish to compute the followings using Minitab and TI-84 Plus calculator.

(a) $p(x)$ and $F(x)$ for $x = 0, 1, 2, 3, 4, 5$

(b) $p(3)$

(c) $F(3) = p(X \leq 3)$

Minitab for Binomial Probability

Ref: Example 6.16

Use the steps below for computing $p(x)$ and $F(x)$ using Minitab.

(1) Enter values of x from 0 to n (here, $n = 5$) in C1 and type $p(x)$ and $F(x)$ for output columns, in C2 and C3, respectively.

(2) From Menu Bar option use: Calc>Probability Distribution>Binomial…

(3) Choose **Probability** for $p(x)$ or **Cumulative probability** for $F(x)$, type **Number of trials:** (here, 5) and **Event probability:** (here, 0.45)

(4) Select **Input column:** C1 for x values, **Optional storage:** C2 for $p(x)$ or C3 for $F(x)$ and click **OK**.

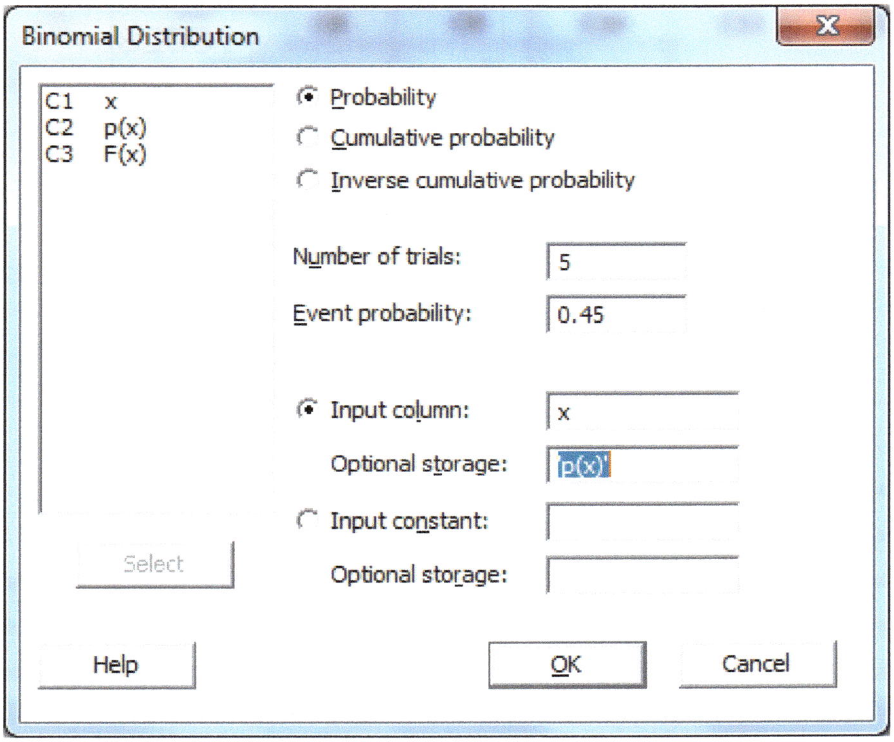

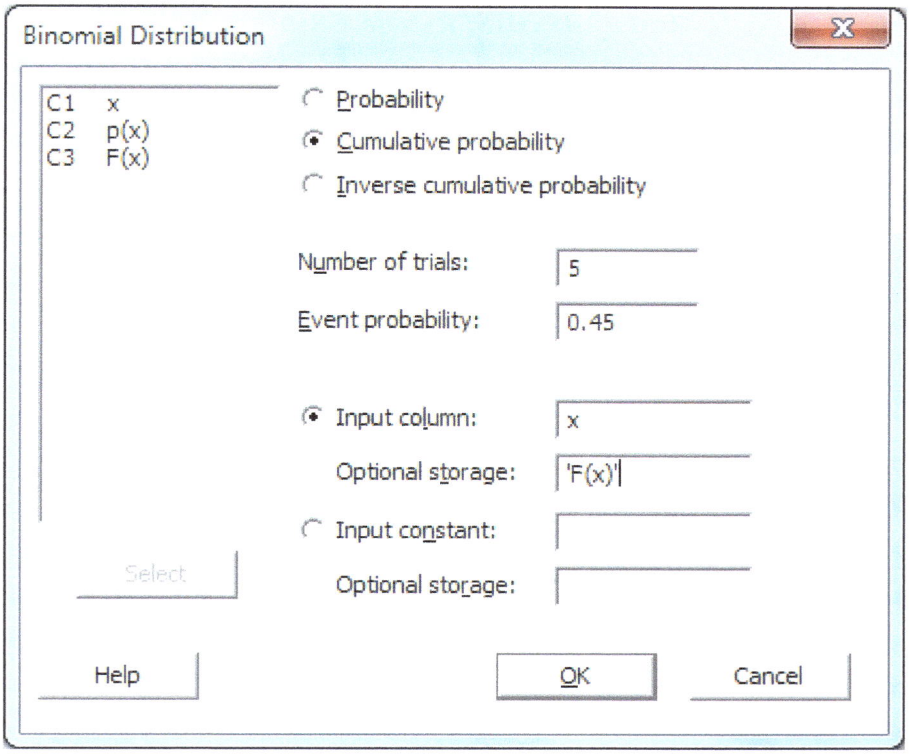

(a) See the output in Minitab Worksheet with $p(x)$ in C2 column and $F(x)$ in C3 column.

Worksheet 1 ***

↓	C1	C2	C3	C4	C5	C6
	x	p(x)	F(x)			
1	0	0.050328	0.05033			
2	1	0.205889	0.25622			
3	2	0.336909	0.59313			
4	3	0.275653	0.86878			
5	4	0.112767	0.98155			
6	5	0.018453	1.00000			
7						

(b) $p(3) = 0.275653$

(c) $F(3) = 0.86878$

TI-84 Calculator for Binomial probability

For computing $p(x) = \binom{n}{x} p^x q^{n-x}$ use the function $\boxed{\boldsymbol{binompdf(n, p, x)}}$ available from

$$\boldsymbol{2nd\ VARS\ (DISTR)}, select\ \boldsymbol{A}: \boldsymbol{binompdf}().$$

For computing $F(x) = P(X \leq x)$ use the function $\boxed{\boldsymbol{binomcdf(n,p,x)}}$ available from

$\mathbf{2nd\ VARS\ (DISTR)}, select\ \mathbf{B}: \mathbf{binomcdf()}$

Use the steps below for computing $p(x)$ and $F(x)$ using TI-84 Plus calculator.

Ref: Example 6.16

(1) Enter values of x from 0 to n (here, $n = 5$) in L1.

(2) To get $p(x)$ in L2, highlight L2 by moving the cursor to L2 and type $\boldsymbol{binompdf(n,p,L1)}$ and press Enter.

(3) To get $F(x)$ in L3, highlight L3 by moving the cursor to L3 and type $\boldsymbol{binomcdf(n,p,L1)}$ and press Enter.

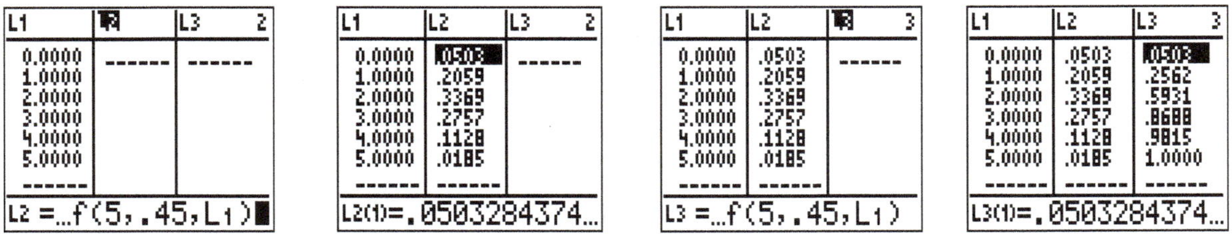

(a) The values of $p(x)$ and $F(x)$ appear in L2 and L3, respectively.

(b) $p(3) = 0.275$, which can be produced by the function $\boldsymbol{binompdf(5, 0.45, 3)}$.

(c) $F(3) = 0.8688$, which can be produced by the function $\boldsymbol{binomcdf(5, 0.45, 3)}$.

Example 6.17

Suppose X has a Poisson distribution with $\theta = 4$. Then, the probability of x is given by

$$p(x) = \frac{e^{-4}4^x}{x!}; \ x = 0, 1, 2, \ldots$$

We wish to compute

(a) $p(x)$ and $F(x)$ for $x = 0, 1, 2, 3, 4, 5$

(b) $p(3)$

(c) $F(3) = p(X \leq 3)$

using Minitab and TI-84 Plus calculator with necessary instructions.

Minitab for Poisson probability

Ref: Example 6.17

Use the steps below for computing $p(x)$ and $F(x)$ using Minitab:

(1) Enter x from 0 to 5 in C1, and type $p(x)$ in C2 and $F(x)$ in C3 for output columns.

(2) From Menu Bar option use: Calc>Probability Distributions>Poisson…

(3) Choose **Probability** for $p(x)$ or **Cumulative probability** for $F(x)$

(4) Specify **Mean:** (here, 4)

(5) For **Input column:** select C1, for **Optional storage:** select C2 for p(x) or C3 for F(x) and click **OK**.

The above steps are illustrated using the snapshot of Minitab dialog box below:

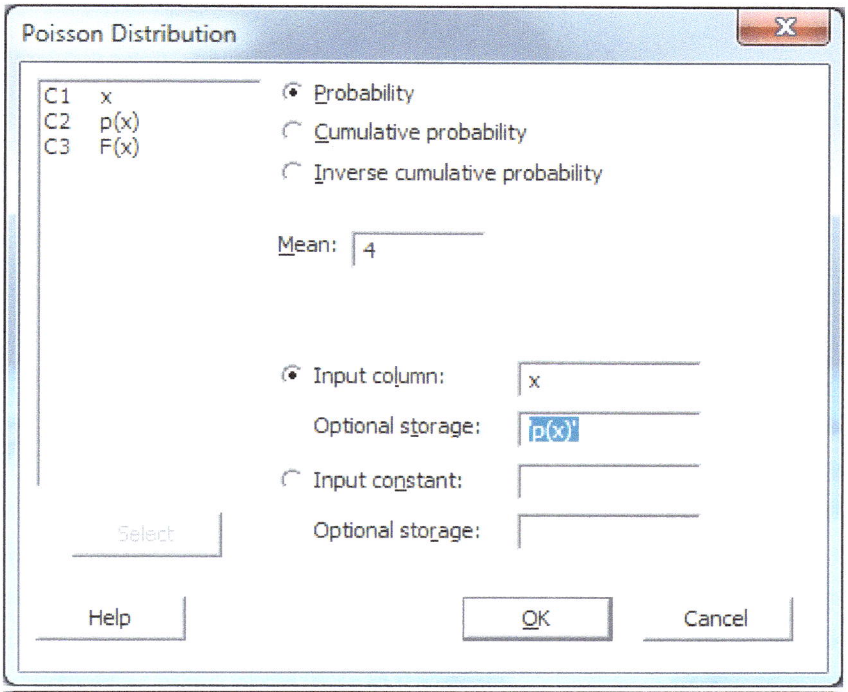

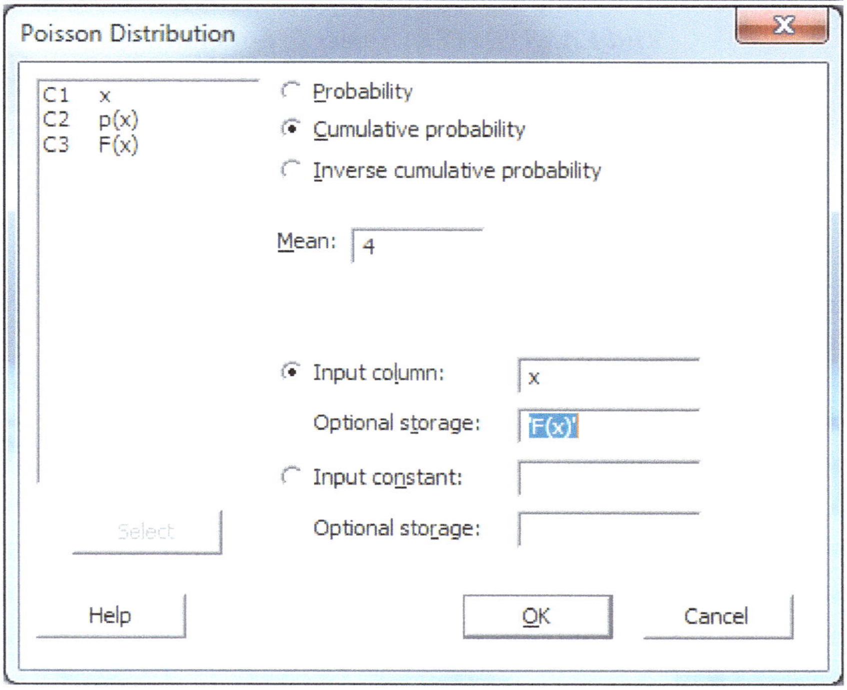

(a) See the output in Minitab Worksheet with $p(x)$ in C2 column and $F(x)$ in C3 column.
(b) $p(3) = 0.1954$
(c) $F(3) = 0.4335$

Worksheet 1 ***

↓	C1	C2	C3	C4	C5
	x	p(x)	F(x)		
1	0	0.018316	0.018316		
2	1	0.073263	0.091578		
3	2	0.146525	0.238103		
4	3	0.195367	0.433470		
5	4	0.195367	0.628837		
6	5	0.156293	0.785130		
7					

TI-84 Calculator for Poisson probability

For computing $p(x)$ use the function $\boxed{\boldsymbol{poissonpdf(\theta, x)}}$ available from

$$\textbf{2nd VARS (DISTR)}, select\ \textbf{C}: \boldsymbol{poissonpdf}()$$

For computing $F(x) = P(X \le x)$ use the function $\boxed{\boldsymbol{poissoncdf(\theta, x)}}$ available from

$$\textbf{2nd VARS (DISTR)}, select\ \textbf{D}: \boldsymbol{poissoncdf}()$$

Use the steps below for computing $p(x)$ and $F(x)$ using TI-84 Plus calculator:

Ref: Example 6.17

(1) Enter values of x from 0 to 5 in L1.

(2) To get $p(x)$ in L2, highlight L2 by moving the cursor to L2 and type $\boldsymbol{poissonpdf(\theta, L1)}$ and press Enter.

(3) To get $F(x)$ in L3, highlight L3 by moving the cursor to L3 and type $\boldsymbol{poissoncdf(\theta, L1)}$ and press Enter.

(4) The output will be in columns L2 and L3.

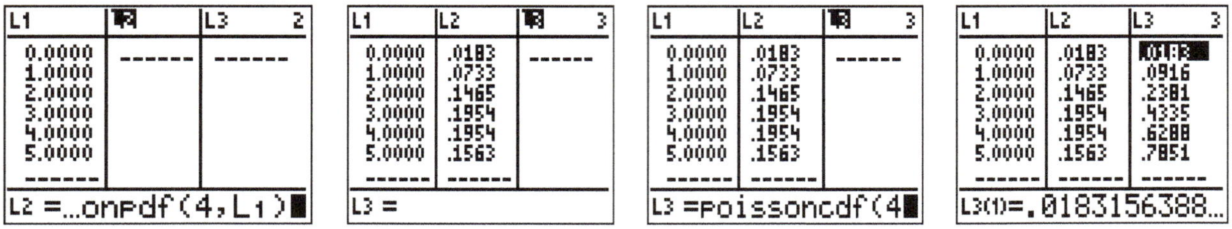

(a) The values of $p(x)$ and $F(x)$ are in L2 and L3, respectively.

(b) $p(3) = 0.1954$, which can also be produced separately by $\boldsymbol{poissonpdf(4, 3)}$.

(c) $F(3) = 0.4335$, which can also be produced separately by $\boldsymbol{poissoncdf(4, 3)}$.

Chapter 6: Practice problems

Sections 6.1-6.2

(6.1) Define random variable and discrete random variable with examples.

(6.2) What do you mean by a discrete probability? Give an example of a discrete probability distribution.

(6.3) Find the value of c from the discrete probability distribution of X given in the table below:

x	$p(x)$
0	0.125
1	c
2	0.375
3	0.125

(6.4) Reference to Exercise (6.3), compute mean and variance of X.

(6.5) A gambler rolls a die in gambling. In each roll, he wins \$30 if it shows up an even number, and loses \$30 if it shows up an odd number. The gambler is using a die that has probability of 0.8 of showing up an even number and 0.2 of showing up an odd number. Let X denote the profit of the gambler from a single roll.

(a) What is the expected value of X?
(b) Find the variance of X.
(c) What is the standard deviation of X?

(6.6) Let X be the number of boys a couple with two children may have. Then, the random variable X takes values 0, 1 and 2 with respective probabilities 1/4, 2/4 and 1/4.

(a) Find the mean, variance and standard deviation of X.
(b) How would you interpret the value of mean?

(6.7) As appears in National Vital Statistics Reports (Volume 62, January 6, 2014), the probability of a 39 year old person dying between ages 39-40 is 0.00164. An insurance company is decided to sell a \$25,000 whole life insurance policy to a person aged 39 years for an annual premium of \$350.

(a) Find the expected profit per policy per year.
(b) What premium would you recommend if the company wants to make an expected profit of \$300 per policy per year?

(6.8) Let us consider rolling two dice and let X denote the sum of two dice. It is easy to see that X takes values 2, 3, 4,…, 12.

(a) Find the probability distribution of X.
(b) Compute $P(X \leq 5)$.
(c) Compute $P(X > 5)$.
(d) Compute the probability that X is an even number.
(e) Compute the probability that X is an odd number.
(f) Find the mean of X.

Section 6.3

(6.9) What assumptions characterize a binomial distribution?

(6.10) Suppose X follows a binomial distribution with two parameters n and p.

(a) Explain what n and p refer to.
(b) What are the mean and variance of X?
(c) Write out the formula for the probability distribution of X.

(6.11) Suppose X has a binomial distribution with $n = 10$ and $p = 0.60$. Compute the followings:

(a) Find mean and variance of X.
(b) Compute $P(x = 2)$.
(c) Find $P(x \geq 2)$.
(d) Compute $P(1 < x < 7)$.
(e) Compute $P(1 \leq x \leq 7)$.

(6.12) A landscaping company claims that 90% of the trees planted by the company survive. The company planted 15 trees for one of its customers.

(a) What is the probability that 10 trees planted will survive?
(b) What is the probability that at most 10 trees planted will survive?
(c) What is the probability that at most 14 trees will survive?
(d) Compute the mean and variance of the number of trees survived.
(e) Suppose 13 trees survived out of 15 trees planted. What do you conclude about the claim made by the company?

(6.13) High blood pressure costs the nation \$47.5 billion annually in direct medical expenses and \$3.5 billion each year in lost productivity. A recent research shows that 30% Americans adults have high blood pressure. We randomly select 20 American adults and let X be the number of adults with high blood pressure.

(a) What is the probability that 8 American adults will have a high blood pressure?
(b) What is the probability that at most 8 American adults will have a high blood pressure?
(c) Suppose that out of 20 adult Americans sampled, 10 Americans are found with a high blood pressure. Do you believe that the research claim is justifiable?
(d) Compute the mean and variance of the adult Americans in the sample with a high blood pressure.

(6.14) About 43% of the U.S. population (including almost 70% of older women) use dietary supplements containing calcium. A sample of 15 people in a small residential area of the U.S. population is considered to see how many of them use dietary supplements containing calcium. Let X be the number of people in the sample who use dietary supplements.

(a) Find the probability that x people use the dietary supplements.
(b) Find $P(0 \leq X \leq 8)$ and $P(X \leq 8)$.
(c) Find the mean, variance and standard deviation of the number of people using the dietary supplements.
(d) Suppose out of 15 people sampled, 9 people are found using dietary supplements. Does this evidence support the claim that 43% people in the U.S. use dietary supplements? Justify your argument. (Hints: Compare the number 9 and the mean).

(e) How many people do you expect use the dietary supplements in the sample to support the claim that 43% of the people in the U.S. use dietary supplements?

(6.15) As per the centers for disease control and prevention, about 12% babies born in the USA are preterm births. A researcher examined 15 newborn babies to identify if they have preterm births or not. Let X be a random variable which denotes the number of preterm born babies.
Answer the following questions:

(a) What is the probability that the researcher will find X babies born with preterm births? What are the possible values of X?
(b) How would you justify that the formula in (a) is a valid probability function?
(c) What assumptions will make your probability function in (a) valid?
(d) What is the probability that no babies are preterm birth?
(e) What is the probability that at least one of them will be a preterm birth?
(f) Find mean and variance of the number of preterm born babies.

Section 6.4

(6.16) What events do you model with Poisson probability? Give some examples of such events.

(6.17) A discrete random variable follows a Poisson distribution with parameter $\theta = 2.5$. What are the mean and variance of the distribution?

(6.18) A random variable X follows a Poisson distribution with parameter $\theta = 2.8$. Answer the following questions:

(a) Find mean and variance of X.
(b) Compute $P(x = 2)$.
(c) Find $P(x \geq 2)$.
(d) Compute $P(1 < x < 7)$.
(e) Compute $P(1 \leq x \leq 7)$.

(6.19) On average, 3 traffic accidents take place at a particular intersection each month. For a given month, compute the followings:

(a) Probability function for x accidents
(b) Probability that no accident will take place
(c) Probability that exactly 3 accidents will take place
(d) Probability that not more than 5 accidents will take place
(e) Probability that at least 3 accidents will take place

(6.20) As per CDC, on average, about 15 out of 1000 children get identified with autism spectrum disorder (ASD) (i.e., about 1 in every 68 children). Let X be the number of children identified with ASD in the next 1000 children inspected. Answer the following questions:

(a) Write out the formula to compute the probability distribution of X.
(b) Find the probability that there will be at least 15 ASD identified children.
(c) Find the probability that there will be not more than 15 ASD identified children.
(d) What are the mean and variance of the number of ASD identified children?

(6.21) On average, a school has 3 days of closing due to heavy snowfall and power outage during the winter. Let X be the number of days the school will be closed during the next winter due to the heavy snowfall and power outage.

(a) What probability model will be suitable for the probability distribution of X?
(b) What is the probability that the school will be closed for exactly 3 days next winter?
(c) What is the probability that the school will not be closed at all next winter?
(d) Find the mean and variance of X.

(6.22) An office, on average, has 3 lightbulb failures every six months. Let X be the number of lightbulbs that will fail in the next six months.

(a) What probability model will be suitable for the probability distribution of X?
(b) What is the probability that there will be no lightbulb failure in the next six months?
(c) What is the probability that there will be exactly 3 lightbulb failures in the next six months?
(d) What is the probability that there will be at most 5 lightbulb failures in the next six months?
(e) What are the mean and variance of the number of lightbulbs that will fail?

(6.23) A life insurance company has figured out that, on average, it receives 8 death claims per day. Let X be the number of claims it receives on a given day.

(a) Write out the formula of probability distribution of X.
(b) Compute $P(x = 0)$.
(c) Compute $P(x \geq 2)$.
(d) Compute $P(x = 8)$.
(e) Compute $P(x \leq 8)$.
(f) Compute $P(3 < x < 7)$.
(g) Compute mean and variance of X.

(6.24) A city, on average, has 3.4 traffic accidents per month at a certain intersection. Let X be the number of traffic accidents reported at the intersection on a given month.

(a) Write out the formula of probability distribution of X.
(b) Compute $P(x = 0)$.
(c) Compute $P(x = 3)$.
(d) Compute $P(x \leq 3)$.
(e) Compute $P(x > 4)$.
(f) Compute $P(0 \leq x \leq 3)$.
(g) Compute variance and standard deviation of X.

(6.25) A city, on average, has 6 traffic accidents per month at a certain intersection. Let X be the number of traffic accidents reported at the intersection on a given month.

(a) Write out the formula of probability distribution of X.
(b) Compute $P(x = 0)$.
(c) Compute $P(x = 6)$.
(d) Compute $P(x \leq 6)$.
(e) Compute $P(x > 7)$.
(f) Compute $P(0 \leq x \leq 6)$.

(g) Compute variance and standard deviation of X.

(6.26) A manufacturing company, on average, produces 5 defective items per day. Let X be the number of defective items it produces on a given day.

(a) What is the probability distribution of X?
(b) What is the probability that it produces no defective item on the day?
(c) What is the probability that it produces at most 5 defective items on the day?
(d) What is the probability that it produces at least 6 defective items on the day?
(e) What are variance and standard deviation of X?

(6.27) A construction company has, on average, 3 workplace injuries per year. Let X be the number of injuries it had on a given year.

(a) What is the probability distribution of X?
(b) What is the probability that it had no injury in the year?
(c) What is the probability that it had at most 4 injuries in the year?
(d) What is the probability that it had at least 5 injuries in the year?
(e) What are variance and standard deviation of X?

Chapter 6: Answer to selected problems

(6.1) Try yourself.
(6.2) Try yourself.
(6.3) $c = 0.375$.
(6.4)

$$\mu = 0p(0) + 1p(1) + 2p(2) + 3p(3) = 0 \times 0.125 + 1 \times 0.375 + 2 \times 0.375 + 3 \times 0.125$$
$$= 1.5$$
$$\sigma^2 = \sum_x (x-\mu)^2 p(x)$$
$$= (0-1.5)^2 p(0) + (1-1.5)^2 p(1) + (2-1.5)^2 p(2) + (3-1.5)^2 p(2)$$
$$= (0-1.5)^2 \times 0.125 + (1-1.5)^2 \times 0.375 + (2-1.5)^2 \times 0.375 + (3-1.5)^2 \times 0.125$$
$$= 0.75$$

(6.5)

$$\mu = 30 \times 0.8 + (-03) \times 0.2 = \$18$$
$$\sigma^2 = \sum_x (x-\mu)^2 p(x)$$
$$= (30-18)^2 \times 0.8 + (-30-18)^2 \times 0.2 = \$576$$
$$\sigma = \sqrt{\sigma^2} = \sqrt{576} = \$24$$

(6.6)

(a) The expected value or mean of X is

$$\mu = 0p(0) + 1p(1) + 2p(2) = 0\left(\frac{1}{4}\right) + 1\left(\frac{2}{4}\right) + 2\left(\frac{1}{4}\right) = \frac{4}{4} = 1$$

The variance of X is

$$\sigma^2 = \sum_x (x-\mu)^2 p(x) = (0-1)^2 p(0) + (1-1)^2 p(1) + (2-1)^2 p(2)$$
$$= (1)\left(\frac{1}{4}\right) + (0)\left(\frac{2}{4}\right) + (1)\left(\frac{1}{4}\right) = 0.50$$

The standard deviation of X is

$$\sigma = \sqrt{\sigma^2} = \sqrt{0.50} = 0.71$$

(b) The mean value of 1 means, a couple with two children, on average will have 1 boy.

(6.7)

(a) Let X be the profit from selling a whole life insurance to a person aged 39 years who dies in the interval 39-40 with probability 0.00164. Then the probability distribution of X is given by:

x	$p(x)$
\$350, if policy holder lives	0.99836
\$350-\$25000, if policy holder dies	0.00164

Then the expected profit is

$$\mu = 350(0.99836) + (350 - 25000)(0.00164)$$
$$= 350(0.99836) + 350(0.00164) - 25000(0.00164)$$
$$= 350 - 41 = 309$$

Therefore, the expected profit per policy per year is \$309.

(b) Let the premium be $\$y$ per policy per year. Then, we will have the following distribution of the profit X.

x	$p(x)$
y, if policy holder lives	0.99836
y-\$25000, if policy holder dies	0.00164

We need to solve for y using the equation

$$300 = \sum_x xp(x)$$
$$300 = y \times 0.99836 + (y - 25000) \times 0.00164$$
$$300 = y \times 0.99836 + y \times 0.00164 - 25000 \times 0.00164$$
$$300 = y \times (0.99836 + 0.00164) - 25000 \times 0.00164$$
$$300 = y \times (1) - 41$$
$$y = 341$$

Therefore, we would recommend the annual price per policy to be \$341.

(6.8)

(a) The probability distribution of X is given by

x	2	3	4	5	6	7	8	9	10	11	12
$p(x)$	1/36	2/36	3/36	4/36	5/36	6/36	5/36	4/36	3/36	2/36	1/36

(b) $P(X \leq 5) = 10/36$.

(c) $(X > 5) = 1 - (X \leq 5) = 1 - \frac{10}{36} = \frac{26}{36}$.

(d) $P(X = even\ number) = \frac{18}{36}$.

(e) $P(X = odd\ number) = \frac{18}{36}$

(f) $\mu = \sum_x x \times p(x)$

$\mu = \frac{2}{36} + \frac{6}{36} + \frac{12}{36} + \frac{20}{36} + \frac{30}{36} + \frac{42}{36} + \frac{40}{36} + \frac{36}{36} + \frac{30}{36} + \frac{22}{36} + \frac{12}{36} = \frac{252}{36} = 7.$

(6.11)

(a) $\mu = np = 10(0.6) = 6, \sigma^2 = np(1-p) = 10(0.6)(0.4) = 2.4.$

(b) $p(2) = 0.0106$

(c) $p(x \geq 2) = 1 - F(1) = 1 - 0.0017 = 0.9983.$

(d) $p(1 < x < 7) = F(6) - F(1) = 0.616$

(e) $p(1 \leq x \leq 7) = F(7) - F(0) = 0.8326$

(6.12)

(a) $p(10) = 0.0105$

(b) $F(10) = p(x \leq 10) = 0.0127.$

(c) $F(14) = p(x \leq 14) = 0.7941.$

(d) $\mu = np = 15(0.9) = 13.5, \sigma^2 = np(1-p) = 15(0.6)(0.9)(0.1) = 1.35.$

(e) Since mean is 13.5, if 13 trees survived it makes sense. Also, 13/15=0.87, close to 0.90, which might be supportive to the claim.

(6.13)

(a) $p(8) = 0.1144$

(b) $F(8) = p(x \leq 8) = 0.8867$

(c) $\mu = np = 20(0.3) = 6$. Therefore, we would expect number of people with high blood pressure close to 6, not 10. Also, $F(8) = P(X \leq 10) = 0.9829$, which is much more than 30%. Therefore, more people might have high blood pressure than the claim. claimed co

(d) $\mu = np = 20(0.3) = 6, \sigma^2 = 4.2$.

(6.14)

(a) $p(x) = \binom{15}{x}(0.43)^x(0.57)^{15-x};\ x = 0,1,\dots,15$

(b) $F(8) = p(x \leq 8) = 0.8573 = p(0 \leq x \leq 8)$, the same for both probabilities.

(c) $\mu = np = 15(0.43) = 6.45$, Variance=3.7

(d) Try yourself.

(e) Close to the average of 6.45. Also, $F(6) = 0.5153$ and $F(5) = 0.3144$. Therefore, we would expect 5 or 6 people using dietary supplement.

(6.15)

(a) $p(x) = \binom{15}{x}(0.12)^x(0.88)^{15-x};\ x = 0,1,\dots,15$

(b) If $p(x) \geq 0$ and $\sum_x p(x) = 1$.

(c) Try yourself

(d)

(e) .

(f) Mean=1.8, Variance=1.58

(6.19) Given $\theta = 3$.

a) $p(x) = \frac{e^{-3}3^x}{x!};\ x = 0, 1, 2, \dots$

b) $p(0) = \frac{e^{-3}3^0}{0!} = 0.0498$

c) $p(3) = \frac{e^{-3}3^3}{3!} = 0.2240$

d) $p(x \leq 5) = F(5) = 0.9161$

e) $p(x \geq 3) = 1 - p(x \leq 2) = 1 - F(2) = 0.5768$

(6.20) Given $\theta = 15$ per 1000 childeren.

(a) $p(x) = \frac{e^{-15}15^x}{x!};\ x = 0, 1, 2, \dots$

(b) $p(x \geq 15) = 1 - F(14) = 0.5343$

(c) $p(x \leq 15) = F(15) = 0.5681$.

(d) Mean=15, variance=15.

(6.21) $X \sim P(\theta = 3)$.

(a) $p(x) = \frac{e^{-3}3^x}{x!};\ x = 0, 1, 2, \dots$

(b) $p(3) = 0.2240$.

(c) $p(0) = 0.0498$.

(d) Mean=3, Variance=3.

(6.23) Given $X \sim P(\theta = 8)$.

(a) $p(x) = \frac{e^{-8}8^x}{x!}$; $x = 0, 1, 2, ...$

(b) $p(0) = \frac{e^{-8}8^0}{0!} = 0.0003$.

(c) $p(x \geq 2) = 1 - p(x \leq 1) = 1 - F(1) = 0.9970$

(d) $p(8) = \frac{e^{-8}8^8}{8!} = 0.1396$.

(e) $p(x \leq 8) = F(8) = 0.5925$

(f) $p(3 < x < 7) = F(6) - F(3) = 0.2710$

(6.25) Given $X \sim P(\theta = 6)$.

(a) $p(x) = \frac{e^{-6}6^x}{x!}$; $x = 0, 1, 2, ...$

(b) $p(0) = \frac{e^{-6}6^0}{0!} = 0.0025$.

(c) $p(6) = \frac{e^{-6}6^6}{6!} = 0.1606$.

(d) $p(x \leq 6) = F(6) = 0.6063$

(e) $p(x > 7) = 1 - p(x \leq 6) = 1 - F(6) = 0.3937$

(f) $p(0 \leq x \leq 6) = F(6) = 0.6063$

CHAPTER 7
NORMAL DISTRIBUTION

Introduction

In everyday life, we deal with a wide variety of variables, such as test-score, height, weight, income, age, life time of electric bulbs, body mass index (BMI), etc. The histogram or the probability distribution of many of these variables has roughly the shape of a bell-shaped curve. If the probability distribution of any variable is bell-shaped and symmetric, then the variable is said to have a ***normal distribution***. The graph of a normal distribution is called the ***normal curve***.
Below is a curve for a variable that is normal with mean 0 and standard deviation 1.

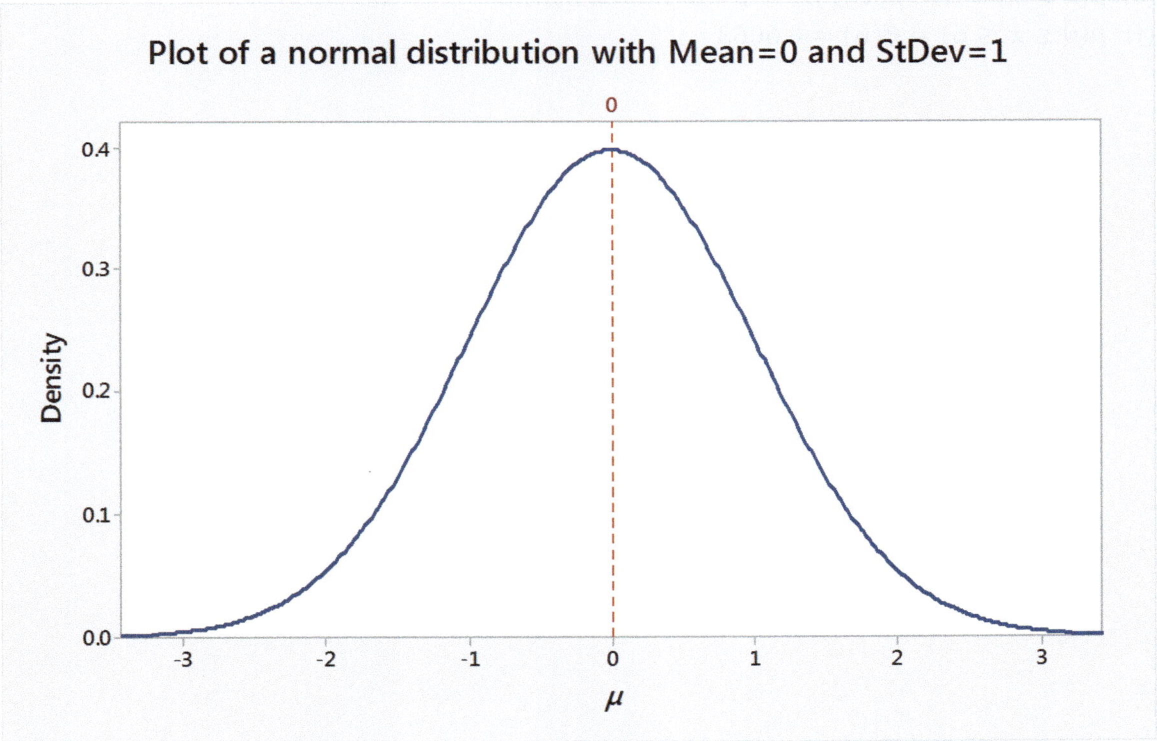

The normal distribution is one of the most useful probability distributions in statistics. Formally, a continuous random variable X is said to have a ***normal probability distribution*** if it has the probability function given by

$$f(x) = \frac{1}{\sigma\sqrt{2\pi}} e^{-\frac{1}{2}\left(\frac{x-\mu}{\sigma}\right)^2}; \ -\infty < x < \infty\,, -\infty < \mu < \infty, \sigma > 0$$

where,
μ and σ are the mean and standard deviation of X, respectively
$\pi \approx 3.14$; the notation $\approx$ stands for "approximately equal to",
$e \approx 2.178$.

Section 7.1: Specifying a normal distribution

If the variable X represents a normal distribution, we sometimes say interchangeably that

- X has a normal distribution,
- the distribution of X is normal,
- X is normally distributed.

A normal distribution is completely specified by two parameters, one is called the mean, denoted by μ and the other is called the standard deviation, denoted by σ. We use the notation $N(\mu, \sigma)$ to refer to a normal distribution that has mean μ and standard deviation σ. A normal distribution with mean 0 and standard deviation 1 will be denoted by $N(0,1)$.

If X is a normally distributed variable with mean μ and standard deviation σ, then we specify it symbolically by $X \sim N(\mu, \sigma)$. For example, if the variable $Exam\ Score$ has a mean of 75 and standard deviation of 5 and if $Exam\ Score$ is normally distributed, we specify it by writing: $Exam\ Score \sim N(75,5)$.

For a $N(\mu, \sigma)$ distribution, the mean μ determines the ***center*** or ***position*** and standard deviation σ determines the ***spread*** or ***shape*** of the distribution. As such, two normal distributions with the same mean and standard deviation refer to an identical distribution or curve. Two normal distributions with different means and standard deviations will have different location and shape.
A normal distribution is symmetric about the mean μ. The larger the standard deviation of a normal curve, the more spread out is the curve.
Below we present several normal curves to show how mean and standard deviation affect the position and shape of a normal curve.

In Case I, we have three normal distributions with means 0, 1.5 and 3, but all with the same standard deviation 1. As such, even though three normal curves have the same shape they have centers at different locations.

In Case II, we have three normal distributions with the same mean 0 and standard deviations 1, 2 and 3. As such, even though three normal curves have the same center, they have different shapes.

In Case III, we have two normal distributions $N(0,1)$ and $N(2,1.5)$. As such, two normal curves have center at two different locations and they have different shapes.

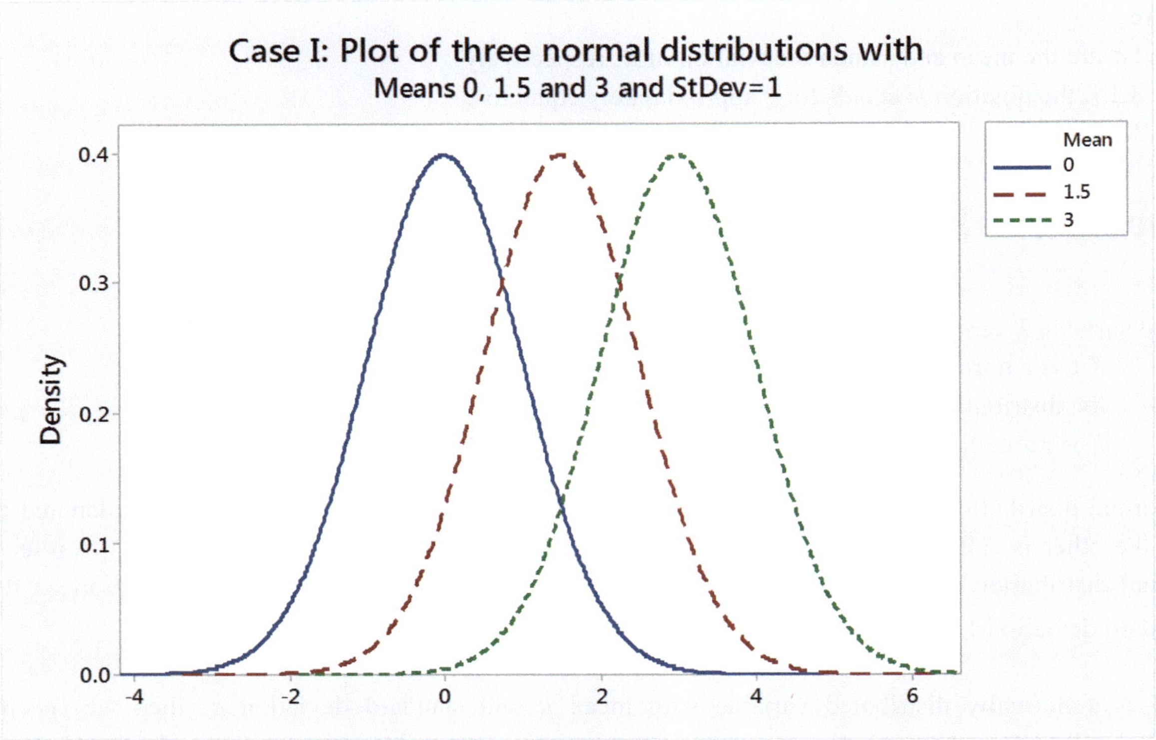
Case I: Plot of three normal distributions with
Means 0, 1.5 and 3 and StDev=1
Density
Mean
0
1.5
3
0.4
0.3
0.2
0.1
0.0
-4
-2
0
2
4
6

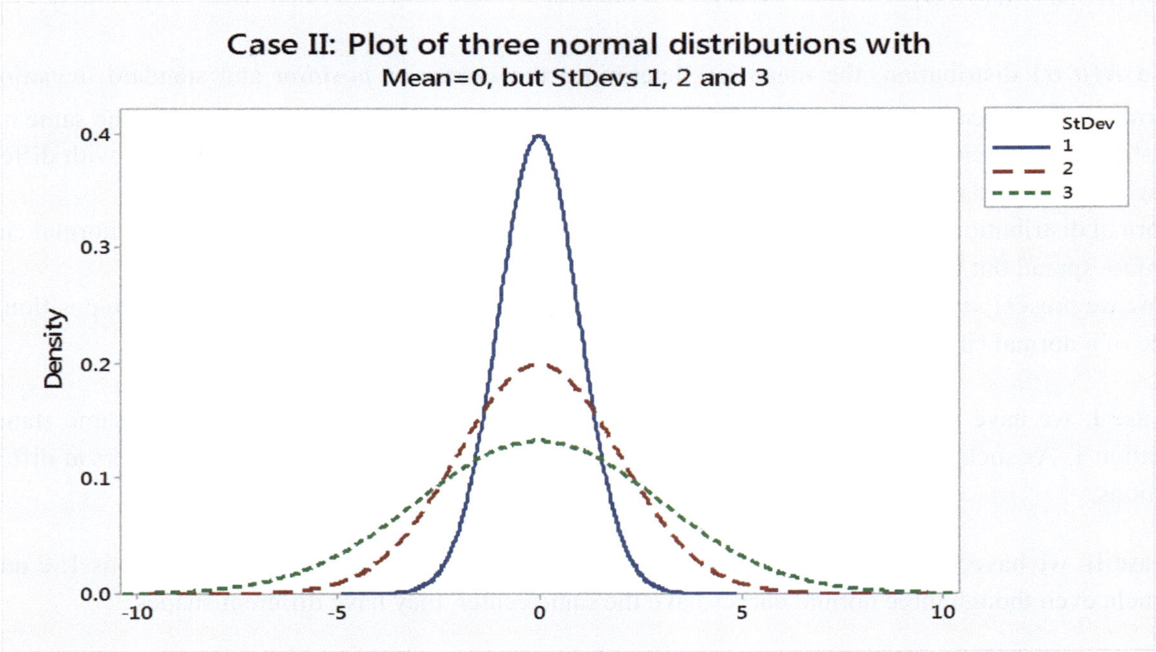
Case II: Plot of three normal distributions with
Mean=0, but StDevs 1, 2 and 3
Density
StDev
1
2
3
0.4
0.3
0.2
0.1
0.0
-10
-5
0
5
10

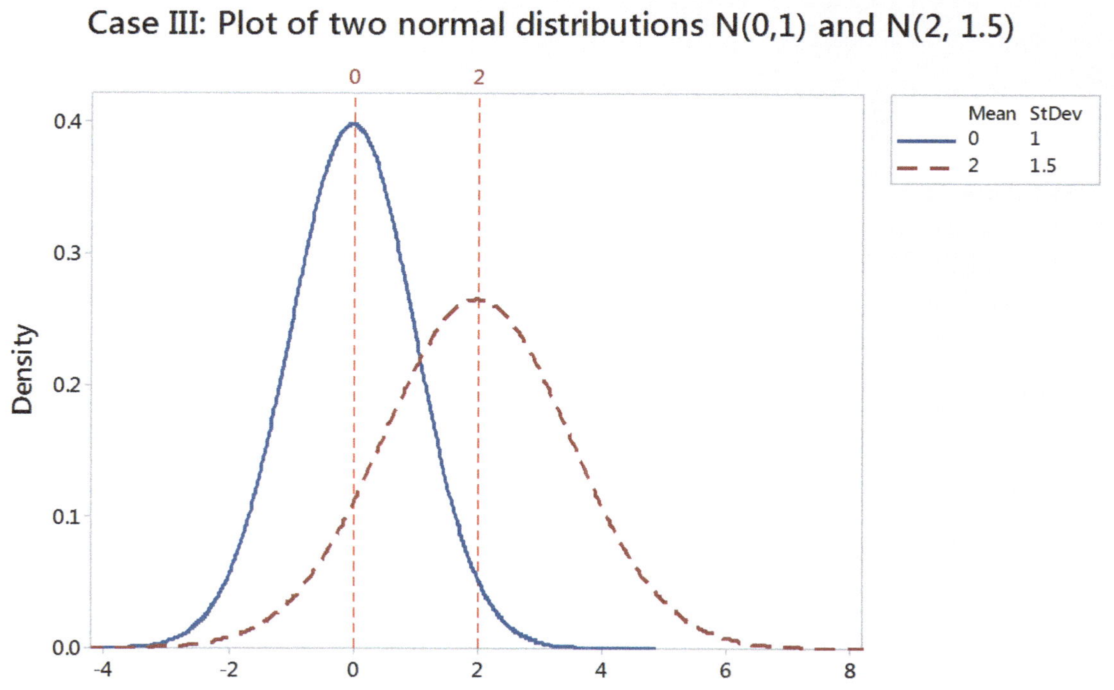

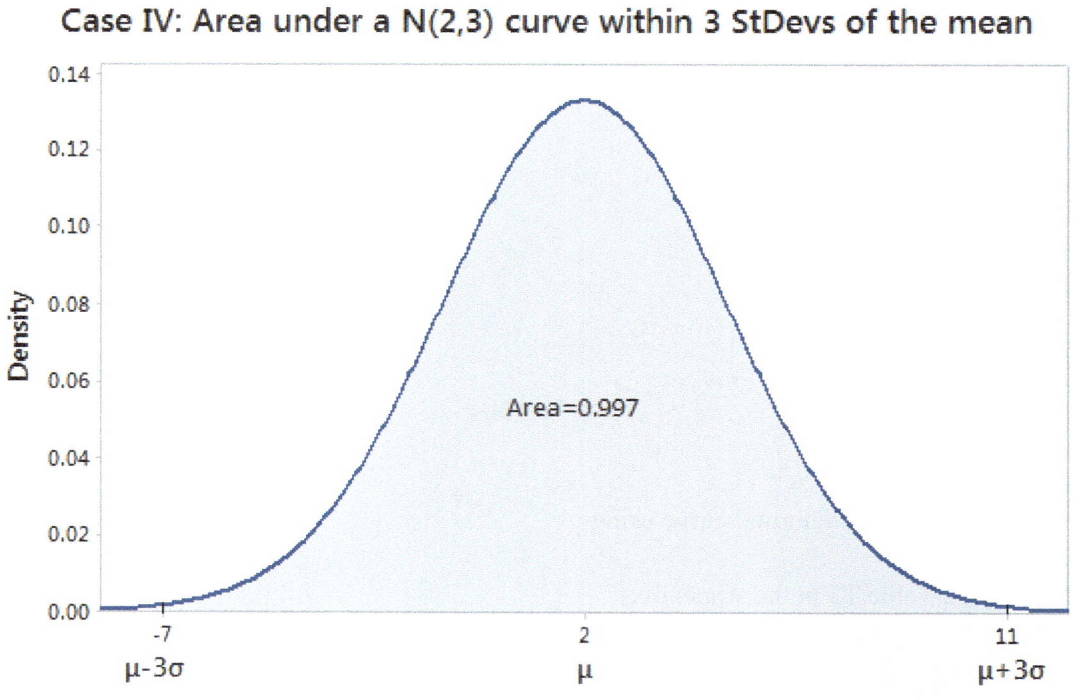

Given a normal $N(\mu, \sigma)$ distribution, the total area under the normal curve is 1 and about 99.7% of the total area is within three standard deviations of the mean. In figure Case IV, the total area between -7 and 11 of a $N(2,3)$ distribution is 99.7%.

Properties of the normal curve

Below are some of the properties of a normal distribution, some of which are readily evident after above discussion.

(1) The curve of a normal distribution is bell-shaped and symmetric about the mean μ.
(2) The center of the curve is at the mean μ.
(3) The standard deviation σ determines the shape of the curve.
(4) The total area under the normal curve is 1, with most of the area within 3 standard deviations of either side of the mean.
(5) The curve never touches the x-axis and always stays above the axis.

Section 7.2: Area under a normal curve

The ***area under a normal curve*** is the probability $P(x_1 \le X \le x_2)$ that a normally distributed variable X takes values between x_1 and x_2, and is represented by the region in figure below:

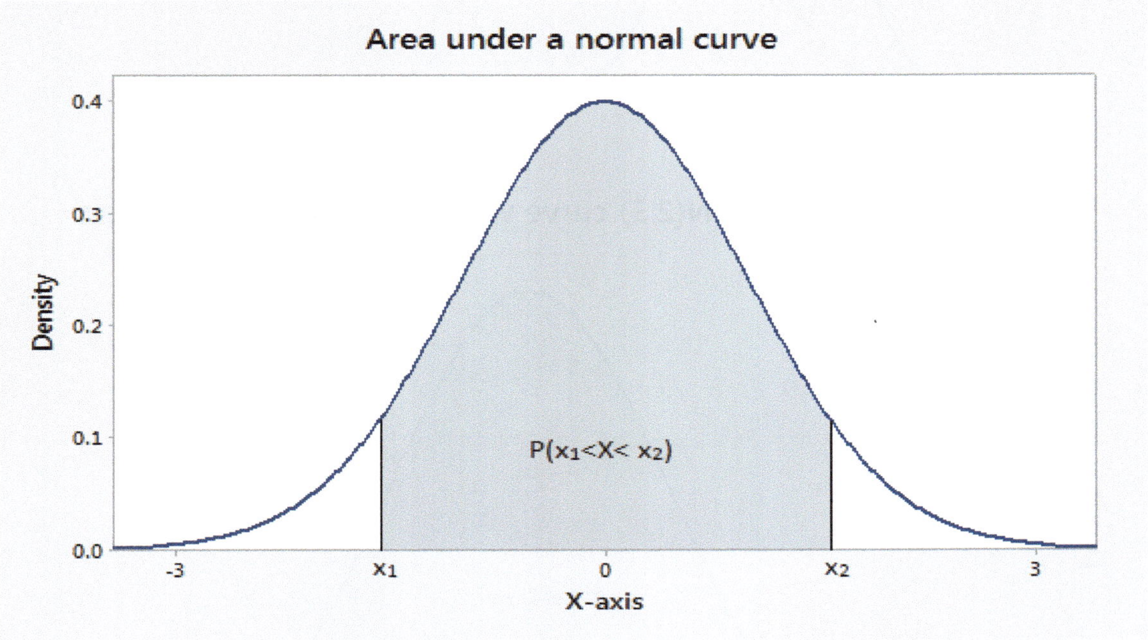

One can find the area under a normal curve using

(1) Empirical Rule
(2) Table (e.g., Table T3 in the Appendix)
(3) Statistical software
(4) Graphing calculator

In this section, we demonstrate how empirical rule can be used for computing area under a normal curve. Computing area using a table has been demonstrated in sections 7.3 and 7.4 after having a substantial knowledge about standard normal distribution and its connection with any given normal distribution. Computing area using statistical software and graphing calculator has been demonstrated in section 7.5.

Empirical Rule

For a $N(\mu, \sigma)$ distribution, an ***empirical rule*** determines approximately what percent of the measurements are within 1, 2 or 3 standard deviations of either side of the mean. The empirical rule is as follows:

(1) Approximately 68.3% of the measurements are within one standard deviation of the mean. That is, $P(\mu - \sigma \leq x \leq \mu + \sigma) \approx 0.683$.

(2) Approximately 95.4% of the measurements are within two standard deviations of the mean. That is, $P(\mu - 2\sigma \leq x \leq \mu + 2\sigma) \approx 0.954$.

(3) Approximately 99.7% of the measurements are within three standard deviations of the mean. That is, $P(\mu - 3\sigma \leq x \leq \mu + 3\sigma) \approx 0.997$.

The above ***empirical rule*** is popularly known as ***68.3-95.4-99.7 rule***.

The figure below illustrates the 68.3-95.4-99.7 rule for a normal $N(0,1)$ distribution.

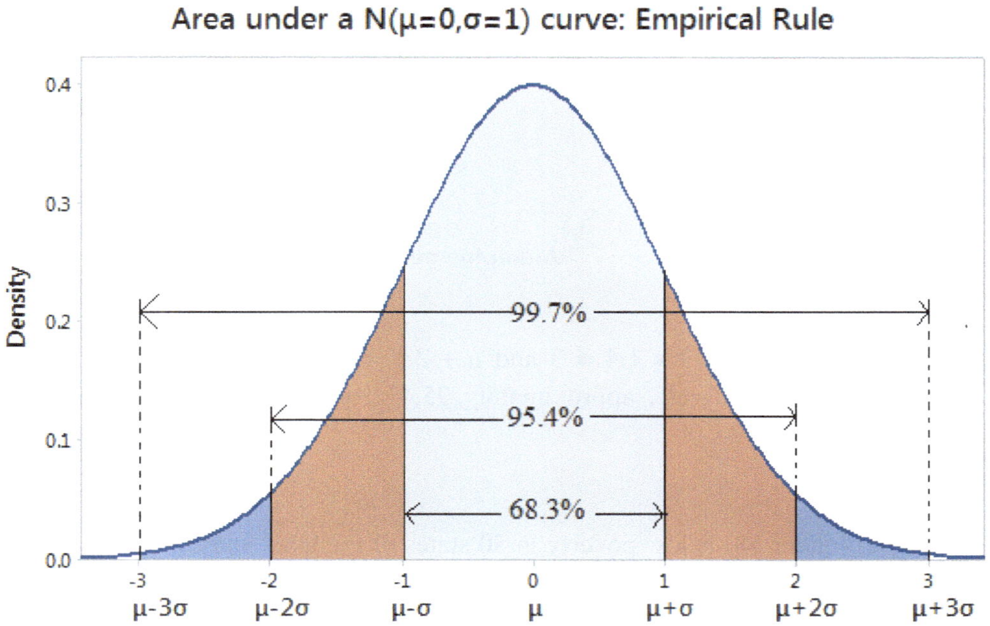

Application of Empirical Rule

Example 7.1

The unemployment rate, expressed as a percent of labor force, for the month of April 2014 in 50 states and the District of Columbia follows approximately a normal distribution with mean 5.8 and standard deviation 1.4. Use empirical rule to answer the following questions.

(a) What percent of states has unemployment rate higher than 4.4?

(b) Evaluate the interval at which the unemployment rate of 95.4% of the states will belong to.

Solution

(a) Here, $\mu - \sigma = 5.8 - 1.4 = 4.4$ and $\mu + \sigma = 5.8 + 1.4 = 7.2$. Therefore, by the 68.3 part of the empirical rule, approximately 68.3% of the states will have unemployment rate between 4.4

and 7.2, and of the remaining 31.7%, 15.85% will have unemployment rate below 4.4 and above 7.2 by the symmetry of normal distribution.

This means, $(68.3 + 15.85)\% = 84.15\%$ of states will have unemployment rate above 4.4. See the figure below for graphical demonstration.

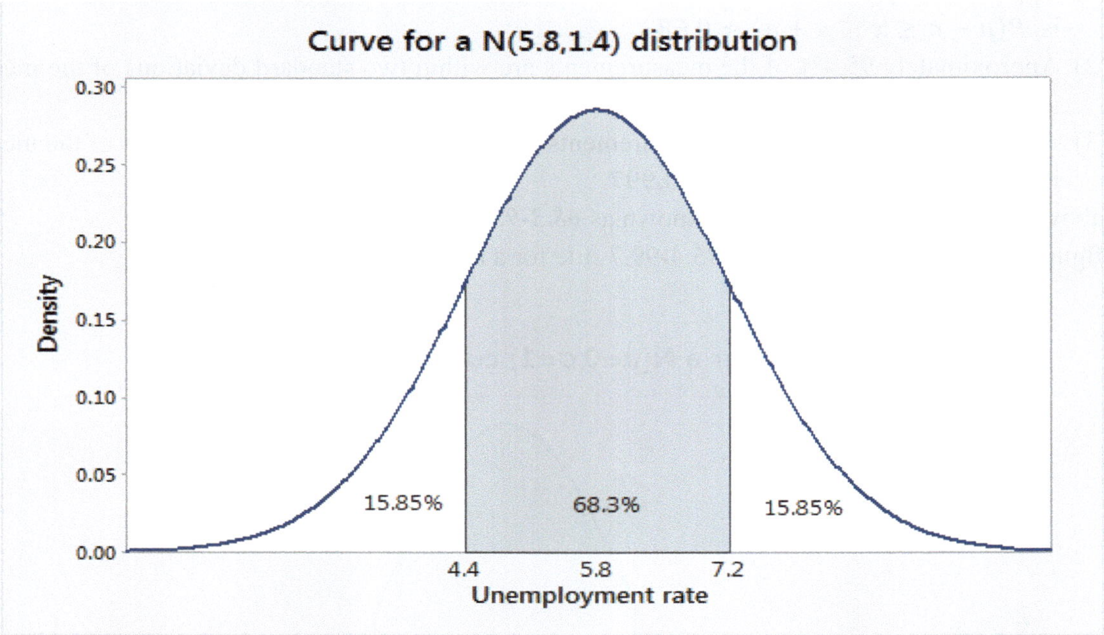

(b) We have, $\mu - 2\sigma = 5.8 - 2 \times 1.4 = 3$ and $\mu + 2\sigma = 5.8 + 2 \times 1.4 = 8.6$. Therefore, by the 95.4 part of the empirical rule, approximately 95.4% of states will have unemployment rate between 3 and 8.6.

Example 7.2

The percent of children (under 18) below poverty in 50 states of the USA and the District of Columbia in 2010 follows approximately a normal distribution with mean 20.5 and standard deviation 5.2. The histogram of percent below the poverty, and a superimposed normal distribution with mean 20.5 and standard deviation 5.2 appear in the figure below:

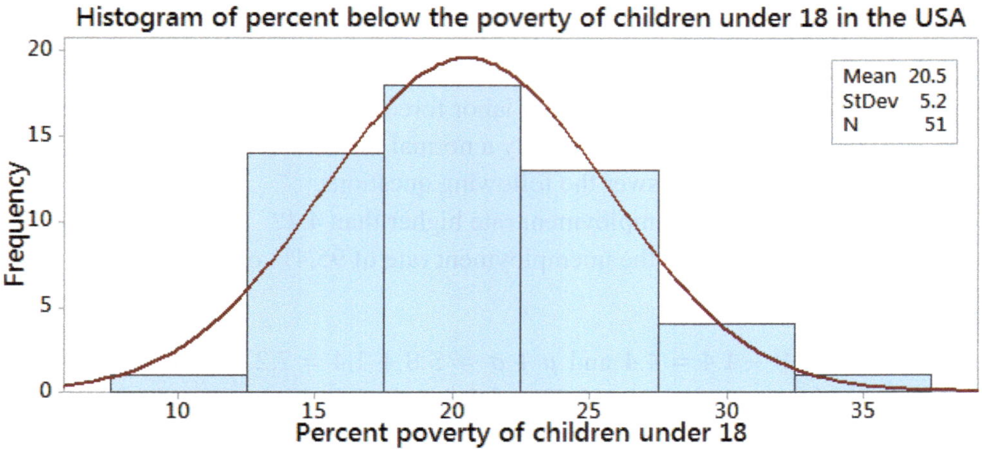

Answer the following questions regarding the distribution of the percent below the poverty of children under 18:

(a) What percent of states has the poverty measurements falling within one standard deviation of the mean?
(b) Find the interval of 95.4% of the poverty measurements.
(c) Find the interval of 99.7% of the poverty measurements.
(d) As you see from the descriptive statistics, the minimum and the maximum value of the percent poverty measurements are equal to 10 and 32.5, respectively. How would you compare and interpret this information with your result in (c)?

Solution

(a) Approximately, 68.3% of the states will have poverty measurements falling within one standard deviation of the mean.
(b) We have, $\mu - 2\sigma = 10.1$ and $\mu + 2\sigma = 30.9$. Therefore, approximately 95.4% of the states have poverty in the interval [10.1, 30.9].
(c) We have, $\mu - 3\sigma = 20.5 - 3 \times 5.2 = 4.9$ and $\mu + 3\sigma = 20.5 + 3 \times 2.5 = 36.1$. By the 99.7 part of the empirical rule, approximately 99.7% of the states have poverty in the interval [4.9, 36.1].
(d) Since the interval [4.9, 36.1] for the approximate 99.7% of the states poverty includes the minimum (10) and the maximum (32.5) of the observed percent poverty, 100% of the observed percent poverty is in the estimated interval. This means the percent poverty data fits a normal distribution reasonably well. However, the existing discrepancy between empirical and exact normal distributions is due to the fact that the state percent poverty may not be exactly normal.

Definition: Tolerance interval

An interval with a specified percentage of the population measurements is called a ***tolerance interval***. For example, the three intervals $[\mu - \sigma, \mu + \sigma]$, $[\mu - 2\sigma, \mu + 2\sigma]$ and $[\mu - 3\sigma, \mu + 3\sigma]$ for a normal population are 68.3%, 95.4% and 99.7% tolerance intervals, respectively.

Example 7.3

Intelligence quotient (IQ) for a population is normally distributed with mean 100 and standard deviation 15. Find 68.3%, 95.4% and 99.7% tolerance intervals and interpret the result of 68.3% tolerance interval.

Solution

We have $\mu = 100$ and $\sigma = 15$.

68.3% tolerance interval is $[\mu - \sigma, \mu + \sigma] = [100 - 15, 100 + 15] = [85, 115]$

95.4% tolerance interval is $[\mu - 2\sigma, \mu + 2\sigma] = [100 - 2 \times 15, 100 + 2 \times 15] = [70, 130]$

95.4% tolerance interval is $[\mu - 2\sigma, \mu + 3\sigma] = [100 - 3 \times 15, 100 + 3 \times 15] = [55, 145]$

Interpretation

Approximately 68.3% of the IQ measurements of the population are in the interval from 85 to 115.

Estimates of tolerance intervals

Given a sample, let $\bar{x}$ be an estimate of the population mean μ and s be an estimate of the population standard deviation σ. As long as the population is normal or approximately normal, the estimates of 68.3%, 95.4% and 99.7% tolerance intervals are $[\bar{x}-s, \bar{x}+s]$, $[\bar{x}-2s, \bar{x}+2s]$ and $[\bar{x}-3s, \bar{x}+3s]$, respectively.

Example 7.4

Suppose that from a sample of IQ scores of 45 people, we have $\bar{x} = 112.5$ and $s = 14.2$. If the population the sample comes from is approximately normal, find estimates of 68.3%, 95.4% and 99.7% tolerance intervals. Interpret the result of 99.7% estimated tolerance interval.

Solution

We have$\bar{x} = 112.5$ and $s = 14.2$.

An estimate of 68.3% tolerance interval is

$$[\bar{x}-s, \bar{x}+s] = [112.5 - 14.2, 112.5 + 14.2] = [98.3, 126.7]$$

An estimate of 95.4% tolerance interval is

$$[\bar{x}-2s, \bar{x}+2s] = [112.5 - 2 \times 14.2, 112.5 + 2 \times 14.2] = [84.1, 140.9]$$

An estimate of 99.7% tolerance interval is

$$[\bar{x}-3s, \bar{x}+3s] = [112.5 - 3 \times 14.2, 112.5 + 3 \times 14.2] = [69.9, 155.1]$$

Interpretation

Given that the sample is approximately normal, 99.7% of the IQ measurements of the population are likely to be included in the interval from 69.9 to 155.1.

Section 7.3: Standard normal distribution

A normal distribution with mean 0 and standard deviation 1 is called the ***standard normal distribution***, and the graph of the distribution is called the ***standard normal curve***. If the random variable Z represents a normal distribution that has a mean 0 and a standard deviation 1, we interchangeably say that

- Z has a standard normal distribution,
- the distribution of Z is standard normal,
- Z is normally distributed.

We use the notation $Z \sim N(0,1)$ to refer to the standard normal distribution. A lowercase letter z refers to a specific value of Z, and is called a z-score. We can easily create a table of the cumulative area under the standard normal curve using an appropriate technology, and use it to determine the area under the curve for specific range of z-scores, to the left of z_1, between z_1 and z_2, etc., as shown in figure below.

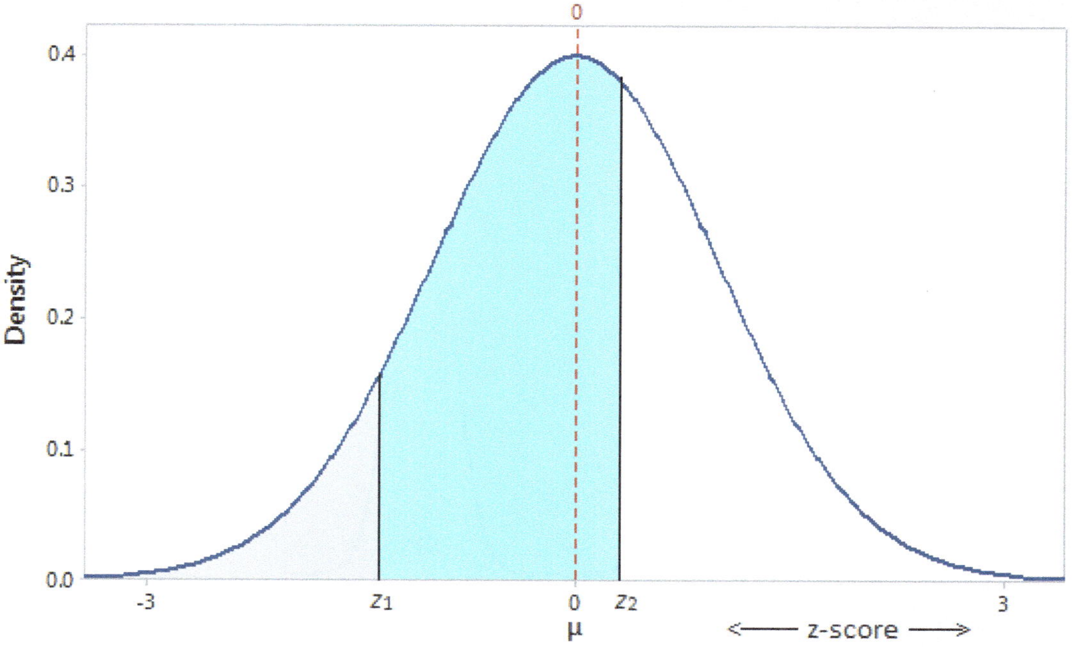

If x is any specified value of a normal distribution with mean μ and standard deviation σ, then we can get a z-score from x using the following conversion formula:

$$z = \frac{x - \mu}{\sigma}$$

The process is called the ***standardization*** of x to z.

Indeed, if X has a $N(\mu, \sigma)$ distribution and x_1 and x_2 are two distinct values of X, $x_1 < x_2$, then

$$P(x_1 < X < x_2) = P\left(\frac{x_1 - \mu}{\sigma} < \frac{X - \mu}{\sigma} < \frac{x_2 - \mu}{\sigma}\right) = P(z_1 < Z < z_2)$$

where Z is a $N(0,1)$ distribution.

The above property allows us to use the standard normal table to find the area under any normal distribution which is not standard normal, when the empirical rule cannot be applied in case we are interested in area beyond the 68.3-95.4-99.7 rule.

However, if we use any technology that supports standardization, we do not need to worry about the standardization process to find the desired area under the curve.

The figure below demonstrates how an original normal curve corresponds to the standard normal curve:

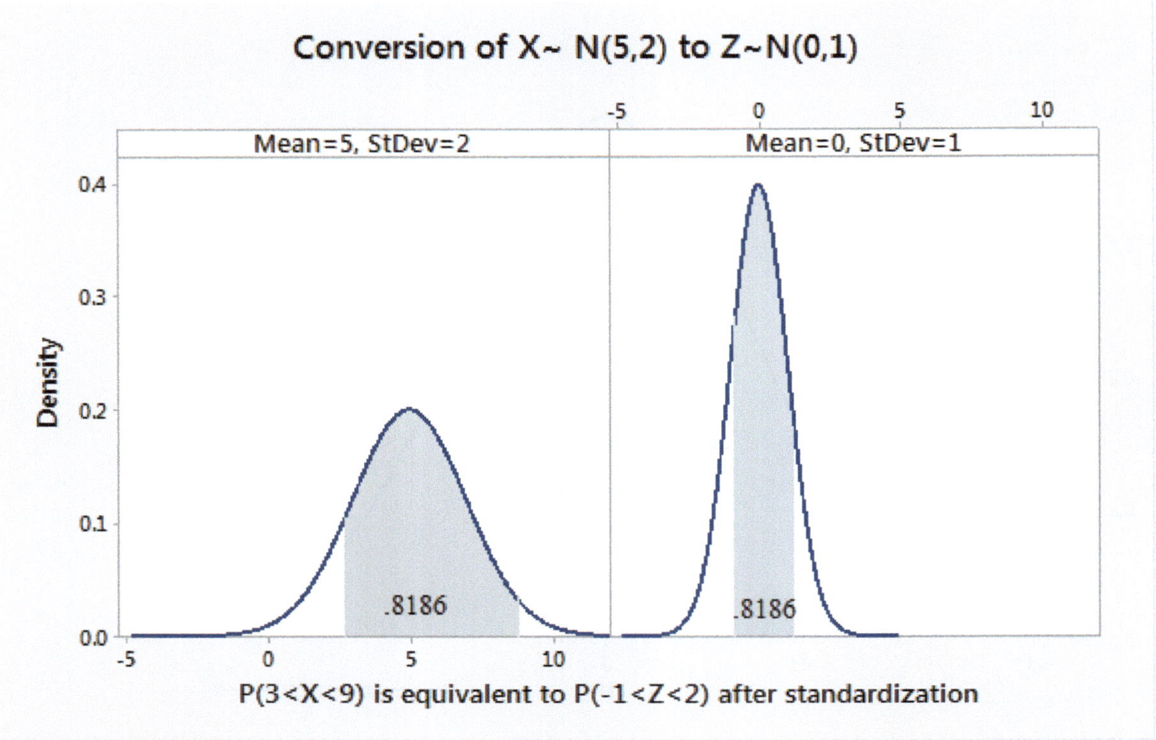

As demonstrated in above figure, the area of $X \sim N(5,2)$ between 2 and 3 is the same as the area of $Z \sim N(0,1)$ between -1 and 2, by the property of standardization.

Example 7.5

Given X is $N(5,2)$ distribution, verify the X conversion of "$3 < X < 9$" to Z of "$-1 < Z < 2$" where Z is $N(0,1)$ distribution.

Solution

Given X is $N(5,2)$, the conversion formula

$$z = \frac{x - \mu}{\sigma}$$

converts $x = 3$ and 9 to equivalent standardized value $z = -1$ and 2, respectively:

$$z = \frac{3-5}{2} = -1$$
$$z = \frac{9-5}{2} = 2$$

Finding area under the standard normal curve

In Appendix Table T3, we have a table of cumulative area under a standard normal curve. Given any value z on the horizontal axis of the curve, the area under the curve to the left of z is the probability $P(Z < z)$.

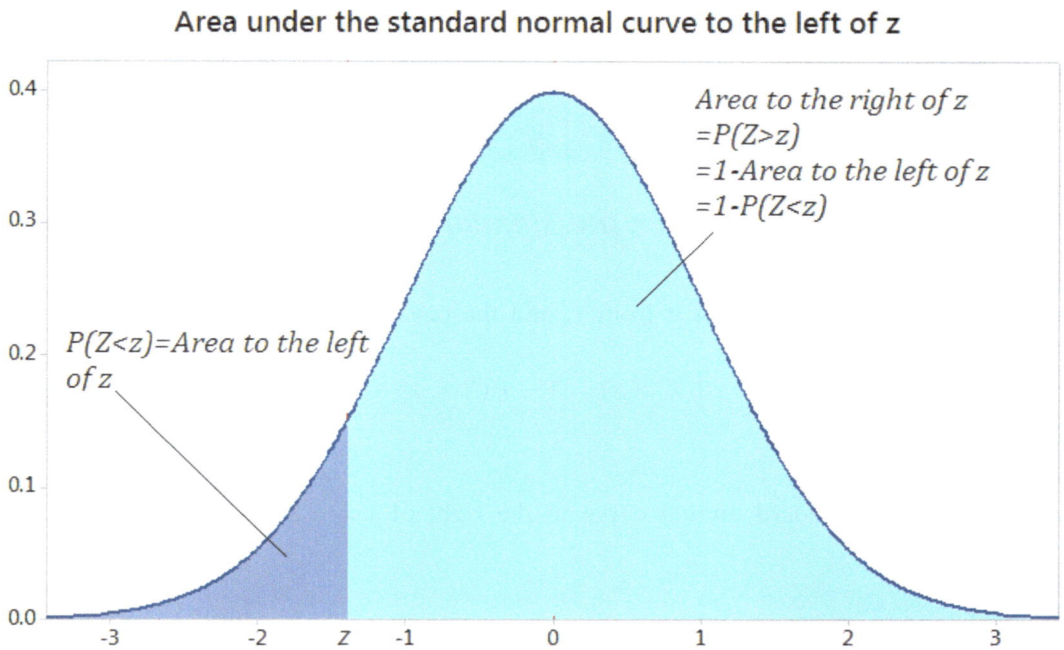

Unless we use any technology, we frequently need to use the Appendix Table T3 for finding area under the standard normal curve. In particular, we need to find area to the left or right of a given z-score or area between two given z scores.

In the following three questions Q1-Q3, we address issues relating to finding area under the standard normal curve.

Q1: How to find the area under the standard normal curve to the left of z using Table T3?

Find ones and tenths places of z from the leftmost column highlighted and hundredths place of z from the topmost row highlighted. The value at the intersection of row and column just obtained corresponds to the area to the left of z under the standard normal curve.

$$Area\ to\ the\ left\ of\ z = P(Z < z).$$

Example 7.6

Find the area under the standard normal curve to the left of $z = 1.35$. In other words, find $P(z \leq 1.35) = P(z < 1.35)$.

Solution

Read 1.3 from leftmost column highlighted and .05 from the topmost row highlighted. Value at the intersection of the selected row and column is 0.91149. Thus, the area to the left of $z = 1.35$ is 0.91149.

z	.00	.01	.02	.03	.04	.05	.06	.07	.08	.09
1.1	0.86433	0.86650	0.86864	0.87076	0.87286	0.87493	0.87698	0.87900	0.88100	0.88298
1.2	0.88493	0.88686	0.88877	0.89065	0.89251	0.89435	0.89617	0.89796	0.89973	0.90147
1.3	0.90320	0.90490	0.90658	0.90824	0.90988	0.91149	0.91309	0.91466	0.91621	0.91774
1.4	0.91924	0.92073	0.92220	0.92364	0.92507	0.92647	0.92786	0.92922	0.93056	0.93189

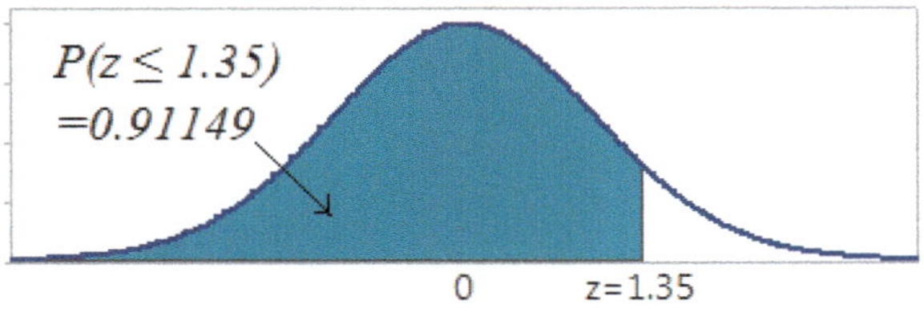

Q2: How to find the area under the standard normal curve to the right of z using Table T3?

Find area to the left of z and subtract it from 1, and the result is the area to the right of z under the standard normal curve.

$$Area\ to\ the\ right\ of\ z = P(Z > z) = 1 - P(Z < z) = 1 - area\ to\ the\ left\ of\ z.$$

Example 7.7

Find the area under the standard normal curve to the right of $z = 1.35$. In other words, find $P(z \geq 1.35) = P(z > 1.35)$.

Solution

The area to the left of $z = 1.35$ is 0.91149.

Thus, the area to the right of $z = 1.35$ is 1 minus the area to the left of $z = 1.35$, i.e., $1 - 0.91149 = 0.08851$.

z	.00	.01	.02	.03	.04	.05	.06	.07	.08	.09
1.1	0.86433	0.86650	0.86864	0.87076	0.87286	0.87493	0.87698	0.87900	0.88100	0.88298
1.2	0.88493	0.88686	0.88877	0.89065	0.89251	0.89435	0.89617	0.89796	0.89973	0.90147
1.3	0.90320	0.90490	0.90658	0.90824	0.90988	0.91149	0.91309	0.91466	0.91621	0.91774
1.4	0.91924	0.92073	0.92220	0.92364	0.92507	0.92647	0.92786	0.92922	0.93056	0.93189

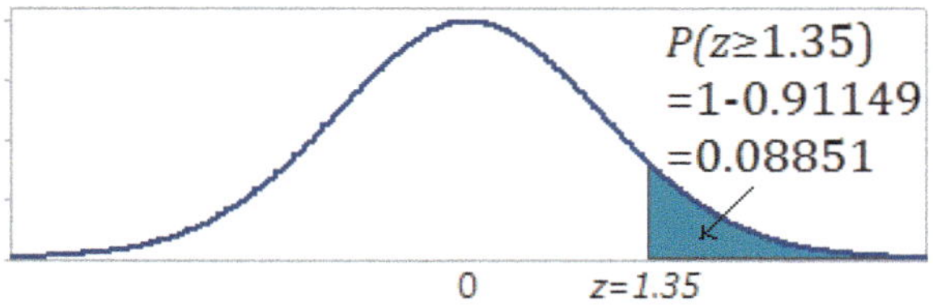

Q3: How to find the area under the standard normal curve between z_1 and z_2, where $z_1 < z_2$, using Table T3?

Subtract area to the left of z_1 from the area to the left of z_2, and the result is the area between z_1 and z_2 under the standard normal curve.

$$Area\ between\ z_1\ and\ z_2 = P(z_1 < Z < z_2) = P(Z < z_2) - P(Z < z_1)$$
$$= Area\ to\ the\ left\ of\ z_2 - Area\ to\ the\ left\ of\ z_1$$

Note that for any two distinct z values

$$Area\ between\ two\ distinct\ z\ values$$
$$= Area\ to\ the\ left\ of\ higher\ value\ of\ z - Area\ to\ the\ left\ of\ lower\ value\ of\ z$$

Example 7.8

Find the area under the standard normal curve between 1.23 and 1.36. In other words, find $P(1.23 \leq z \leq 1.36)$.

Solution

Note that $P(1.23 \leq z \leq 1.36)$ or the area under the standard normal curve between 1.23 and 1.36 is equal to area to the left of 1.36 minus the area to the left of 1.23.

Here, the area to the left of 1.36 is 0.91309, and the area to the left of 1.23 is 0.89065. Therefore, $P(1.23 \leq z \leq 1.36) = 0.91309 - 0.89065 = 0.02244$.

z	.00	.01	.02	.03	.04	.05	.06	.07	.08	.09
1.1	0.86433	0.86650	0.86864	0.87076	0.87286	0.87493	0.87698	0.87900	0.88100	0.88298
1.2	0.88493	0.88686	0.88877	0.89065	0.89251	0.89435	0.89617	0.89796	0.89973	0.90147
1.3	0.90320	0.90490	0.90658	0.90824	0.90988	0.91149	0.91309	0.91466	0.91621	0.91774
1.4	0.91924	0.92073	0.92220	0.92364	0.92507	0.92647	0.92786	0.92922	0.93056	0.93189

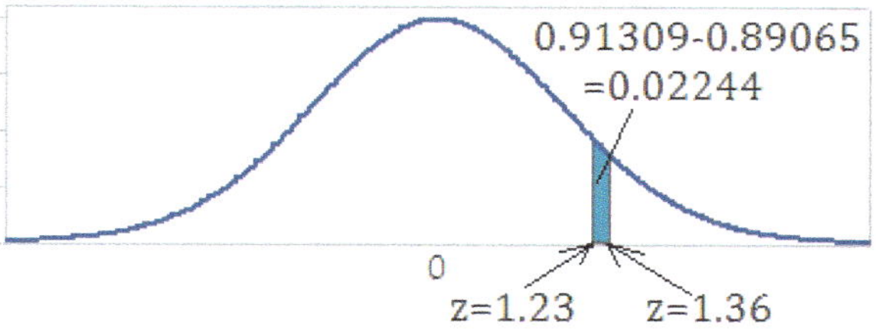

Important remarks regarding finding the area

(1) For any continuous variable, like a normally distributed variable Z, the probability at any particular point is zero. Therefore, $P(Z > z) = P(Z \geq z)$ or $P(Z < z) = P(Z \leq z)$.

(2) Following two formulas are very useful for computing area under the standard normal curve:

(a) $P(Z > z) = 1 - P(Z < z)$, i.e., to compute the area to the right of a given score, just compute 1 minus the area to the left of the score.

(b) $P(z_1 < Z < z_2) = P(Z < z_2) - P(Z < z_1)$, i.e., to compute the area between two distinct scores, just compute the area to the left of the higher score minus the area to the left of the lower score.

Example 7.9

Find the following using Table T3.

(a) The area to the left of -1.45, i.e., $P(z < -1.45)$.

(b) The area to the right of 1.45, i.e., $P(z > 1.45)$ and compare it with the result in (a).

(c) The area to the left of 1.45.

(d) The area between -1.45 and 1.45, i.e., $P(-1.45 < z < 1.45)$.

Solution

Remember in order to answer each of questions (a)-(d), we read ones and tenths places of z from leftmost column highlighted and hundredths from the topmost row highlighted from Table T3.

- The value at the intersection of row and column is the area to the left of z.

- The area to the right of z is just 1 minus the area to the left of z.
- The area between two distinct z values is just the area to the left of the higher z value minus the area to the left of the lower z value.

In light of the above discussion, we have:

(a) $P(z < -1.45) = 0.07353$.
TI-84 Plus Calculator ***normalcdf***() function can be used to compute the area under the normal curve (see section 7.6 for detail instruction).

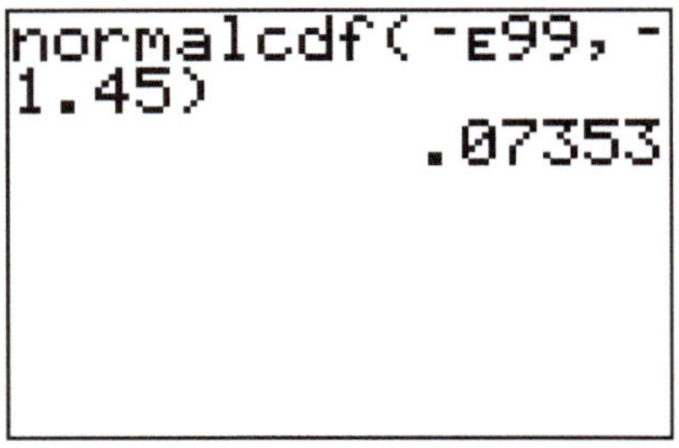

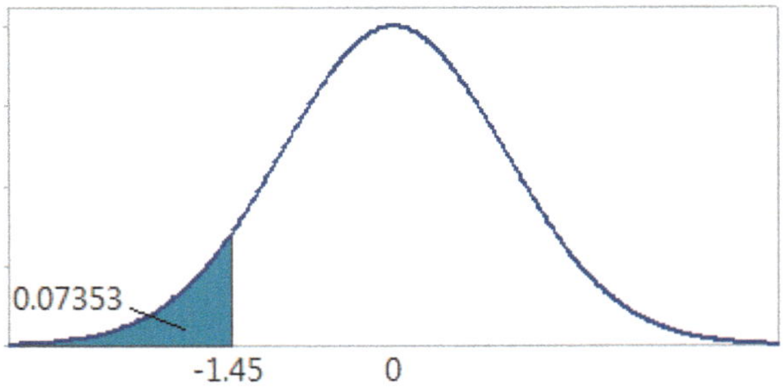

(b) $P(z > 1.45) = 1 - P(z < 1.45) = 1 - 0.92647 = 0.07353$.

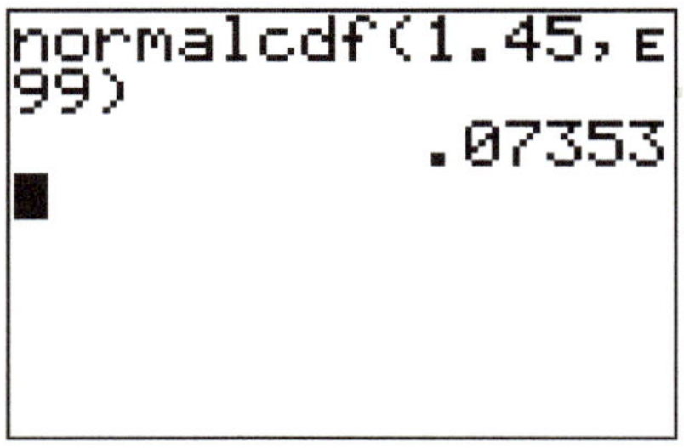

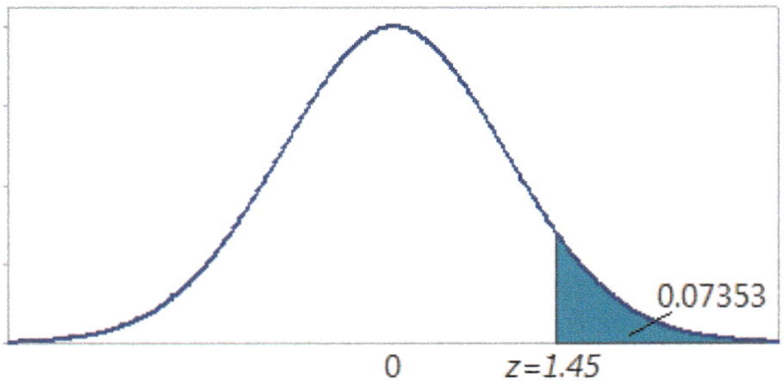

Because the standard normal curve is symmetric about the mean 0, results in (a) and (b) are the same.

(c) $P(Z < 1.45) = 0.92647$.

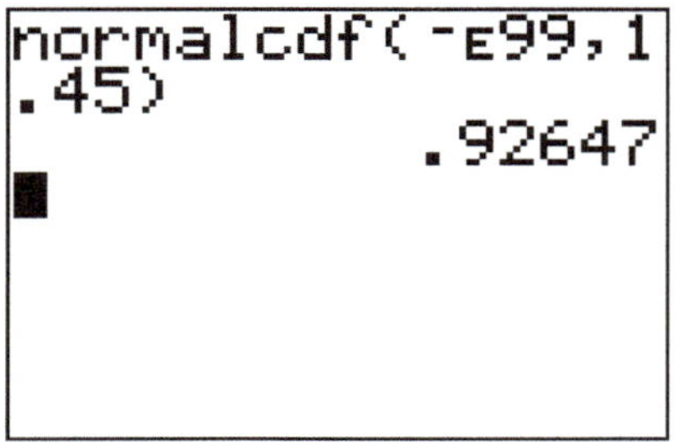

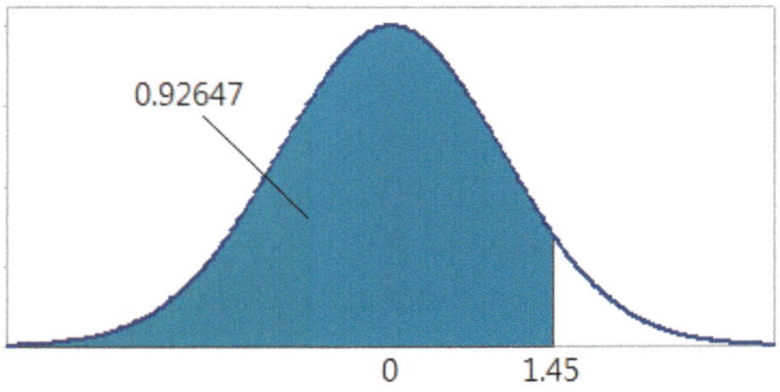

(d) $P(-1.45 < Z < 1.45) = P(Z < 1.45) - P(Z < -1.45)$
$= 0.92647 - 0.07353 = 0.85294$.

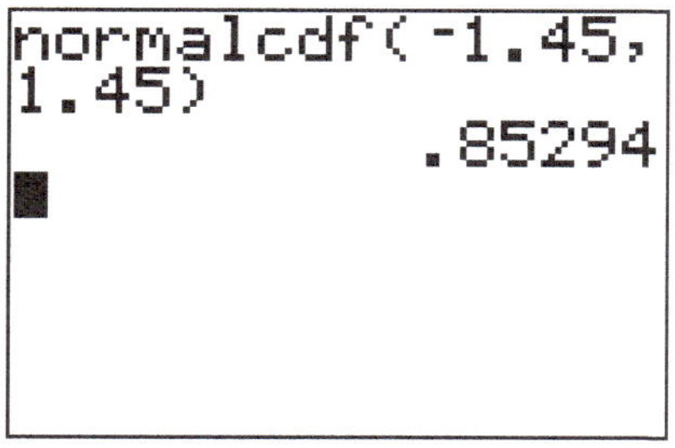

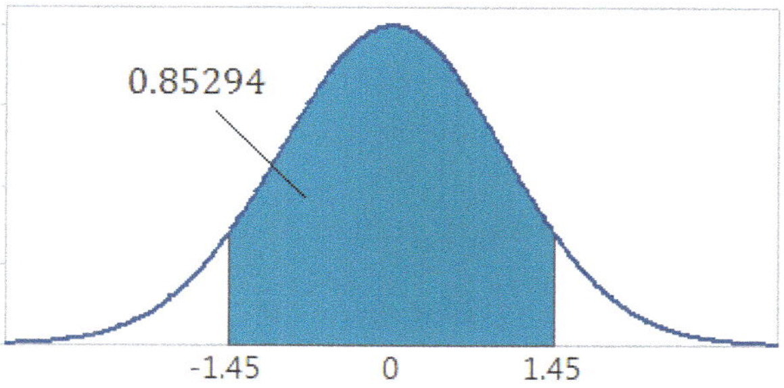

Finding area under any normal curve

Suppose we have a normally distributed variable X which has mean μ and standard deviation σ, i.e., $X \sim N(\mu, \sigma)$. We wish to find area under the curve of $X \sim N(\mu, \sigma)$. In other words, for fixed values of X, say x, x_1 and x_2, we wish to compute probability of the form

- $P(X < x)$
- $P(X > x)$
- $P(x_1 < X < x_2)$

We wish to compute above probabilities using Table T3.

Note that Table T3 is applicable for a normal distribution that has mean 0 and standard deviation 1. In other words, we can use Table T3 to compute area under a standard normal variable. Therefore, before computing the probability or area under the curve of X, we have to covert x, x_1 and x_2 into z, z_1 and z_2, respectively, by using the relations:

$$z = \frac{x - \mu}{\sigma}$$

$$z_1 = \frac{x_1 - \mu}{\sigma}$$

$$z_2 = \frac{x_2 - \mu}{\sigma}$$

Once z scores are obtained, we can compute the area under the curve of X in terms of the area under the standard normal curve as follows:

- $P(X < x) = P(Z < z)$
- $P(X > x) = P(Z > z)$
- $P(x_1 < X < x_2) = P(z_1 < Z < z_2)$

Example 7.10

Suppose that the annual salary for workers in a company is approximately normal with mean \$26,362 and standard deviation \$6,500. Let X be the salary for workers in the company. Answer the following questions using the standard normal table T3.

(a) What percent of people earn below \$35,000, i.e., $P(x < 35{,}000) =$?
(b) What percent earn between \$35,000 and \$50,000, i.e., $P(35{,}000 < x < 50{,}000) =$?
(c) What percent earn below \$15,000, i.e., $P(x < 15{,}000) =$?
(d) What percent earn above \$40,000, i.e., $P(x > 40{,}000) =$?

Solution

Note that we have $X \sim N(\mu = \$26{,}362, \sigma = \$6{,}500)$. Therefore, for each of the problems (a)-(d), we have to convert x-value to the equivalent z-score and then use Table T3 of area under the standard normal curve.

(a) We wish to compute $P(x < 35{,}000)$.
For $x = 35000$,

$$z = \frac{35{,}000 - 26{,}362}{6{,}500} = 1.33$$

Then, $P(x < 35{,}000) = P(z < 1.33) = 0.9082$.
Technology such as TI-84 Graphing Calculator, Minitab or other software can be used directly to compute the probability. These technology has built-in function to convert x into z. See the computation below in TI-84 Plus Calculator:

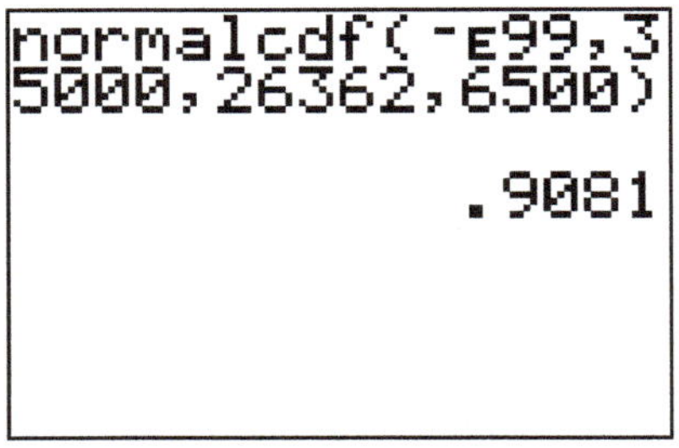

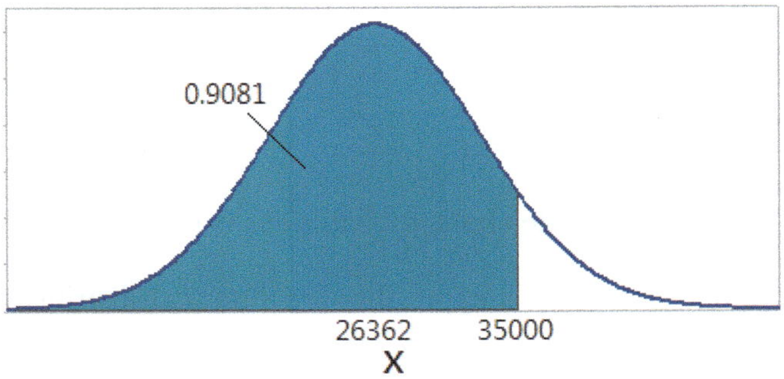

(b) We wish to compute $P(35{,}000 < x < 50{,}000)$
For $x_1 = 35000$,

$$z_1 = \frac{35{,}000 - 26{,}362}{6{,}500} = 1.33$$

For $x_2 = 50000$,

$$z_2 = \frac{50000 - 26362}{6{,}500} = 3.64$$

Therefore, $P(35{,}000 < x < 50{,}000) = P(1.33 < z < 3.64) = 0.99986 - 0.908 = 0.09166$.

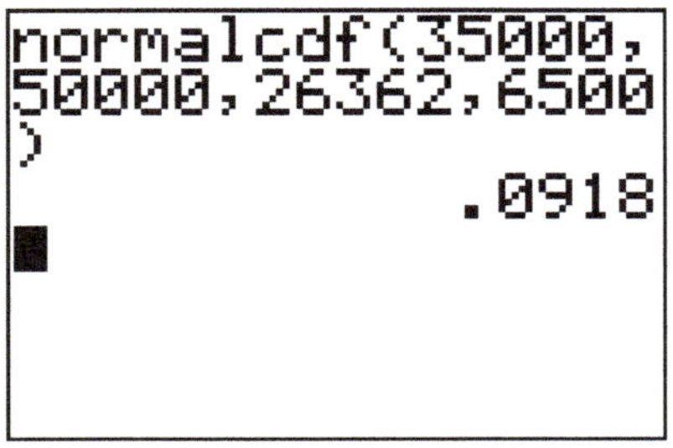

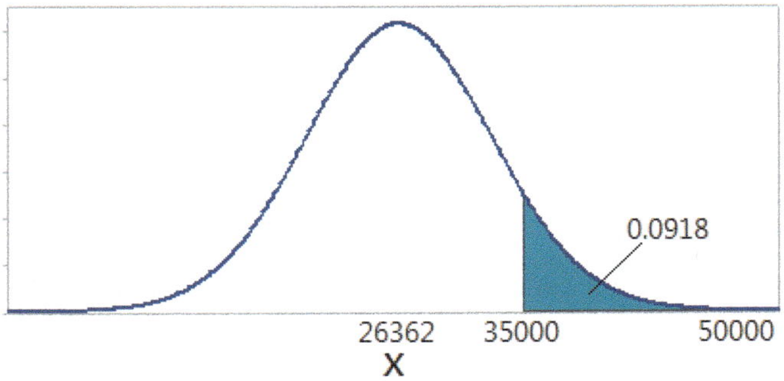

(c) We wish to compute $P(x < 15{,}000)$.
For $x = 15000$,

$$z = \frac{15000 - 26362}{6500} = -1.75$$

Therefore, $P(x < 15{,}000) = P(z < -1.75) = 0.0401$

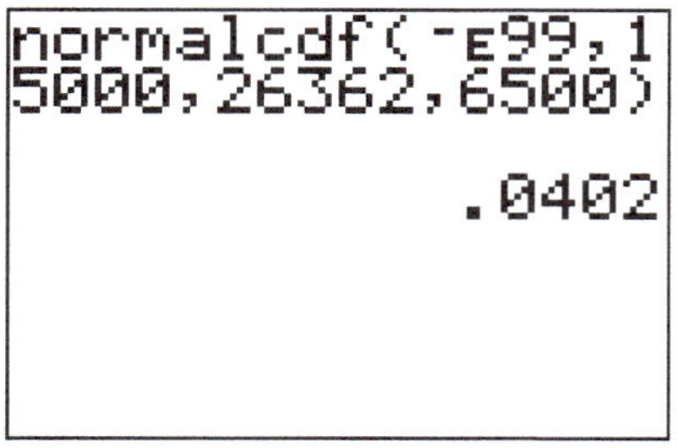

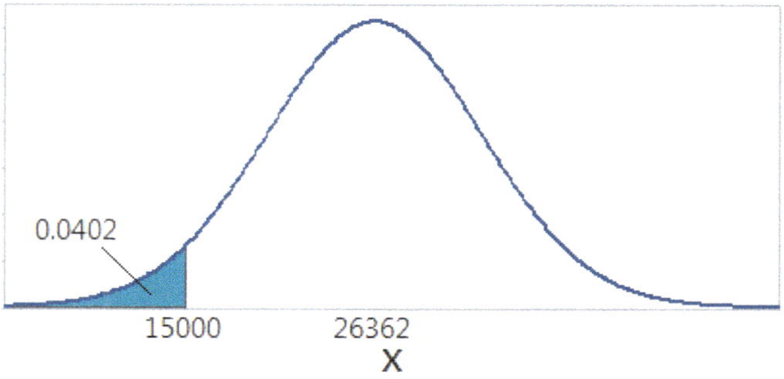

(d) We wish to compute $P(x > 40{,}000)$
For $x = 40000$,

$$z = \frac{40000 - 26362}{6500} = 2.10$$

Therefore, $P(x > 40{,}000) = P(z > 2.10) = 1 - P(z < 2.10) = 1 - 0.9821 = 0.0179$.

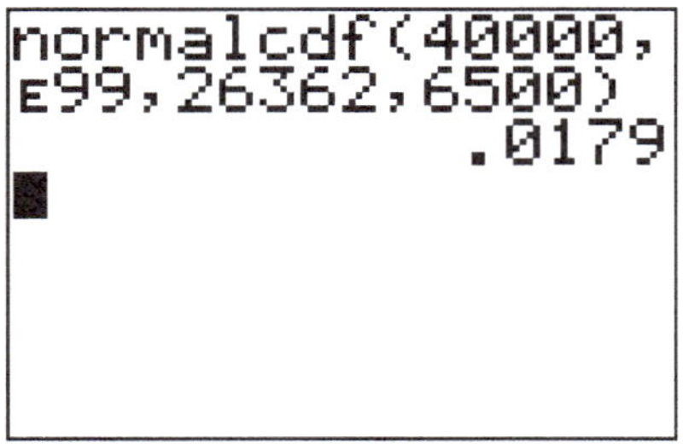

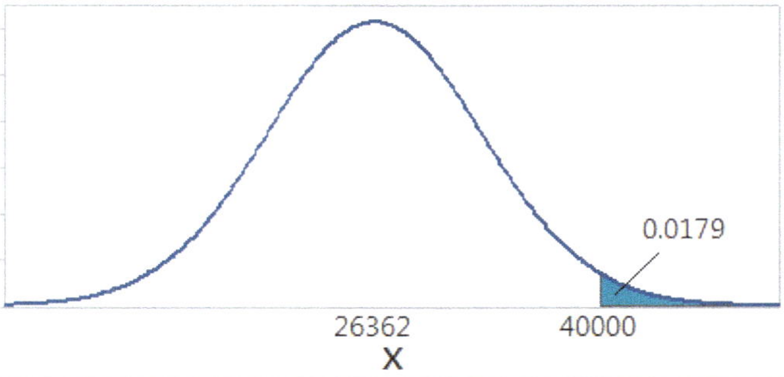

Section 7.4: z_α and $-z_\alpha$ notations

The notations z_α and $-z_\alpha$ have wide applications in confidence interval estimation and test of hypothesis regarding the parameters mean and proportion (to be discussed later).

(1) z_α is the z-score or value of the standard normal variable with area α to its right or equivalently, an area 1- α to its left.

(2) $-z_\alpha$ is the value of the standard normal variable with area α to its left or equivalently, an area 1- α to its right.

Simply speaking, z_α and $-z_\alpha$ are values of standard normal variables with area α to its right and left, respectively.

For any given value of α, technology can be used to find the value of the z_α. Below we find some values of z_α using Table T3, TI-84 Plus Calculator and Minitab, for selected values of α.

How to find z_α using Table T3?

Find the area of 1-α in Table T3 and read the corresponding z value using the leftmost column and the topmost row highlighted. The value so obtained is the z_α with area α to its right.

How to find $-z_\alpha$ using Table T3?

Find the area of α in Table T3 and read the corresponding z value using the leftmost column and the topmost row highlighted. The value so obtained is the $-z_\alpha$ with area α to its left.

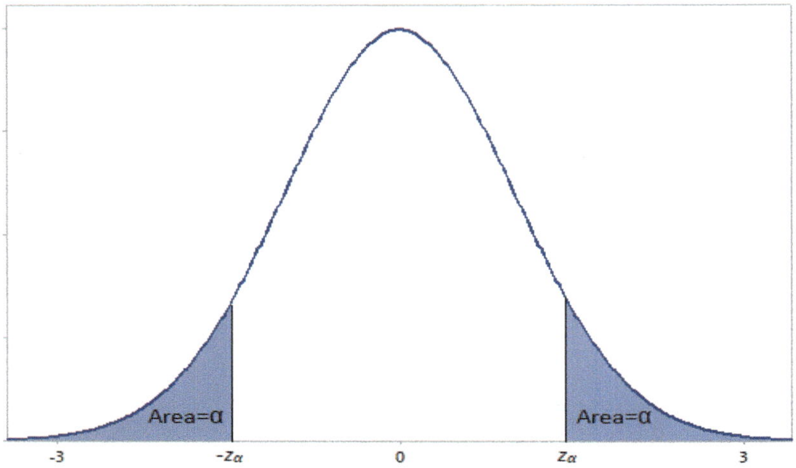

Example 7.11

Find the value of

(a) $z_{0.025}$
(b) $z_{0.975}$
(c) $-z_{0.025}$
(d) $z_{0.05}$
(e) $-z_{0.05}$
(f) $-z_{0.0505}$
(g) $z_{0.0505}$

Solution

(a) $z_{0.025}$ refers to a z-score with area 0.025 to its right, i.e., an area of 0.975 to its left. Therefore, to find the value of $z_{0.025}$, we find the cumulative area of 0.975, and the corresponding z-score from Table T3, which is 1.96. Therefore, $z_{0.025} = 1.96$.

TI-84 Plus Calculator ***invNorm***() function can be used to compute z_α value (see section 7.6 for detail instruction).

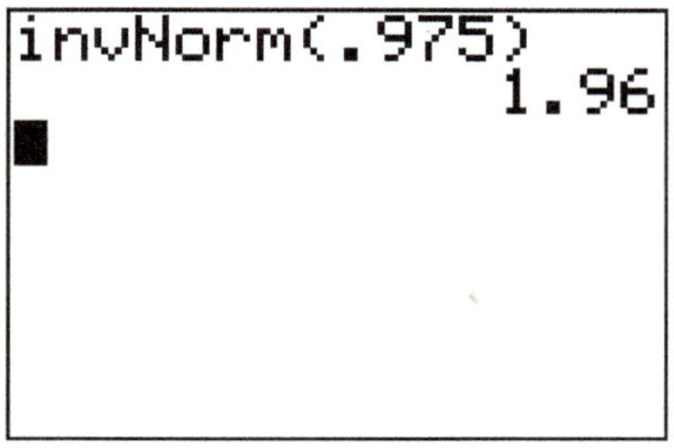

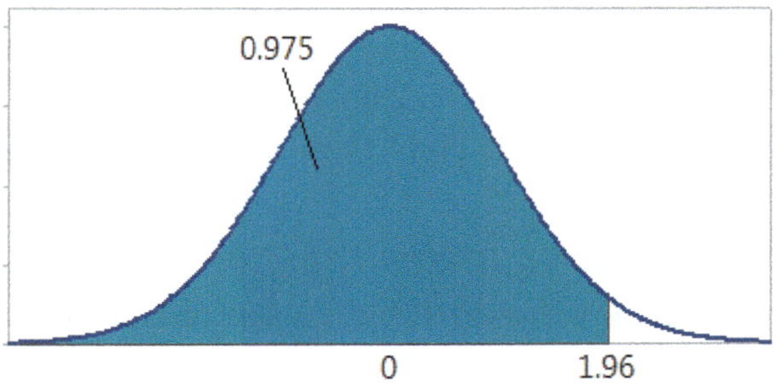

(b) To $z_{0.975}$ find cumulative area of 0.025 in Table T3 and read the corresponding z score, which is -1.96. Therefore, $z_{0.975} = -1.96$.

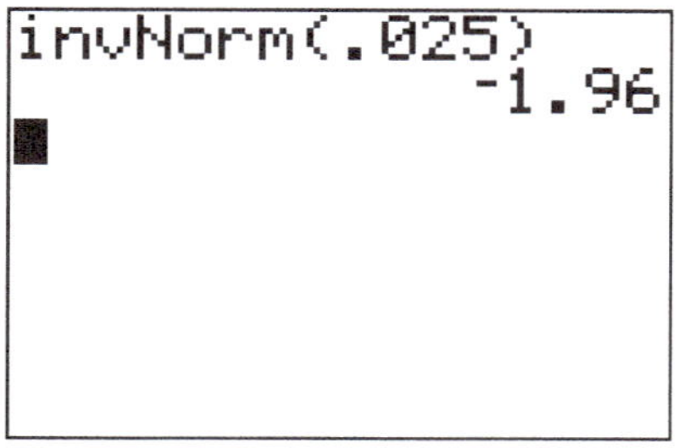

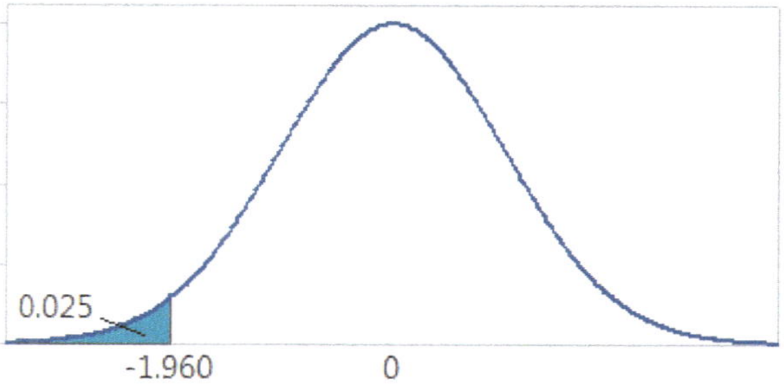

(c) To find $-z_{0.025}$, simply find cumulative area of 0.025 and read the corresponding z score, which is -1.96. Therefore, $-z_{0.025} = -1.96$

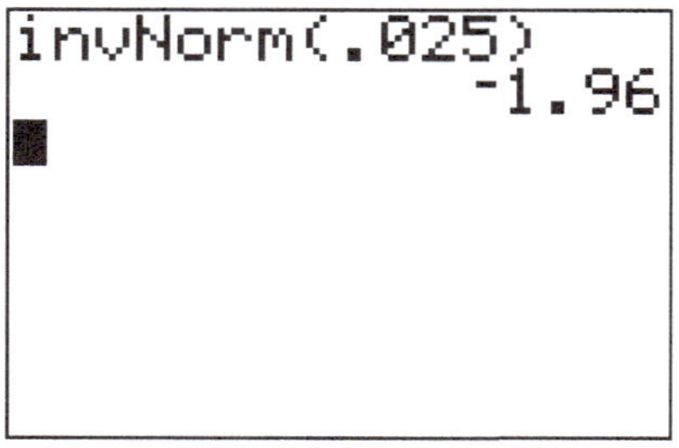

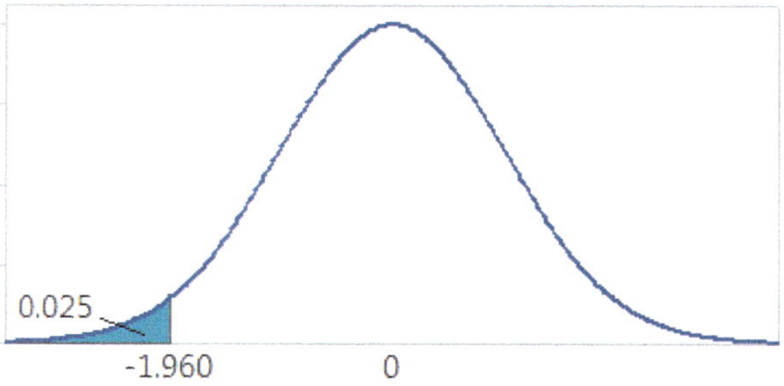

Note that due to the symmetry of the normal curve about 0, $z_{0.975=} - z_{0.025} = -1.96$.

(d) To find $z_{0.05}$, simply find cumulative area of $1 - 0.05 = 0.95$ and read the corresponding z-score, which is 1.645, approximately. Therefore, $z_{0.05} = 1.645$.

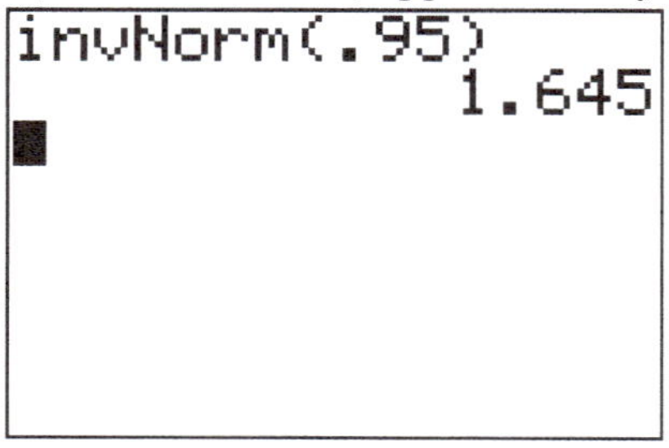

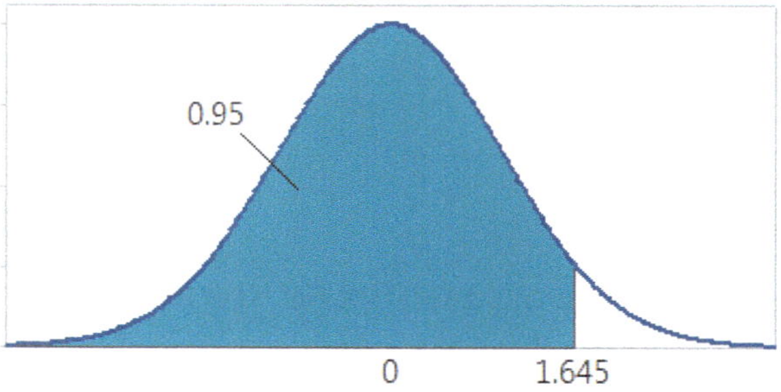

(e) To find $-z_{0.05}$, simply find cumulative area of 0.05 and read the corresponding z-score, which is -1.645, approximately. Therefore, $-z_{0.05} = -1.645$.

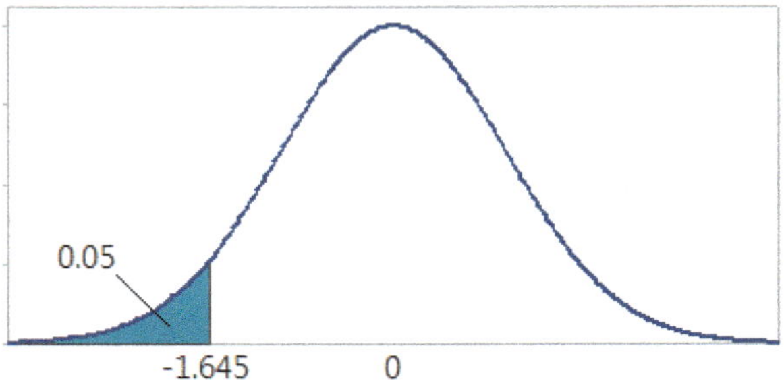

(f) To find $-z_{0.0505}$, simply find cumulative area of 0.0505 and read the corresponding z-score, which is -1.64, approximately. Therefore, $-z_{0.0505} = -1.64$.

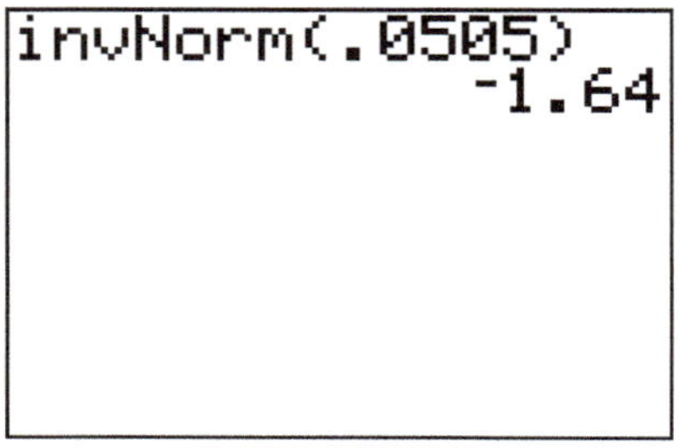

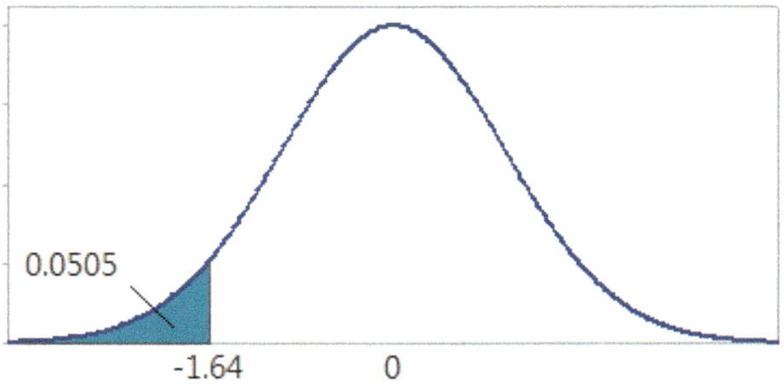

(g) To find $z_{0.0505}$, simply find cumulative area of $1 - 0.0505 = 0.9495$ and read the corresponding z-score, which is 1.64, approximately. Therefore, $z_{0.0505} = 1.64$.

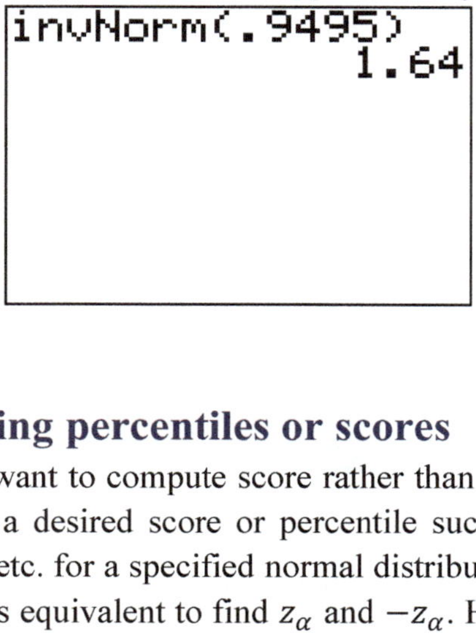

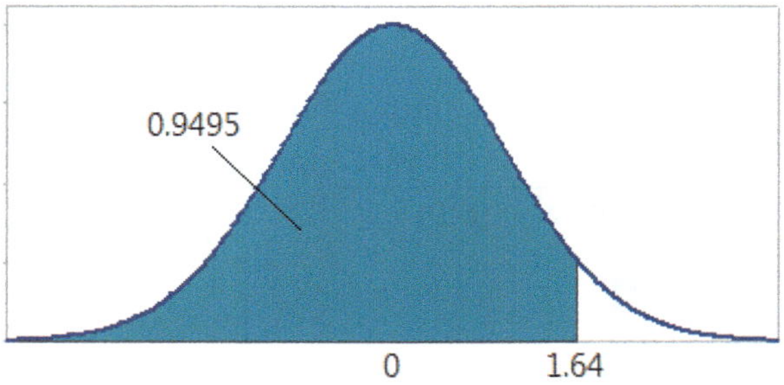

Computing percentiles or scores

Often, we want to compute score rather than area or probability. For example, we might be interested in computing a desired score or percentile such as top 5% score, top 10% score, lower 85% score, 75th percentile, etc. for a specified normal distribution. For a standard normal distribution, finding the score or percentile is equivalent to find z_α and $-z_\alpha$. However, for any normal distribution other than the standard normal distribution, we follow the steps below to compute a desired score or percentile for a given probability, percent or area:

(1) Determine the lower percent, probability or area, for which the score is being asked.
(2) Find z-score using similar techniques as z_α and $-z_\alpha$ for the answer in step (1).
(3) Compute the desired score x by using the relation $z = \frac{x-\mu}{\sigma}$ or $x = \mu + z\sigma$.

Example 7.12

Find the followings for a standard normal distribution using Table T3 or any technology:

(a) The top 5% score
(b) The bottom 5% score
(c) The 2.5 percentile
(d) The upper 2.5 percentile
(e) The 97.5 percentile
(f) The 50 percentile

Solution

(a) The top 5% score is the z-score $z_{0.05} = -z_{0.95} = 1.645$ (we have done this in Example 7.11).
(b) The bottom 5% score is the z-score $-z_{0.05} = -1.645$.
(c) The 2.5 percentile is the z-score $-z_{0.025} = -1.96$
(d) The upper 2.5 percentile is the z-score $z_{0.025} = -z_{0.975} = 1.96$
(e) The 97.5 percentile is the z-score $-z_{0.975} = 1.96$
(f) The 50 percentile is the z-score $z_{0.5} = -z_{0.5} = 0$

Example 7.13

An office will increase salary to its top 5% employees on the basis of a performance score the office created for each employee. The performance score is approximately normal with mean 80 and standard deviation 10. How high must an employee score in order to be in the top 5% of all employee scores?

Solution

Note that the top 5% area under the standard normal curve lies to the right of the is above the z-score $z = 1.645$. What is the x equivalent of $z = 1.645$?

Top 5% area under the standard normal curve

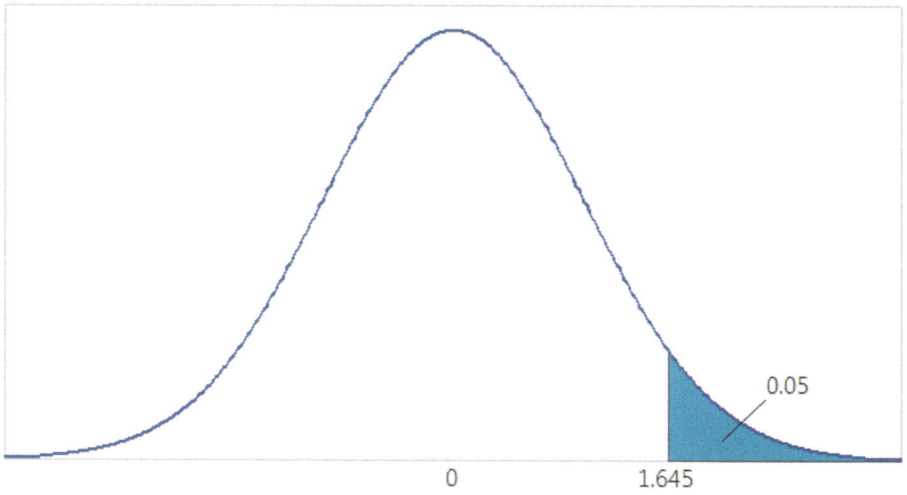

(1) Step 1: The top 5% score corresponds to a score with an area 0.95 to its left.
(2) Step 2: Find z score for which the area to the left is 0.95 by consulting Appendix Table T.3, which is $z = 1.645$.
(3) Step 3: Using the relation $x = \mu + z\sigma$, $x = 80 + 1.645 \times 10 = 96.45$

Thus, an employee must get a score of 96.45 or higher to fall in the top 5% of all employees' score to qualify for an increase in salary.

The computation can be done in TI-84 Plus Calculator:

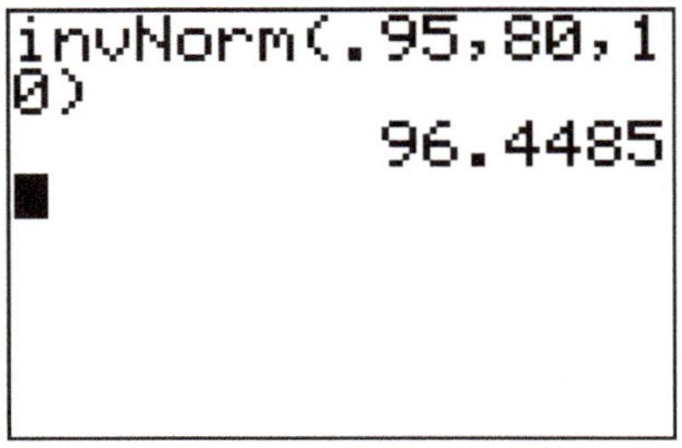

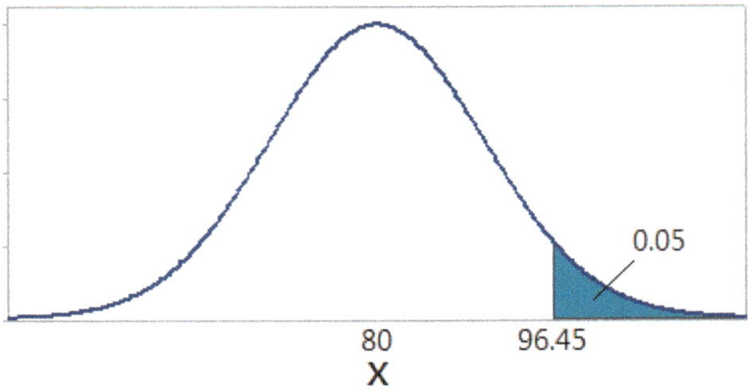

Section 7.5: Normal approximation to binomial distribution

If the number of trials n in a binomial distribution is very large, the table of binomial probability may not be available for all possible values of x. Can we approximate a binomial distribution by the normal distribution? Indeed, when n is large and the success probability p is close to 0.5, the graph of a binomial distribution approximately resembles the normal curve. Therefore, we may approximate a binomial distribution by a normal distribution. However, in order to approximate a discrete binomial distribution legitimately by a continuous normal distribution, we need some adjustment or correction due to the normal approximation to the binomial distribution, what we call the **continuity correction**.

The idea of continuity correction is to compute the probability at an integer x, by the probability of all values within 0.5 units of the integer x. In other words, the binomial probability $P(X = a)$ is computed as $P(a - 0.5 < X < a + 0.5)$ using the normal approximation.

In general, the following rules apply to the normal approximation to the binomial distribution.

Binomial Probability	**Normal Approximation**
(1) $P(X = a)$	(1) $P(a - 0.5 < X < a + 0.5)$
(2) $P(X \geq a)$	(2) $P(X > a - 0.5)$
(3) $P(X > a)$	(3) $P(X > a + 0.5)$
(4) $P(X < a)$	(4) $P(X < a - 0.5)$
(5) $P(X \leq a)$	(5) $P(X < a + 0.5)$
(6) $P(a \leq X \leq b)$	(6) $P(a - 0.5 \leq X \leq b + 0.5)$
(7) $P(a < X < b)$	(7) $P(a + 0.5 < X < b - 0.5)$

In normal approximation, we use $\mu = n \times p$, $\sigma = \sqrt{n \times p \times q}$. The applicability of normal approximation also requires $n \times p \geq 5$ and $n \times q \geq 5$.

Section 7.6: Technology for computing normal probability

In this section, we will briefly demonstrate how to use Minitab and TI 83/84 Plus calculator for computing probability or area and percentiles for a normal distribution. We will solve **Examples 7.14 and 7.15** with necessary instructions for the type of technology used.

Example 7.14

Suppose Z is a $N(0,1)$ variable. Compute the followings:

(a) $P(z < 1.5)$
(b) $P(-1 < z < 2.5)$
(c) $P(z > 2.5)$
(d) z value for the top 10% area
(e) z value for bottom 7% area

Example 7.15

Suppose X is a $N(\mu = 40, \sigma = 7)$ variable. Compute the followings

(a) $P(x < 27)$
(b) $P(30 < x < 50)$
(c) $P(x > 45)$
(d) x value for the top 10% area
(e) x value for bottom 7% area

Minitab Instruction and Solution for Example 7.14

For (a) follow the steps below in Minitab:

(1) Select **Graph>Probability Distribution Plot…**
(2) Select **View Probability** and click **OK**.
(3) For Distribution: Choose **Normal**, **Mean: 0**, **Standard deviation: 1**.
(4) Choose Shaded Area, and type **X-value: 1.5** and click picture for **Left Tail** and OK.

See the dialog box and options selection in figures below together with the produced graph and area.

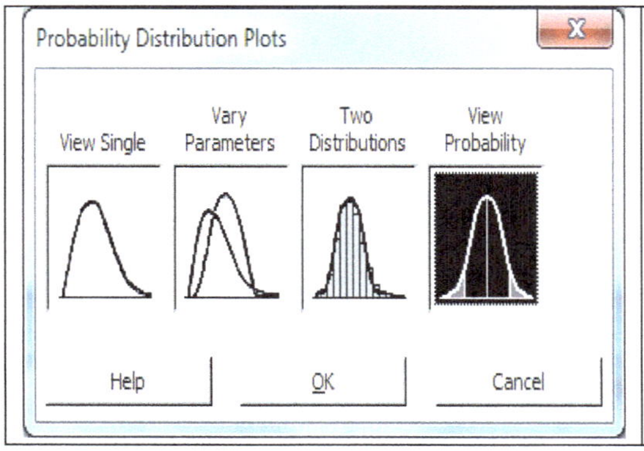

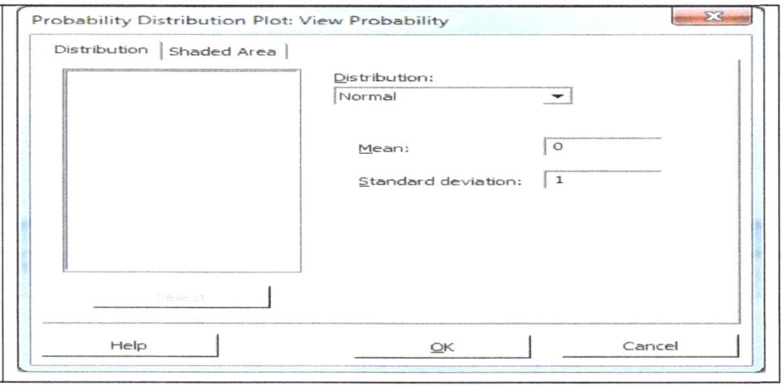

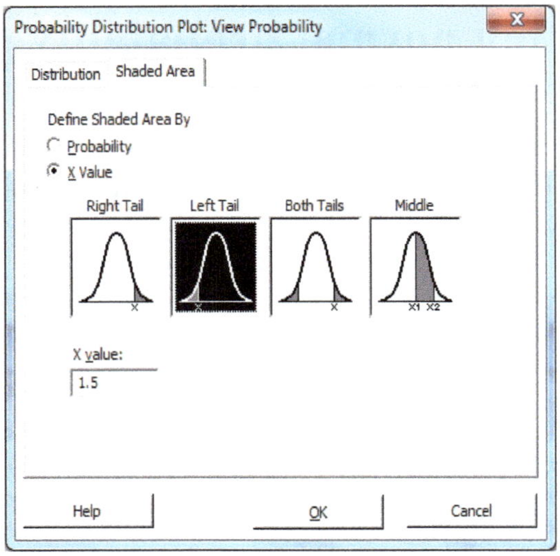

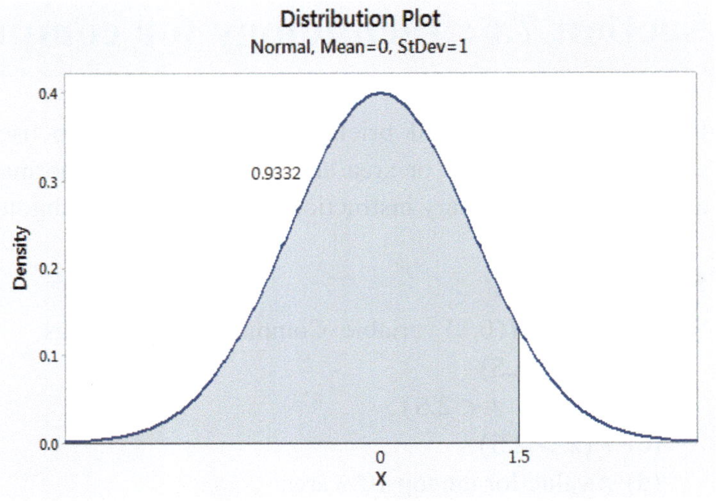

For (b),

(1) Select Edit>Edit Last Dialog,
(2) Select Shaded Area and click picture for Middle, type for **X value 1: -1** and **X value 2: 2.5** and click OK.

See below for option selection and resultant graph.

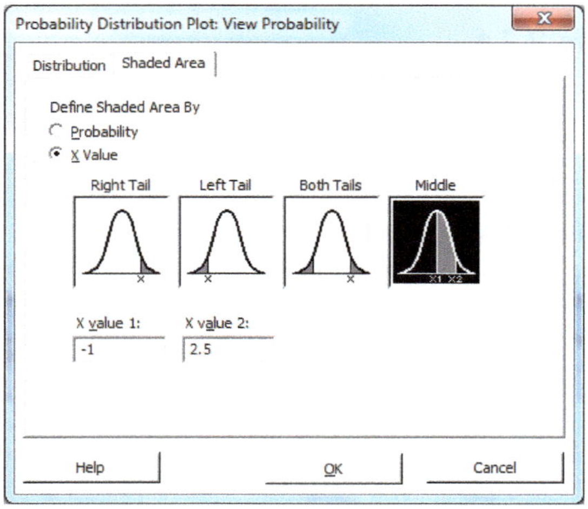

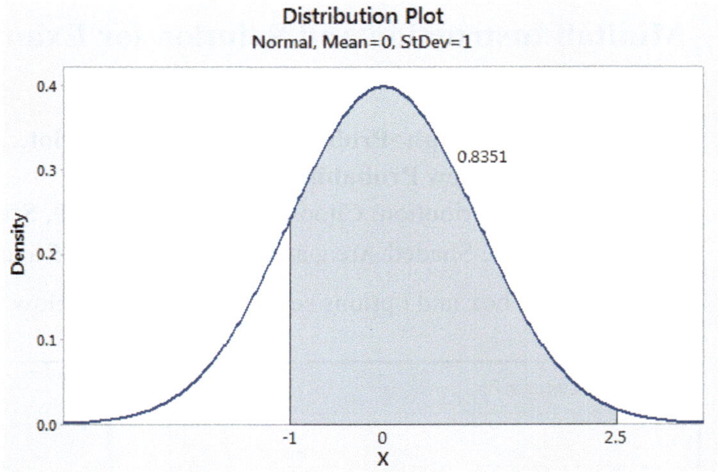

For (c) follow the steps below:

(1) Select Edit>Edit Last Dialog,
(2) Select Shaded Area and click picture for Right Tail, type for **X value: 2.5** and click OK.

See below for option selection and resultant graph.

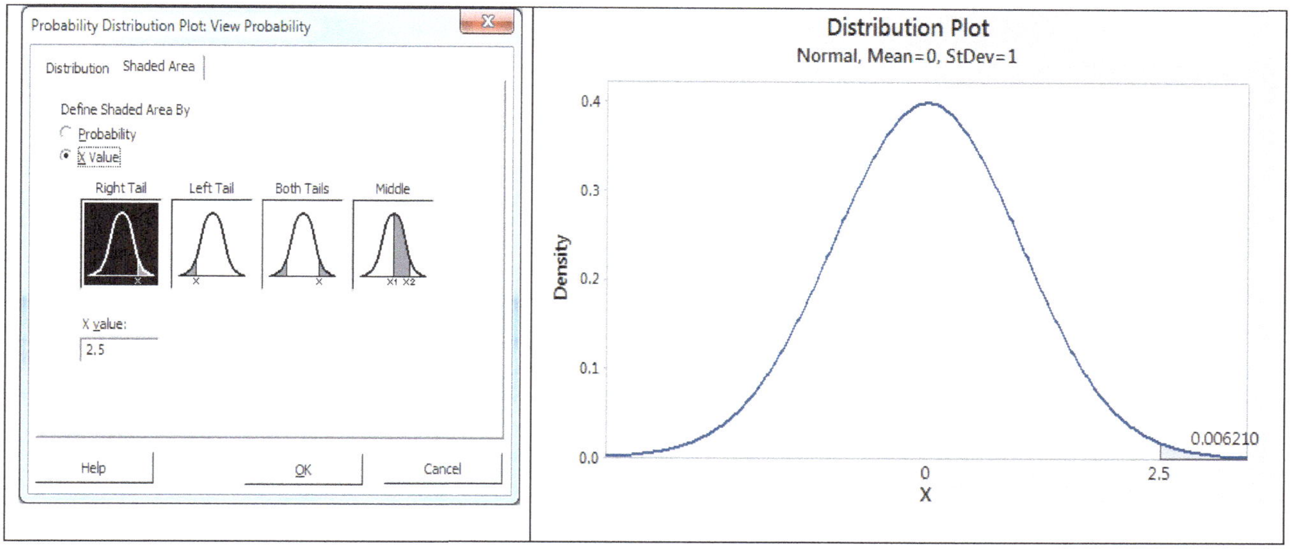

For (d) follow the steps below:

(1) Select Edit>Edit Last Dialog,
(2) Select **Shaded Area**, option **Define Shaded Area By Probability:** and click picture for Right Tail, type for **Probability: 0.1** and click OK.

See below for option selection and resultant graph.

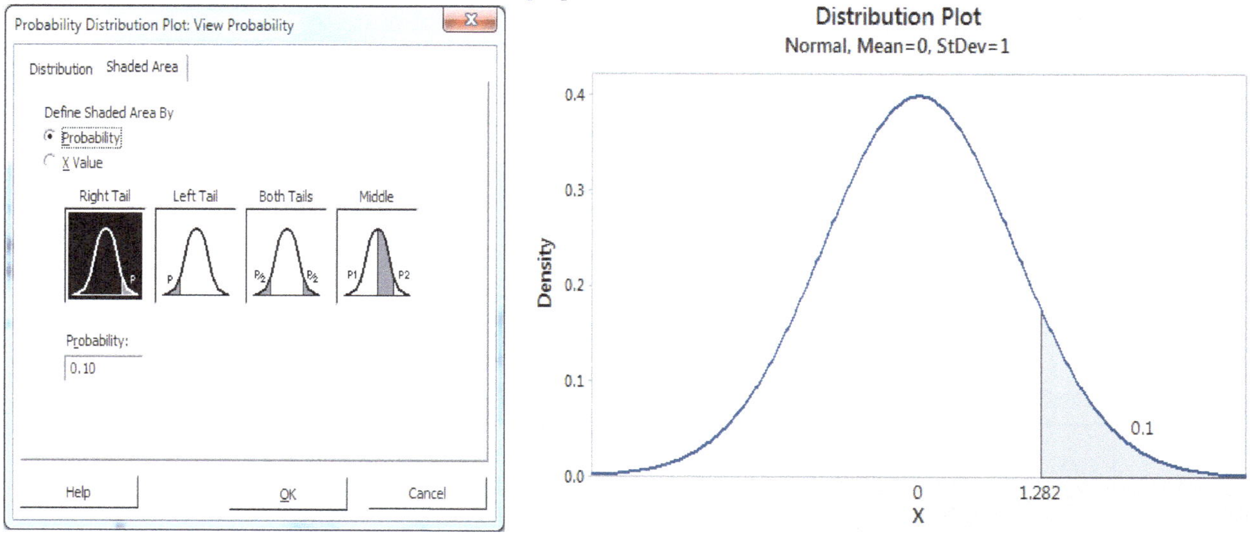

For (e) follow the steps below:

(1) Select Edit>Edit Last Dialog,
(2) Select **Shaded Area**, and click picture for Left Tail, type for **Probability: 0.07** and click OK.

See below for option selection and resultant graph.

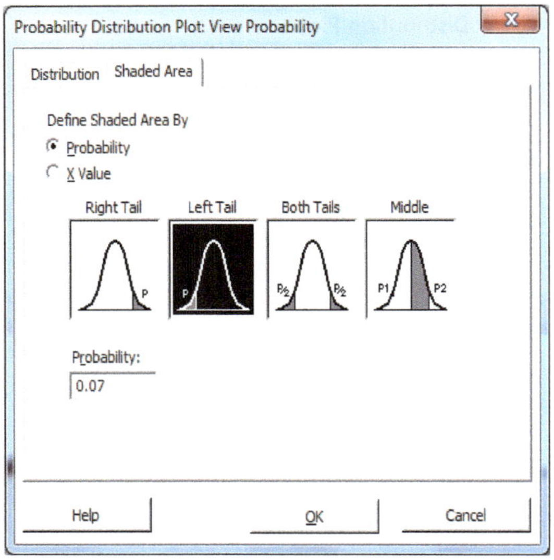

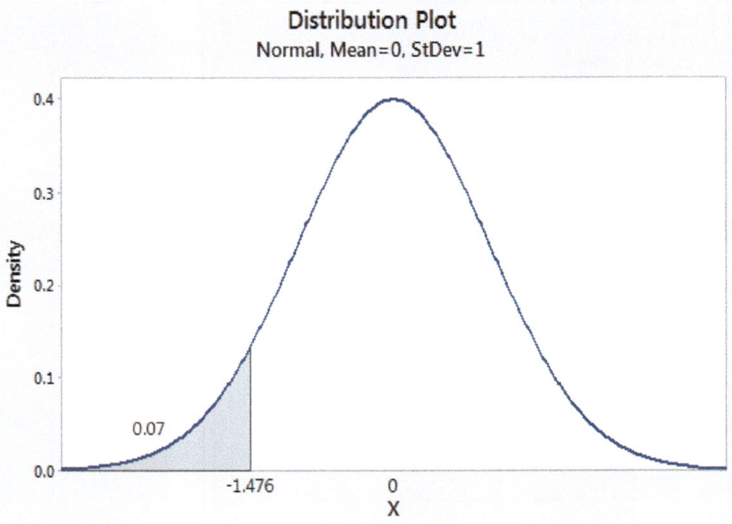

Minitab Instruction and Solution for Example 7.15

For (a) follow the steps below in Minitab:

(1) Select **Graph>Probability Distribution Plot…**

(2) Select **View Probability** and click **OK**.

(3) For Distribution: Choose **Normal**, **Mean: 40**, **Standard deviation: 7**.

(4) Choose Shaded Area, and type **X-value: 27** and click picture for **Left Tail** and OK.

See the dialog box and options selection in figures below together with the produced graph.

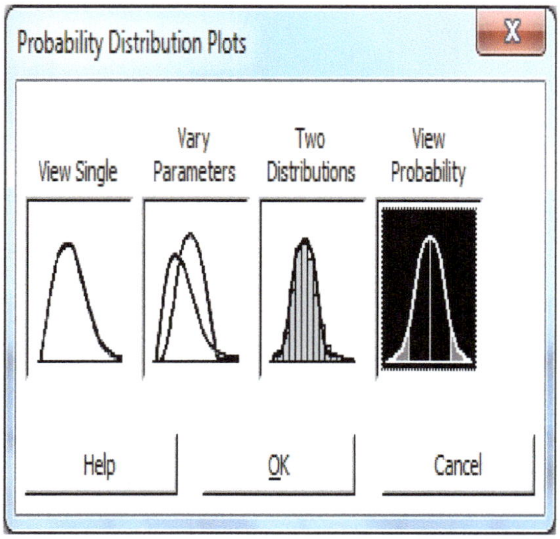

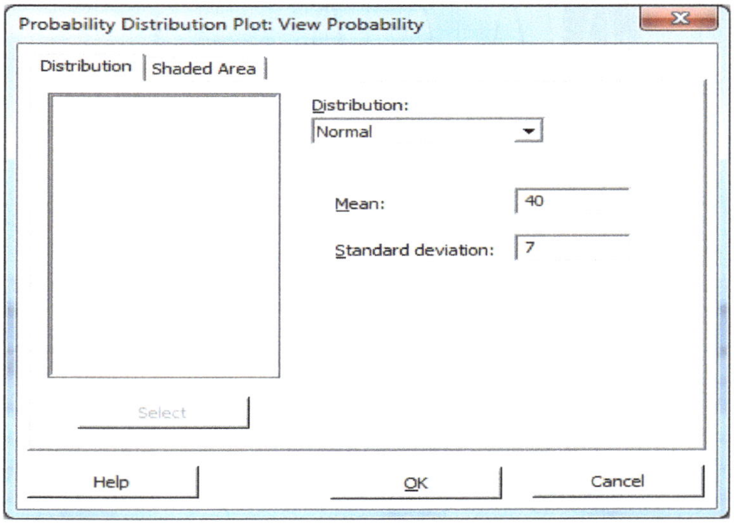

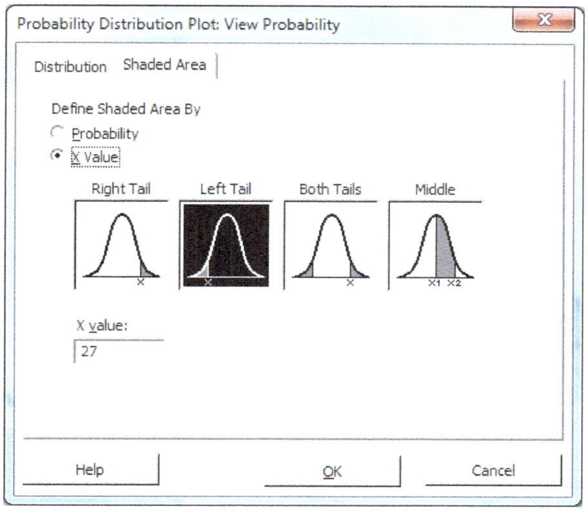

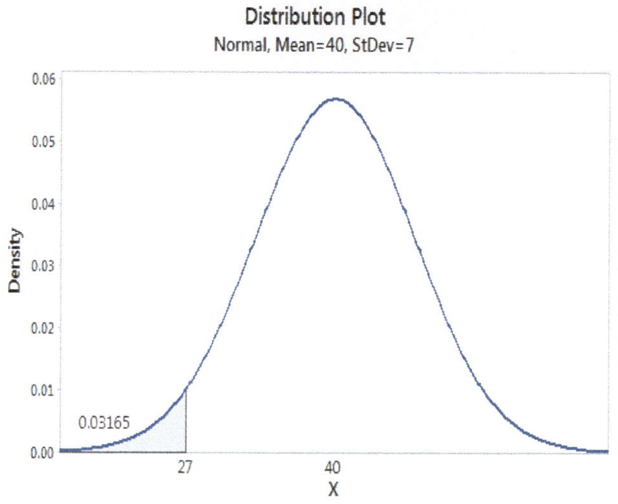

For (b),

(1) Select Edit>Edit Last Dialog,

(2) Select Shaded Area and click picture for Middle, type for **X value 1: 30** and **X value 2: 50** and click OK.

See below for option selection and resultant graph.

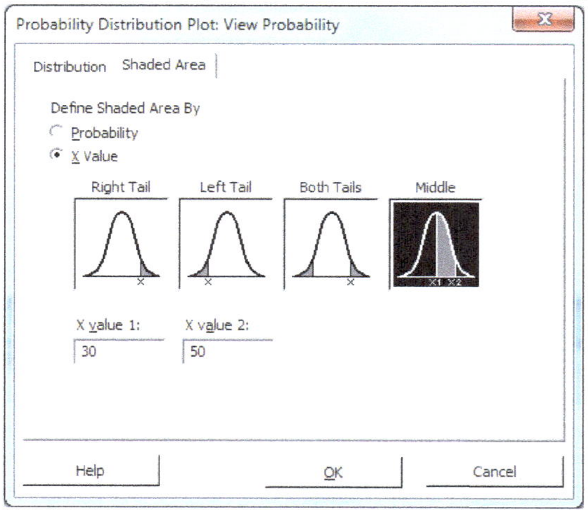

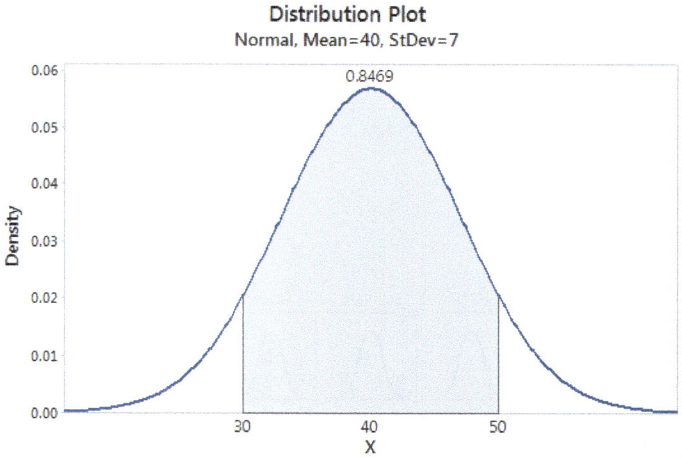

For (c) follow the steps below:

(1) Select Edit>Edit Last Dialog,

(2) Select Shaded Area and click picture for Right Tail, type for **X value: 45** and click OK.

See below for option selection and resultant graph.

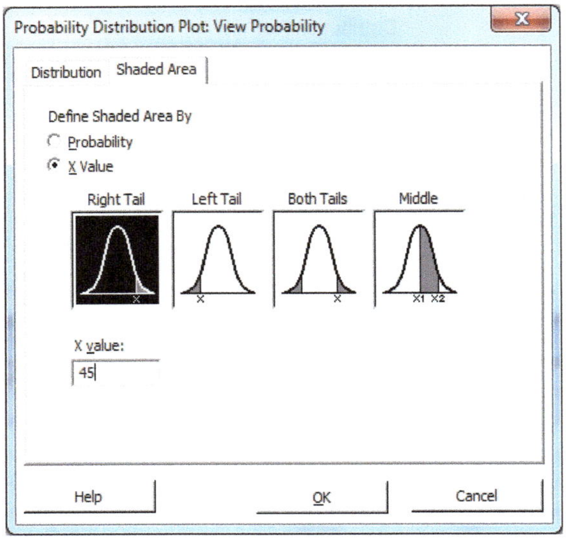

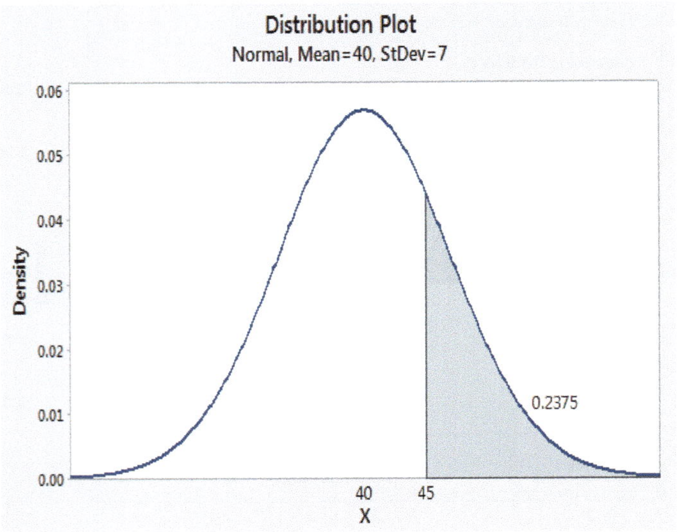

For (d) follow the steps below:

(1) Select Edit>Edit Last Dialog,

(2) Select **Shaded Area**, option **Define Shaded Area By Probability:** and click picture for Right Tail, type for **Probability: 0.1** and click OK.

See below for option selection and resultant graph.

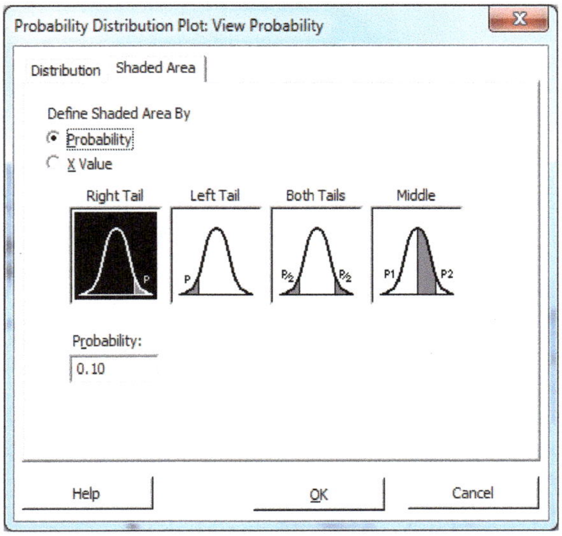

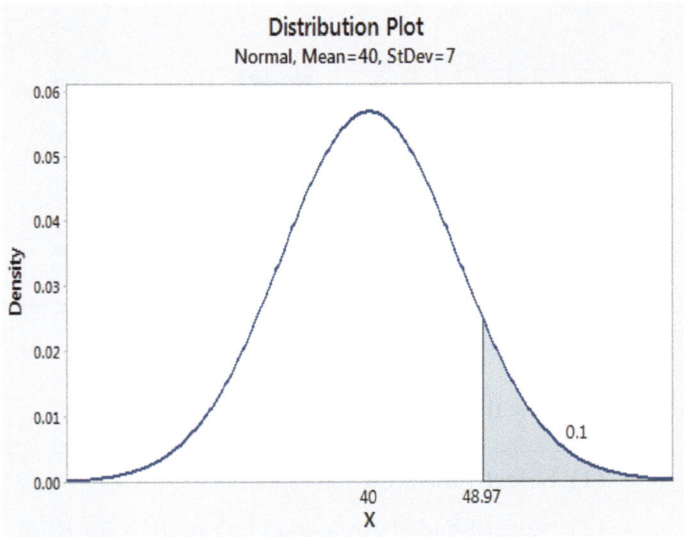

For (e) follow the steps below:

(1) Select Edit>Edit Last Dialog,

(2) Select **Shaded Area**, and click picture for Left Tail, type for **Probability: 0.07** and click OK.

See below for option selection and resultant graph.

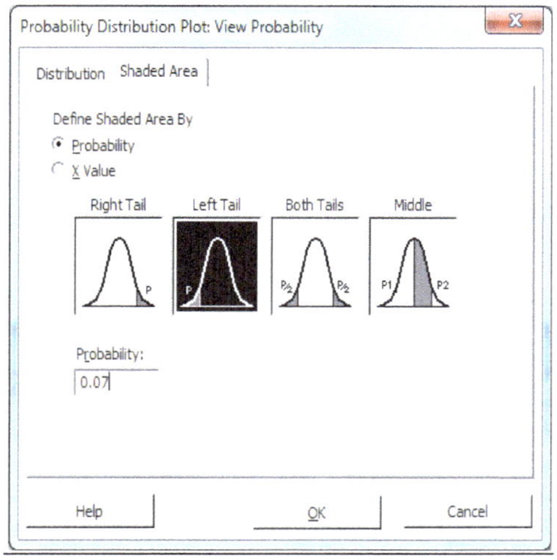

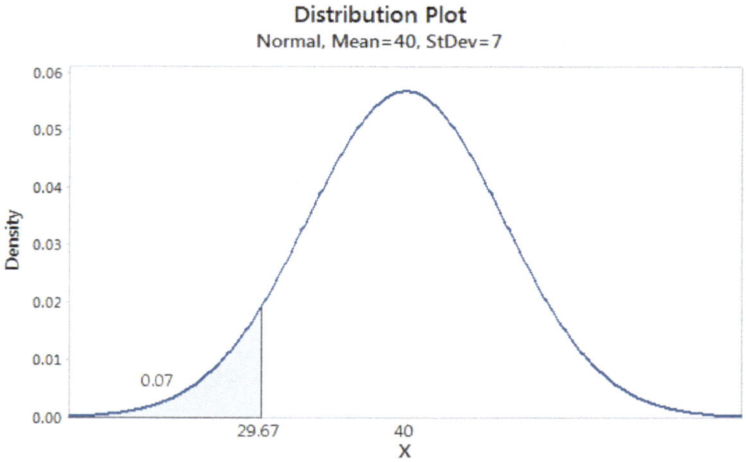

TI-84 Plus Instruction and Solution for Examples 7.14 and 7.15

In TI-84 Plus calculator, the functions

(1) ***normalcdf***$(lowerX, upperX, mean, sd)$ computes ***probability, percent, proportion or area***.

(2) ***invNorm***$(p, mean, sd)$ computes ***inverse cumulative probability*** **or score** (such as top 5% score, top 10% score, lower 85% score, etc.), where p is the left area or probability for which score or percentile is being sought for.

For a standard normal distribution, the mean 0 and standard deviation 1 in above functions are optional. Functions ***normalcdf***() and ***invNorm***() are accessed by pressing **2nd VARS[DISTR]** and then selecting options **2** and **3,** respectively (see the figure on the right).
Use **E99** for ∞, typed using **2nd ,99** and **-E99** for $-\infty$, typed using as **(-)2nd ,99**.

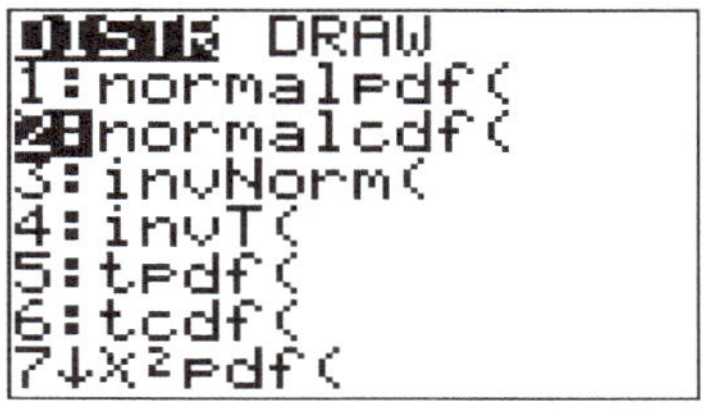

TI-84 Plus Solution for Example 7.14

The following functions with specified arguments as we have described above will solve the problem:

(a) normalcdf(-E99,1.5).
(b) normalcdf(-1,2.5).
(c) normalcdf(2.5,E99).
(d) invNorm(0.90).
(e) invNorm(0.07).

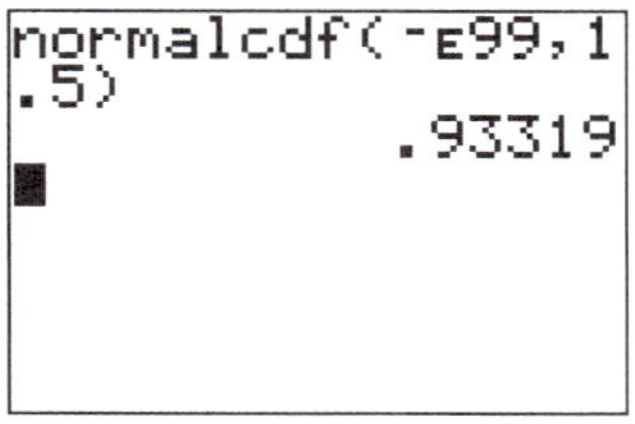

(a)

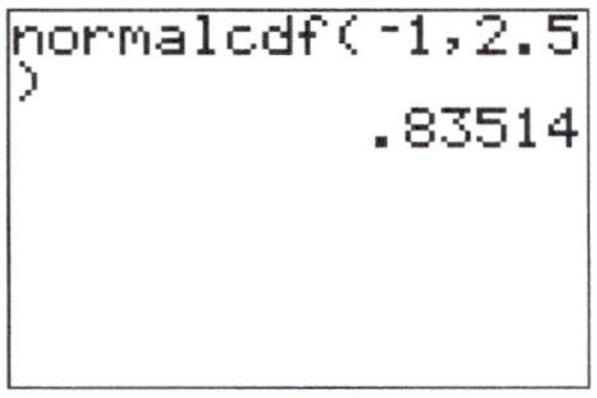

(b)

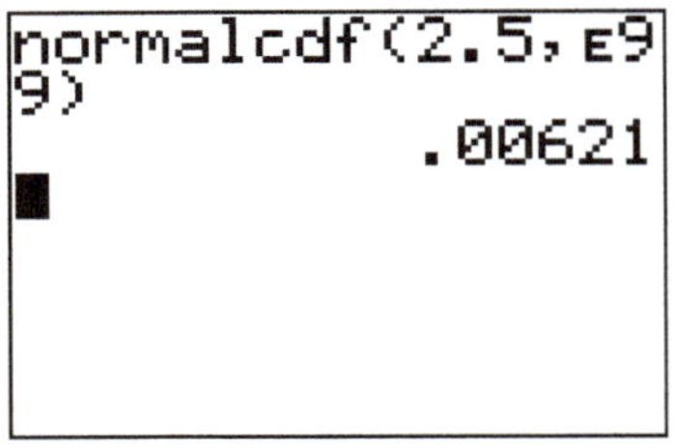

(c)

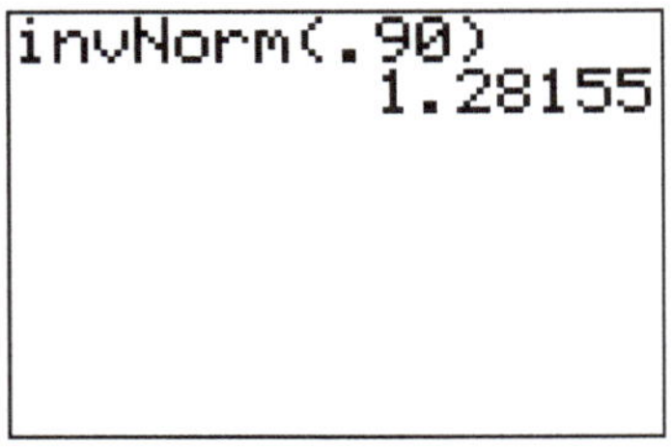

(d)

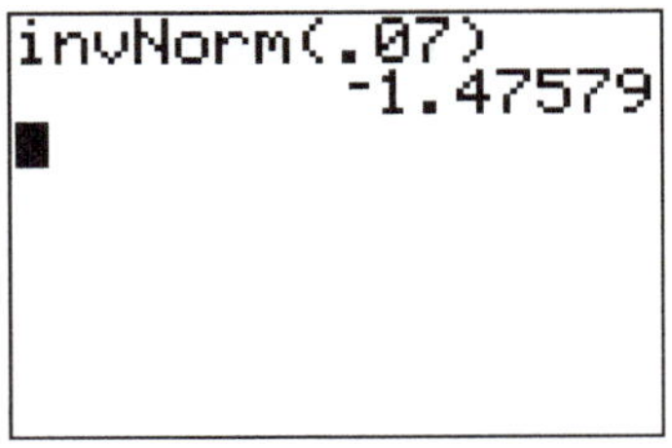

(e)

TI-84 Plus Solution for Example 7.15

The following functions with specified arguments as we have described already will solve the problem:

(a) normalcdf(-E99,27,40,7).
(b) normalcdf(30,50,40,7).
(c) normalcdf(45,E99,40,7).
(d) invNorm(0.90,40,7).
(e) invNorm(0.07,40,7).

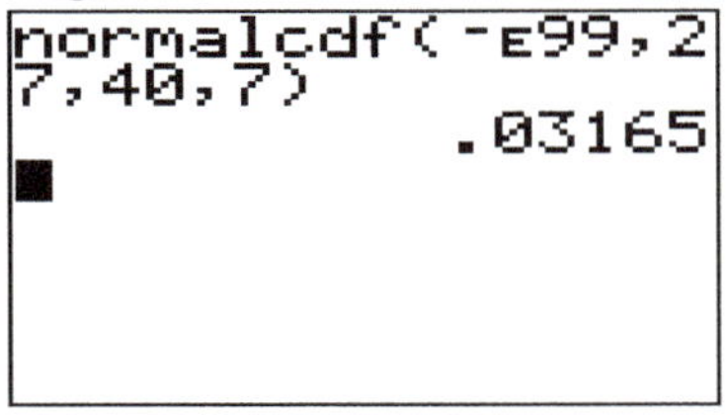

(a)

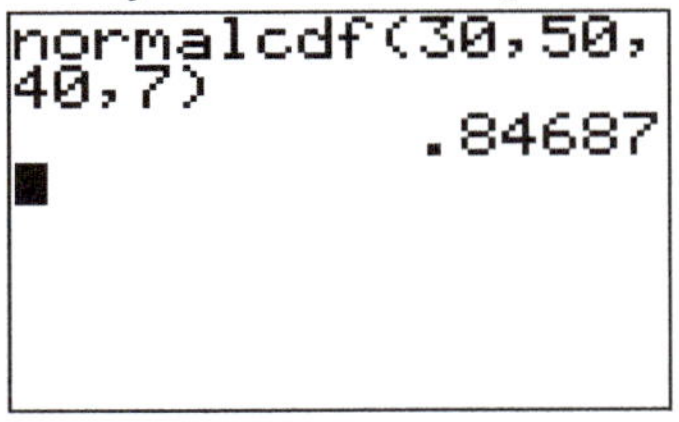

(b)

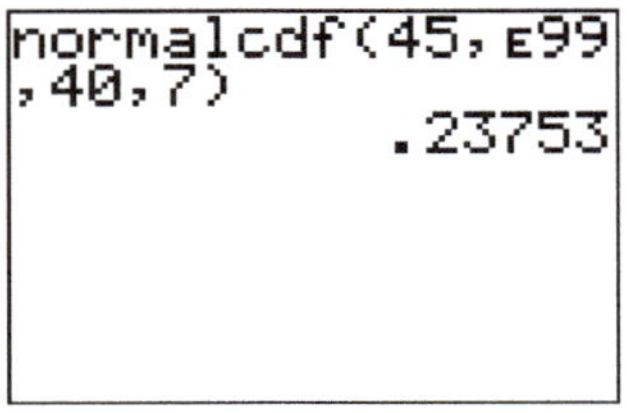

(c)

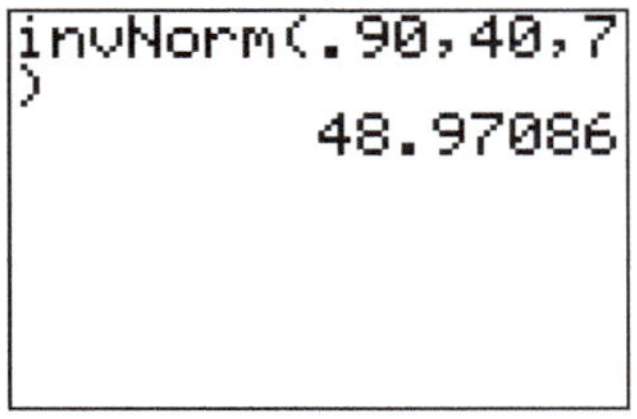

(d)

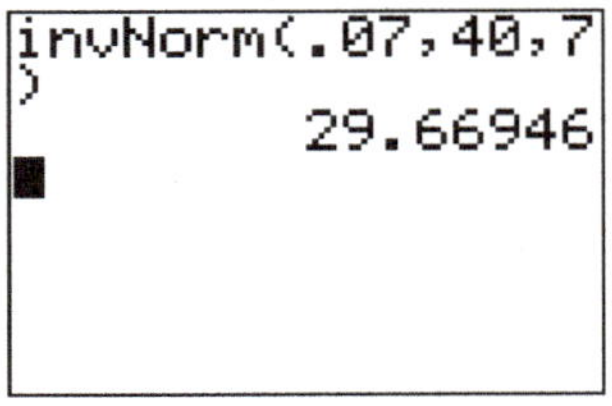

(e)

Chapter 7: Practice problems

(7.1) Answer the following questions using plain sentences:

(a) What do you mean by a normal distribution?
(b) What are two parameters of a normal distribution?
(c) What do the two parameters refer to?
(d) Which parameter determines the center of a normal curve?
(e) Which parameter determines the shape of a normal curve?
(f) How can you specify a normal distribution completely?
(g) What do you mean by the area under the standard normal curve?

(7.2) If a distribution is approximately normal with a given mean and standard deviation, then

(a) What percent of measurements fall within one standard deviation of the mean?
(b) What percent of measurements fall within two standard deviations of the mean?
(c) What percent of measurements fall within three standard deviations of the mean?
(d) What do you call the rule that applies to the answer to questions in parts (a)-(c)?

(7.3) The variable X follows a $N(5,1.5)$ distribution. Find intervals that contain

(a) the 68.3% of observed values of X.

(b) the 95.4% of observed values of X.
(c) the 99.7% of observed values of X.

(7.4) Sketch the two normal distributions as specified below:
(a) N(0,1) and N(5,2)
(b) N(5,1) and N(5,2)
(c) N(3,2) and N(5,2)
(d) N(0,1) and N(2,1)

(7.5) Let X be a N(30,10) distribution. Find equivalent Z-value for each of the following observed values of X and interpret the value of Z.
(a) $x = 15$ (b) $x = 20$ (c) $x = 25$ (d) $x = 30$ (e) $x = 40$ (f) $x = 45$

(7.6) Sketch the following area under the standard normal curve as specified in questions (a)-(e) below and find the area using the standard normal table T3:
(a) to the right of $z = 1.54$
(b) to the left of $z = 1.54$
(c) between $z = -1.54$ and $z = 1.54$
(d) between z = -1.96 and z = 1.96
(e) between z = -2.56 and z = 2.56

(7.7) Let z_α be the value of a $N(0,1)$ distribution, which has area of α to its right. Find the value of z_α for (a)-(i) below:
(a) $z_{0.5}$, (b) $z_{0.05}$, (c) $z_{0.025}$, (d) $z_{0.1}$, (e) $z_{0.01}$, (f) $z_{0.95}$, (g) $z_{0.975}$, (h) $z_{0.90}$, (i) $z_{0.99}$

(7.8) According to a recent study, US adults spent on average 34 hours per month using the mobile internet on Smartphone. If the time spent on the mobile internet on Smartphone follows a normal distribution with mean 34 hours per month and standard deviation 12 hours per month, find the followings:
(a) What percent of US adults spend between 20 and 45 hours per month?
(b) What percent of US adults spend over 45 hours per month?
(c) What is the top 10% of time spent on the mobile internet?
(d) What is the bottom 10% of time spent on mobile internet?

(7.9) The time it takes to finish an exam is approximately normal with mean 50 minutes and standard deviation 10 minutes. Let X denote the time it takes to finish the exam. Then,
(a) Find the percent of students finishing between 50 and 70 minutes.
(b) Find the percent of students finishing between 60 and 75 minutes.
(c) What is the second quartile of the finishing time?
(d) What is the third quartile of the finishing time?

(7.10) According to a recent study, the average payment for a vaginal childbirth in the USA in 2010 was \$12,520. Assume that the distribution of payment for a vaginal childbirth is approximately normal with mean \$12, 520 and standard deviation \$3,000.
(a) What percent of payment is less than \$8,000?

(b) What percent of payment is between \$10,000 and \$13,000?
(c) What percent of payment is above \$12,520?
(d) What percent of payment is above \$10,000?
(e) What is the minimum payment for the top 10% of the payment distribution?
(f) What is the third quartile of the payment distribution?
(g) How low the payment needs to be to fall in the lower 15% of the payment distribution?

(7.11) According to a recent study, US adults spent on average 27 hours per month using the PC internet. If the time spent on the PC internet follows a normal distribution with mean 27 hours per month and standard deviation 10 hours per month, find the followings:
(a) What percent of US adults spend between 20 and 45 hours per month on PC internet?
(b) What percent of US adults spend over 45 hours per month on PC internet?
(c) What is the top 10% of time spent on PC internet?
(d) What is the bottom 10% of time spent on PC internet?

(7.12) According to a recent study, the average payment for a cesarean childbirth in the USA in 2010 was \$16,673. Assume that the distribution of payment for a cesarean childbirth is approximately normal with the mean \$16,673 and standard deviation \$3,500.
(a) What percent of payment is less than \$10,000?
(b) What percent of payment is between \$10,000 and \$17,000?
(c) What percent of payment is above \$16,673?
(d) What percent of payment is above \$15,000?
(e) What is the minimum payment for the top 10% of the payment distribution?
(f) What is the third quartile of the payment distribution?
(g) How low the payment needs to be to fall in the lower 15% of the payment distribution?

(7.13) Intelligence quotient (IQ) for a population is normally distributed with mean 100 and standard deviation 15. Answer the following questions:
(a) What would be the lowest value of IQ score to fall in the top 5% of the score distribution?
(b) What would be the lowest value of IQ score to fall in the top 2.5% of the score distribution?
(c) What is the top 5 percentile of IQ score distribution?
(d) What is the top 2.5 percentile of IQ score distribution?
(e) What would be the IQ score to fall in the bottom 5% of the score distribution?
(f) What would be the IQ score to fall in the bottom 2.5% of the score distribution?

(7.14) Intelligence quotient (IQ) for a population is normally distributed with mean 100 and standard deviation 15. Answer the following questions:
(a) What percent of people will have IQ below 75?
(b) What percent of people will have IQ above 75?
(c) What percent of people will have IQ between 75 and 85?
(d) What percent of people will have IQ between 60 and 75?
(e) Find 68.3%, 95.4% and 99.7% tolerance intervals and interpret the result of 95.4% tolerance interval.

Chapter 7: Answer to selected problems

(7.1) (b) mean μ and standard deviation σ, (d) μ, (e) σ, (f) By mean μ and standard deviation σ.

(7.3)

(a) $[\mu - \sigma, \mu + \sigma] = [5 - 1.5, 5 + 1.5] = [3.5, 6.5]$
(b) $[\mu - 2\sigma, \mu + 2\sigma] = [5 - 2 \times 1.5, 5 + 2 \times 1.5] = [2, 8]$
(c) $[\mu - 3\sigma, \mu + 3\sigma] = [5 - 3 \times 1.5, 5 + 3 \times 1.5] = [0.5, 9.5]$

(7.7)

(a) 0, (b) 1.645, (c) 1.96, (d) 1.285, (e) 2.325, (f) -1.645, (g) -1.96, (h) -1.285, (i) -2.325

(7.8) Let $X \sim N(\mu = 34, \sigma = 12)$ distribution.

(a) $P(20 < x < 45) = 0.6987$.
(a) $P(x > 45) = 0.17966$
(b) 49.4 hours or more per month.
(c) 18.6 hours or less per month.

(7.9) Let $X \sim N(\mu = 50, \sigma = 10)$ distribution.

(a) $P(50 < x < 70) = 0.477$, i.e., 47.7%.
(b) $P(60 < x < 75) = 0.1524$, i.e., 15.2%
(c) 50 minutes
(d) 56.7 minutes

(7.12) Let $X \sim N(\mu = 16673, \sigma = 3500)$ distribution.

(a) $P(x < 10000) = 0.0283$, i.e., 2.83%
(b) $(10000 < x < 17000) = 0.5089$, i.e., 50.9%.
(c) $P(x > 16673) = 0.5000$, i.e., 50%.
(d) $P(x > 15000) = 0.6837$, i.e., 68.4%
(e) \$21,158.43
(f) \$19,033.71
(g) \$13,045.48

(7.13) Let $IQ \sim N(\mu = 100, \sigma = 15)$ distribution.

(a) 124.67
(b) 129.4
(c) 124.67
(d) 129.4
(e) 75.33 or less.
(f) 70.6 or less.

CHAPTER 8
SAMPLING DESIGNS AND SAMPLING DISTRIBUTIONS

Source: CDC

Introduction

Sampling designs and sampling distributions are essential components of a statistical study. Statisticians need adequate knowledge of sampling designs and sampling distributions for making an appropriate statistical inference. For example, statistical inference via a confidence interval estimate or test of the hypothesis (to be discussed later) depends to a great extent on sampling designs and sampling distributions. Sampling designs take into account how samples are taken from a population. Sampling distribution talks about the probability distribution of a statistic. Suppose we wish to know the average cost of a child birth nationwide. How should we proceed? Should we study the entire population or should we rely on a sample for the average cost of a child birth? If we consider a sample, how should we take a sample? How good is the undertaken sample or the estimate from the sample? Answers to all such questions are covered within the domain of the sampling designs and sample distributions.

This chapter briefly covers important sampling designs and sampling distributions. In particular, section 8.1 devotes to define various sampling designs and examples, section 8.2 covers sampling distribution of the sample mean and related properties and applications, and section 8.3 covers the sampling distribution of the sample proportion and related properties and applications. Other sampling distributions such as Chi-square and F distributions will be introduced in Chapter 12.

Section 8.1: Sampling designs

It is well-known that studying the entire population is statistically impractical and inefficient. But we need to know various descriptive measures of a population characteristic, called ***parameters***. Therefore, a representative part of the population, called a ***sample***, is desirable. A descriptive measure of a sample characteristic, called a ***statistic***, is used for making an inference about the parameter. Therefore, selection of an adequate sample is essential for making a meaningful inference about the population or the parameter. ***Sampling design*** refers to how samples are taken from a population. Popular sampling designs are simple random sampling, stratified random sampling, cluster sampling and multistage sampling. Other sampling designs such as volunteer sampling, convenience sampling, etc. also exist in literature.
In this section, we mostly define various sampling designs and some common terminologies associated with them together with examples.

Definition: Sampling frame

A ***sampling frame*** is a list of all population units we sample from.

Example 8.1

Suppose a study is about an opinion of students of a college. Then, the sampling frame is simply the list of all registered students of that college. If the study is about voters of a city, then list of all registered in the city is a sampling frame. For a class of students, list of all enrolled students in the class is the sampling frame.

Definition: Bias

A sampling design is said to be ***biased*** if it systematically favors one sampling unit over others.

Example 8.2

A researcher decides to conduct a survey about the intake of dietary supplement in a big residential area. He will use the list of all landline phone users living in that area as a sampling frame. This design will incur ***sampling bias*** because those in the area without a phone or users of cell phone only will be missed out completely from the study.

Definitions: Simple random sampling and simple random sample

A sampling design in which all possible samples of a given size are equally probable is called a ***simple random sampling***. A sample is called a ***simple random sample*** (SRS) if it is drawn following the design of the simple random sampling. In this text, whenever we specify a sample, it refers to a simple random sample (SRS), unless it is specified otherwise.

Definition: Sampling with replacement

A sampling design in which the repetition of a unit in the sample is allowed is called a ***sampling with replacement***. In a sampling with replacement, a unit already been drawn is returned to the population and thus it has a chance of reselection in the sample and thereby allows repetition of the unit in a given sample.

Example 8.3

Consider a population with only even numbers between 1 and 8, consisting of the set $\{2,4,6,8\}$. Note that we have four units in the population, and all four units are equally likely. We wish to sample two units with replacement. The first pick may be any of the four units. Suppose that the first pick is 6 and it has the probability of 1/4 of being chosen. Now, imagine replacing it back to the population. Then, the second pick has the same probability of 1/4 of being chosen. Indeed, there are exactly $4^2 = 16$ different possibilities: (2,2), (2,4), (2,6), (2,8), (4,2), (4,4), (4,6), (4,8), (6,2), (6,4), (6,6), (6,8), (8,2), (8,4), (8,6), (8,8). In a sampling with replacement, the two sample values are independent. In other words, the first pick does not affect the outcome of the second pick.

In this example, all the 16 different possibilities are equally probable and thus represent 16 simple random samples, each of size 2, with replacement.

Definition: Sampling without replacement

A sampling design in which the repetition of a unit in the sample is not allowed is called a ***sampling without replacement***. In a sampling without replacement, a unit already been drawn is out from the population and thus it has no reselection chance. A sampling without replacement allows the maximum possible look at the population by including different units in the sample.

Example 8.4

Consider the population of the previous example: $\{2,4,6,8\}$. Initially, we have four units in the population, and all four units are equally likely. We wish to sample two units without replacement. The first pick may be any of the four units. Suppose that the first pick is 6 and therefore, it has the probability of 1/4 of being chosen. Now, that 6 is out of the population, the second pick will be any of the remaining 3 units, each with the probability of 1/3 of being chosen. Indeed, there are exactly $4 \times 3 = 12$ different possibilities: (2,4), (2,6), (2,8), (4,2), (4,6), (4,8), (6,2), (6,4), (6,8), (8,2), (8,4), (8,6). In a sampling without replacement, the two sample values are not independent. In other words, the first pick affects the second pick.

In this example, all the 12 different possibilities are all equally probable and thus represent 12 simple random samples, each of size 2, without replacement.

A sampling procedure in which every individual in the population has an equal chance of being included is called a ***probability sampling***. An SRS is a probability sample.

How to draw an SRS?

In this text, we will consider any given sample to be an SRS, unless it is specified otherwise. However, given a sampling frame, we can draw an SRS using statistical software or a computer generated table of random numbers.

In general, almost every statistical software has a built-in-procedure of its own to draw an SRS. The idea is to take a random number and then select an individual matching this random number from the sampling frame as a sampling unit. Continue this procedure until a sample of a desired size n is obtained.

Suppose a class consists of 40 students and we wish to take a sample of 10 students from the class. We can follow the steps below to draw a sample of 10 students in R (an open source software).

(1) For an SRS with replacement, we can type the following line of code in R:

```
> sample(c(1:40), 10, replace = TRUE)
```

An output of this code is:

```
[1]   13   1   30   23   2   13   39   36   2   8
```

(2) For an SRS without replacement, we can type the following line of code in R:

```
> sample(c(1:40), 10, replace = FALSE)
```

An output of this code is:

```
[1]   3   28   22   33   34   14   36   38   4   24
```

In above code, the function sample () draws a sample of size 10 from a list of the first 40 natural numbers, with replacement if replace=TRUE option is used or without replacement if replace=FALSE option is used. The students bearing the observed numbers as identification numbers constitute the desired sample of size 10.

Often, we take a sample with the help of a table of random numbers. A table of random numbers can be created using various procedures or commands. For example, we use the R command below to create following table of 250 random numbers.

matrix(round(runif(250, min=12345, max=99999), digits=0), 25, 10)

73300	90541	16310	13508	79002	64696	86388	58150	30273	69923
26032	40567	54526	94218	75290	63250	31228	44877	59528	19164
81115	82057	82952	14849	36745	29854	79875	33587	40939	82204
68116	44122	83442	43967	30676	39433	61451	50627	74204	43222
14476	87089	91595	52948	48061	57294	60992	32861	94337	80297
20397	81172	20465	82177	38667	14759	20338	28764	16745	74210
88787	18459	43268	64634	90706	99896	35617	91685	61866	57829
13175	88550	59254	74378	87715	56204	19268	23366	13717	71970
35696	80581	55522	41622	63472	68700	64173	54886	74717	95709
50096	43331	29902	62239	71384	85095	90868	76648	87601	81175
70096	50865	12881	91809	32198	69814	56498	92568	29491	18774
30447	57276	76812	38909	14052	99767	34157	26173	58496	23364
61852	67539	30351	59455	90450	77994	12366	23855	79529	33324
20195	62012	70689	89754	96623	24463	12405	76680	79821	75984
52323	81023	80837	13954	33327	90753	22551	56703	82127	23901
36142	82226	68105	20334	94618	65212	66258	28853	37952	93728
80426	32612	92618	84776	84731	98485	13704	46988	95710	20559
32182	22193	87722	90650	42987	32050	37637	72009	18649	17590
78903	56507	42763	78213	78088	53221	39019	54624	68974	21568
16156	16893	82769	80163	86933	53608	69974	92902	52039	14901
35979	93958	19635	40552	77284	39801	45122	32570	97304	27394
27298	62172	24565	52451	95537	92337	62321	90810	70191	86872

15344	47711	58863	98141	47409	58835	97048	63529	76873	89561
15907	67143	36795	91145	95142	94445	70614	21108	15293	76212
85217	91570	28569	84706	47533	94855	30735	77513	69269	69344

We can take the help of this table to select a sample of 10 students from a population of 40 students of the class. Given the fact that the class has 40 students, we can make the sampling frame with a list of two-digit numbers from 01 to 40.

Now, that we have table of random numbers, we can start at any row or column and continue taking random numbers systematically. For example, let us start from the first row:

73300	90541	16310	13508	79002	64696	86388	58150	30273	69923

The two-digit numbers to this row are as follows:

73 **30** **09** **05** 41 **16** **31** **01** **35** **08** 79 00 **26**
46 96 86 **38** 85 81 50 **30** **27** **36** 99 **23**

These two-digit numbers are equally likely and can be used to select the 10 sample units. Starting the first number, the sample of size 10 could be the students bearing identification or serial numbers:

30 09 05 16 31 01 35 08 26 38

The two digit-numbers greater than 40 or less than 01 are ignored from considering in the sample because the sample frame consists of the list from 01 to 40. Using any other row or column of our choice, we could end up with a different sample of 10 students. Note that should we have a sampling frame with the list from 00 to 40, we could use 00 as a sampling unit.

Similarly, we could consider a three-digit, four-digit or any higher-digit number sequence from the table of the random numbers to select a sample of desired size depending on the size of the population. For a population with higher size, we need a larger table of random numbers, which we can create easily using a technology. Indeed, due to the modern computational facility, it is good to use a computer software package to generate random numbers of any desired size and select a sample of a desired size thereby.

Example 8.5

Suppose a special education class has 9 students, with each student assigned a number between 1 and 9 as shown below:

1. Amy	4. Elizabeth	7. Emily
2. Arianna	5. Lucie	8. Alan
3. Kimberly	6. Cecilia	9. Melissa

Then the sampling frame is the list of numbers between 1 and 9.

Select an SRS of size four without replacement to perform an activity using following sequence of computer generated random numbers starting

(a) left to right
(b) right to left:

73300	90541	16310	13508

Solution

(a) Starting at the beginning of the random number sequence and reading each single digit from left to right, we have:
7 3 3 0 0 9 0 5 4 1…

Therefore, a sample of size four without replacement would consists of students bearing numbers: 7, 3, 9, 5. An SRS with this scheme would include Emily, Kimberly, Melissa and Lucie.

(b) Starting at the end of the random number sequence and reading each single digit from right to left, we have:

8 0 5 3 1 0 1 3 6 1 …

Therefore, a sample of size of four without replacement would consists of students bearing numbers: 8, 5, 3, 1. Following this scheme, an SRS would consist of the four students, namely: Alan, Lucie, Kimberly and Amy.

Definition: Stratified random sample

Often, the population as a whole is not homogeneous. Therefore, an SRS cannot provide a reliable estimate of any desired population characteristic, unless some special means are taken into account to adjust for the heterogeneity of the population. The first step to this adjustment is to divide the entire population into several relatively homogeneous subpopulations, called ***strata***, and the second step is to select an SRS from each subpopulation, called ***stratum***. A sampling design that incorporates these two steps is called a ***stratified random sampling***.
A sample is called a ***stratified random sample*** if it is made up of all SRSs, each of which is an SRS from each stratum.

Example 8.6

We wish to find the average income of all faculty members of a university. Several students are taking it as a group project. Students consider income information from SRSs of 30 Professors, 35 Associate Professors and 35 Assistant Professors. Finally, they combine income information from these three SRSs to form a sample of 100 faculty members to be used to estimate the unknown average income of all faculty members of the university. The final sample of size 100 is an example of a stratified random sample.

Definition: Cluster Sampling

A sampling design in which an SRS of clusters or subgroups are considered first and then all subjects in selected clusters are considered together as a sample, is called a ***cluster sampling***.
A sample is called a ***cluster sample*** if it consists of all subjects from a sample of clusters.

Example 8.7

In order to sample from a big neighborhood, all the streets of the neighborhood are considered as clusters. Suppose an SRS of three clusters (streets) are then selected and then all the people living in these three streets are considered to be included in the sample. This is an example of a cluster sample.

Definition: Multistage Sampling

A sampling design in which sampling is done in stages, and at each stage sampling is performed from the units chosen at previous stage, is called a ***multistage sampling*** and the sample so obtained is called a ***multistage sample***.

Example 8.8

In order to estimate the average cost of education in the United States, a sample is intended from all undergraduate students in the United States. To this end, a group of researchers selects an SRS of five states. From each of these five states, they take an SRS of three colleges or universities. Finally, from each of these fifteen colleges or universities, they select an SRS of 30 undergraduates. The final sample consists of 450 undergraduate students. This sample is an example of multistage sampling.

Definition: Voluntary Response Sample

A sampling design in which participants self-nominate themselves is called a ***voluntary response sampling*** and the sample of self-nominated participants is called a ***voluntary response sample***. Generally, the participants of a voluntary response sample select themselves by personal contact out of their own interest, a voluntary response sample is biased and non-representative.

Example 8.9

Suppose a university wish to gather opinion about the on-campus student parking problem. Instead of contacting all students individually, the university is asking students to provide their opinion about the parking problem by responding to an e-mail notification, an automated phone call response or by personal contact. Therefore, the university is relying on students who are self-selecting and giving their opinion. Suppose the university received opinion about the problem from a self-nominated sample of 345 students. This is an example of a voluntary response sample.

Definition: Convenience Sample

A sampling design in which units of the sample are chosen only if it is convenient is called a ***convenience sampling***, and the sample so obtained is called a ***convenience sample***.
A convenience sample is biased and non-representative.

Example 8.10

As a part of a student project, a student is collecting information about the students' use of library by asking questions to whoever he meets in the library on a given day. The sample so obtained is an example of convenience sample.

Definition: Undercoverage

Undercoverage refers to the situation in which some eligible members of the population are excluded from the sampling frame, and thereby from the process of sampling.

Example 8.11

A researcher decides to conduct a survey about the intake of dietary supplement in a big residential area. He will use the list of all landline phone users in the area as a sampling frame. Naturally, the residence without a landline phone or only users of a cell phone will be ignored in the process of sampling. This design is subject to an undercoverage.

Definition: Nonresponse

Nonresponse refers to a situation in which information from one or more sampling units are unavailable either because the individuals in the sample refuse to respond to the survey question or cannot be contacted due to time constraint, lack of personnel or other reasons.

Example 8.12

Suppose a research scientist mails a questionnaire to 500 randomly selected patients as a part of the follow-up study. In one month, responses were obtained from 350 patients. This situation is an example of a nonresponse.

Definition: Sampling Error

Sampling error occurs due to the use of the sample and is the difference of a sample result from the corresponding true population result.

Example 8.13

Suppose a sample of size 10 from a normal distribution with mean $\mu = 2.5$ and standard deviation $\sigma = 0.5$ reveals that the sample mean $\bar{x} = 2.13$. Then the sampling error associated with this sample is

$$\mu - \bar{x} = 2.5 - 2.13 = 0.37$$

Section 8.1: Practice problems

Every three months, a doctor sends survey questionnaires to a randomly selected 100 patients out of all eligible patients treated by the doctor, to better understand the treatment outcome. In one occasion, 78 patients sent back the survey questionnaires to the doctor with the information sought for.
Answer questions (8.1)-(8.3) using above scenario.

(8.1) The population in this problem is
- (a) 100 patients who received survey questionnaires.
- (b) 78 patients who returned the survey questionnaires.
- (c) all eligible patients treated by the doctor.
- (d) all visited the doctor office.

(8.2) The sample in this problem is
- (a) 100 patients who received survey questionnaires.
- (b) 78 patients who returned the survey questionnaires.
- (c) all eligible patients treated by the doctor.
- (d) all visited the doctor office.

(8.3) The sample is subject to

(a) a bias.
(b) nonresponse.
(c) an undercoverage.
(d) none of the above.

(8.4) In order to estimate the average cost of education in the United States, a sample is intended from all undergraduate students in the United States. To this end, a group of researchers select a simple random sample of five states. From each of these states, they take a simple random sample of three colleges or universities. Finally, from each of these fifteen colleges or universities, they select a simple random sample of 30 undergraduates. The final sample consists of 450 undergraduate students. This sample is an example of
(a) simple random sample.
(b) multistage sample.
(c) stratified random sample.
(d) cluster sample.

(8.5) A college is gathering opinion about the on-campus student parking problem. It may be impractical to contact all students. By a notice, the university is decided to collect information from whoever chooses to provide opinion by personal contact. A sample to be collected this way is an example of a
(a) simple random sample.
(b) biased sample.
(c) voluntary response sample.
(d) both (b) and (c).

(8.6) As a part of a student research project, a graduate student is planning to collect information about the students' use of library by asking questions to whoever he meets in the library on a given day. The sample to be obtained this way is an example of
(a) simple random sample.
(b) voluntary response sample.
(c) convenience sample.
(d) cluster sample.

(8.7) A researcher decides to conduct a survey in a large residential area about residents' Facebook usage. He will use the list of all landline phone users living in that area as a sampling frame. It is expected that sampling this way will miss out those without a phone or users of cell phone. A sample to be drawn this way must be subject to a
(a) Nonresponse.
(b) Undercoverage.
(c) Sampling error.
(d) All of the above.

(8.8) A physician mails a survey form to 150 randomly selected patients as a part of the follow-up study. It turns out that responses were obtained from 130 patients. This situation must be an example of a
(a) non-response.
(b) Undercoverage.

(c) sampling error.
(d) none of the above.

(8.9) Political analysts believe that Race and Ethnicity is going to be a big factor in the upcoming presidential election. Using a call-in-poll survey conducted by a local radio station, an assessment of the belief is being made on the basis of 530 calls of volunteer radio listeners. The sampling design being used in this case is an example

(a) simple random sampling.
(b) stratified random sampling.
(c) multistage sampling.
(d) volunteer sampling.

(8.10) The average income of all faculty members in a university is unknown to students. Several students decided to carry out a sample survey to evaluate the average income of all faculty members of the university. They considered income information from SRSs of 40 Professors, 35 Associate Professors and 35 Assistant Professors. Finally, they combined income information from three SRSs to form a sample of 110 faculty members to be used to estimate the unknown average income of all faculty members. The final sample of size 110 is an example of

(a) simple random sampling.
(b) stratified random sampling.
(c) voluntary response sampling.
(d) convenience sampling.

(8.11) A student is interested in doing a research on "Campus Safety". He decides to interview whoever willing to give their opinion while leaving the central library between 6:00pm and 9:pm. The student decided to interview users of library on three consecutive days of his choice. The sampling design of this problem would be an example of

(a) simple random sampling.
(b) convenience sampling.
(c) volunteer sampling.
(d) both (b) and (c) above.

(8.12) A campus organization in a large university has 320 student members and 45 faculty members. They take a random sample of 10 faculty members and 45 student members for a study. The combined sample of 55 members in the study is an example of a

(a) simple random sample.
(b) stratified random sample.
(c) volunteer response sample.
(d) all of the above.

(8.13) How much trust and confidence do you have in our government handling of the immigration reform system? An analyst used an online system to get response from the people on this issue. During a certain period of time, 840 people participated with an answer of the type: great deal, fair amount, not very much, not at all, no opinion. The sampling design being used in this case is a

(a) simple random sampling.

(b) stratified random sampling.
(c) multistage sampling.
(d) volunteer sampling.

(8.14) In order to sample from all High Schools in Michigan Pittsfield Charter Township, a simple random sample of two schools are considered and then all students in these two schools are considered for an interview as a part of the study. Then all students in these two high schools constitute a
(a) simple random sample.
(b) stratified random sample.
(c) cluster sample.
(d) multistage sample.

(8.15) A simple random sample of 10 female professors and a simple random sample of 15 male professors are selected by a proportional allocation (proportional allocation roughly means take more sample from larger group) from a university. The two samples are combined together to form an overall sample of 25 professors for a study. The overall sample forms a
(a) cluster sample.
(b) stratified random sample.
(c) volunteer response sample.
(d) all of the above.

Section 8.2: Sampling distribution of sample mean

The main objective of this section is to address the probability distribution of all possible sample means of a given size, called the ***sampling distribution of sample mean***. In particular, we will learn about sampling distribution of sample mean, central limit theorem for sample mean, and related applications.

Let us start with a brief review of terminologies such as population, parameter, sample and statistic as they are related to the sampling distribution of mean.

Population and Parameter

Population is the collection of all individuals of a study. A ***parameter*** is any descriptive measure of a population characteristic. For example, a population could be all students in a college, and their average height (μ) and standard deviation of height (σ) are parameters.

For a population of a set of N values $\{X_1, X_2, \dots, X_N\}$, the population mean (μ) and population variance (σ^2) are defined respectively by

$$\mu = \frac{X_1 + X_2 + \cdots + X_N}{N} = \frac{\sum X}{N}$$

$$\sigma^2 = \frac{\sum (X - \mu)^2}{N}$$

Generally speaking, the parameters μ and σ^2 are unknown and estimated from a sample.

Sample and statistic

A ***sample*** is a representative part of the population. A ***statistic*** is any descriptive measure of a sample characteristic. For example, 50 students selected at random from all students of a college is a sample, and their mean height ($\bar{x}$) and standard deviation of height (s) are statistics.

For a sample of n values $\{x_1, x_2, \dots, x_n\}$, the sample mean and sample variance, denoted by $\bar{x}$ and s^2, respectively are defined by

$$\bar{x} = \frac{x_1 + x_2 + \cdots + x_n}{n} = \frac{\sum x}{n}$$

$$s^2 = \frac{\sum (x - \bar{x})^2}{n - 1}$$

Given a sample, the statistics $\bar{x}$ and s^2 are known and used to estimate the parameters μ and σ^2, respectively.

Example 8.14

Given a population of 4 natural numbers $\{1,2,3,4\}$,

(a) Compute population mean (μ) and population variance (σ^2).
(b) Write out all possible samples of size 2 with replacement.
(c) Compute sample means of all possible samples in (b).
(d) Compute the mean of the population of sample means, denoted by $\mu_{\bar{x}}$.
(e) Compute the variance of the population of sample means, denoted by $\sigma_{\bar{x}}^2$.
(f) Compute σ^2/n.
(g) What connection can you make by comparing results in (a) and (d)?
(h) What connection can you make by comparing results in (e) and (f)?

Solution

(a) The population mean is

$$\mu = \frac{1 + 2 + 3 + 4}{4} = \frac{10}{4} = 2.5$$

The population variance is

$$\sigma^2 = \frac{(1 - 2.5)^2 + (2 - 2.5)^2 + (3 - 2.5)^2 + (4 - 2.5)^2}{4} = \frac{5}{4} = 1.25$$

(b) From a population of 4 numbers, we get a total of $2^4 = 16$ samples of size 2, with replacement. The 16 samples, together with their serial numbers (Sample #) and means (Mean) appear in table below:

Sample #	Sample	Mean	Sample #	Sample	Mean
1	1,1	1	9	3,1	2
2	1,2	1.5	10	3,2	2.5
3	1,3	2	11	3,3	3
4	1,4	2.5	12	3,4	3.5
5	2,1	1.5	13	4,1	2.5
6	2,2	2	14	4,2	3
7	2,3	2.5	15	4,3	3.5
8	2,4	3	16	4,4	4

(c) The sample means are reported in above table.

(d) Let $\mu_{\bar{x}}$ denote the mean of the population of all possible sample means.

Then, $\mu_{\bar{x}} = \frac{1+1.5+2+2.5+\cdots+2.5+3+3.5+4}{16} = \frac{40}{16} = 2.5$

(e) Let $\sigma_{\bar{x}}^2$ be the variance of the population of all possible sample means. Then,

$$\sigma_{\bar{x}}^2 = \frac{(1-2.5)^2 + (1.5-2.5)^2 + (2-2.5)^2 + \cdots + (3.5-2.5)^2 + (4-2.5)^2}{16}$$

$$= \frac{10}{16} = 0.625$$

(f) $\frac{\sigma^2}{n} = \frac{1.25}{2} = 0.625$

(g) By comparing results in (a) and (d), it follows that:

$$\mu = 2.5 = \mu_{\bar{x}}$$

(h) By comparing results in (e) and (f), it follows that

$$\sigma_{\bar{x}}^2 = 0.625 = \frac{\sigma^2}{n}$$

Note that 16 samples of size 2 are equally probable and thus they represent SRSs. Also, note that 16 samples correspond to 16 sample means, and thus "sample" varies and so does the "sample mean". Therefore, **the sample mean is a variable**.

Now, that we know that the sample mean is a variable, take a moment and think about answering the following four questions:

(1) What should we call the collection of all possible sample means?
(2) What is the distribution of all possible sample means?
(3) What is the mean ($\mu_{\bar{x}}$) of all possible sample means?
(4) What is the variance $(\sigma_{\bar{x}}^2)$ of all possible sample means?

Of course,

(1) By the definition of a population, the collection of all possible sample means is best described by saying "**the population of sample means**".
(2) The distribution of the population of sample means is called "**the sampling distribution of mean**".
(3) The mean of all possible sample means ($\mu_{\bar{x}}$) is the same as the population mean (μ), that is:

$$\mu_{\bar{x}} = \mu$$

(4) The variance of all possible sample means $(\sigma_{\bar{x}}^2)$ is the variance of the population variance (σ^2) divided by the sample size n, that is:

$$\sigma_{\bar{x}}^2 = \frac{\sigma^2}{n}$$

The last two expressions or equations are what we have evidenced from Example 8.14 and are true for any population and SRSs.

A histogram of 16 sample means along with a superimposed normal curve with mean $\mu = 2.5$ and standard deviation $\sigma_{\bar{x}} = \sqrt{0.625} = 0.791$ appears in Figure 8.1.

Figure 8.1: Histogram of sample means, and a normal N(2.5, 0.791) distribution superimposed

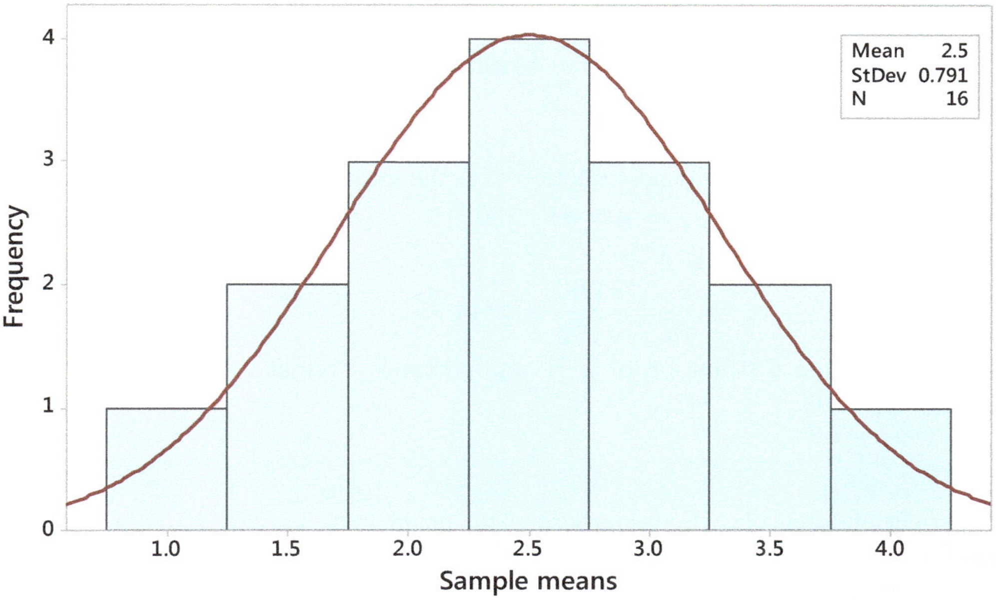

The histogram in Figure 8.1 is the sampling distribution of the sample mean which tells what different values the variable $\bar{x}$ takes in all possible samples of the same size. We also see that the histogram is symmetric and is approximated by a normal curve with mean $\mu = 2.5$ and standard deviation $\sigma_{\bar{x}}$.

Properties of the sampling distribution of sample mean

What we have observed in the histogram of Figure 8.1 are called properties of $\bar{x}$, which we can summarize as follows:

Given a population with mean μ and standard deviation σ,

(1) *the mean and standard deviation of the sampling distribution of the sample mean $\bar{x}$ over all possible samples of size n are given by*

$$\mu_{\bar{x}} = \mu$$
$$\sigma_{\bar{x}} = \sigma/\sqrt{n}$$

(2) *if the sample size n is large, the sampling distribution of the sample mean is approximately normal $N\left(\mu, \frac{\sigma}{\sqrt{n}}\right)$. This property is called the* ***central limit theorem*** *(CLT) for the sampling distribution of the sample mean. In general, n is considered large if $n \geq 30$.*

Example 8.15

According to a recent study, the average payment for a vaginal childbirth in the USA in 2010 was \$12,520. Suppose that the payment for a vaginal childbirth varies with a standard deviation of \$3,000. We randomly select 150 payments from the population of all vaginal childbirth in the USA in 2010. Let $\bar{x}$ represent the sample mean of 150 payments of vaginal childbirth.

(a) What are the mean and standard deviation of the distribution of $\bar{x}$?

(b) What is the sampling distribution of $\bar{x}$?

Solution

(a) Following the property of the sampling distribution of $\bar{x}$, the mean of $\bar{x}$ is

$$\mu_{\bar{x}} = \mu = \$12{,}520$$

The standard deviation of the average $\bar{x}$ is

$$\sigma_{\bar{x}} = \frac{\sigma}{\sqrt{n}} = \frac{3{,}000}{\sqrt{150}} \cong \$244.95$$

(b) By the CLT, the sampling distribution of $\bar{x}$ is approximately normal with mean \$12,520 and standard deviation \$244.95.

Example 8.16

A recent study reveals that children classified appropriate for gestational age (AGA) have mean birth weight of 3503 grams with a standard deviation of 358 grams. A sample of 40 AGA children is considered and let $\bar{x}$ represent their mean birth weight.

(a) What are the mean and standard deviation of the sampling distribution of $\bar{x}$?

(b) What is the sampling distribution of $\bar{x}$?

(c) What makes your answer in (b) justifiable?

(d) Compute $P(\bar{x} > 3500)$.

Solution

(a) The mean of the sampling distribution of $\bar{x}$ is

$$\mu_{\bar{x}} = \mu = 3503\ grams$$

The standard deviation of the sampling distribution of $\bar{x}$ is

$$\sigma_{\bar{x}} = \frac{\sigma}{\sqrt{n}} = \frac{358}{\sqrt{40}} = 56.6$$

(b) The sampling distribution of $\bar{x}$ is approximately normal $N(3503, 56.6)$.

(c) The answer in (b) is justifiable by the CLT because the sample size is large ($n = 40$).

(d) $P(\bar{x} > 3500) = P\left(z > \frac{3500-3503}{56.6}\right) = P(z > -0.05) = 1 - 0.4801 = 0.5199.$

Sampling distribution of $\bar{x}$ for a normal population

The CLT requires that the sample size be large for the distribution of $\bar{x}$ to be approximately normal; nothing is required about the distribution of the population the samples come from.

What happens if the population the samples come from is normal?

The answer is that we no longer require the "large sample" condition for the distribution of $\bar{x}$ to be normal! The following statement best specifies the sampling distribution of $\bar{x}$ in this situation:

Given a population that is normal with mean μ and finite standard deviation σ, the sampling distribution of all possible sample means $\bar{x}$ of size n, is exactly normal $N\left(\mu, \frac{\sigma}{\sqrt{n}}\right)$.

Example 8.17

A recent study reveals that for children classified smaller for gestational age (SGA), the mean onset of rolling is 5.1 months with a standard deviation of 2.6 months. Assume that the time it takes in month (x) for the onset of rolling for the population of SGA children is approximately normal. A researcher takes a sample of 20 SGA children and let $\bar{x}$ be the mean onset of rolling.

(a) What are the mean and standard deviation of the sampling distribution of $\bar{x}$?
(b) What is the sampling distribution of $\bar{x}$?
(c) Compute $P(x > 4)$.
(d) Compute $P(\bar{x} > 4)$.

Solution

(a) The mean of the sampling distribution of $\bar{x}$ is

$$\mu_{\bar{x}} = \mu = 5.1 \text{ months}$$

The standard deviation of the sampling distribution of $\bar{x}$ is

$$\sigma_{\bar{x}} = \frac{\sigma}{\sqrt{n}} = \frac{2.6}{\sqrt{20}} = 0.58 \text{ months}$$

(b) The sampling distribution of $\bar{x}$ is approximately normal $N(5.1, 0.58)$.

(c) $P(x > 4) = P\left(z > \frac{4-5.1}{2.6}\right) = P(z > -0.42) = 1 - 0.3372 = 0.6628$

(d) $P(\bar{x} > 4) = P\left(z > \frac{4-5.1}{0.58}\right) = P(z > -1.90) = 1 - 0.0287 = 0.9713$

Law of large number (LLN)

Let $\bar{x}$ be the mean of an SRS of size n from a population with unknown mean μ. As the size of the sample n increases, the sample mean $\bar{x}$ gets closer and closer to the population mean μ.

Example 8.18

Suppose a sample of size 10 from a normal distribution with mean $\mu = 2.5$ and standard deviation $\sigma = 0.5$ reveals that the sample mean $\bar{x} = 2.13$. The law of large number says that as n gets higher and higher, the mean $\bar{x}$ gets closer and closer to $\mu = 2.5$.

To justify the LLN, we select various samples for selected size n and compute corresponding mean as reported in table below and plotted in a scatterplot in Figure 8.2.

Sample size (n)	10	25	50	100	200	300	400	500	1000	1500	2000
Sample mean ($\bar{x}$)	2.2	2.5	2.51	2.44	2.46	2.46	2.55	2.52	2.51	2.52	2.5

Figure 8.2: Scatterplot showing the effect of sample size on sample means

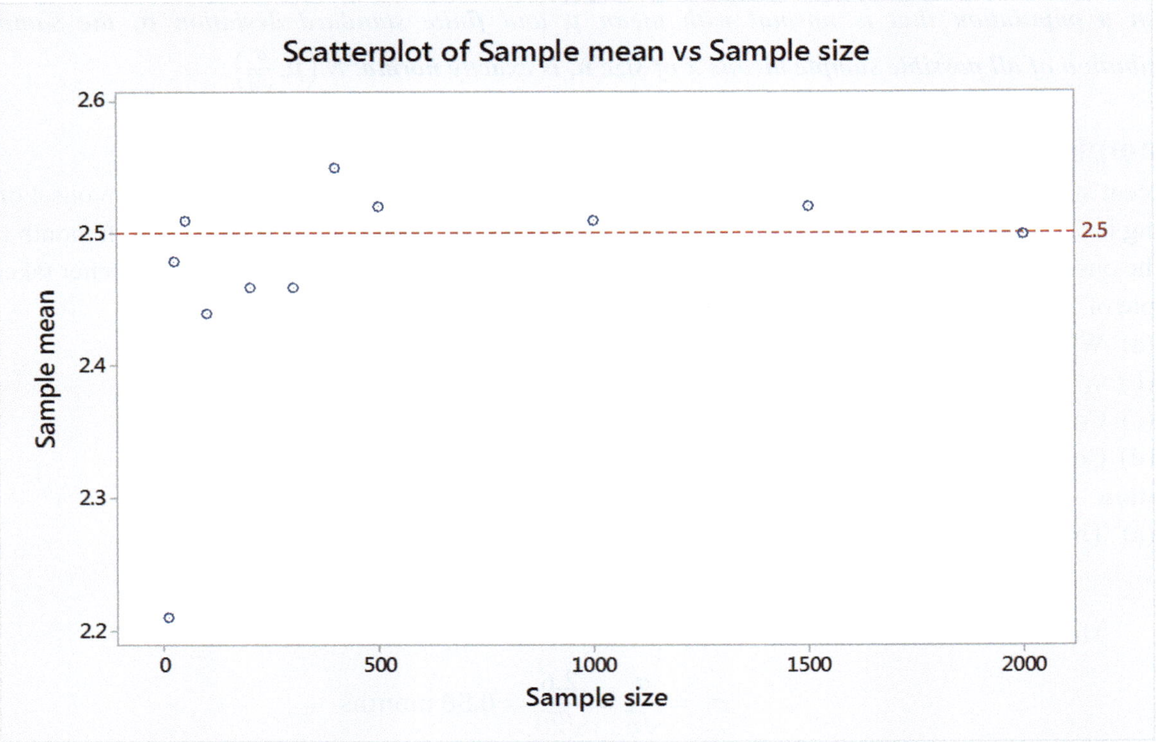

As we see in the table and the scatterplot, as n goes higher, the sample mean gets closer to the population mean. However, it should be noted here that any sample mean is subject to the sampling error. That is why we should not expect sample mean to be exactly equal to the population mean for any sample size. Of course, we could expect the sample mean to be closer to the population mean for large n.

Section 8.2: Practice problems

(8.16) The sample mean $\overline{x}$ has approximately a normal distribution because of

(a) the central limit theorem.
(b) the law of large number.
(c) the empirical rule of a normal distribution.
(d) the normal population.

(8.17) The sample mean $\overline{x}$ has an exact normal distribution if

(a) the sample size is large.
(b) the empirical rule applies to the sample mean.
(c) the population the samples come from is normal.
(d) the central limit theorem can be applied to the mean.

(8.18) Using a sample survey of 40 students, it appears that, on average, they spend 38 hours per month using mobile internet on Smartphone. Which of the followings is true with respect to this scenario?

(a) 40 is the size of the population being studied.
(b) 38 is a statistic that represents an estimate of the unknown value of a parameter of interest.
(c) 38 is a parameter that represents an estimate of the unknown value of a statistic of interest.

(8.19) The sampling distribution of a statistic is

(a) the probability the statistic takes in repeated samples.
(b) for which the histogram looks like normal.
(c) the probability distribution of the population of the statistic of the same size from the same population.
(d) the extent to approximation to the normality.

(8.20) According to a recent study, the average payment for a vaginal childbirth in the USA in 2010 was \$12,520. Suppose that the payment for a vaginal childbirth varies with a standard deviation of \$3,000. We randomly select 150 payments from the population of all vaginal childbirth in the USA in 2010. Let $\bar{x}$ represent the sample mean of 150 payments for a vaginal childbirth. Find the mean and standard deviation of the distribution of $\bar{x}$.

(8.21) In reference to the problem **(8.20)**, find the probability that the mean sample payment is less than \$12,000.

(a) 0.5279.
(b) 0.0169.
(c) 0.1894.
(d) 0.4721.

(8.22) The finishing time of a quiz is N(15,5). A sample of 10 students is considered and sample mean of sample is computed. What is the sampling distribution of the sample mean $\bar{x}$?

(a) is exactly N(15, 5).
(b) is approximately N(15, 5).
(c) is exactly N(15, 1.58).
(d) is approximately N(15, 1.58).

(8.23) A manufacturing process produces bags of chips whose weight is N(16 oz, 1.5 oz). On a given day, the quality control officer takes a sample of 45 bags and computed the mean weight of these bags. What is standard deviation of $\bar{x}$?

(a) 0.224.
(b) 0.444.
(c) 0.664.
(d) 0.884.

(8.24) For question in (8.23), what is the probability that the sample mean weight is below 16 oz?

(a) 0.1000
(b) 0.3000
(c) 0.5000

(d) 0.7000

(8.25) The central limit theorem says that when a sample of size n is drawn from any population with mean μ and standard deviation σ, then for any large n

(a) the standard deviation of the sample mean is σ^2/n.
(b) the distribution of the population is approximately Normal.
(c) the distribution of the sample mean is approximately Normal.
(d) the distribution of the sample mean is exactly Normal.

(8.26) Suppose a large population has mean μ and standard deviation σ, and a sample of size n with replacement is taken. The sampling distribution of the sample mean $\bar{x}$ has mean and variance respectively equal to

(a) μ/n and σ^2/n .
(b) μ and σ/n.
(c) μ/n and σ^2/n^2.
(d) μ and σ^2/n.

(8.27) The average exam score of students of a large class is 70 with a standard deviation of 10. A sample of 35 students is selected, and the mean score $\bar{x}$ of these students is computed. The sampling distribution of the sample mean $\bar{x}$ has approximately a normal distribution because of

(a) the central limit theorem.
(b) the 68–95–99.7 rule.
(c) the population we're sampling from has a Normal distribution.
(d) all of the above.

(8.28) In question **(8.27)**, the probability that the mean score $\bar{x}$ of 35 students is more than 73 is (choose the closest possible answer)

(a) 0.958.
(b) 0.058.
(c) 0.038.
(d) 0.068.

(8.29) A recent study reveals that, on average, US adults spend 27 hour per month on PC internet. Suppose, the time that US adults spend on PC internet follows approximately a normal distribution with a mean of 27 hours per month and a standard deviation of 10 hours per month. A sample of 100 US adults is considered. The sampling distribution of the sample mean $\bar{x}$ is

a) exactly Normal, mean 27, standard deviation 10.
b) exactly Normal, mean 27, standard deviation 1.
c) approximately Normal, mean 27, standard deviation 10.
d) approximately Normal, mean 27, standard deviation 10.

(8.30) In reference to problem (8.29), find the probability that mean $\bar{x}$ is greater than 28.

(8.31) The law of large number states that as the sample size gets larger and larger, the value of the sample mean $\bar{x}$

(a) gets larger and larger.
(b) gets smaller and smaller.
(c) gets closer and closer to the population mean μ.
(d) takes the same value as the population mean μ.

Section 8.3: Sampling distribution of sample proportion

Given a population, the proportion of subjects with a particular characteristic is called the **population proportion**. For example, the proportion of people with salmonella infection in the USA, proportion of people buying a dietary supplement in a big city, proportion of smokers in a university, etc. are examples of population proportions.
In general, the population proportion is unknown, and given a sample, we wish to estimate the unknown population proportion. In order to estimate the proportion, we need a sample of subjects and count how many of the subjects in the sample possess the characteristic of interest.

Let p be the unknown population proportion with a specified characteristic.

Definition: Sample proportion

$\hat{p} = \frac{x}{n}$ is the sample proportion or a point estimate of p, where n is the sample size and x is the number of individuals in the sample with a specified characteristic.

Example 8.19

Among the included sources, shell eggs are responsible for the highest proportion of foodborne Salmonella serotype Enteritidis (SE) infections during the baseline period (2007-2009). In a sample of 83 SE infections, 34 SE infections are due to the shell eggs. Then, the sample proportion of SE infections attributable to shell eggs is

$$\hat{p} = \frac{34}{83} = 0.41$$

Source: Due to a CDC study: http://www.cdc.gov/foodsafety/ifsac/projects/completed.html.

Sampling distribution of sample proportion

Generally, the sample proportion $\hat{p}$ varies from sample to sample, and therefore, the sample proportion $\hat{p}$ is a random variable. Then, what would be probability distribution of sample proportions over all possible samples?
Let us consider the following example to illustrate the idea a little more.

Example 8.20

As per CDC, the most common source of SE infection is shell eggs. Other sources include raw milk, pork, beef, sprouts, raw almonds, international travel, contact with reptiles, etc. Given a population of SE infection, suppose that each person has a probability $p = 0.5$ of having an SE infection.

Suppose that four people are diagnosed with SE infection in a large community. We want to know the source of SE infection of these four patients. Particularly, we would like to estimate the proportion of SE infection due to shell eggs. By an inspection, we assign an S to a patient if the source is shell egg and an N if the source is not a shell egg.

Note that every outcome of the inspection represents a sample of size $n = 4$, and we will get a total of $2^4 = 16$ possible samples. Consider estimating the proportion of SE infection for each of 16 possible samples. Table below provides sample proportions ($\hat{p}$) from 16 samples, a histogram of $\hat{p}$ and a normal curve with mean 0.5 and standard deviation 0.25 superimposed on the histogram.

Sample #	Sample	$\hat{p}$	Sample #	Sample	$\hat{p}$
1	NNNN	0.00	9	SNNS	0.50
2	SNNN	0.25	10	NSSN	0.50
3	NSNN	0.25	11	SSNN	0.50
4	NNSN	0.25	12	SSSN	0.75
5	NNNS	0.25	13	SSNS	0.75
6	NNSS	0.50	14	SNSS	0.75
7	NSNS	0.50	15	NSSS	0.75
8	SNSN	0.50	16	SSSS	1

Figure 8.3: Histogram of sample proportions with a N(0.5, 0.25) distribution superimposed

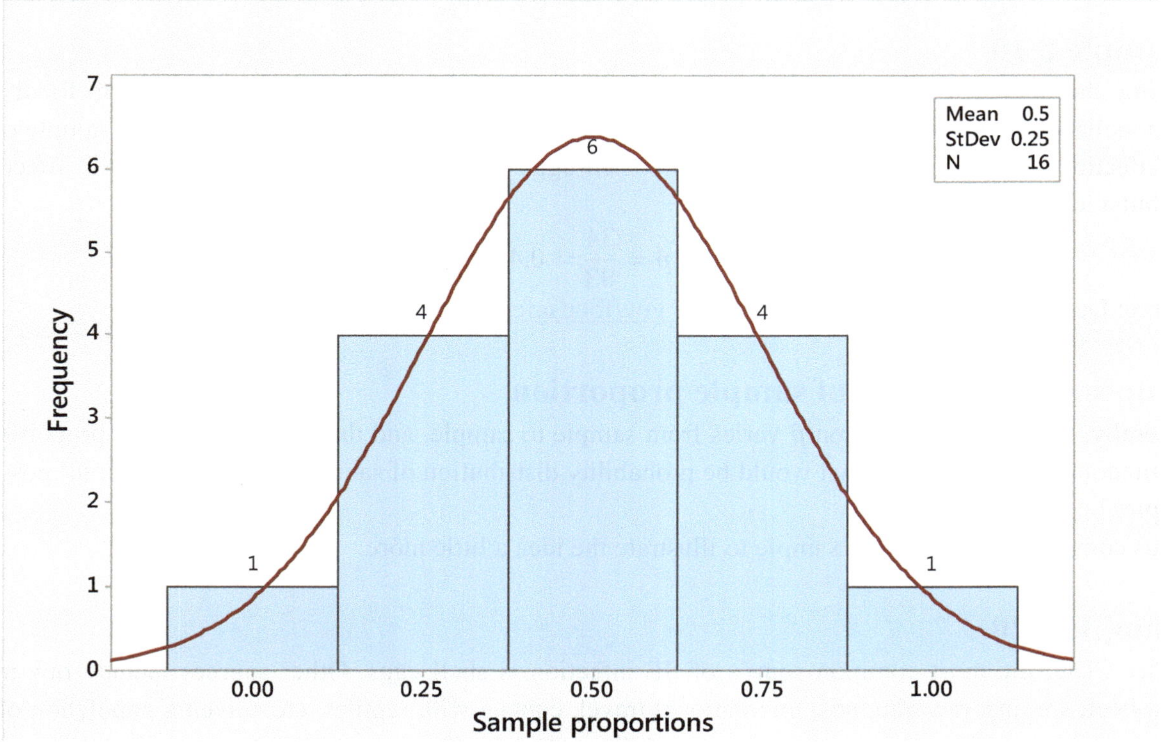

Looking at the table and histogram in Figure 8.3, it is evident that

(1) $\hat{p}$ varies from sample to sample and hence $\hat{p}$ is a random variable.

(2) The histogram of $\hat{p}$ is symmetric and it represents the distribution of $\hat{p}$.
(3) The normal curve approximates the histogram of $\hat{p}$.

Definition: Sampling distribution of sample proportion

The probability distribution of $\hat{p}$ over all possible samples of a given size is called the **sampling distribution of $\hat{p}$**. In other words, the sampling distribution of $\hat{p}$ is the probability distribution of the population of $\hat{p}$.

In order to understand some important properties of the sampling distribution of the sample proportion, let us consider the example below:

Example 8.21

Use 16 sample proportions of Example 8.20 and answer questions (a)-(e) below.

For answering questions (a)-(e), wherever needed, use the fact that each person has a probability $p = 0.5$ of having an SE infection in the population.

Sample #	Sample	$\hat{p}$	Sample #	Sample	$\hat{p}$
1	NNNN	0.00	9	SNNS	0.50
2	SNNN	0.25	10	NSSN	0.50
3	NSNN	0.25	11	SSNN	0.50
4	NNSN	0.25	12	SSSN	0.75
5	NNNS	0.25	13	SSNS	0.75
6	NNSS	0.50	14	SNSS	0.75
7	NSNS	0.50	15	NSSS	0.75
8	SNSN	0.50	16	SSSS	1

(a) Compute mean of 16 sample proportions ($\mu_{\hat{p}}$).
(b) Establish a relationship between your answer in (a) and the value of p.
(c) Compute variance of 16 sample proportions ($\sigma_{\hat{p}}^2$).
(d) Establish a relationship between your answer in (c) and the value of p.
(e) What formulas do you seem to have from the results of (b) and (d)?

Solution

(a) The mean of 16 sample proportion is

$$\mu_{\hat{p}} = \frac{0+4\times0.25+6\times0.50+4\times0.75+1}{16} = 0.50$$

(b) By comparing result of (a) and the value of p, it follows that $\mu_{\hat{p}} = 0.5 = p$
(c) The variance of the 16 sample proportions is

$$\sigma_{\hat{p}}^2 = \frac{(0-0.5)^2+4\times(0.25-0.50)^2+6\times(0.50-0.50)^2+4\times(0.75-0.5)^2+(1-.5)^2}{16} = 0.0625$$

(d) By comparing result of (c) and the value of p, it follows that

$$\sigma_{\hat{p}}^2 = 0.0625 = \frac{0.5 \times (1-0.5)}{4} = \frac{p \times (1-p)}{n}$$

(e) The results of (b) and (d) lead to the following formulas

$$\mu_{\hat{p}} = p$$
$$\sigma_{\hat{p}}^2 = \frac{p \times (1-p)}{n}$$

The formula established in Example 8.21 hold for any value of n and p, and are called the properties of the sampling distribution of $\hat{p}$.

Properties of the sampling distribution of $\hat{p}$

For a population with proportion p

(1) *the mean and standard deviation of the sampling distribution of* $\hat{p}$ *over all possible samples of size n are given by*

$$\mu(\hat{p}) = p$$
$$\sigma(\hat{p}) = \sqrt{\frac{p \times (1-p)}{n}}$$

(2) *Sampling distribution of* $\hat{p}$ *is approximately normal with mean* $\mu(\hat{p}) = p$ *and standard deviation* $\sigma(\hat{p}) = \sqrt{\frac{p \times (1-p)}{n}}$ *provided* $np \geq 10$ *and* $n(1-p) \geq 10$. *This property is called the* ***central limit theorem*** *(CLT) for the sampling distribution of the sample proportion.*

Standard error (S.E.) of $\hat{p}$

In sampling distribution terminology, the standard deviation of any statistic is called the standard error of the statistic. The standard error (S.E.) of $\hat{p}$ is the same as the standard deviation of $\hat{p}$. That is,

$$\sigma(\hat{p}) = S.E.(\hat{p}) = \sqrt{\frac{p \times (1-p)}{n}}$$

When p is unknown, which is true in most cases, an estimate of $\sigma(\hat{p})$ is

$$\hat{\sigma}(\hat{p}) = s.e.(\hat{p}) = \sqrt{\frac{\hat{p} \times (1-\hat{p})}{n}}$$

Example 8.22

According to CDC, of all single etiology foodborne disease outbreaks, 42% is due to the bacterial infection. Let a sample of 49 patients with foodborne illness be considered, and let $\hat{p}$ be the sample proportion with a bacterial infection.

Find the mean and standard deviation of $\hat{p}$.

Verify that the conditions of the normality of $\hat{p}$ by the CLT are satisfied.

What is the sampling distribution of $\hat{p}$?

Solution

(a) The population proportion of bacterial foodborne illness is $p = 0.42$ and the sample size is $n = 49$. Therefore,

$$\mu_{\hat{p}} = p = 0.42$$

$$\sigma_{\hat{p}} = \sqrt{\frac{0.42\times(1-0.42)}{49}} = 0.0705$$

(a) Note that $np = 49 \times 0.42 = 20.6 > 10$ and $n(1-p) = 49 \times (1-0.42) = 28.4 > 10$. Thus the conditions of CLT are satisfied.

(b) By the sampling distribution of $\hat{p}$, $\hat{p}$ is approximately normal with mean 0.42 and standard deviation 0.0705.

Example 8.23

A recent CDC study shows that 46% of children identified with autism spectrum disorder (ASD) had average or above average intellectual ability (IQ greater than 85). Suppose a sample of 30 ASD identified children is considered and let $\hat{p}$ be the proportion of children in the sample with an average or above average intellectual ability.

(a) Find the mean and standard deviation of $\hat{p}$.
(b) What is the sampling distribution of $\hat{p}$?
(c) What is the probability that the sample proportion of ASD identified children is greater than 0.46?
(d) What is the probability that the sample proportion of ASD identified children is greater than 0.80?

Solution

Given, the population proportion of ASD identified children is $p = 0.46$ and the sample size is $n = 30$.

(a) $\mu_{\hat{p}} = p = 0.46$ and $\sigma_{\hat{p}} = \sqrt{\frac{0.46\times(1-0.46)}{30}} = 0.091$

(b) Note that $np = 30(0.46) = 13.8$ and $n(1-p) = 30(0.54) = 16.2$. Therefore, by the CLT, the sampling distribution of $\hat{p}$ is approximately normal with mean 0.46 and standard deviation 0.091.

(c) $P(\hat{p} > .46) = P\left(z > \frac{0.46-0.46}{0.091}\right) = P(z > 0) = 0.5.$

(d) $P(\hat{p} > .80) = P\left(z > \frac{0.80-0.46}{0.091}\right) = P(z > 3.74) = 0.00009.$

Example 8.24

According to a recent study, 85% (rounded to the nearest percentage) of all flights by Delta Airlines Inc. (DL) have on-time arrivals in April 2014.
SOURCE: http://www.transtats.bts.gov/OT_Delay/OT_DelayCause1.asp?pn=1; Bureau of Transportation Statistics, Airline Service Quality Performance 234.

Arrival time of a sample of 100 DL flights are considered. Let $\hat{p}$ be sample proportion of on-time arrivals.

(a) Find the mean and standard deviation of $\hat{p}$.
(b) Justify if the conditions of CLT applies to evaluate the sampling distribution of $\hat{p}$.
(c) What is the sampling distribution of proportions of DL on-time flights?
(d) What is the probability that the sample proportion of on-time arrival of DL flights would be greater than 0.75? Does this probability make sense to you?

(e) Is it unusual if the sample proportion of DL on-time arrival is less than 75% ?
(f) Is it unusual if the sample proportion of DL on-time arrival is less than 84% ?

Solution

Let $\hat{p}$ be the proportion of on-time arrivals of DL flights in April 2014.

(a) Then the mean and standard deviation of $\hat{p}$ are:

$$\mu_{\hat{p}} = p = 0.85 \text{ and } \sigma_{\hat{p}} = \sqrt{\frac{0.85\times(1-0.85)}{100}} = 0.0357.$$

(b) We have: $np = 100(0.85) = 85 > 10$ and $n(1-p) = 100(0.15) = 15 > 10$. Thus the conditions of CLT are satisfied for the sampling distribution of $\hat{p}$.

(c) By the CLT, the sampling distribution of $\hat{p}$ is approximately normal with mean 0.85 and standard deviation 0.0357.

(d) To find $P(\hat{p} > 0.75)$, we compute z score of 0.75 and use Table T3 to find the area to the right of it:

$$z = \frac{0.75 - 0.85}{0.0375} = -2.67$$

$$P(\hat{p} > .75) = P(z > -2.67) = 1 - 0.0038 = 0.9962$$

Since the distribution of $\hat{p}$ is approximately normal with a mean of 0.85, we expect $P(\hat{p} > .85) = 0.50$. In other words, we expect $P(\hat{p} > .75) > 0.5$, and indeed we have $P(\hat{p} > .75) = 0.9962$, which makes sense.

(e) In answer to the question (d) we found that $P(\hat{p} > .75) = 0.9962$. Then,

$$P(\hat{p} < .75) = 1 - P(\hat{p}(> 0.75) = 1 - 0.9962 = 0.0038$$

By the similar argument of (d), we expect $P(\hat{p} \leq .85) = 0.5$. In other words, we expect $P(\hat{p} < .75) < 0.5$, and indeed we have $P(\hat{p} < .75) = 0.0038 < 0.5$, which makes sense.

(f) To find $P(\hat{p} < .84)$, we compute z score of 0.84 and use Table T3 to find the area to the left of it:

$$z = \frac{0.84 - 0.85}{0.0375} = -0.27$$

$P(\hat{p} < -0.27) = 0.3936$, which is close to 0.5, and it makes sense.

Section 8.3: Practice problems

(8.32) Define sample proportion and the sampling distribution of the sample proportions.

(8.33) Suppose that in a sample of 45 subjects, 30 subjects recycle their household recyclable items. What is the sample proportion of subjects who recycle their household items?

(8.34) The sampling distribution of $\hat{p}$ is approximately normal with mean p and standard deviation $\sqrt{p(1-p)/n}$ provided_______________ and ____________________.

(8.35) If n is the size of a sample from a population with a proportion p, then the mean and standard deviation of the sampling distribution of $\hat{p}$ are _________________ and _________________.

(8.36) Let p be the population proportion of a certain characteristic, n be the sample size and $\hat{p}$ be the sample proportion. For each of the following situations, specify if the central limit theorem of sampling distribution of $\hat{p}$ is applicable or not:

(a) $n = 45, p = 0.25$
(b) $n = 100, p = 0.15$
(c) $n = 200, p = 0.5$

(8.37) Let p be the population proportion of a certain characteristic, n be the sample size and $\hat{p}$ be the sample proportion. For each of the following situations, specify the sampling distribution of $\hat{p}$ and evaluate the indicated probability, if applicable:

(a) $n = 100, p = 0.25;\ P(\hat{p} > 0.35)$
(b) $n = 49, p = 0.50;\ P(\hat{p} > 0.65)$
(c) $n = 81, p = 0.25;\ P(\hat{p} > 0.35)$

(8.38) According to the National Institutes of Health (NIH), about 43% of the U.S. population use dietary supplements containing calcium. Let a sample of 45 U.S. people be considered, and let $\hat{p}$ be the sample proportion of people using dietary supplements containing calcium.

(a) Find the mean of $\hat{p}$.
(b) Find the standard deviation of $\hat{p}$.
(c) Can you apply central limit theorem of proportions to determine the sampling distribution of $\hat{p}$ for this problem?
(d) What is the sampling distribution of $\hat{p}$?
(e) What is the probability that the proportion of people in the sample buying the dietary supplements is greater than 0.40?
(f) What is the probability that the proportion of people in the sample buying the dietary supplements is between 0.40 and 0.70?

Source: Based on NIH Calcium Dietary Supplement Fact Sheet.

(8.39) According to the NIH, about 70% of older women in the USA use dietary supplements containing calcium. Suppose a sample of 35 older women in the USA is considered, and let $\hat{p}$ be the sample proportion of older women who use dietary supplements containing calcium.

(a) Find the mean and standard deviation of $\hat{p}$.
(b) Verify how the conditions of central limit theorem of proportions are satisfied to determine the sampling distribution of $\hat{p}$.
(c) What is the sampling distribution of $\hat{p}$?
(d) What is the probability that the proportion of people in the sample using the dietary supplements is greater than 0.50?
(e) What is the probability that the proportion of people in the sample using the dietary supplements is between 0.50 and 0.85?

Source: Based on NIH Calcium Dietary Supplement Fact Sheet.

(8.40) According U.S. Census Bureau, 32% (rounded up to the nearest integer) Americans aged 25 years and over have a Bachelor's degree. Let $\hat{p}$ be the sample proportion of Bachelor's degree from a sample of 45 Americans aged 25 years and over.

(a) Find the mean of $\hat{p}$.
(b) Find the standard deviation of $\hat{p}$.
(c) Can you apply central limit theorem of proportions to determine the sampling distribution of $\hat{p}$?
(d) What is the sampling distribution of $\hat{p}$?
(e) Find the probability that the sample proportion of Americans aged 25 and over with a Bachelor's degree is above 0.30.
(f) What is probability that the sample proportion of Americans aged 25 and over with a Bachelor's degree is between 0.30 and 0.50?

Source: U.S. Census Bureau, Educational Attainment 2013.

(8.41) A recent study shows that 22% Americans aged 25 years and over are not in the civilian labor force in 2013. Let $\hat{p}$ be the sample proportion of Americans not in the civilian labor force in a sample of 50 Americans aged 25 years and over.

(a) Find the mean of $\hat{p}$.
(b) Find the standard deviation of $\hat{p}$.
(c) Justify if the sampling distribution of $\hat{p}$ is determinable.
(d) What is the sampling distribution of $\hat{p}$?
(e) Find the probability that the sample proportion of Americans aged 25 and over not in the labor force is less than 0.20.
(f) Find the probability that the sample proportion of Americans aged 25 and over not in the labor force is between 0.15 and 0.45.

Source: U.S. Census Bureau, Educational Attainment 2013.

(8.42) A recent study of Employee Job Satisfaction and Engagement by the Society for Human Resource Management (SHRM) shows that 30% of employees are very satisfied with the job security. Suppose the proportion of employees very satisfied with the job security, $\hat{p}$, is estimated from a sample of 45 employees.

(a) Find the mean of $\hat{p}$.
(b) Find the standard deviation of $\hat{p}$.
(c) Justify if the requirements of the CLT of the sampling distribution of $\hat{p}$ are met.
(d) Specify the sampling distribution of $\hat{p}$.
(e) Find the probability that the sample proportion of employees very satisfied with the job security is above 0.50.
(f) Find the probability that the sample proportion of employees very satisfied with the job security is less than 0.30.

Source: 2012 Employee Job Satisfaction and Engagement: A Research Report by SHRM

(8.43) A recent study by the Society for Human Resource Management reveals that 36% of employees agree that the job specific training is very important. Suppose the proportion of employees who agree that the job specific training is very important, $\hat{p}$, is estimated from a sample of 30 employees.

(a) Find the mean of $\hat{p}$.
(b) Find the standard deviation of $\hat{p}$.
(c) Justify if the requirements of the CLT of the sampling distribution of $\hat{p}$ are met.
(d) What is the sampling distribution of $\hat{p}$?

(e) Find the probability that the sample proportion of employees who agree that the job specific training is very important is less than 0.50.
(f) Find the probability that the sample proportion of employees who agree that the job specific training is very important is between 0.25 than 0.50.

Source: 2012 Employee Job Satisfaction and Engagement: A Research Report by SHRM

(8.44) What do you use your iPhone-for? Those who respond to this question, 61% uses iPhone for games, while 21% uses for shopping. Suppose a sample of 100 iPhone users are considered and let $\hat{p}$ be the sample proportion of iPhone users who use it games.

(a) Find the mean of $\hat{p}$.
(b) Find the standard deviation of $\hat{p}$.
(c) Justify the requirements of the sampling distribution of $\hat{p}$.
(d) What is the sampling distribution of $\hat{p}$?
(e) Find the probability that the sample proportion of people who use iPhone for games is less than 0.25.
(f) Find the probability that the sample proportion of people who use iPhone for games is greater than 0.25.
(g) Find the probability that the sample proportion of people who use iPhone for games is between 0.25 and 0.60.

Source: http://jamesdalesio.hubpages.com/hub/What-do-you-use-your-iPhone-for.

(8.45) What do you use your iPhone-for? Those who respond to this question, 49% uses iPhone for social networking. Suppose a sample of 100 iPhone users are considered and let $\hat{p}$ be the sample proportion of iPhone users who use it for social networking.

(a) Find the mean of $\hat{p}$.
(b) Find the standard deviation of $\hat{p}$.
(c) Justify if the sampling distribution of $\hat{p}$ is determinable.
(d) What is the sampling distribution of $\hat{p}$?
(e) Find the probability that the sample proportion of people using iPhone for social networking is more than 0.40.
(f) Find the probability that the sample proportion of people using iPhone for social networking is not more than 0.45.
(g) Find the probability that the sample proportion of people using iPhone for social networking is between 0.45 and 0.70.

Source: http://jamesdalesio.hubpages.com/hub/What-do-you-use-your-iPhone-for.

(8.46) According to 2014 Work Stress Survey, 80% Americans workers aged 18 years and over are stressed out at workplace. Suppose a sample of 100 American workers aged 18 and over is considered. Let $\hat{p}$ be the sample proportion of American workers stressed out at workplace.

(a) Find the mean of $\hat{p}$.
(b) Find the standard deviation of $\hat{p}$.
(c) What is the sampling distribution of $\hat{p}$?
(d) Find out the probability that the sample proportion of stressed out workers is between 0.75 and 0.85.

(e) What is the probability that the proportion of stressed out workers is above 0.80?

Source: 2014 Work Stress Survey

(8.47) According to a recent study, 83% (rounded to the nearest percentage) of all flights by US Airways Inc. (US) have on-time arrivals in April 2014. If arrival times of 100 US flights were sampled, then

(a) What is the sampling distribution of proportion of on-time arrival US flights?

(b) What is the probability that the sample proportion of on-time arrival US flights would be greater than 0.70? Does this probability make sense to you?

(c) What assumptions do you need to verify in order to determine the sampling distribution of proportion of on-time arrival US flights?

(d) Is it unusual if the sample proportion of on-time arrival US flights is less than 70% ?

(e) Is it unusual if the sample proportion of on-time arrival US flights is less than 82% ?

Source: Bureau of Transportation Statistics, Airline Service Quality Performance 234

(8.48) It appears that 85% (rounded to the nearest percentage) of all flights by AirTran Airways (FL) have on-time arrivals in April 2014. If arrival time of 100 FL flights were sampled, then

(a) What is the sampling distribution of proportions of on-time arrival FL flights?

(b) What is the probability that the sample proportion of on-time arrival FL flights would be greater than 0.70? Does this probability make sense to you?

(c) What assumptions do you need to verify in order to determine the sampling distribution of proportion of on-time arrival FL flights?

(d) Is it unusual if the sample proportion of on-time arrival FL flights is less than 70% ?

(e) Is it unusual if the sample proportion of on-time arrival FL flights is less than 82% ?

Source: Bureau of Transportation Statistics, Airline Service Quality Performance 234

Chapter 8: Answer to selected problems

Section 8.1

(8.1) (c); (8.2) (a); (8.3) (b); (8.4) (b); (8.5) (c); (8.6) (c); (8.7) (b); (8.8) (a); (8.9) (d); (8.10) (b); (8.11) (d); (8.12) (b); (8.13) (d); (8.14) (c); (8.15) (b).

Section 8.2

(8.16) (a); (8.17) (c); (8.18) (b); (8.19) (c)
(8.20) mean = \$12,520 and standard deviation = \$245; (8.21) (b); (8.22) (c); (8.23) (a); (8.24) (c) (8.25) (c); (8.26) (d); (8.27) (a); (8.28) (c); (8.29) (b); (8.30) 0.1587 (8.31) (c)

Section 8.3

(8.33) 0.667; **(8.34)** $np \geq 0$ and $n(1-p) \geq 0$; **(8.35)** p and $\sqrt{\frac{p(1-p)}{n}}$

(8.36) Verify that $np \geq 0$ and $n(1-p) \geq 0$. Yes, for (a)-(c), the CLT is applicable.

(8.37) For (a)-(c), $\hat{p}$ is $N\left(p, \sqrt{\frac{p(1-p)}{n}}\right)$.

(a) $P(\hat{p} > 0.35) = 0.0105$
(b) $P(\hat{p} > 0.65) = 0.0161$
(c) $P(\hat{p} > 0.35) = 0.019$

(8.38) (a) 0.43 (b) 0.074 (c) Yes (d) approximately normal $N(0.43, 0.074)$
(e) $P(\hat{p} > 0.4) = 0.6574$ (f) $P(0.40 < \hat{p} < 0.70) = 0.6573$

(8.39) (a) 0.7 and 0.0775 (b) Verify that $np \geq 0$ and $n(1-p) \geq 0$. (c) approximately normal $N(0.7, 0.0775)$ (d) $P(\hat{p} > 0.50) = 0.9951$ (e) $P(0.50 < \hat{p} < 0.85) = 0.9686$.

(8.40) (a) 0.32 (b) 0.0695 (c) Verify that $np \geq 0$ and $n(1-p) \geq 0$; Yes, we can apply CLT.
(d) approximately normal $N(0.32, 0.0695)$ (e) $P(\hat{p} > 0.30) = 0.6132$ (f) $(0.30 < \hat{p} < 0.50) = 0.6084$.

(8.41) (a) 0.22 (b) 0.059 (c) Verify that $np \geq 0$ and $n(1-p) \geq 0$; Yes, we can apply CLT.
(d) approximately normal $N(0.22, 0.059)$ (e) $P(\hat{p} < 0.20) = 0.3673$ (f) $(0.15 < \hat{p} < 0.45) = 0.8822$.

(8.42) (a) 0.30 (b) 0.0683 (c) Verify that $np \geq 0$ and $n(1-p) \geq 0$; Yes, we can apply CLT.
(d) approximately normal $N(0.30, 0.0683)$ (e) $P(\hat{p} > 0.50) = 0.0017$ (f) $(\hat{p} < 0.30) = 0.50$.

(8.43) (a) 0.36 (b) 0.0876 (c) Verify that $np \geq 0$ and $n(1-p) \geq 0$; Yes, we can apply CLT.
(d) approximately normal $N(0.36, 0.0876)$ (e) $P(\hat{p} < 0.50) = 0.945$ (f) $(0.25 < \hat{p} < 0.50) = 0.8404$.

(8.44) (a) 0.61 (b) 0.0487 (c) Verify that $np \geq 0$ and $n(1-p) \geq 0$; Yes, we can apply CLT.
(d) approximately normal $N(0.61, 0.0487)$ (e) $P(\hat{p} < 0.25) = 0.0000$ (f) $P(\hat{p} > 0.25) = 1.0000$ (f) $(0.25 < \hat{p} < 0.60) = 0.4187$.

(8.45) (a) 0.49 (b) 0.0499 (c) Verify that $np \geq 0$ and $n(1-p) \geq 0$; Yes, we can apply CLT. (d) approximately normal $N(0.49, 0.0499)$ (e) $P(\hat{p} > 0.40) = 0.9644$ (f) $P(\hat{p} < 0.45) = 0.2114$ (f) $(0.45 < \hat{p} < 0.70) = 0.7886$.

(8.46) (a) 0.80 (b) 0.04 (c) approximately normal $N(0.80, 0.04)$ (d) $(0.75 < \hat{p} < 0.85) = 0.7887$ (e) $P(\hat{p} > 0.80) = 0.50$.

(8.47) (a) approximately normal $N(0.83, 0.0375)$ (b) $P(\hat{p} > 0.70) = 0.9997$ (c) Verify that $np \geq 0$ and $n(1-p) \geq 0$ (d) May be because mean 0.83>0.70. (e) Yes: $P(\hat{p} < 0.82) = 0.394$.

(8.48) (a) approximately normal $N(0.85, 0.0357)$ (b) $P(\hat{p} > 0.70) = 0.9999$ (c) Verify that $np \geq 0$ and $n(1-p) \geq 0$ (d) Try yourself (e) Try yourself.

CHAPTER 9
ESTIMATING MEANS

Introduction

Generally speaking, a population parameter is usually unknown and we want to know about it. What is the way out of this problem? Of course, we estimate the parameter using a sample. Often, we wish to test a specified hypothesis about the parameter for making a decision. The estimation of a parameter and test of a desired hypothesis about it is collectively called a statistical inference. In other words, the first step of a statistical inference is to estimate the parameter of interest. In this chapter, we will estimate one population mean and compare two population means. The specific objectives of the chapter include the followings:

- Estimate an unknown population mean by a point and confidence interval estimates
- Compute margin of error of an estimate
- Determine sample size for a specified margin of error
- Estimate difference of two population means
- Estimate the difference of means in paired samples

Section 9.1: Estimating mean when population standard deviation is known

Estimation is the first step of any statistical inference. In this section, we want to estimate the mean μ of a population given a sample $(x_1, x_2, \dots, x_n)$ of size of n from the population. There are two types of estimate of μ- one is called the point estimate and the other confidence interval estimate.

Definition: Point estimate of the population mean μ

A **point estimate** of the population mean μ is the sample mean $\bar{x}$.
Indeed, a point estimate of a parameter is a single number computed from the sample to estimate the parameter. Let us consider the following example:

Example 9.1

A researcher wants to know the average time an US adult spends on the internet per month. To this end, the researcher randomly considers 50 US adults out of all US adults who use internet. On the basis of the sample, an observed value of the sample mean is found to be 39 hours per month. Then, the point estimate of the unknown mean μ is $\bar{x} = 39$ hours per month.

If the researcher is happy with the point estimate, then there is nothing to be worried about. However, the researcher knows that the sample mean $\bar{x}$ is a random variable, and as such, a different sample will end up with a different value of the point estimate. Generally, the variability resulting from different point estimates is something the researcher is not happy about.

Example 9.2

How much, on an average, a data analyst earns? To be more specific, suppose we want to know the mean salary μ of data analysts working in all branch offices of a big company. Suppose on the basis of a sample of salaries of 35 employees, the mean salary turns out to be \$68,690/year.
Then, a point estimate of μ is $\bar{x} = \$68{,}690$/year.

Q: If a different sample is considered, would the point estimate of μ be the same?

Of course, we might get a different estimate. So, the variability of a point estimate is again a concern.

Q: How could we get an estimate by taking the variability of point estimates into account?

The answer to this question leads to what is called a ***confidence interval estimate*** of μ.

From chapter 8, we know that the standard deviation of all possible sample means of size n is

$$\sigma_{\bar{x}} = \sigma/\sqrt{n}$$

Also, by the CLT for large n, or if the population is normal, the distribution of $Z = \frac{\bar{x}-\mu}{\sigma_{\bar{x}}}$ is a $N(0,1)$ variable. By taking this piece of information into account, the confidence interval estimate of the mean μ not only takes care of the variability of the point estimate, but also gives certain confidence that the interval would contain the unknown mean μ.

Definition: Confidence interval estimate

A $100(1-\alpha)\%$ confidence interval (CI) of μ, with confidence level $C = 1 - \alpha$, is the interval

$$\left[\bar{x} - z_{\alpha/2} \times \frac{\sigma}{\sqrt{n}}, \bar{x} + z_{\alpha/2} \times \frac{\sigma}{\sqrt{n}}\right]$$

where $z_{\alpha/2}$ is a z-value with area $\alpha/2$ to its right under the standard normal curve.

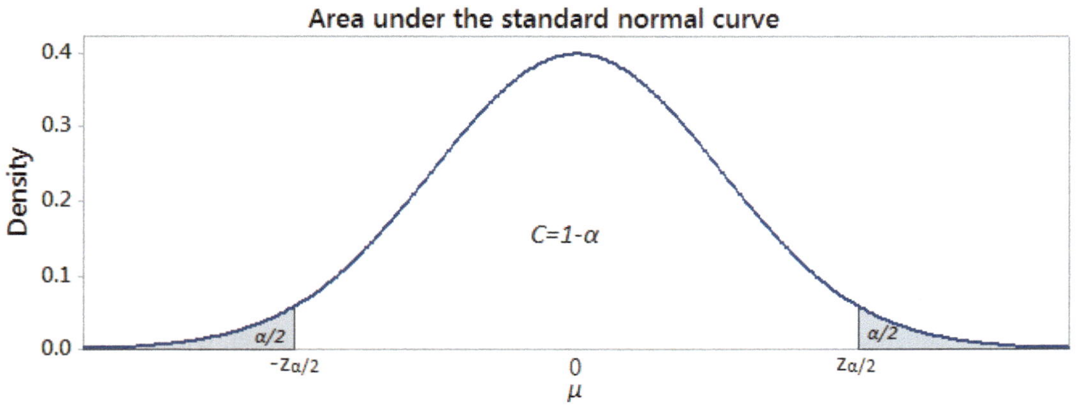

For a given confidence level $C = 1 - \alpha$, one can get the value of α by subtracting the confidence level from 1. If $C = 1 - \alpha = 0.90$, 0.95 and 0.99, the interval $\left[\bar{x} - z_{\alpha/2} \times \frac{\sigma}{\sqrt{n}},\ \bar{x} + z_{\alpha/2} \times \frac{\sigma}{\sqrt{n}}\right]$ is called a 90%, 95% and 99% confidence interval, respectively, with α taking values 0.1, 0.05 and 0.01, respectively. Once α is known, one can find the value of $z_{\alpha/2}$ from the area under the standard normal curve using Table 3 in the Appendix, calculator or any statistical software.

For a 90% CI of μ, α=0.1 and $z_{\alpha/2} = z_{0.05} = 1.65$, or more accurately, 1.645.
For a 95% CI of μ, α=0.05 and $z_{\alpha/2} = z_{0.025} = 1.96$.
For a 99% CI of μ, α=0.01 and $z_{\alpha/2} = z_{0.005} = 2.58$, or more accurately, 2.575.

The table below provides values of $z_{\alpha/2}$ for common choices of confidence level C:

C	90%	95%	99%
$z_{\alpha/2}$	1.645	1.96	2.575

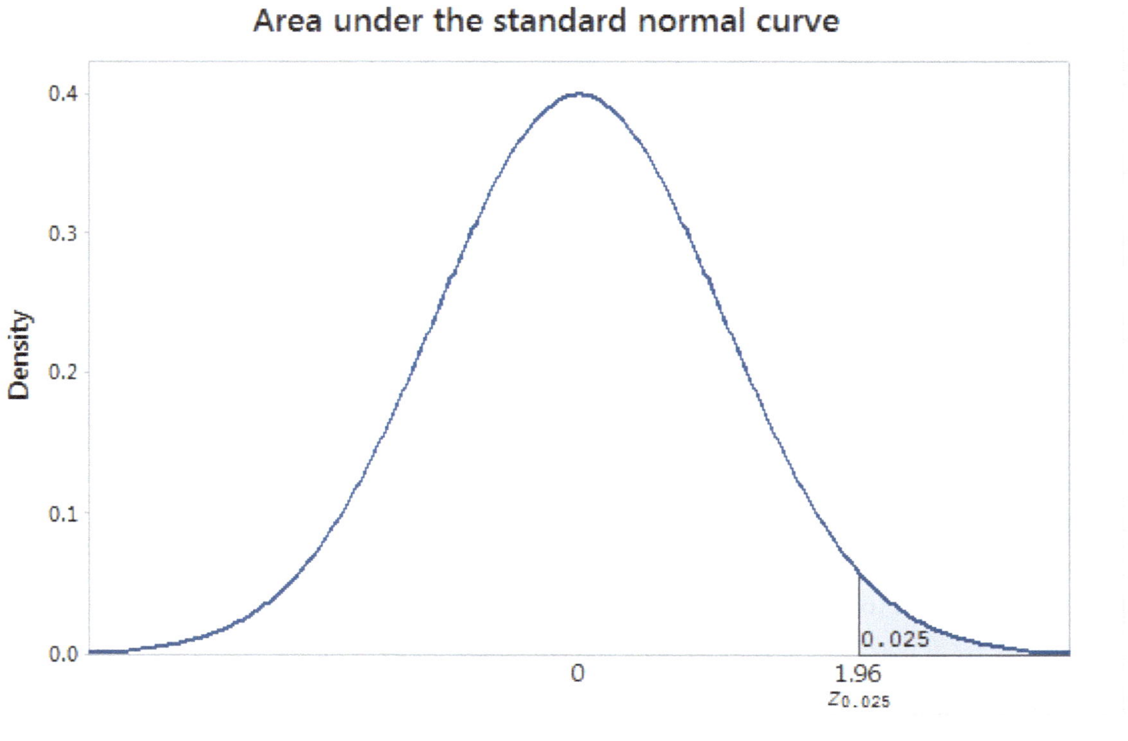

The idea of a confidence interval estimate is to use the fact that

$$P\left(-z_{\alpha/2} \leq \frac{\bar{x} - \mu}{\sigma/\sqrt{n}} \leq z_{\alpha/2}\right) = 1 - \alpha$$

or equivalently,

$$P\left(\bar{x} - z_{\alpha/2} \times \frac{\sigma}{\sqrt{n}} \leq \mu \leq \bar{x} + z_{\alpha/2} \times \frac{\sigma}{\sqrt{n}}\right) = 1 - \alpha$$

where α is a number between 0 and 1.

Then, the interval $\left[\bar{x} - z_{\alpha/2} \times \frac{\sigma}{\sqrt{n}},\ \bar{x} + z_{\alpha/2} \times \frac{\sigma}{\sqrt{n}}\right]$ contains the unknown mean μ with probability $1-\alpha$.

The quantity $1-\alpha$, often denoted by C, is called the confidence level of the interval.

Generally, the confidence level is expressed as percentage and the interval $\left[\bar{x} - z_{\alpha/2} \times \frac{\sigma}{\sqrt{n}},\ \bar{x} + z_{\alpha/2} \times \frac{\sigma}{\sqrt{n}}\right]$ is called $100(1-\alpha)\%$ confidence interval of μ and is popularly known as the Z-interval.

Assumptions for Z-interval

The way it has been derived, the Z-interval requires following three assumptions:

(1) Data represents an SRS,

(2) Either the population is normal or the sample is large ($n \geq 30$),

(3) σ is known.

Computation of Z-interval and interpretation of the interval

Let us start with the following example:

Example 9.3

Suppose we want to know the mean salary μ of data analysts working in all branch offices of a big company. On the basis of a sample of salaries of 35 employees, the mean salary turns out to be \$68,690/year. Assume that the population of all salaries has a standard deviation $\sigma = \$9480$.

Compute a 95% confidence interval of the mean salary μ of all data analysts working for the company, and interpret the result.

Solution

The formula for Z-interval is $\left[\bar{x} - z_{\alpha/2} \times \frac{\sigma}{\sqrt{n}},\ \bar{x} + z_{\alpha/2} \times \frac{\sigma}{\sqrt{n}}\right]$.

We have $\bar{x} = \$68{,}690$ and $\sigma = \$9480$. For a 95% CI, $\alpha = 1 - 0.95 = 0.05$ and $z_{\alpha/2} = z_{0.025} = 1.96$.

Then, we can compute: $z_{\alpha/2} \times \frac{\sigma}{\sqrt{n}} = 1.96 \times \frac{9480}{\sqrt{35}} = 3140.73$.

Then,

$$\left[\bar{x} - z_{\alpha/2} \times \frac{\sigma}{\sqrt{n}},\ \bar{x} + z_{\alpha/2} \times \frac{\sigma}{\sqrt{n}}\right]$$

$$= [68{,}690 - 3140.73, 68{,}690 + 3140.73] = [65549.27, 71830.73]$$

Therefore, a 95% CI of μ is $[65549.27, 71830.73]$.

See the TI-84 Plus Calculator option selection and output given below:

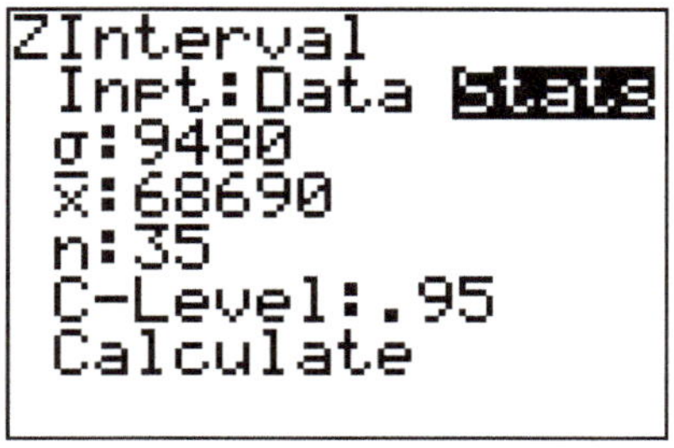

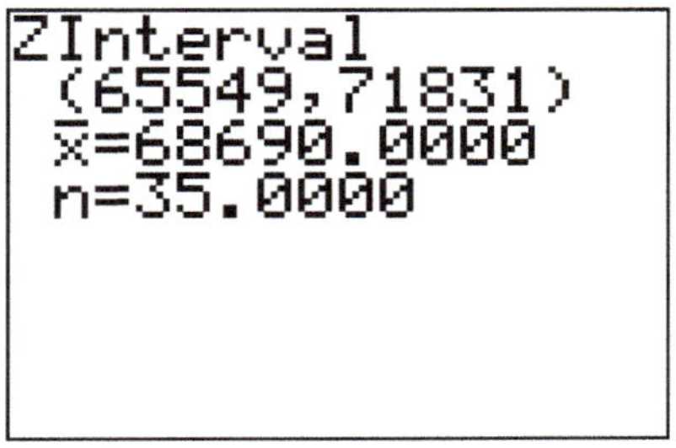

See the Minitab Software option selection and output below:

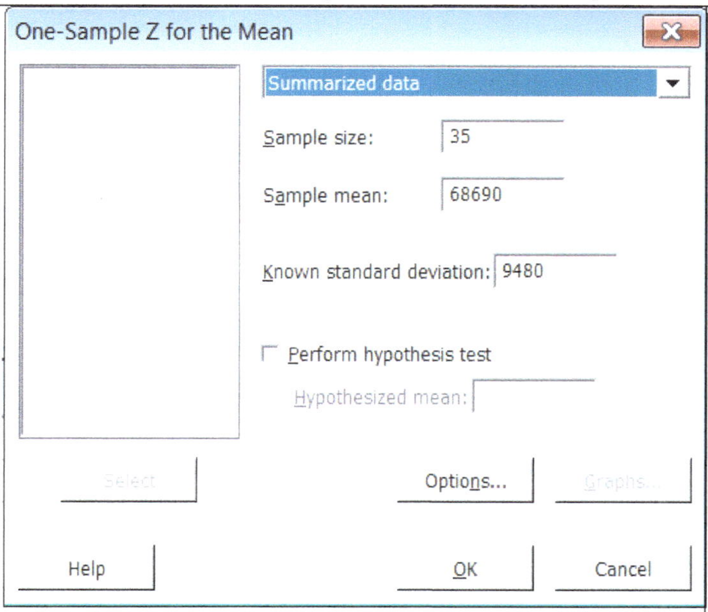

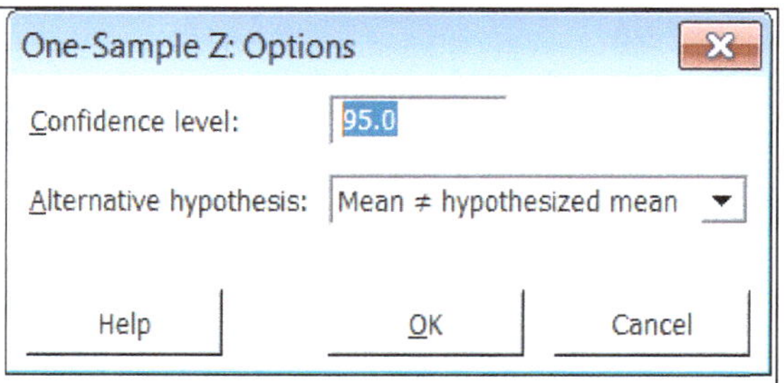

One-Sample Z

The assumed standard deviation = 9480

N	Mean	SE Mean	95% CI
35	68690	1602	(65549, 71831)

Interpretation

We are 95% confident that the interval $[65549, 71831]$ would capture the population mean salary μ of all data analysts working in all branch offices of the company. In other words, there is a 95% chance that the interval from 65549 to 71831 contains the true mean salary of all data analysts working for the company.

Generally speaking, a 95% confidence interval means that in repeated sampling, 95% of the times the confidence intervals would actually capture the unknown population mean.

Therefore, the interval $[65549, 71831]$ is one of all 95% CIs that would capture the true mean salary μ of all data analysts working for the company.

Example 9.4

The posted speed limit of vehicles in a neighborhood is 25 miles per hour. It is believed that the outsiders passing through the neighborhood exceeds the speed limit, which is a risk for the children playing in the neighborhood. Therefore, some speed bumps were requested to the neighborhood association. As a part of the process, the speed of all cars passing through the neighborhood on a given day was recorded. The table below contains the sample data of speeds for 24 cars.

25.7	30.1	27.3	28.6	32.2	28.6	24.1	27.3	32.2	30.4	25.5	29.7
28.4	31.2	28.4	24.5	32.1	25.8	25.5	27.4	31.7	29.7	26.1	25.3

Assume that the speed of cars follows a normal distribution with $\sigma = 4.1$ miles/hour.

(a) Find a point estimate of the mean speed (μ) of all cars passing through the neighborhood.
(b) Construct a 95% CI (C=0.95) for the mean speed of all cars passing through the neighborhood.
(c) How would you interpret your result in (b)?

Solution

Given $n = 24$ and $\sigma = 4.1$.

(a) Verify that $\bar{x} = 28.24$, which is the point estimate of all cars passing through the neighborhood.

(b) For a 95% CI, $\alpha = 0.05$ and $z_{\alpha/2} = z_{0.025} = 1.96$; $z_{\alpha/2}\left(\frac{\sigma}{\sqrt{n}}\right) = 1.96\left(\frac{4.1}{\sqrt{24}}\right) = 1.64$.

Then, $\left[\bar{x} - z_{\alpha/2}\left(\frac{\sigma}{\sqrt{n}}\right), \bar{x} + z_{\alpha/2}\left(\frac{\sigma}{\sqrt{n}}\right)\right]$ = [28.24-1.64, 28.24+1.64=[26.60, 29.88], which is the 95% CI of μ.

(c) We can be 95% confident that the mean speed of all cars passing through the neighborhood is somewhere between 26.60 and 29.88 miles per hour.

Definition: Margin of error

The quantity $z_{\alpha/2}\left(\frac{\sigma}{\sqrt{n}}\right)$ in the $100(1-\alpha)\%$ CI $\left[\bar{x} - z_{\alpha/2} \times \frac{\sigma}{\sqrt{n}},\ \bar{x} + z_{\alpha/2} \times \frac{\sigma}{\sqrt{n}}\right]$ is called the margin of error, usually denoted by m. In other words, $m = z_{\alpha/2}\left(\frac{\sigma}{\sqrt{n}}\right)$. Therefore, for the computational convenience, one can write the $100(1-\alpha)\%$ CI of μ as

$$[\bar{x} - m, \bar{x} + m]$$

where $m = z_{\alpha/2}\left(\frac{\sigma}{\sqrt{n}}\right)$ is the margin of error of the interval and $\bar{x}$ is the point estimate of μ.

Note that the margin of error depends on the sample size n. As sample size n increases, the margin of error m decreases.

Let L be the length of the interval $\left[\bar{x} - z_{\alpha/2}\left(\frac{\sigma}{\sqrt{n}}\right), \bar{x} + z_{\alpha/2}\left(\frac{\sigma}{\sqrt{n}}\right)\right]$. Then, one can verify that

$$m = z_{\alpha/2}\left(\frac{\sigma}{\sqrt{n}}\right) = \frac{L}{2}$$

That is, the margin of error is half of the length of the confidence interval.

Example 9.5

Revisit Example 9.4, where we found the point estimate of μ to be $\bar{x} = 28.24$, using a sample of size $n = 24$. We are interested in computing a 90% and 99% CIs of μ, the mean speed of all cars passing through the neighborhood. To this end, compute the followings:

(a) Find the margin of error for a 90% CI of μ.

(b) Find the margin of error of 99% CI of μ.

(c) Compare the results in (a) and (b).

(d) Compute a 90% CI of μ.

(e) Compute a 99% CI of μ.

(f) Compare CIs in (d) and (e) and interpret the result.

Solution

Given, $n = 24$, $\sigma = 4.1$, and $\bar{x} = 28.24$.

(a) For a 90% CI, $\alpha = 0.1$; $z_{\alpha/2} = z_{0.05} = 1.645$; $m = z_{\alpha/2}\left(\frac{\sigma}{\sqrt{n}}\right) = 1.645\left(\frac{4.1}{\sqrt{24}}\right) = 1.38$.

(b) For a 99% CI, $\alpha = 0.01$; $z_{\alpha/2} = z_{0.005} = 2.58$; $m = z_{\alpha/2}\left(\frac{\sigma}{\sqrt{n}}\right) = 2.58\left(\frac{4.1}{\sqrt{24}}\right) = 2.16$.

(c) We see that the margin of error for a 99% CI is higher than the one from the 90% CI.
(d) The 90% CI is $[28.24 - 1.38, 28.24 + 1.38] = [26.86, 29.62]$.
(e) The 99% CI is $[28.24 - 2.16, 28.24 + 2.16] = [26.08, 30.40]$.
(f) The 99% CI in (c) is wider, this is because the margin of error for the 99% CI is higher than the 90% CI. It means that the higher the confidence level, the wider is the CI.

Sample size determination

Often, researchers might have a desired specification for a margin of error. In other words, they would like to control the margin of error for a given confidence level. How does it work? Well, from the formula of margin of error, $m = z_{\alpha/2}\left(\frac{\sigma}{\sqrt{n}}\right)$, it is evident that one can control the margin of error by choosing different values of the sample size n. Indeed, by increasing the sample size, one can reduce the margin of error, and vice versa. Given these circumstances, one might be interested to know the answer to the following question:

Q: What is the sample size for a desired margin of error m for a level C confidence interval?

To answer this question, all we need is to solve the equation $m = z_{\alpha/2}\left(\frac{\sigma}{\sqrt{n}}\right)$ for n. This leads to the formula:

$$n = \left(\frac{z_{\alpha/2}\sigma}{m}\right)^2$$

Therefore, given a desired level of confidence and margin of error, we can determine the sample size to be considered.

A useful note for determining n

Since by increasing the sample size one can decrease the margin of error, n should be rounded up to the nearest integer in order to ensure the specified margin of error. Rounding n down to the nearest integer will make the margin of error slightly larger than what is being specified.

Example 9.6

In reference to the Examples 9.4 and 9.5, suppose the researcher wants a margin of error of 2 in the estimation of 90%, 95%, and 99% CIs of μ. As a part of the process, the researcher wants to investigate what sample size will be appropriate to guarantee 90%, 95% and 99% CIs.

Given the fact that $\sigma = 4.1$ and $m = 2$, answer the following questions:
(a) Determine the sample size to guarantee a 90% CI of μ.
(b) Determine the sample size to guarantee a 95% CI of μ.
(c) Determine the sample size to guarantee a 99% CI of μ.
(d) What conclusion can you make from your results in (a)-(c).
(e) What sample size will you recommend for a margin of error within 1.25 for a 95% CI?

Solution

Given $\sigma = 4.1$ and $m = 2$. We want to find sample size for 90%, 95% and 99% CIs using the formula $n = \left(\frac{z_{\alpha/2}\sigma}{m}\right)^2$.

(a) For a 90% CI, $\alpha = 0.1$ and $z_{\alpha/2} = z_{0.05} = 1.645$. Then, $n = \left(\frac{1.645\times4.1}{2}\right)^2 = 11.37$.
By rounding it up to the nearest integer, we get $n = 12$, which will provide 90% confidence level for estimating mean.

(b) For a 95% CI, $\alpha = 0.05$ and $z_{\alpha/2} = z_{0.025} = 1.96$. Then, $n = \left(\frac{1.96\times4.1}{2}\right)^2 = 16.14$.
By rounding it up to the nearest integer, we get $n = 17$.

(c) For a 99% CI, $\alpha = 0.01$ and $z_{\alpha/2} = z_{0.005} = 2.58$. Then, $n = \left(\frac{2.58\times4.1}{2}\right)^2 = 27.97$.
By rounding it up to the nearest integer, we get $n = 28$.

(d) Given a fixed margin of error, we will have to sample more subjects to achieve a higher confidence level.

(e) Here, $m = 1.25$ and we want to determine the sample size for a 95% CI. Since for a 95% CI, $\alpha = 0.05$ and $z_{\alpha/2} = z_{0.025} = 1.96$, we have $n = \left(\frac{1.96\times4.1}{1.25}\right)^2 = 41.3$.
By rounding it up to the nearest integer, we get $n = 42$. Therefore, we would recommend a sample size of 42 to ensure a margin of error of 1.25 for a 95% CI of μ.

Example 9.7

As of January 2014, 58% of American adults have a Smartphone and on average, spend 34 hours per month using the mobile internet on Smartphone. We wonder what would be the average use of the mobile internet on Smartphone for students taking Elementary statistics at a given semester in a university. Is this average different than the national average? As a first step of the study, a simple random sample of 20 students taking Elementary Statistics in a given semester is considered for estimating the unknown average time (μ) spent on the mobile internet on Smartphone. The sample data on the amount of time spent (in hours per month) on mobile internet on Smartphone is tabulated below.

23	33	51	41	31	42	43	36	23	46
39	42	47	45	32	45	37	34	33	29

Assume that the population data follows a normal distribution with unknown mean and standard deviation $\sigma = 10$.

Answer the following questions:

(a) What is the point estimate of μ?

(b) Find a 90% CI of μ and interpret your result.

(c) Find a 95% confidence interval of μ. How is this interval different from your result in (b)?

(d) What sample size would you recommend for a 95% confidence interval of μ within 2.5 hours per month?

Solution

(a) The point estimate of μ is $\bar{x} = 37.6$.

(b) For a 90% CI, $\alpha = 0.1$; $z_{\alpha/2} = z_{0.05} = 1.645$; $m = z_{\alpha/2}\left(\frac{\sigma}{\sqrt{n}}\right) = 1.645\left(\frac{10}{\sqrt{20}}\right) = 3.68$. Then the 90% CI is $[37.6 - 3.68, 37.6 + 3.68] = [33.92, 41.28]$. This means that the interval between 33.92 and 41.28 hours per month is one of the 90% of all CIs that would capture the unknown mean time μ that students spent on the internet using the Smartphone. In other words, we can be 90% confident that the unknown mean time μ that students spent on the internet using the Smartphone is somewhere between 33.92 and 41.28 hours per month.

(c) For a 95% CI, $\alpha = 0.05$; $z_{\alpha/2} = z_{0.025} = 1.96$; $m = z_{\alpha/2}\left(\frac{\sigma}{\sqrt{n}}\right) = 1.96\left(\frac{10}{\sqrt{20}}\right) = 4.38$. Then the 95% CI is $[37.6 - 4.38, 37.6 + 4.38] = [33.22, 41.98]$. This interval is wider than the interval in (b), because this interval provides more confidence level with higher margin of error.

(d) For a 95% CI, $z_{\alpha/2} = z_{0.025} = 1.96$. Then, $n = \left(\frac{1.96\times 10}{2.5}\right)^2 = 61.47$. By rounding it up to the nearest integer, we recommend the value of the sample size be $n = 62$.

Random sample and Z-Interval estimate in Minitab

Example 9.8

In this example, we apply Minitab to compute 95% CI of the population mean using Z-interval with necessary instructions for drawing samples with and without replacement. Assume that a population consists of the following 40 measurements:

23	13	21.3	23.1
27	30.4	26.5	28.2
22.2	28.3	26.4	28.4
19.6	24.3	14.5	28.7
23.7	29.6	19.9	18
18.5	20	16.9	26.8
26.2	33.9	29	25.9
28.5	31.3	23.8	24.7
19.2	35.5	24.9	15.8
17.8	24.4	22.9	28.6

(a) Draw a sample of size 25 with replacement.
(b) Draw a sample of size 25 without replacement.
(c) Using sample in (a), compute a 95% confidence interval for population mean assuming that the population is normally distributed with standard deviation 5.
(d) Compute the population mean of the original 40 measurements as well and comment on whether or not the confidence interval in (c) contains the population mean.

Solution

Below we have Minitab instruction for drawing samples with and without replacement:

(1) Select **Calc>Random Data>Sample From Columns...**
(2) Copy 40 observations in column **C1** and give it a label X.
(3) For **Number of rows to sample**: type 25 and for **From columns:** select X

(4) For Store Samples in: type SWR to store sample with replacement (SWR) and check on **Sample with replacement** option box and click OK.

(5) Select Edit>Edit Last Dialog, for Store Samples in: type SWOR to store sample without replacement, uncheck on **Sample with replacement** option box and click OK.

See the Minitab dialog box below with option selections as specified above for drawing sample with and without replacement.

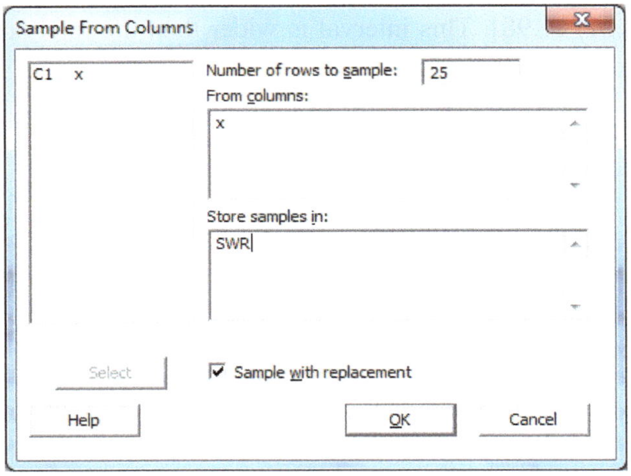

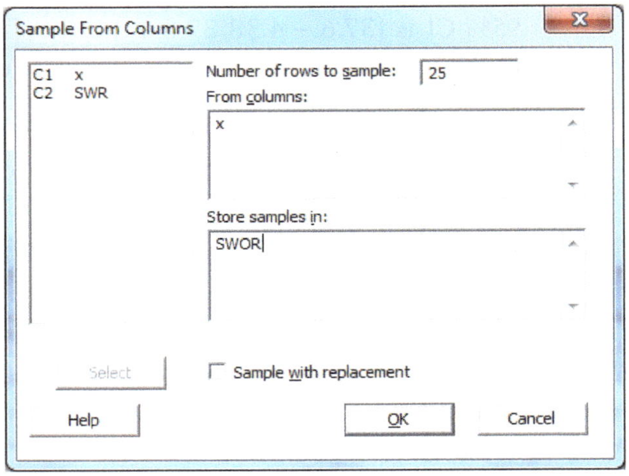

(a) A portion of the sample with replacement, SWR, appers in column C2 of Minitab worksheet snapshot.

(b) A portion of the sample without replacement, SWOR, appers in column C3 of Minitab worksheet snapshot.

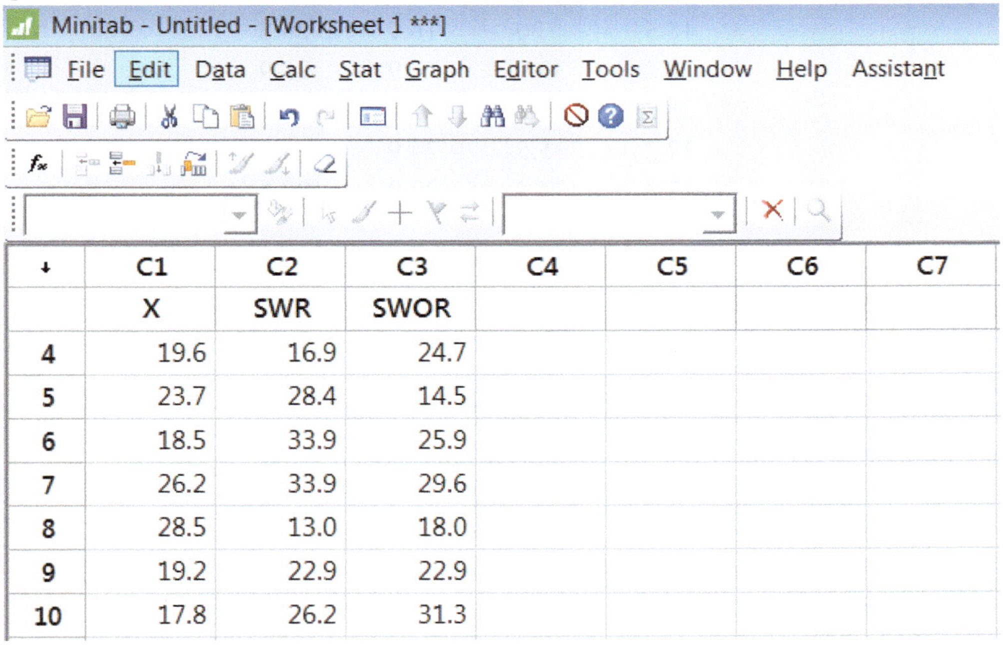

↓	C1	C2	C3	C4	C5	C6	C7
	X	SWR	SWOR				
4	19.6	16.9	24.7				
5	23.7	28.4	14.5				
6	18.5	33.9	25.9				
7	26.2	33.9	29.6				
8	28.5	13.0	18.0				
9	19.2	22.9	22.9				
10	17.8	26.2	31.3				

(c) For a 95% CI of mean using sample in (a), we can follow the steps below:

(1) Select Stat>Basic Statistics>1-Sample Z…
(2) For **One or more samples, each in a column**, Select C2 (SWR) and for ***Known standard deviation***: type 5
(3) Use **Options…** to specify ***Confidence level***: 95 for a 95% CI and Click [OK] two times.
See the dialog boxes below with the necessary option selections for computing Z-Interval for the sample.

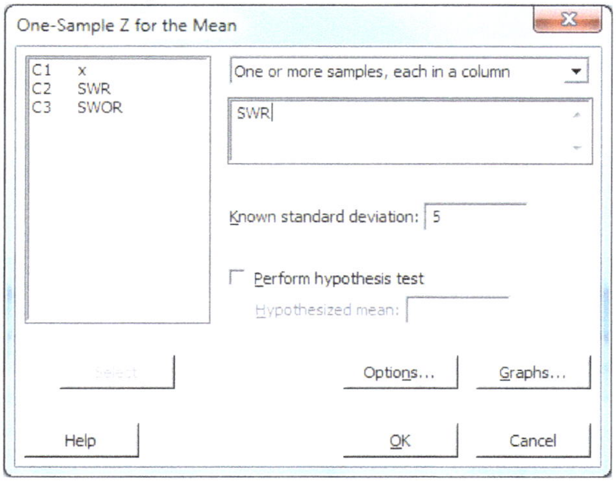

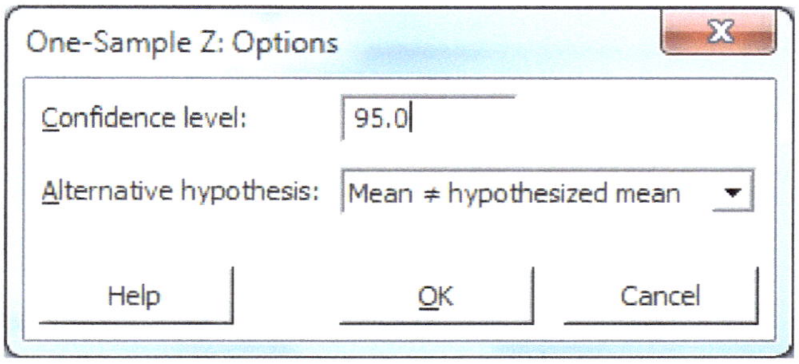

Below is the Minitab output from the session window for 95% CI of Mean:
One-Sample Z: SWR

The assumed standard deviation = 5

Variable	N	Mean	StDev	SE Mean	95% CI
SWR	25	24.41	4.57	1.00	(22.45, 26.37)

(d) For the population of the 40 measurements, the mean μ =24.2675. As we see, the 95% confidence interval (22.45, 26.37) contains the population mean μ =24.2675.

In general, the sample is assumed given, as in the example below.

Example 9.9

Suppose we have a sample of size 6 from a normal distribution with unknown mean and a known standard deviation 7.

3.2	13.3	15.1	13.4	9.1	16.4

Compute a 95% CI interval for the unknown population mean.

Solution

For Minitab we use the steps below:
(1) Enter the sample data in C1

(2) Select Stat>Basic Statistics>1-Sample Z…
(3) For Samples in Columns, Select C1 and for ***Known standard deviation***: type 7
(4) Use option to specify ***Confidence level***: 95.0 for a 95% CI and Click [OK].

See the dialog boxes below with the necessary options selection:

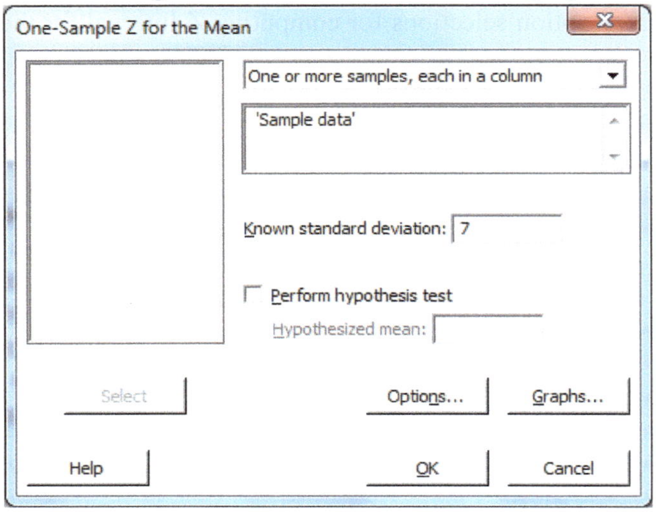

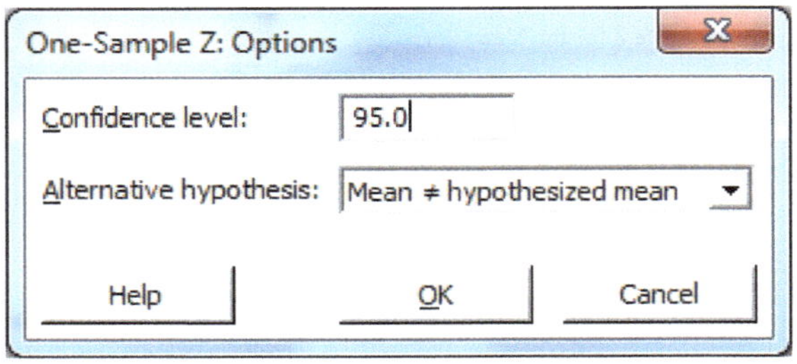

Below is the Minitab output from the session window:
One-Sample Z: Sample data

The assumed standard deviation = 7

Variable	N	Mean	StDev	SE Mean	95% CI
Sample data	6	11.75	4.86	2.86	(6.15, 17.35)

TI-84 Plus instruction for Z-Interval

Data Entry

1. STAT>EDIT>1:Edit… (Enter data in any of L1,L2,…L6), we enter data in Example 9.7 in L1.
2. 2nd >MODE[:Quit] (to get back to the home screen after the data entry)

Getting Z-Interval

1. STAT> TESTS> Press 7 for ZInterval
2. For Inpt: Choose **Data**, specify $\boldsymbol{\sigma}$: 7, **List:** L1, **Freq:**1, **C-Level:** 0.95 as appropriate.
3. Move the cursor to Calculate and press ENTER.

See the figure below for option selections and result:

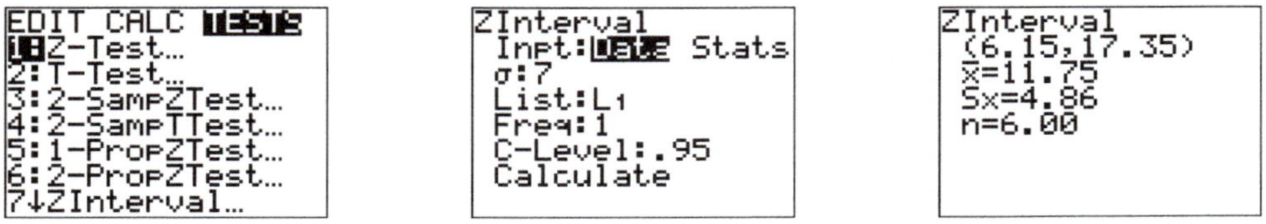

Section 9.2: Estimating mean when population standard deviation is not known

In section 9.1, we have the confidence interval estimate of μ, called the Z-interval given by the formula

$$\left[\bar{x} - z_{\alpha/2}\left(\frac{\sigma}{\sqrt{n}}\right), \bar{x} + z_{\alpha/2}\left(\frac{\sigma}{\sqrt{n}}\right)\right]$$

where σ is assumed to be known.

What if σ is unknown in above expression?

In most real life problems, the population standard deviation σ is unknown. As a result, we no longer can use the Z-Interval for estimating μ.

It is very intuitive that if σ is unknown, we have to estimate it by the sample standard deviation S, that leads to a different confidence interval of μ, we call T-Interval. Before we talk about T-interval, we will now address a new distribution, called the t-distribution.

The t-distribution

A t-distribution is a probability distribution similar to the standard normal distribution. Both t- and standard normal distributions are single-peaked, bell-shaped curve symmetric about 0. However, unlike standard normal curve, the t-curve is little more thick at both tails, occupying more area at tails and less area at the center as compared to the standard normal curve.

Definition: The t-distribution

The probability distribution of the statistic

$$T = \frac{\bar{x} - \mu}{s/\sqrt{n}}$$

is called the t-distribution with $(n-1)$ degrees of freedom, usually denoted by the symbol $t(n-1)$. The t-distribution is characterized by its degrees of freedom which is one less than the number of observations in the sample. As the degrees of freedom of t-distribution gets larger, the t-distribution approaches the standard normal distribution. See the graph below to compare Z and T distributions.

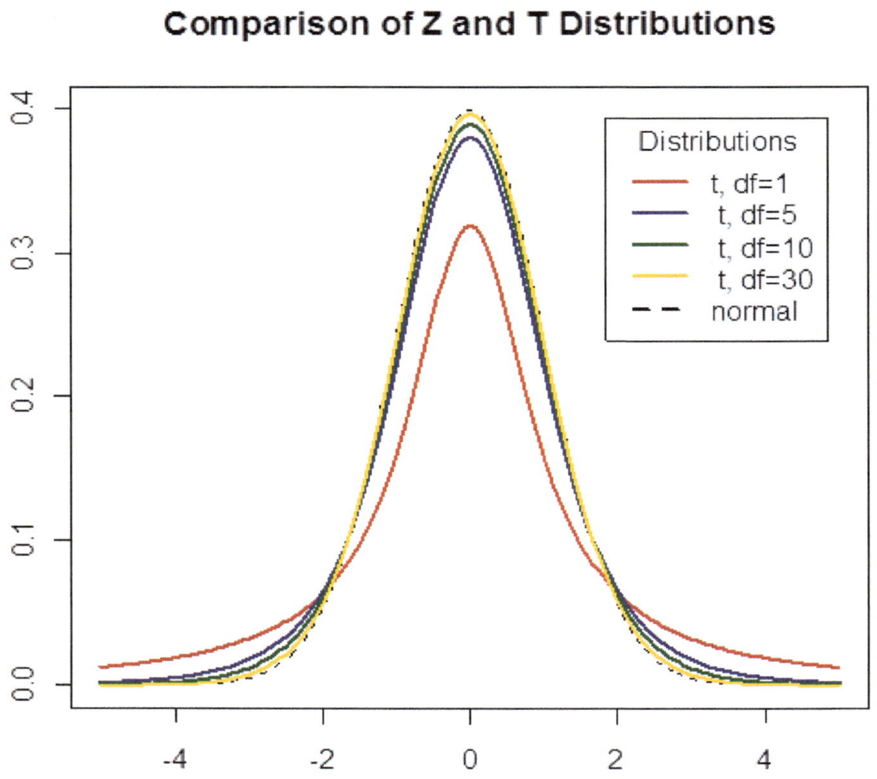

T- Interval

A $100(1-\alpha)\%$ confidence interval (CI) of μ when σ is unknown, is the interval

$$\left[\bar{x} - t_{\alpha/2}\left(\frac{s}{\sqrt{n}}\right), \bar{x} + t_{\alpha/2}\left(\frac{s}{\sqrt{n}}\right)\right]$$

where $t_{\alpha/2}$ is T-value with an area $\alpha/2$ to its right under the curve of a t-distribution with $(n-1)$ degrees of freedom. This interval is often called the one sample T-Interval and provides a confidence level of $C = 1 - \alpha$.

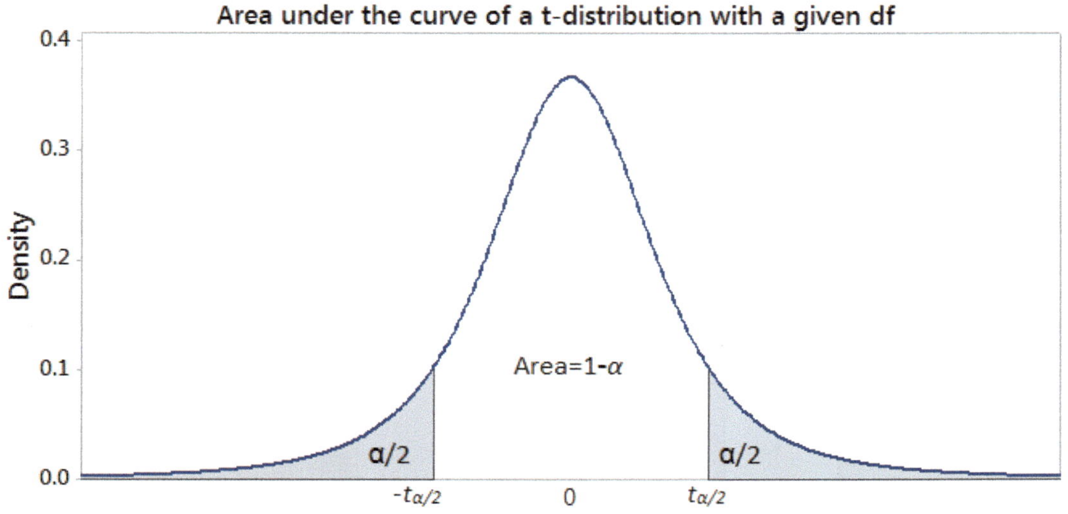

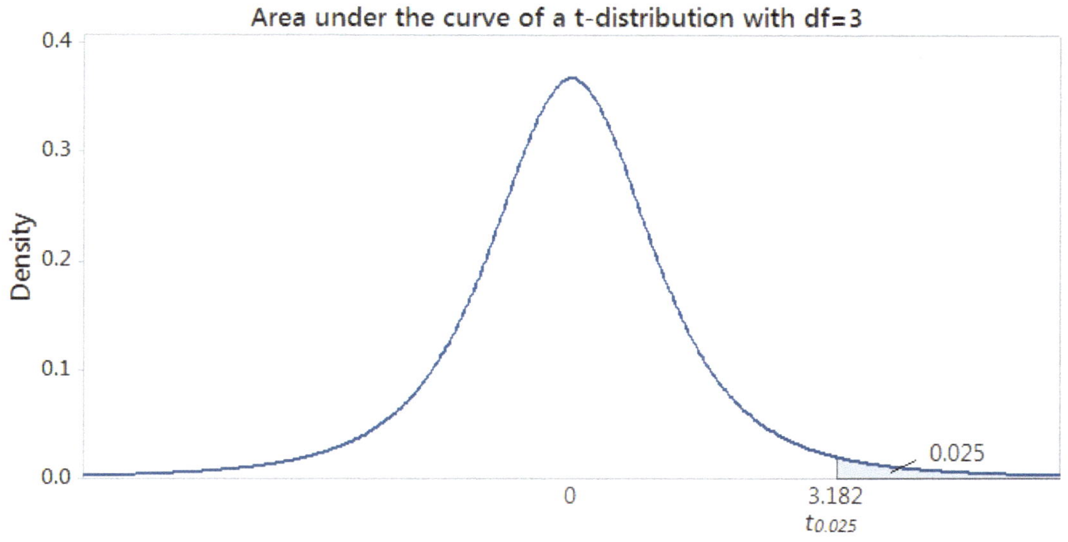

Example 9.10

Data below refers to time spent, in hours per month, on mobile internet using Smartphone of 10 randomly selected students in a class. Suppose the sample comes from a distribution that is approximately normal with unknown mean and standard deviation.

39	42	47	45	32	45	37	34	33	29

(a) What is the point estimate of average time spent on mobile internet using Smartphone for all students in the class?

(b) Find a 90% CI of μ and conclude your interval.

Solution

For the given sample, it follows that $\bar{x} = 38.3$, s= 6.27, $n = 10$, $df = 9$. For 90% CI, with $df = 9$, $t_{\alpha/2} = t_{0.05} = 1.833$. Then, $t_{\alpha/2}\left(\frac{s}{\sqrt{n}}\right) = 1.833\left(\frac{6.27}{\sqrt{10}}\right) = 3.63$

(a) The point estimate of average time spent on internet using Smartphone by all students in the class is 38.30 hours per month.

(b) The 90% CI is

$$\left[\bar{x} - t_{\alpha/2}\left(\frac{s}{\sqrt{n}}\right), \bar{x} + t_{\alpha/2}\left(\frac{s}{\sqrt{n}}\right)\right] = [38.3 - 3.63, 38.3 + 3.63] = [34.67, 41.93].$$

Therefore, we can say with 90% confidence that the mean time spent on mobile internet using the Smartphone by all students in the class is somewhere between 34.67 and 41.93 hours per month.

See the TI-84 Plus Calculator option selections and output given below:

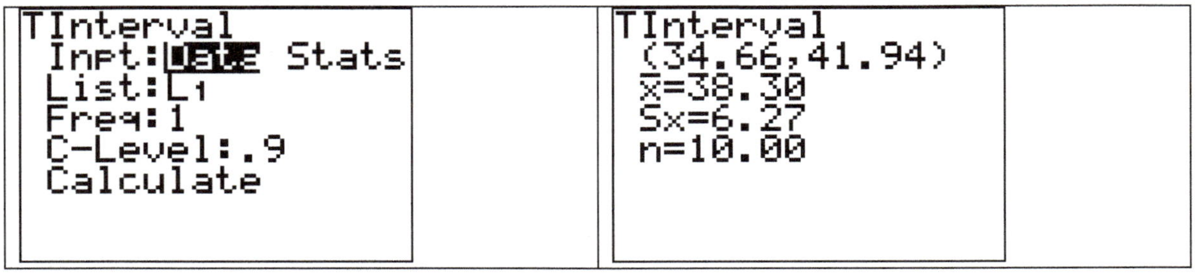

See the Minitab option selection and output given below:

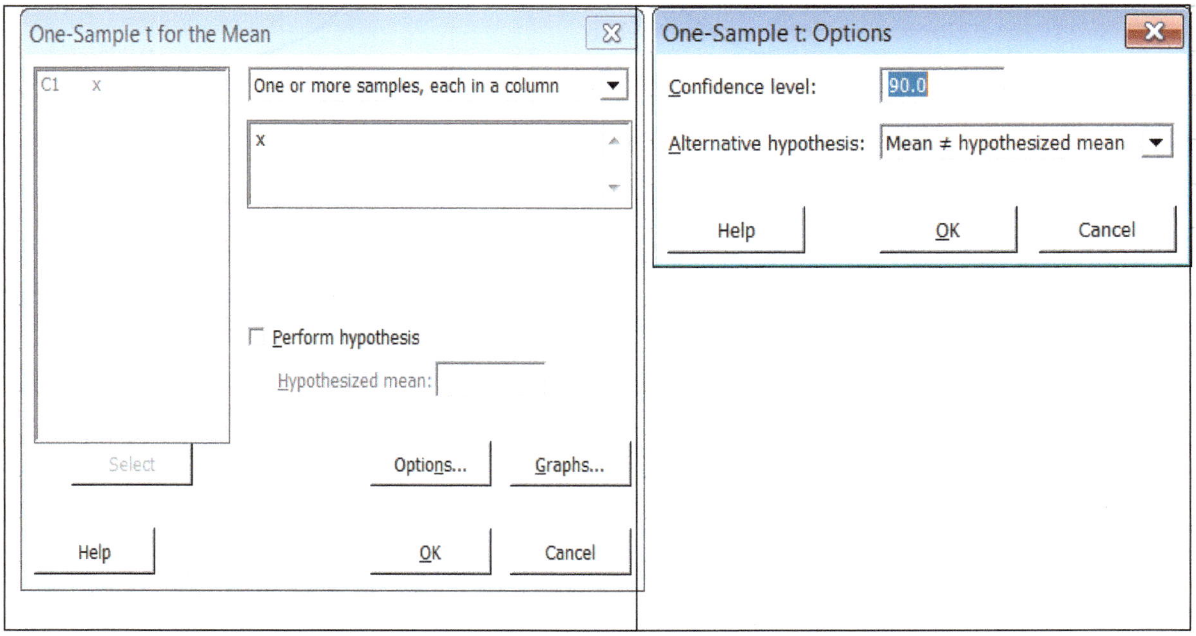

One-Sample T:

Variable	N	Mean	StDev	SE Mean	90% CI
x	10	38.30	6.27	1.98	(34.66, 41.94)

Use of Technology for one-sample T-interval

Below we provide an example where we apply Minitab and TI 84 Plus calculator to find a 95% CI of an unknown population mean using sample data and summary statistics from a sample.

Example 9.11

Consider the following two problems for estimating a 95% CI for a population mean:

(a) Suppose we have a sample of size 6 from a normal distribution with unknown mean and standard deviation:

3.2	13.3	15.1	13.4	9.1	16.4

(b) Suppose, given a sample of size 30 from a population with unknown mean and standard deviation, we have $\bar{x} = 6.5$ and $s = 2.5$.

We will use Minitab and TI 84 Plus calculator to compute one sample 95% T-interval using necessary instructions for (a) and (b).

Minitab Instruction for one-sample T-interval (for sample data)

For sample data in (a), we follow the steps below:

(1) Enter the sample data in C1
(2) Select Stat>Basic Statistics>1-Sample t…
(3) Choose option **One or more samples, each in a column,** Select Sample data or C1.
(4) Use **Options…** to specify ***Confidence level***: 95 and Click [OK].

See the dialog boxes below with the necessary option selections:

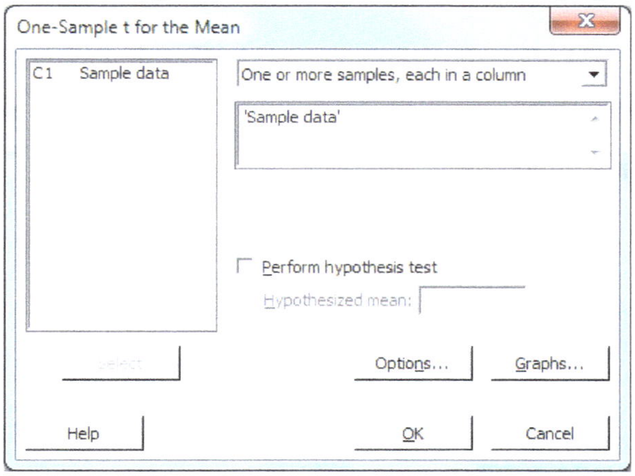

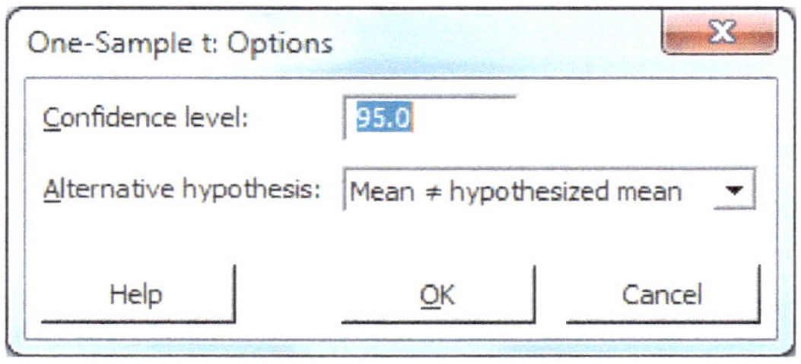

Below is the Minitab output from the session window:

One-Sample T: Sample data

Variable	N	Mean	StDev	SE Mean	95% CI
Sample data	6	11.75	4.86	1.98	(6.65, 16.85)

For sample summary Stats in (b), we follow the steps below:

(1) Select Stat>Basic Statistics>1-Sample t…
(2) Choose option **Summarized data**, and specify **sample size:** 30, **sample mean:** 6.5, **standard deviation:** 2.5..
(3) Use **Options**… to specify ***Confidence level***: 95 and Click [OK].

See the dialog boxes below with the necessary option selections:

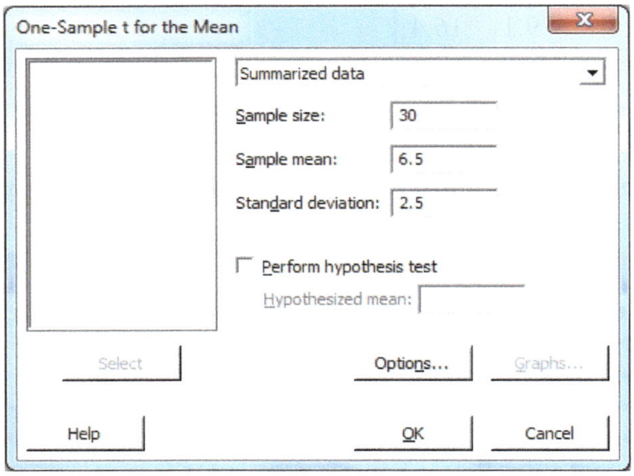

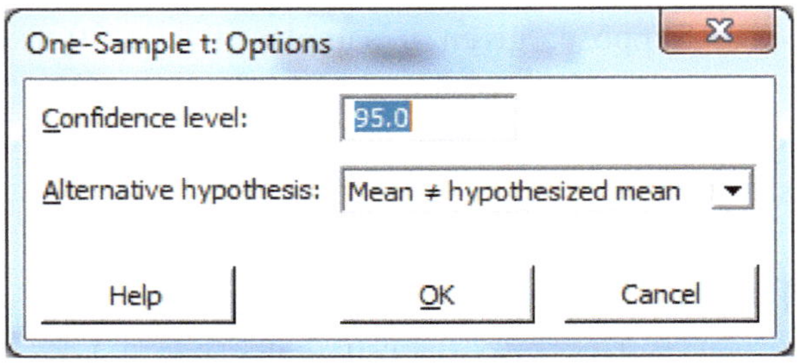

Below is the Minitab output from the session window:

One-Sample T

N	Mean	StDev	SE Mean	95% CI
30	6.500	2.500	0.456	(5.566, 7.434)

TI-84 Plus for one-sample T-Interval (for sample data)

Given the sample in (a) we follow the steps below to get a 95% CI for mean:

1) Enter sample data in L1.
2) Press STAT, select TESTS, choose **8: TInterval**
3) Move the cursor to **Data** and press **ENTER**, Specify **List:** L1, **Freq:** 1, C-Level: 0.95.
4) Move the cursor to Calculate and **press** ENTER

The above steps and result are demonstrated in figures below:

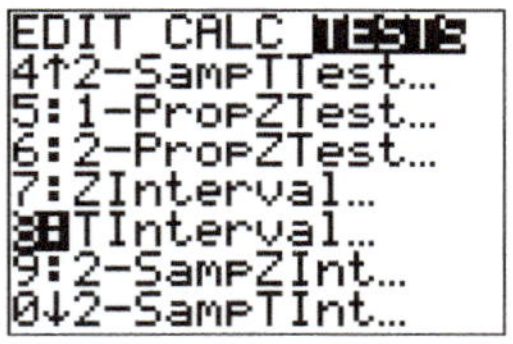

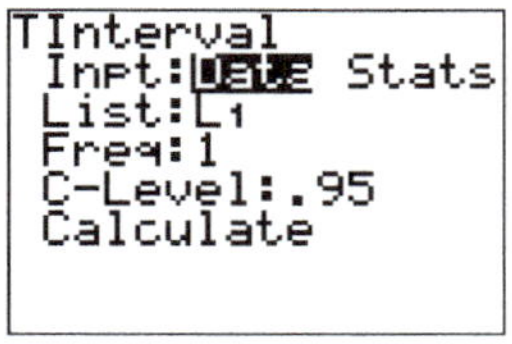

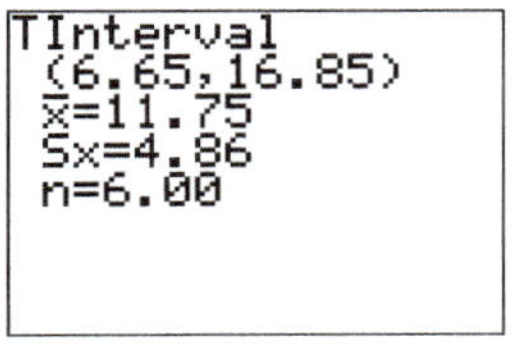

TI-84 Plus for one-sample T-Interval (for Stats)

Given summary Stats in (b), we follow the steps below to get a 95% CI of mean:

1) Press STAT, select TESTS, choose **8: TInterval**
2) Move the cursor to **Stats** and press **ENTER**.
3) Specify appropriate Stats: $\bar{x}$: 6.5, s_x: 2.5 , n: 30, C-Level: 0.95.
4) Move the cursor to Calculate and press ENTER

The above steps and result are demonstrated in figures below:

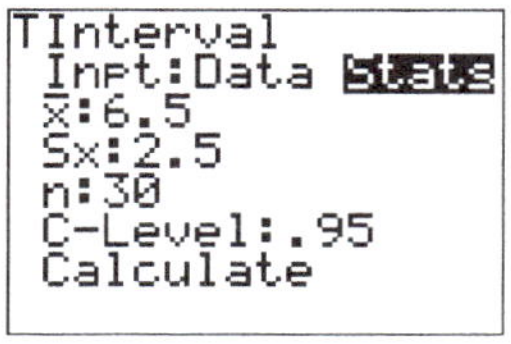

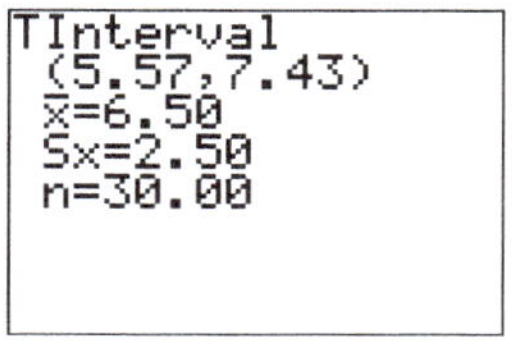

Section 9.3: Estimating difference of two population means using independent samples

Suppose we have two independent samples of sizes n_1 and n_2 from two populations, one with mean μ_1 and standard deviation σ_1, and the other with mean μ_2 and standard deviation σ_2, respectively. We want to estimate the difference of two population means, $\mu_1 - \mu_2$.
The estimation procedure is similar to what have been applied in sections 9.1 and 9.2. Let us provide some formula that will be useful in this section.

(1) The point estimate of $\mu_1 - \mu_2$ is $\bar{x}_1 - \bar{x}_2$, where $\bar{x}_1$ and $\bar{x}_2$ are means of two samples.
(2) The mean of $\bar{x}_1 - \bar{x}_2$ is $\mu_{\bar{x}_1 - \bar{x}_2} = \mu_1 - \mu_2$.
(3) The standard deviation of $\bar{x}_1 - \bar{x}_2$ is $\sigma_{\bar{x}_1 - \bar{x}_2} = \sqrt{\frac{\sigma_1^2}{n_1} + \frac{\sigma_2^2}{n_2}}$, which reduces to the expression

$\sigma_{\bar{x}_1 - \bar{x}_2} = \sqrt{\sigma^2\left(\frac{1}{n_1} + \frac{1}{n_2}\right)}$ if $\sigma_1^2 = \sigma_2^2 = \sigma^2$, say.
(4) When σ_1^2 and σ_2^2 are equal ($\sigma_1^2 = \sigma_2^2$), but unknown, an estimate of the standard deviation of $\bar{x}_1 - \bar{x}_2$ is $\hat{\sigma}_{\bar{x}_1 - \bar{x}_2} = \sqrt{s_p^2\left(\frac{1}{n_1} + \frac{1}{n_2}\right)}$, where s_p^2 is the pooled estimator of the common variance σ^2 and is given by

$$s_p^2 = \frac{(n_1 - 1)s_1^2 + (n_2 - 1)s_2^2}{n_1 + n_2 - 2}$$

and s_1^2 and s_2^2 in above expression are sample variances of two samples.
(5) When σ_1^2 and σ_2^2 are unknown and unequal, an estimate of the standard deviation of $\bar{x}_1 - \bar{x}_2$ is $\hat{\sigma}_{\bar{x}_1 - \bar{x}_2} = \sqrt{\frac{s_1^2}{n_1} + \frac{s_2^2}{n_2}}$, where s_1^2 and s_2^2 are sample variances of two samples.

Confidence Interval of $\mu_1 - \mu_2$

In order to estimate the confidence interval of $\mu_1 - \mu_2$, we need the following three assumptions:
(1) Two samples are independent,
(2) Sample sizes are either large or the populations have normal or approximately normal distributions, and
(3) Any of the followings:
(a) Two population variances σ_1^2 and σ_2^2 are known (Case I: Two-sample Z-interval).
(b) Two population variances σ_1^2 and σ_2^2 are equal but unknown (Case II: Two-sample pooled T-interval).

(c) Two population variances σ_1^2 and σ_2^2 are unequal and unknown (Case III: Two-sample non-pooled T-interval).

Case I: Two-sample *Z*-interval

When two population variances σ_1^2 and σ_2^2 are known, a $100(1-\alpha)\%$ Z-interval of $\mu_1 - \mu_2$, with confidence level $C = 1 - \alpha$, is

$$\left[(\bar{x}_1 - \bar{x}_2) - z_{\alpha/2}\sqrt{\frac{\sigma_1^2}{n_1} + \frac{\sigma_2^2}{n_2}}, (\bar{x}_1 - \bar{x}_2) + z_{\alpha/2}\sqrt{\frac{\sigma_1^2}{n_1} + \frac{\sigma_2^2}{n_2}}\right]$$

where $z_{\alpha/2}$ is Z-value with area $\alpha/2$ to its right under the standard normal curve, as described in section 9.1.

Case II: Two-sample pooled T-interval

When two population variances σ_1^2 and σ_2^2 are unknown but equal, a $100(1-\alpha)\%$ Pooled T- interval of $\mu_1 - \mu_2$, with confidence level $C = 1 - \alpha$, is

$$\left[(\bar{x}_1 - \bar{x}_2) - t_{\alpha/2}\sqrt{s_p^2\left(\frac{1}{n_1} + \frac{1}{n_2}\right)}, (\bar{x}_1 - \bar{x}_2) + t_{\alpha/2}\sqrt{s_p^2\left(\frac{1}{n_1} + \frac{1}{n_2}\right)}\right]$$

where $t_{\alpha/2}$ is the T-value with an area $\alpha/2$ to its right under the curve of a t-distribution with $(n_1 + n_2 - 2)$ degrees of freedom.

Case III: Two-sample non-pooled T-interval

When two population variances σ_1^2 and σ_2^2 are unknown and unequal, a $100(1-\alpha)\%$ Non-pooled T-interval of $\mu_1 - \mu_2$, with confidence level $C = 1 - \alpha$, is

$$\left[(\bar{x}_1 - \bar{x}_2) - t_{\alpha/2}\sqrt{\left(\frac{s_1^2}{n_1} + \frac{s_2^2}{n_2}\right)}, (\bar{x}_1 - \bar{x}_2) + t_{\alpha/2}\sqrt{\left(\frac{s_1^2}{n_1} + \frac{s_2^2}{n_2}\right)}\right]$$

where $t_{\alpha/2}$ is T-value with an area $\alpha/2$ to its right of an approximate t-distribution with an approximate degrees of freedom (df) calculated using the formula:

$$df = \frac{\left(\frac{s_1^2}{n_1} + \frac{s_2^2}{n_2}\right)^2}{\left(\frac{1}{n_1 - 1}\right)\left(\frac{s_1^2}{n_1}\right)^2 + \left(\frac{1}{n_2 - 1}\right)\left(\frac{s_2^2}{n_2}\right)^2}$$

The common practice is to round down the quantity to the nearest integer.

Implementing Technology for Case I, II and III

Example 9.12

Suppose two independent samples, one of size 22 males, and the other of size 20 females, reveal that the mean age of diagnosis of diabetes for male and female are 44 and 41 years and standard deviations are

9.45 and 8.53 years, respectively. Suppose that the age of diagnosis of diabetes among males and females approximately follow normal distributions with unknown means and unknown standard deviations (σ_1 and σ_2 are unknown for two populations).

(a) Find a 90% CI of $\mu_1 - \mu_2$ assuming that $\sigma_1 = \sigma_2$.

(b) Find a 90% CI of $\mu_1 - \mu_2$ assuming that $\sigma_1 \neq \sigma_2$.

(c) Find a 90% CI of $\mu_1 - \mu_2$ assuming that $\sigma_1 = 10$ and $\sigma_2 = 8$.

Show every computational steps of your work in (a)-(c) and use appropriate tables.

Solution

We will solve problems (a)-(c) first manually by means of tables and then apply Minitab and TI 84 Plus calculator to get output so that we can verify our results.

(a) We have two populations with equal but unknown standard deviations. Therefore, a Pooled T-interval is appropriate for this problem, which is given by:

$$\left[(\bar{x}_1 - \bar{x}_2) - t_{\alpha/2}\sqrt{s_p^2\left(\frac{1}{n_1} + \frac{1}{n_2}\right)}, (\bar{x}_1 - \bar{x}_2) + t_{\alpha/2}\sqrt{s_p^2\left(\frac{1}{n_1} + \frac{1}{n_2}\right)}\right]$$

We have, $\bar{x}_1 - \bar{x}_2 = 44 - 41 = 3$,

$df = 22 + 20 - 2 = 40$

For 90% CI, with $df = 40$, $t_{\alpha/2} = t_{0.05} = 1.684$

$$s_p^2 = \frac{(n_1 - 1)s_1^2 + (n_2 - 1)s_2^2}{n_1 + n_2 - 2} = \frac{(22 - 1)(9.45)^2 + (20 - 1)(8.53)^2}{22 + 20 - 2} = 81.45$$

$$t_{\alpha/2}\sqrt{s_p^2\left(\frac{1}{n_1} + \frac{1}{n_2}\right)} = (1.684)\sqrt{81.45\left(\frac{1}{22} + \frac{1}{20}\right)} = 4.70$$

Therefore, the 90% CI is $[3 - 4.70, 3 + 4.70] = [-1.7, 7.7]$.

Minitab steps for Pooled T-Interval

We follow the steps below:

(1) Select Stat>Basic Statistics>2-Sample t…

(2) Choose option **Summarized data**, and specify **sample sizes** 22 and 20, **sample means** 44 and 41, **Standard deviations** 9.45 and 8.53,

(3) Use **Options**... to specify ***Confidence level***: 90

(4) Check **Assume equal variances** and Click [OK].

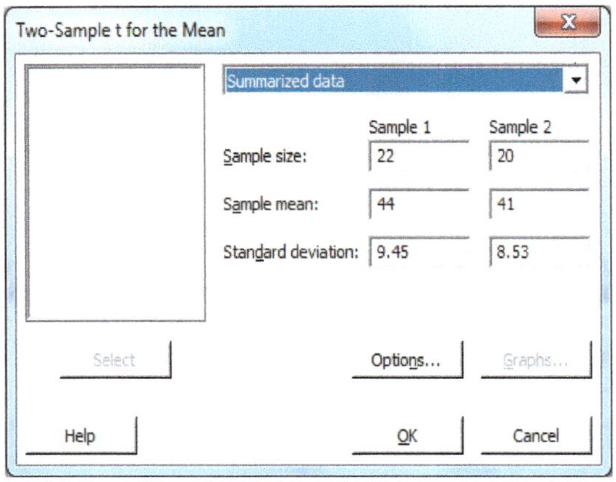

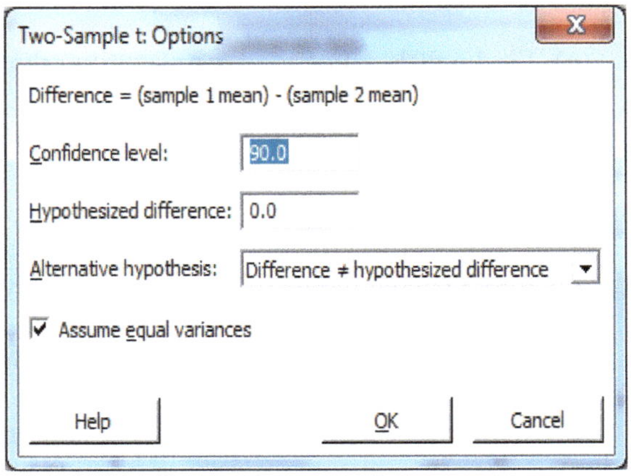

Below is the Minitab output which is fairly close to the value obtained manually before.

Two-Sample T-Test and CI

Sample	N	Mean	StDev	SE Mean
1	22	44.00	9.45	2.0
2	20	41.00	8.53	1.9

Difference = μ (1) - μ (2)
Estimate for difference: 3.00
90% CI for difference: (-1.69, 7.69)
T-Test of difference = 0 (vs ≠): T-Value = 1.08 P-Value = 0.288 DF = 40
Both use Pooled StDev = 9.0247

(b) This is the case with two populations with unknown and unequal standard deviations. Therefore, a non-pooled T-interval is appropriate, which is given by:

$$\left[(\bar{x}_1 - \bar{x}_2) - t_{\alpha/2}\sqrt{\left(\frac{s_1^2}{n_1} + \frac{s_2^2}{n_2}\right)}, (\bar{x}_1 - \bar{x}_2) + t_{\alpha/2}\sqrt{\left(\frac{s_1^2}{n_1} + \frac{s_2^2}{n_2}\right)}\right]$$

$$df = \frac{\left(\frac{s_1^2}{n_1}+\frac{s_2^2}{n_2}\right)^2}{\left(\frac{1}{n_1-1}\right)\left(\frac{s_1^2}{n_1}\right)^2+\left(\frac{1}{n_2-1}\right)\left(\frac{s_2^2}{n_2}\right)^2} = \frac{\left(\frac{89.30}{22}+\frac{72.76}{20}\right)^2}{\frac{1}{21}\left(\frac{89.30}{22}\right)^2+\frac{1}{19}\left(\frac{72.76}{20}\right)^2} = \frac{59.25}{0.78+0.70} = 40.03$$

Rounding down to the nearest integer, $df = 40$, for which $t_{\alpha/2} = t_{0.05} = 1.684$.

$$t_{\alpha/2}\sqrt{\left(\frac{s_1^2}{n_1}+\frac{s_2^2}{n_2}\right)} = 1.684 \times \sqrt{\left(\frac{89.30}{22}+\frac{72.76}{20}\right)} = 4.67$$

Then, a 90% CI is $[3-5.61, 3+5.61] = [-1.67, 7.67]$

Minitab steps for non-Pooled T-Interval

Follow the steps (1)-(3) for Pooled T-Interval mentioned in part (a), and then uncheck **Assume equal variances** option (see the dialog box option selection below to uncheck equal variances option).

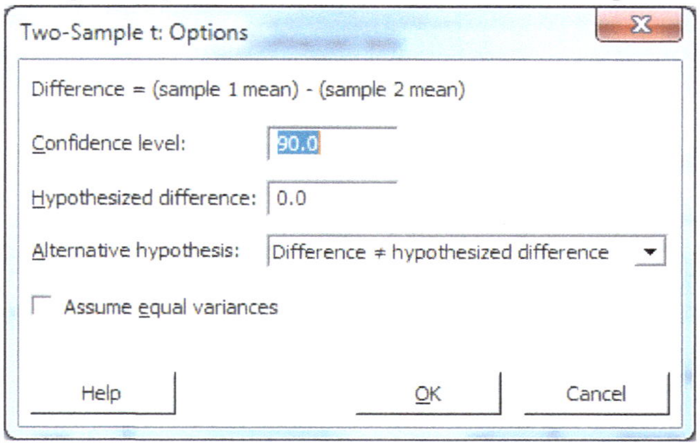

Two-Sample T-Test and CI

Sample	N	Mean	StDev	SE Mean
1	22	44.00	9.45	2.0
2	20	41.00	8.53	1.9

Difference = μ (1) - μ (2)
Estimate for difference: 3.00
90% CI for difference: (-1.67, 7.67)
T-Test of difference = 0 (vs ≠): T-Value = 1.08 P-Value = 0.286 DF = 39

c) This case is solved by Z-Interval:

$$\left[(\bar{x}_1-\bar{x}_2) - z_{\alpha/2}\sqrt{\frac{\sigma_1^2}{n_1}+\frac{\sigma_2^2}{n_2}}, (\bar{x}_1-\bar{x}_2) + z_{\alpha/2}\sqrt{\frac{\sigma_1^2}{n_1}+\frac{\sigma_2^2}{n_2}}\right]$$

For 90% CI, $z_{\alpha/2} = z_{0.05} = 1.645$. Then, $z_{\alpha/2}\sqrt{\frac{\sigma_1^2}{n_1}+\frac{\sigma_2^2}{n_2}} = 1.645 \times \sqrt{\frac{100}{22}+\frac{64}{20}} = 4.58.$

So, the 90% CI is: $[3-4.58, 3+4.58]=[-1.58, 7.58]$

TI 84 Plus instruction for solving above problems

We follow the steps below to get a 90% CI of mean difference:

1) Press STAT, select TESTS, choose **0: 2-SampTInt…**and Press **Enter.**
2) Move the cursor to **Stats** and press **ENTER**.
3) Specify appropriate Stats:: $\bar{x}_1$:44, sx_1: 9.45, n1:22, $\bar{x}_2$: 41, sx_2: 8.53, n2:20, C-level:.9.
4) Select **Pooled**: **YES** for pooled T-interval or **Pooled: NO** for non-pooled T-interval.
5) Move the cursor to Calculate and press ENTER

The above steps and result are demonstrated in figures below:

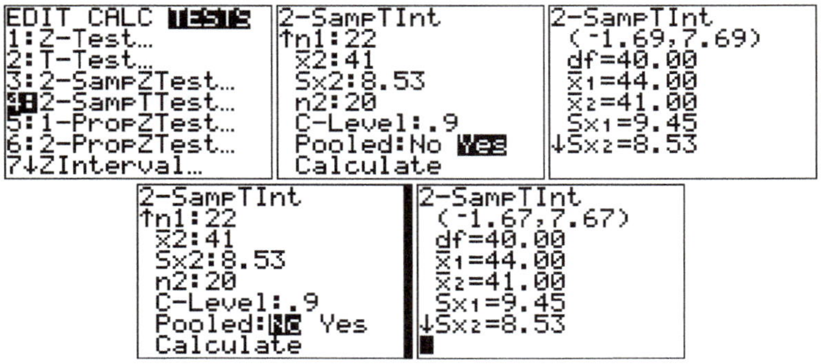

As we see, manually computed values of CIs and those obtained by Minitab and TI 84 Plus calculator give the same result.

Section 9.4: Estimating difference of two population means using dependent samples

In this section, we study two populations whose measurements are paired. For example, let μ_1 and μ_2 be the means of two populations that represent before and after measurements or measurements of effects under two different treatments, etc. Indeed, we want to estimate difference of two population means $\mu_1-\mu_2$ by a confidence interval estimate. This interval estimate is called a ***Paired T-Interval***.

Definition: Paired T-Interval

Let us have a sample of n paired measurements: $(x_1, y_1), (x_2, y_2), \dots, (x_n, y_n)$.

Let $d_i=x_i-y_i$; $i=1,2,\dots,n$; $\bar{d}=\bar{x}-\bar{y}$ and $s_d^2=\frac{\sum(d_i-\bar{d})^2}{n-1}$.

Under above notations, a $100(1-\alpha)\%$ paired T-interval of $\mu_1-\mu_2$, with confidence level $C=1-\alpha$, is given by

$$\left[\bar{d}-t_{\alpha/2}\left(\frac{s_d}{\sqrt{n}}\right), \bar{d}+t_{\alpha/2}\left(\frac{s_d}{\sqrt{n}}\right)\right]$$

where $t_{\alpha/2}$ is T-value with an area $\alpha/2$ to its right of a t-distribution with $(n-1)$ degrees of freedom. In above expression, the quantity $t_{\alpha/2}\left(\frac{s_d}{\sqrt{n}}\right)$ is the margin of error of $100(1-\alpha)\%$ CI. By writing $m = t_{\alpha/2}\left(\frac{s_d}{\sqrt{n}}\right)$, the above CI can equivalently be expressed as

$$[\bar{d} - m\,, \bar{d} + m]$$

Assumptions for Paired T-Interval

A paired T-interval is valid under the fulfilment of the following assumptions:

(1) The observations of paired differences represent a SRS,
(2) Either the sample is large ($n \geq 30$) or the population of paired differences follows a normal distribution,
(3) Standard deviation of the population of paired differences is unknown.

Example 9.13

A paired sample of 10 domestic itinerary fares (in dollars) is considered from the top 100 ranked US Airports for Quarter 4 of 2012 (x) and 2013 (y). The sample data appears in table below:

Rank	Airport	Q42012 (x)	Q42013 (y)	$d = x - y$
83	Knoxville, TN	470.19	402.89	67.30
84	White Plains, NY	396.17	414.21	-18.04
31	Kansas City, MO	356.73	363.36	-6.62
55	Palm Beach, FL	316.14	337.82	-21.68
34	Oakland, CA	302.99	304.06	-1.08
46	Hartford, CT	408.09	381.76	26.33
41	Cleveland, OH	461.46	458.53	2.93
33	Raleigh/Durham, NC	354.24	372.33	-18.09
17	Baltimore, MD	323.29	328.25	-4.96
58	Oklahoma City, OK	411.52	400.42	11.10
Average		380.08	376.36	3.72

Source: Bureau of Transportation Statistics, Origin and Destination Survey: DB1BTicket
http://www.rita.dot.gov/bts/airfares/programs/economics_and_finance/air_travel_price_index/html/table_09.html

Suppose that the difference of itinerary fares in Quarter 4 of 2012 and 2013 in top 100 Airports follow a normal distribution with unknown mean and standard deviation. Also, from the sample it follows that $\bar{d} = 3.72$ and $s_d = 26.7$.

Find a 95% CI estimate of unknown mean difference μ_d. Using the 95% CI of μ_d, can you conclude whether the mean difference is significantly different than 0?

Solution

We have, $\bar{d} = 3.72$ and $S_d = 26.7$. For $df = 9$, $t_{\alpha/2} = t_{0.025} = 2.262$. Then, $m = t_{\alpha/2}\left(\frac{S_d}{\sqrt{n}}\right) = 2.262 * \frac{26.7}{\sqrt{10}} = 19.1$. Therefore, a 95% CI is $[3.72 - 19.1, 3.72 + 19.1] = [-15.38, 22.82]$.

The 95% CI of μ_d includes 0, which implies than the mean difference may not be significantly different from 0.

<u>Minitab steps for paired T-Interval</u>

We follow the steps below:

(1) Select Stat>Basic Statistics>Paired t…

(2) Choose option **Summarized data**, and specify **sample size:10**, **sample mean:** 3.72, **Standard deviation:** 26.7,

(3) Use option to specify ***Confidence level***: 95

(4) Click [OK] two times.

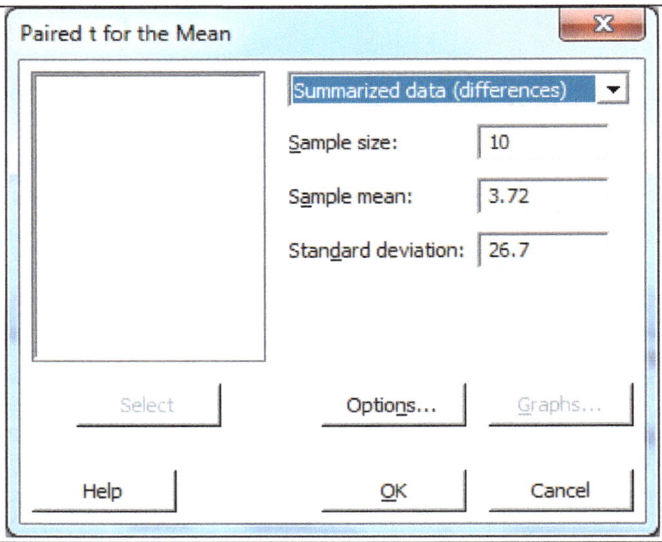

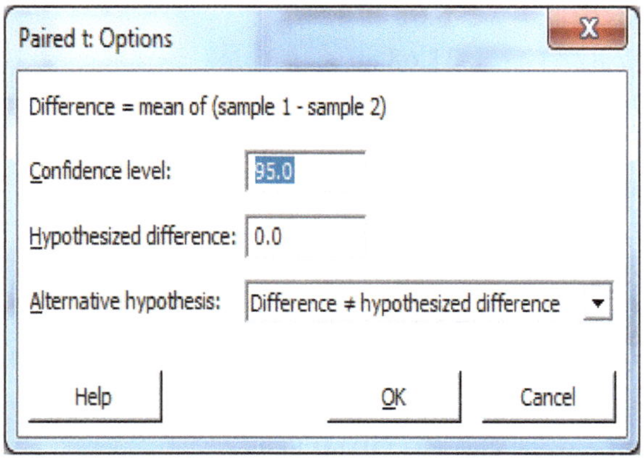

Paired T-Test and CI

	N	Mean	StDev	SE Mean
Difference	10	3.72	26.70	8.44

95% CI for mean difference: (-15.38, 22.82)

T-Test of mean difference = 0 (vs ≠ 0):
T-Value = 0.44 P-Value = 0.670

Chapter 9: Practice Problems

Section 9.1

(9.1) Suppose the data below represents a sample of salary (in thousands of dollars) of 12 doctors. Suppose the salary of the doctors approximately follows a normal distribution with an unknown mean and standard deviation 30.

127	132	138	141	144	146	152	154	162	171	177	192

(a) What is the point estimate of average salary μ of all medical doctors?
(b) Find the margin of error for a 90% CI of μ.
(c) Find a 90% CI of μ and interpret the result.
(d) Find the margin of error for a 99% CI of μ.
(e) Find a 99% CI of μ.
(f) How is the interval different in (e) different from the interval in (c)?
(g) What sample size would you recommend for a 95% confidence interval of μ with a margin of error of 3?

(9.2) What is the average time to finish the Final Exam in a large class? In order to answer this question, the instructor takes a random sample of 20 students and records their time in minutes to finish the exam. Suppose that the time to finish the exam follows a normal distribution with a standard deviation of 10.5 minutes. The data as observed from the sample appears in table below:

48	61	54	64	40	41	58	33	45	60
58	71	64	74	55	61	58	53	71	75

Let μ be the mean time to finish the Exam.
(a) Find a point estimate of μ.
(b) Find a 95% CI of μ. Interpret your result.
(c) What sample size will be required for a 95% confidence interval for μ within 2.5 minutes?

(9.3) Suppose the data below refers to a sample of gas mileage, measured as miles/gallon, of 30 cars of a particular model.

14	22	20	19	20	18	18	24	16	22	25	22	28	14	14
34	19	18	19	18	18	17	12	21	12	16	16	22	24	19

Suppose that the gas mileage of cars of this particular model has approximately a normal distribution with an unknown mean and a standard deviation 5.2.
Let μ be the unknown mean of gas mileage of all cars of the model.
(a) Find a point estimate of μ.
(b) Find a 90% CI of μ. Interpret your result.
(c) Explain how the assumptions for the validity of the interval in (b) are satisfied by the problem.
(d) What would be the recommended sample size for a 90% CI of μ with a margin of error of 1.5 miles/gallon?

(9.4) We want to estimate the average age of students taking an Elementary Statistics course in a large class. To this end, an SRS of size 25 is considered. The data below refers to age, in years, of 25 students in the sample:

18	18	18	20	21	19	22	21	21	18	18	28	24
18	20	24	32	18	18	18	18	32	19	22	18	

Assume that the age of all students taking the Elementary Statistics class follows a normal distribution with an unknown mean and standard deviation $\sigma = 5.1$ years. Let μ be the mean age of all students taking the class.

(a) Find a point estimate of μ.
(b) Find a 90% CI of μ and interpret the result.
(c) Compute the margin of error for a 95% confidence interval estimate.
(d) Find the required sample size for a 95% confidence interval with a margin of error of 2.25 years.

(9.5) Suppose that on the basis of a sample of 25 payments for Sprint customers, the average monthly bill is found to be \$143, including taxes and fees. Assume that the population of monthly bills of all Sprint customers follows a normal distribution with standard deviation of \$9.50, including taxes and fees. Let μ be the mean of monthly bills of all Sprint customers.

(a) Find a point estimate of μ.
(b) What CI method should you consider for a CI estimate of μ?
(c) Find a 90% confidence interval estimate of μ and interpret the result.
(d) Are the assumptions of CI estimate met? How?

(9.6) We wish to estimate the unknown mean speed of all cars passing through a neighborhood on a given day. On the basis of a sample 30 cars, the average speed is found to be 28.5 miles/hour. Suppose that speed of all cars passing through the neighborhood follows a normal distribution with standard deviation $\sigma = 4$ miles/hour.

Let μ be the mean speed of all cars passing through the neighborhood.

(a) What is the point estimate of μ?
(b) Compute the margin of error for a 95% CI estimate of μ.
(c) Compute a 95% CI estimate of μ and interpret the result.
(d) Compute a 99% CI estimate of μ. How is this result different from the result in (c)?
(e) If a 95% CI estimate is desired with a margin of error of 1.5, what sample size should you recommend?

Section 9.2

(9.7) Suppose the data below represents a sample of salary (in thousands of dollars) of 10 doctors. Suppose the salary of the doctors approximately follows a normal distribution with an unknown mean and unknown standard deviation.

132	138	141	144	146	152	154	162	171	177

(a) What is the point estimate for average salary μ of all medical doctors?
(b) What CI is appropriate for μ if a CI estimate is desired and why?
(c) Find a 90% CI of μ and interpret the result.
(d) Find a 99% CI of μ. How is this interval different from the interval in (c)?

(9.8) The data below is a sample of time (in minutes) to finish an exam by 12 students in a class. It is assumed that the time to finish the exam is approximately normal with unknown mean and unknown standard deviation.

61	54	64	40	41	58
71	64	74	55	61	58

(a) What is the point estimate of the average time, μ, it takes to finish the exam of all students in the class?
(b) What CI method should you use if a CI estimate is desired? Justify how assumptions of this method are satisfied.
(c) Find a 90% CI of μ.
(d) Find a 95% CI of μ and interpret your result.
(e) How is the interval in (d) different from the interval in (c)?

(9.9) Below is a sample of gas mileage, in miles/gallon, of 10 domestic automobiles of a particular model. Suppose that the gas mileage of automobiles of this particular model follows approximately a normal distribution with unknown mean and unknown standard deviation.

20	18	18	24	16
18	18	17	12	21

Let μ be the unknown mean of gas mileage of all cars of the model.

(a) Find a point estimate of μ.
(b) What CI method is appropriate for a CI estimate? How are the assumptions of the method satisfied?
(c) Find a 90% CI of μ and interpret the interval obtained.
(d) Find a 95% CI estimate of μ.
(e) Do the CI estimates in (c) and (d) differ? Why?

(9.10) Suppose that on the basis of a sample of 20 payments for Sprint customers, the mean and standard deviation of monthly bill is found to be \$140 and \$10, respectively, including taxes and fees. Assume that the population of monthly bills of all Sprint customers follows a normal distribution with an unknown mean and unknown standard deviation.
Let μ be the mean of monthly bills of all Sprint customers.

(a) What is the point estimate of μ?
(b) If a CI estimate is desired, what CI method should you consider?
(c) Are the assumptions of CI estimate required in (b) met? How?
(d) Find a 90% CI estimate of μ and interpret the result.
(e) Compute a 99% CI estimate of μ and explain how this result is different from the result in (d).

(9.11) In order to estimate the mean speed of all cars passing through a neighborhood, a sample of 25 cars is considered. On the basis of the sample of 25 cars, the mean and standard deviation of speed of cars are 28.5 miles/hour and 5 miles/hour, respectively. Suppose that speed of all cars passing through the neighborhood follows approximately a normal distribution with an unknown standard deviation.
Let μ be the mean speed of all cars passing through the neighborhood.

(a) What is the point estimate of μ?
(b) Compute a 95% CI estimate of μ and interpret the result.
(c) Compute a 99% CI estimate of μ. How is this result different from the result in (b)?

Section 9.3

(9.12) Suppose on the basis of two independent samples, we have $\bar{x}_1 = 44$, $\bar{x}_2 = 41$, $s_1 = 9$, $s_2 = 8$, $n_1 = 15$ and $n_2 = 17$. Assume that the two populations are approximately normally distributed with unknown but equal variances.
Answer the following questions:
(a) Find a point estimate of $\mu_1 - \mu_2$.
(b) Find a 90% CI of $\mu_1 - \mu_2$ and interpret the result.
(c) Find a 95% CI of $\mu_1 - \mu_2$.
(d) Compare the results in (b) and (c).
(e) Now, assume that the two populations are approximately normally distributed with unknown and unequal variances. What CI should you consider for a 90% CI? Find a 90% CI of $\mu_1 - \mu_2$ in this new scenario.

(9.13) Suppose on the basis of a sample of 15 payments of Verizon customers, the mean and standard deviation of monthly bills are \$147 and \$9.5, respectively, including taxes and fees. On the basis of a sample of 20 payments for Sprint customers, the mean and standard deviation of monthly bills are found to be \$143 and \$10.9, respectively, including taxes and fees. Suppose the monthly bills for Verizon and Sprint customers approximately follow normal distributions with unknown means and unknown but equal standard deviations.
(a) Find a point estimate of $\mu_1 - \mu_2$.
(b) Find a 90% CI of $\mu_1 - \mu_2$.
(c) Find a 95% CI of $\mu_1 - \mu_2$.
(d) Compare the results in (b) and (c).
(e) Are the assumptions for a CI estimate met by the problem? How?

(9.14) In reference to the Exercise (9.13), suppose the standard deviations of bills of all customers for Verizon and Sprint are known to be \$11 and \$9.5, respectively.
(a) Find a point estimate of $\mu_1 - \mu_2$.
(b) What CI method should you use for a CI estimate?
(c) Find a 95% CI of $\mu_1 - \mu_2$.
(d) Find a 99% CI of $\mu_1 - \mu_2$.
(e) Compare the results in (c) and (d).
(f) Are the assumptions for a CI estimate met by the problem? How?

(9.15) Suppose on the basis of two independent samples, we have $n_1 = 25$, $n_2 = 20$, $\bar{x}_1 = 30$ and $\bar{x}_2 = 28$. Assume that the two populations are approximately normally distributed with unknown means and standard deviations $\sigma_1 = 9$ and $\sigma_2 = 10$.
Answer the following questions:
(a) Find a point estimate of $\mu_1 - \mu_2$.

(b) Find a 90% CI of $\mu_1 - \mu_2$ and interpret the result.
(c) Find a 95% CI of $\mu_1 - \mu_2$.
(d) Compare the results in (b) and (c).

(9.16) Given two independent samples, suppose we have the followings:

$$n_1 = 20, n_2 = 20, \bar{x}_1 = 25 \text{ and } \bar{x}_2 = 27.$$

Assume that the two populations are approximately normally distributed with unknown means and standard deviations $\sigma_1 = 6$ and $\sigma_2 = 7$.
Answer the following questions:

(a) Find a point estimate of $\mu_1 - \mu_2$.
(b) Find a 90% CI of $\mu_1 - \mu_2$ and interpret the result.
(c) Find a 95% CI of $\mu_1 - \mu_2$.
(d) Compare the results in (b) and (c).

Section 9.4

(9.17) Suppose that for a sample of 36 paired differences, we have $\bar{d} = 1.54$ and $s_d = 2.21$. Let μ_d be the mean of the paired differences. Suppose the population of paired differences follows approximately a normal distribution with unknown mean and standard deviation.

(a) What is the point estimate of μ_d?
(b) Find a 95% CI estimate of μ_d.
(c) Find a 99% CI estimate of μ_d.
(d) Do the assumptions of CI estimate meet? How?

(9.18) Suppose that a sample of 9 video game players rated a game on a 10-point scale after playing the game first and second time. The scale consists of numbers 1 through 10, with a lower number indicating the lower satisfaction rate. The 9 ordered pair data appears in the table below, where x and y refer to the rating after playing the game first and second time, respectively.

x	8	7	10	6	8	7	8	7	4
y	7	7	8	6	7	7	9	6	5
$d = x - y$	1	0	2	0	1	0	-1	1	-1

Suppose the population of paired differences follows approximately a normal distribution with unknown mean and standard deviation. Let μ_d be the mean of all paired difference of ratings in the population.
Verify that $\bar{d} = 0.33$ and $s_d = 1$.

(a) Find a point estimate of μ_d.
(b) Find a 95% CI estimate of μ_d.
(c) Find a 99% CI estimate of μ_d.
(d) Do the assumptions of CI estimate meet? How?

Chapter 9: Answer to selected problems

(9.1)

(a) Point estimate of μ is $\bar{x} = 153$ thousand dollars.
(b) $m = 14.24$.
(c) A 90% CI of μ is [138.76,167.24].
(d) $m = 22.31$
(e) A 99% CI of μ is [130.69,175.31].
(f) The interval in (e) is wider than in (c).
(g) $n = 385$.

(9.2) (a) $\bar{x} = 57.2$ minutes (b) A 95% CI of μ is [52.6,61.8] (c) The sample size required is 68.
(9.3) (a) 19.6 miles/gallon (b) A 90% CI of μ is [17.8,20.9] (c) $n = 33$.
(9.4) (a) 20.92 years (b) 90% CI: [19.24,22.60] (c) $m = 2.0$.
(9.5) (a) $143 (b) Z-interval (c) A 90% CI of μ is [139.87,146.13] (d) Yes, try yourself.
(9.6)

(a) 28.5 miles/hour.
(b) $m = 1.431$.
(c) A 95% CI of μ is [27.069,29.931].
(d) A 99% CI of μ is [26.62, 30.38]. A 99% CI is wider than 95% CI.
(e) $n = 28$.

(9.7)

(a) The point estimate is $\bar{x} = 151.7$ thousand dollars.
(b) One sample T-Interval because population standard deviation σ is unknown. Also, the assumptions of t-interval are satisfied (normal population, random sample and unknown population standard deviation).
(c) A 90% CI of μ is [143.26, 160.14]. Try yourself.
(d) A 99% CI of μ is [136.74, 166.66].

(9.8)

(a) The point estimate is $\bar{x} = 58.42$ minutes.
(b) One sample T-interval. Exam time is approximately distributed as normal, we have a random sample and population standard deviation is unknown.
(c) A 90% CI of μ is [(53.11, 63.72].
(d) A 95% CI of μ is [51.92, 64.92]. Try yourself.
(e) The interval in (d) is wider than interval in (c).

(9.9)

(a) $\bar{x} = 18.2$ miles/gallon.
(b) One sample T-interval. Try yourself.
(c) A 90% CI of μ is [(16.371, 20.029].
(d) A 95% CI of μ is [15.94, 20.46]. Try yourself.
(e) Try yourself.

(9.10)

(a) $\bar{x} = \$140.00$
(b) One sample T-interval.

(c) Yes. Try yourself.
(d) A 90% CI of μ is [136.13, 143.87].
(e) A 99% CI of μ is [133.6, 146.4].

(9.11)

(a) $\bar{x} = 28.5$ miles/hour.
(b) A 95% CI of μ is [26.4, 30.6].
(c) A 99% CI of μ is [25.7, 31.3].

(9.12)

(a) A point estimate of $\mu_1 - \mu_2$ is $\bar{x}_1 - \bar{x}_2 = 44 - 41 = 3$.
(b) A 90% CI of $\mu_1 - \mu_2$ is [-2.1, 8.1].
(c) A 95% CI of $\mu_1 - \mu_2$ is [-3.14, 9.14].
(d) Try yourself.
(e) Two-sample non-pooled T-interval. Try yourself.

(9.13)

(a) A point estimate of $\mu_1 - \mu_2$ is $\bar{x}_1 - \bar{x}_2 = 147 - 143 = 4$.
(b) A 95% CI of $\mu_1 - \mu_2$ is [-1.97, 9.97].
(c) A 95% CI of $\mu_1 - \mu_2$ is [-3.18, 11.18].
(d) Try yourself.
(e) Try yourself.

(9.14)

(a) A point estimate of $\mu_1 - \mu_2$ is $\bar{x}_1 - \bar{x}_2 = 147 - 143 = 4$.
(b) Try yourself.
(c) A 95% CI of $\mu_1 - \mu_2$ is [-2.95, 10.95].
(d) A 95% CI of $\mu_1 - \mu_2$ is [-5.14, 13.14].
(e) Try yourself.
(f) Try yourself.

(9.15)

(a) A point estimate of $\mu_1 - \mu_2$ is $\bar{x}_1 - \bar{x}_2 = 30 - 28 = 2$.
(b) A 90% CI of $\mu_1 - \mu_2$ is [-2.7, 6.7].
(c) A 95% CI of $\mu_1 - \mu_2$ is [-3.6, 7.6].
(d) Try yourself.

(9.17)

(a) A point estimate of μ_d is $\bar{d} = 1.54$.
(b) A 95% CI of μ_d is [0.79, 2.29].
(c) A 99% CI of μ_d is [0.54, 2.54].
(d) Try yourself.

(9.18)

(a) A point estimate of μ_d is $\bar{d} = 0.33$.
(b) A 95% CI of μ_d is [-0.44, 1.1].
(c) A 99% CI of μ_d is [-0.79, 1.45].
(d) Try yourself.

CHAPTER 10
TEST OF HYPOTHESES ABOUT MEANS

Introduction

An essential component of a statistical inference is a test of hypothesis. A ***test of hypothesis*** is a statistical procedure for making a decision about rejection or acceptance of a statistical hypothesis on the basis of sample information. For example, it is believed that internet users (about 80% of the total population) spend, on average, 13 hours per week on internet. A researcher wonders if the average hours spent on internet by people in a large community is different from the national average.

Suppose, on the basis of a sample of 40 people in the community, the average hours spent on internet is found to be 15.2 hours/week with a standard deviation of 5.4 hours/week. Should it then be concluded that the mean hours spent on internet by the people in the community is higher or same as the national average?

Solving this kind of problems involves the technique of test of hypotheses.

This chapter devotes to basics of test of hypotheses and carry out tests about means. The specific objectives of this chapter are as follows:

- Basics of test of hypothesis
- Test of hypothesis about single mean
- Test of hypothesis about two population means
- Test of hypothesis using paired T-test

Section 10:1 Defining key terms for a test of hypothesis

In this section, we define the following terms:

- Null and alternative hypotheses
- Type I and Type II errors
- Level of significance

The above terms play a vital role in the test of hypothesis. We apply these concepts in subsequent sections of this chapter for test of hypotheses about means, and will carry out throughout the rest of the book.

Definition: Null and alternative hypotheses

Any statement of "equality", "no difference" or "no effect" about the true parameter that is being tested for possible **rejection** is called a ***null hypothesis*** and is usually denote by H_0 (read as "H-naught"). We use an equal sign between the unknown parameter and the hypothesized value in this statement. In other words, a null hypothesis generally takes the form:

$$H_0: unknown\ parameter = a\ hypothesized\ or\ specified\ value\ of\ the\ parameter$$

Any sensible hypothesis other than the null hypothesis which researchers try to establish is called an ***alternative hypothesis*** and is usually denoted H_a (read as "H-a"). In general, a less than, a greater than or an unequal sign is placed between the unknown parameter and the hypothesized value in this statement. Thus, an alternative hypothesis takes any of the following forms:

$$H_a: unknown\ parameter < a\ hypothesized\ or\ specified\ value\ of\ the\ parameter$$
$$H_a: unknown\ parameter > a\ hypothesized\ or\ specified\ value\ of\ the\ parameter$$
$$H_a: unknown\ parameter \neq a\ hypothesized\ or\ specified\ value\ of\ the\ parameter$$

Suppose we want to test a hypothesis about the unknown population mean μ. Let μ_0 be the hypothesized or specified value of μ. Therefore, the null hypothesis for a test of hypothesis regarding a value of μ will be specified as:

$$H_0: \mu = \mu_0$$

The alternative hypothesis could be any of the following three forms:

(a) $H_a: \mu < \mu_0\ (left\ tailed\ alternative)$
(b) $H_a: \mu > \mu_0\ (right\ tailed\ alternative)$
(c) $H_a: \mu \neq \mu_0\ (two\ tailed\ alternative)$

A test of a hypothesis is called a ***left tailed***, ***right tailed*** or ***two tailed*** test if the alternative is specified as (a), (b) or (c) above, respectively. A test of a hypothesis is called a ***one-tailed*** test if it is either a left-tailed or a right-tailed test.

Note that, given a sample, the value of the statistic is always known; we do not need to make any hypothesis about it. Therefore, ***the statements in null and alternative hypotheses are always about the unknown parameter, not about any statistic***.

Example 10.1

Suppose we want to test a claim that the population of all students in a college, on average, sleeps for 6.30 hours per day. How would you set up the null and alternative hypotheses for the test?

Solution

The null hypothesis would be $H_0: \mu = 6.30$ (**Recall**: this is a statement of equality).

The alternative hypothesis can take any of the following three possible forms:

$H_a: \mu < 6.30$ (left- tailed)
$H_a: \mu > 6.30$ (right- tailed)
$H_a: \mu \neq 6.30$ (two-tailed)

Example 10.2

A researcher believes that foreign students spend more time in library than native students. The researcher will be using samples from native and foreign students to test the belief. Let the unknown mean time spent in library by the foreign and native students be denoted by μ_1 and μ_2, respectively. How should the researcher set up the null and alternative hypotheses for the test?

Solution

The null hypothesis should be set to $H_0: \mu_1 = \mu_2$ (**Recall**: this is a statement of equality or no difference). The alternative hypothesis could be: $H_a: \mu_1 > \mu_2$ (right-tailed), which the researcher wishes to establish.

Example 10.3

We want to test if there is any gender difference in average hours of sleep for all students in a college. Let the unknown mean hours of sleep by male and female students be denoted by μ_m and μ_f, respectively. How should we set up the null and alternative hypotheses for the test? Specify whether your choice refers to a one-tailed test or a two-tailed test.

Solution

The null hypothesis is $H_0: \mu_m = \mu_f$ (**Recall**: this is a statement of no difference).
The alternative hypothesis is $H_a: \mu_m \neq \mu_f$ (two-tailed).

Definitions: Type I error and level of significance α

An error due to the rejection of H_0 when it is true, is popularly known as the ***Type I*** error. The ***level of significance***, denoted by α, is the probability of Type I error. That is,

$$\alpha = P(Reject\ H_0\ when\ H_0\ is\ true) = P(Type\ I\ error)$$

Generally speaking, we do not want to reject $\mathbf{H_0}$ which is actually true. In other words, we wish to make a wrong decision as rarely as possible. Therefore, the value of $\boldsymbol{\alpha}$ is generally set to 10%, 5% and 1%, which under the usual notation, we write as $\boldsymbol{\alpha = 0.10}$, $\boldsymbol{\alpha = 0.05}$ and $\boldsymbol{\alpha = 0.01}$, respectively.

Definition: Type II error

An error due to the acceptance of H_0 when it is indeed false, is called the ***Type II*** error. The probability of making a Type II error is denoted by β. That is,

$$\beta = P(Accept\ H_0\ when\ H_0\ is\ false) = P(Type\ II\ error)$$

Section 10.2: Z-test for a mean

Let the mean μ of a population be unknown. We wish to test $H_0: \mu = \mu_0$ when σ is known. In order to make a decision regarding the rejection or acceptance of the null hypothesis, we need a test statistic.

Definition: Test statistic

A ***test statistic*** is a formula that consists of a statistic, a hypothesized value of the parameter specified in the null statement and a standard deviation of the statistic (often, referred to as the standard error).
The test statistic to test $H_0: \mu = \mu_0$ when σ is known is given by

$$Z = \frac{\bar{x} - \mu_0}{\sigma/\sqrt{n}}$$

and is called a ***Z-test statistic***, which follows a $N(0,1)$ distribution.

Given a sample, the observed value of the test statistic Z is generally denoted by z. Unless otherwise specified, the upper and lower case distinction between Z and z are assumed while doing a test of hypothesis. Often, the decision for the rejection or acceptance of the null hypothesis is made by comparing the observed z with a standard value of the distribution of Z, which we call a critical value.

Definition: Critical value

Suppose we wish to test $H_0: \mu = \mu_0$ vs $H_a: \mu > \mu_0$ when σ is known. The test statistic for the test is $= \frac{\bar{x}-\mu_0}{\sigma/\sqrt{n}}$.

Suppose given a sample, the observed value for the test statistic is z. Now, what would be the decision regarding the rejection or acceptance of $H_0: \mu = \mu_0$?

Intuitively, if z is close to zero, we accept $\mathbf{H_0: \mu = \mu_0}$, and if z is large, we reject $\mathbf{H_0: \mu = \mu_0}$ in favor of $\mathbf{H_a: \mu > \mu_0}$.

The question is how close is close or how large is large? This requires defining a cut-off value z_α, called the critical value at significance level α. If z is less than the critical value z_α, we accept $H_0: \mu = \mu_0$ vs $H_a: \mu > \mu_0$ or if z is greater than critical value z_α we reject $H_0: \mu = \mu_0$ vs $H_a: \mu > \mu_0$.

A ***critical value*** is a cut-off value of the distribution of the test statistic with which the observed value of the test statistic is compared for the decision of rejection or acceptance of the null hypothesis. A critical value depends on the level of significance α at which the test is performed, and whether the test is one-tailed or two-tailed.

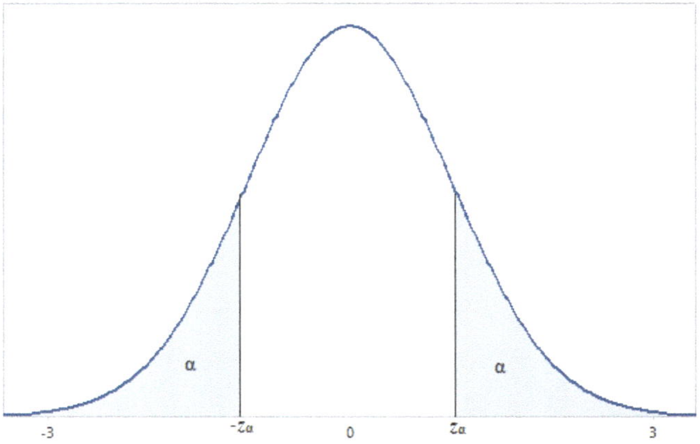

At significance level α, the critical value of the test $H_0: \mu = \mu_0$ (when σ is known) is

(1) z_α if $H_a: \mu > \mu_0$ (i.e., right-tailed test).
(2) $-z_\alpha$ if $H_a: \mu < \mu_0$ (i.e., left-tailed test).
(3) $\pm z_{\alpha/2}$ if $H_a: \mu \neq \mu_0$

where z_α is a value under the standard normal curve with area α to its right, and $-z_\alpha$ is a value under the standard normal curve with area α to its left. Note that, by the symmetry of the normal curve, $-z_\alpha = z_{1-\alpha}$.

The table below provides critical values (approximated up to two decimal places) for α set to 0.1, 0.05. and 0.01

α	0.10	0.05	0.01
Critical value (1-tailed test)	± 1.28	± 1.65	± 2.33
Critical value (2-tailed test)	± 1.65	± 1.96	± 2.58

Decision rule using a critical value

(1) Reject $H_0: \mu = \mu_0$ if $z < -z_\alpha$, when $H_a: \mu < \mu_0$ (i.e., left-tailed test).

(2) Reject $H_0: \mu = \mu_0$ if $z > z_\alpha$, when $H_a: \mu > \mu_0$ (i.e., right-tailed test).

(3) Reject $H_0: \mu = \mu_0$ if $z > z_{\alpha/2}$ or $z < -z_{\alpha/2}$, when $H_a: \mu \neq \mu_0$ (i.e., two-tailed test). In other words, reject $H_0: \mu = \mu_0$ if $|z| > z_{\alpha/2}$, when $H_a: \mu \neq \mu_0$.

(4) Accept $H_0: \mu = \mu_0$ if we fail to reject it.

The above choices of critical values are shown in following graphs:

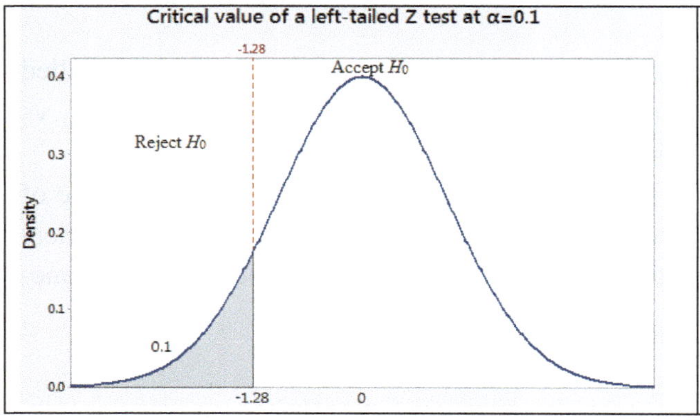

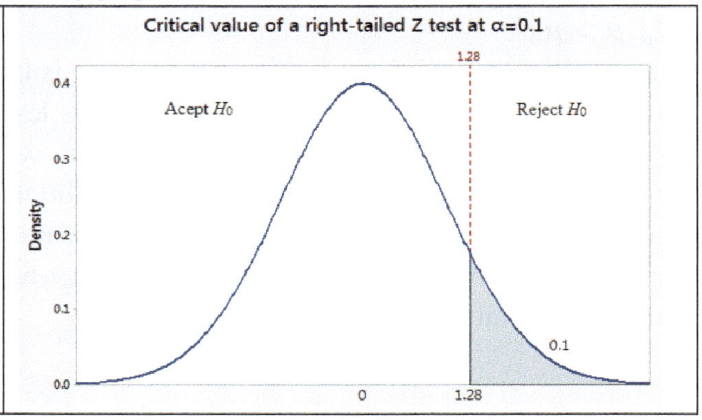

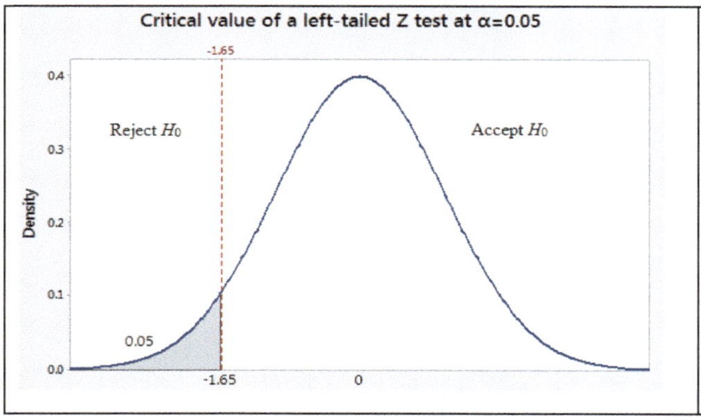

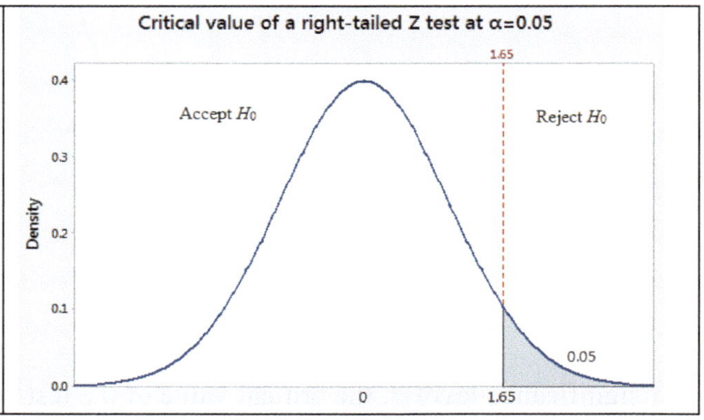

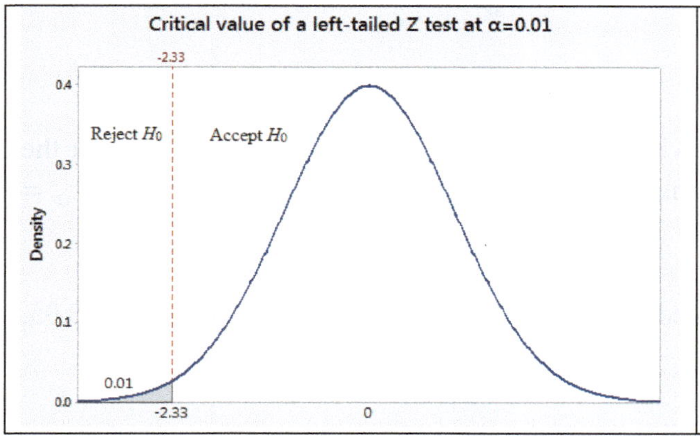

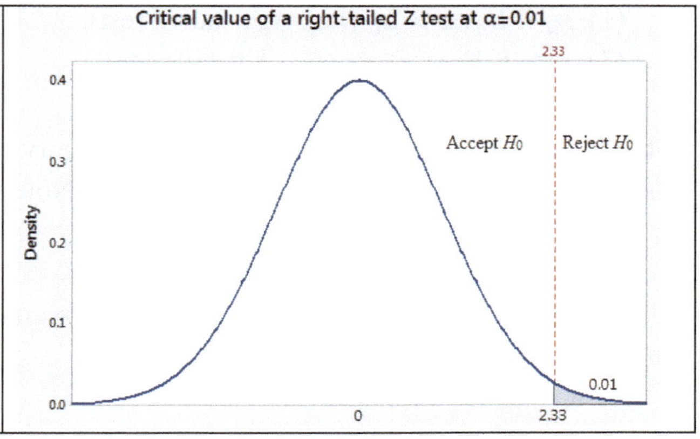

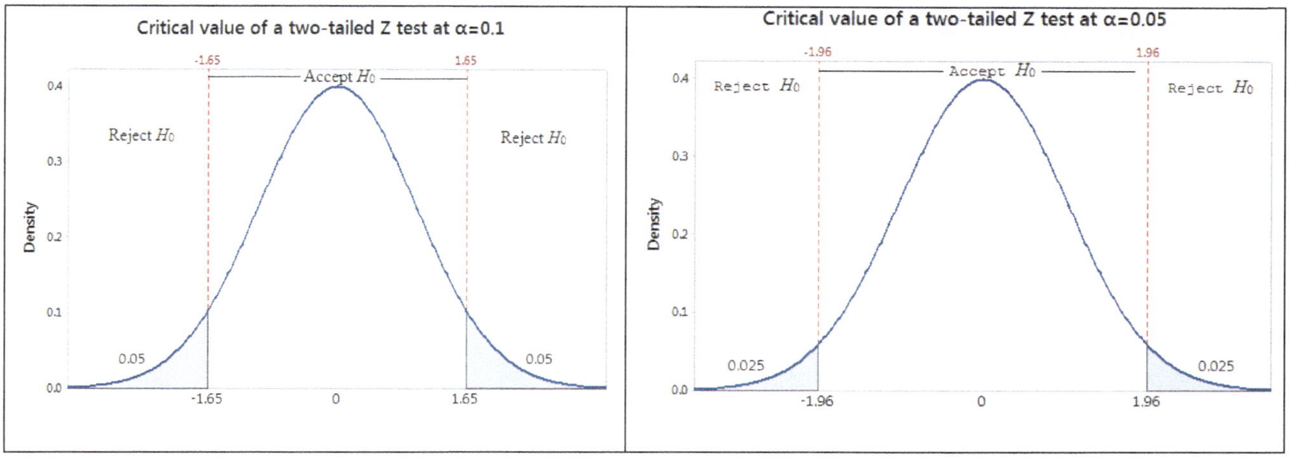

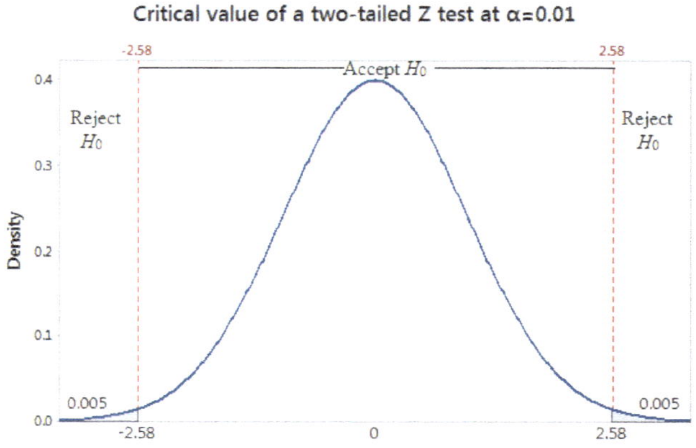

Assumptions for Z-test

The Z-test requires following three assumptions:

(1) Data represents an SRS,

(2) Either the population is normal or the sample is large ($n \geq 30$),

(3) σ is known.

Example 10.4

Let μ be the unknown mean gas mileage of a certain model of compact SUV. Suppose that the gas mileage of this SUV model approximately follows a normal distribution with a population standard deviation of 4 mpg. On the basis of a sample of 45 compact SUVs, the mean gas mileage is found to be 28 mpg. Test H_0: $\mu = 30$ mpg versus H_a: $\mu > 30$ mpg by answering following questions:

(a) Write out the test statistic and value of the test statistic for the test.

(b) Find the critical value of the test at the significance level $\alpha = 0.05$.

(c) What is your conclusion regarding rejection or acceptance of the test at $\alpha = 0.05$?

Solution

(a) The test statistic is $Z = \frac{\bar{x}-\mu_0}{\sigma/\sqrt{n}}$.

Here, we have $\bar{x} = 28$ mpg, $\mu_0 = 30$, $\sigma = 4$ mpg and $n = 45$ and hence the value of the test statistic is $z = \frac{28-30}{4/\sqrt{45}} = -3.35$.

(b) Since $H_a: \mu > 30$, the critical value of the test at $\alpha = 0.05$ is $z_\alpha = z_{0.05} = 1.645$ (using Table T3).

In TI – 84 Plus calculator, we can compute critical value using ***invNorm***() function with the left area of 0.95 as argument of the function:

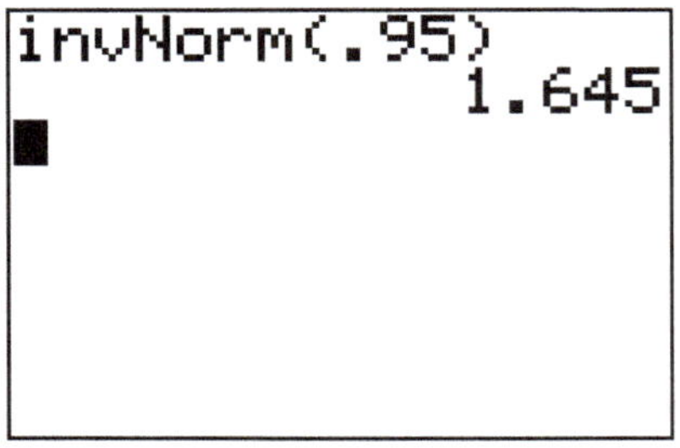

(c) Since the value of the test statistic $z = -3.35$ is less than 1.645, we fail to reject $H_0: \mu = 30$ mpg versus $H_a: \mu > 30$ mpg.

Example 10.5

In reference to Example 10.4, test $H_0: \mu = 30$ mpg versus $H_a: \mu < 30$ mpg and make a conclusion at $\alpha = 0.05$.

Solution

We have $z = -3.35$. Since $H_a: \mu < 30$, the critical value of the test at $\alpha = 0.05$ is $-z_\alpha = -z_{0.05} = -1.645$ (using Table T3).

In TI – 84 Plus calculator, we can compute critical value using ***invNorm***() function with the left area of 0.05 as argument of the function:

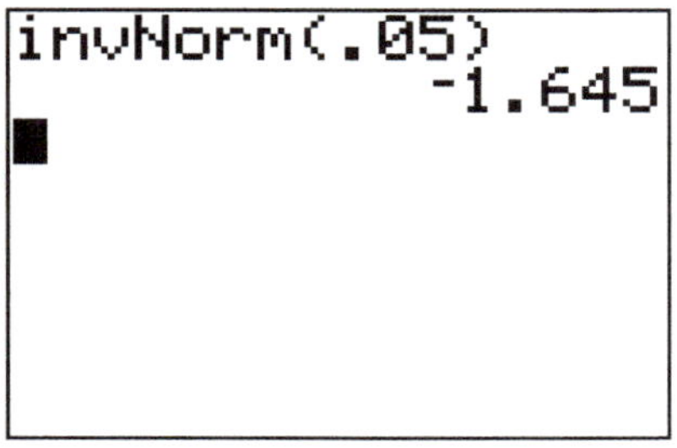

Conclusion: Since $z < -z_\alpha$, we reject $H_0: \mu = 30$ mpg versus $H_a: \mu < 30$ mpg.

Example 10.6

In reference to Example 10.4, test $H_0: \mu = 30$ mpg versus $H_a: \mu \neq 30$ mpg and make a conclusion at $\alpha = 0.05$.

Solution

We have $z = -3.35$. Since $H_a: \mu \neq 30$, the right-tail critical value of the test at $\alpha = 0.05$ is $z_{\alpha/2} = z_{0.025} = 1.96$ (using Table T3).

In TI – 84 Plus calculator, we can compute critical value using ***invNorm***() function with the left area of 0.975 as argument of the function:

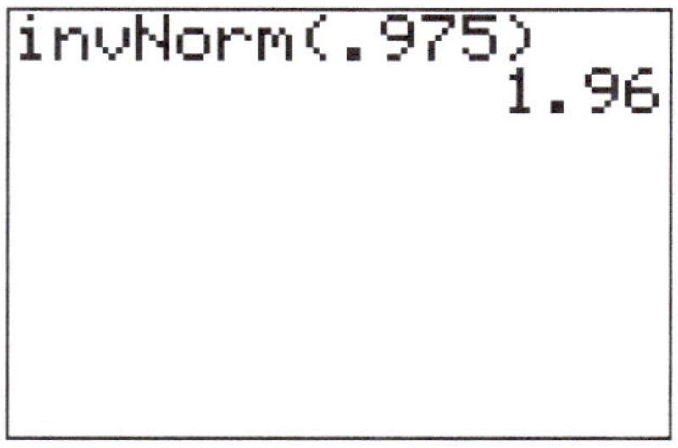

Conclusion: Since $|z| = 3.35 > z_{\alpha/2} = 1.96$, we reject $H_0\!: \mu = 30$ mpg versus $H_a\!: \mu \neq 30$ mpg.

Use of the p-value for the decision of the Z-test

Like comparing z with critical value, we can compare p-value with α for making decision regarding the test, and we reach an identical conclusion with either of the approaches.

Given an observed value of the test statistic z, the **p-value** is the probability that the test statistic Z takes values beyond z in the ***direction of alternative***, assuming that the null hypothesis is true. Therefore, p-value is a measure of evidence for the rejection of H_0 when it is true.

Indeed, p-value is an estimate of the level of the significance α. A p-value higher than α provides evidence in favor of H_0 and p-value lower than α provides evidence against H_0.

Computation of the p-value

The computation of p-value for the test $H_0\!: \mu = \mu_0$ when σ is known is given by

p-value$= P(Z > z)$ if $H_a\!: \mu > \mu_0$ (right-tailed test)

p-value$= P(Z < z)$ if $H_a\!: \mu < \mu_0$ (left-tailed test)

p-value$= P(Z < -z) + P(Z > z) = 2 \times P(Z > |z|)$ if $H_a\!: \mu \neq \mu_0$ (two-tailed test)

where Z is a N(0,1) distribution and z is the observed value of the test statistic given a sample.

Decision rule using a p-value

Using the p-value approach,

(1) Reject H_0 if p-value $\leq \alpha$.

(2) Accept H_0 if we fail to reject H_0 (i.e., p-value$> \alpha$).

Example 10.7

In reference to the Example 10.4, consider testing $H_0\!: \mu = 30$ mpg versus $H_a\!: \mu > 30$ mpg using the test statistic $Z = \frac{\bar{x}-\mu_0}{\sigma/\sqrt{n}}$, where Z has a $N(0,1)$ distribution. Compute the p-value and make a conclusion about the test at $\alpha = 0.05$.

Solution

Here, we have $\bar{x} = 28$ mpg, $\mu_0 = 30$, $\sigma = 4$ mpg and $n = 45$ and hence the observed value of the test statistic is $z = \frac{28-30}{4/\sqrt{45}} = -3.35$.

Since $H_a: \mu > 30$, p-value = P(Z > z) = $P(Z > -3.35) = 1 - P(Z \leq -3.35) = 1 - 0.0004 = 0.9996$ (using Table T3).
In TI – 84 Plus calculator, we can compute p-value using $\boldsymbol{normalcdf}(lv, uv, mean, sd)$ function, where lv =lower value, uv =upper value, $mean = 0$ and $sd = 1$.

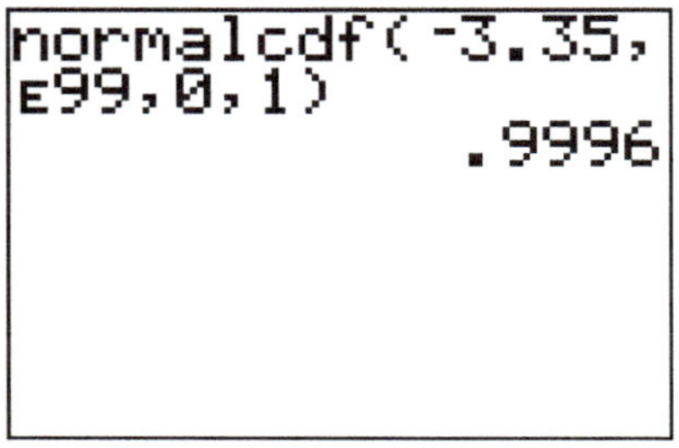

Conclusion: Since p-value $= 0.9996 > \alpha$, we fail to reject $H_0: \mu = 30$ at $\alpha = 0.05$.

Note that $mean = 0$ and $sd = 1$ are default values for a standard normal distribution, and can be ignored as arguments in the $\boldsymbol{normalcdf}()$ function. Also, -E99 and E99 are used to denote $-\infty$ and ∞ in the $\boldsymbol{normalcdf}()$ function.

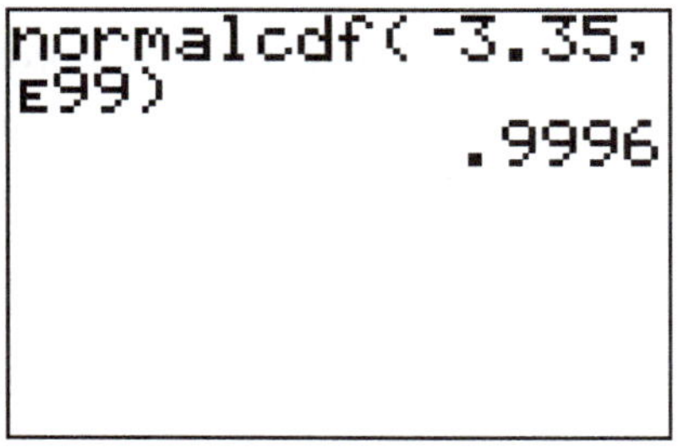

Example 10.8

In reference to the Example 10.4, compute p-value for the test $H_0: \mu = 30$ mpg versus $H_a: \mu < 30$ mpg and make a conclusion about the test at $\alpha = 0.05$.

Solution

We have $z = -3.35$. Since $H_a: \mu < 30$, p-value $= P(Z < z) = P(Z < -3.35) = 0.0004$ (using Table T3). Using TI-84 Plus $\boldsymbol{normalcdf}()$ function we have:

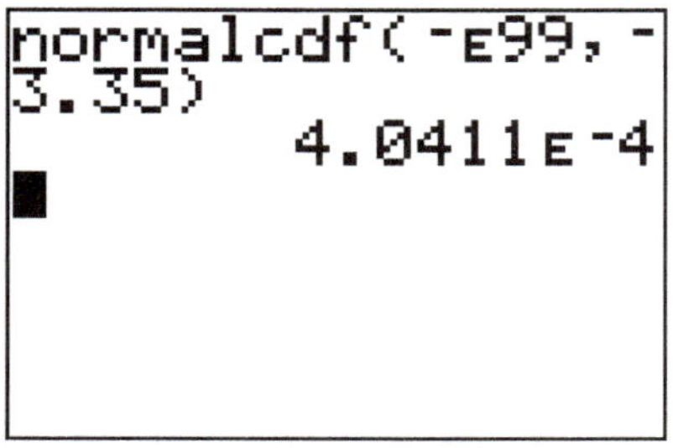

Note that $4.0411E^{-4} = 4.0411 \times 10^{-4} \cong 0.0004$.
Conclusion: Since p-value $< \alpha$, we reject $H_0: \mu = 30$ at $\alpha = 0.05$.

Example 10.9

In reference to the Example 10.4, compute p-value for the test $H_0: \mu = 30$ mpg versus $H_a: \mu \neq 30$ mpg and make a conclusion about the test at $\alpha = 0.05$.

Solution

The observed value of Z is $z = -3.35$. Since $H_a: \mu \neq 30$, p-value is given by
p-value$=2 \times P(Z > |z|) = 2 \times P(Z > |-3.35|) = 2 \times P(Z > 3.35) = 2 \times (1 - 0.9996) = 0.0008$ (using Table T3).
See TI-84 Plus ***normalcdf*()** function below for computing p-value:

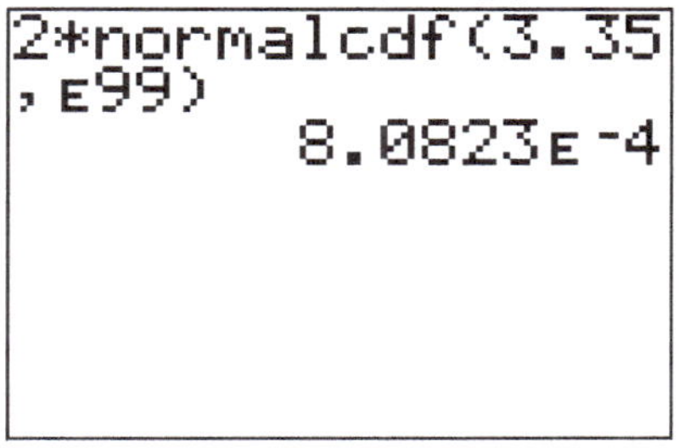

Note that $8.0823E^{-4} = 8.0823 \times 10^{-4} \cong 0.0008$.
Conclusion: Since p-value $< \alpha$, we reject $H_0: \mu = 30$ at $\alpha = 0.05$.

Example 10.10

Suppose the average payment for a cesarean childbirth is found to be \$15,500 on the basis of a sample of 45 payments. Assume that population of payment for a cesarean childbirth has a standard deviation of \$3,500. A researcher believes that the average payment for all cesarean childbirth is actually greater than \$16,500. Answer the questions below:

(a) Set up appropriate null and alternative hypotheses for the test. Identify your hypothesis to be one-tailed or two-tailed.
(b) Write up appropriate test statistic for the test and observed value of the test statistic.
(c) Use critical value approach to make a conclusion of your test at $\alpha = 0.05$.
(d) Use p-value approach to make a conclusion of your test at $\alpha = 0.05$.
(e) Comment on your conclusions from (c) and (d).

Solution

(a) $H_0: \mu = 16{,}500$ vs $H_a: \mu > 16{,}500$; a right-tailed test.
(b) The test statistic is $Z = \frac{\bar{x} - \mu_0}{\sigma/\sqrt{n}}$.
The observed value of the test statistic is $z = \frac{15500 - 16500}{3500/\sqrt{45}} = -1.92$
(c) At $\alpha = 0.05$ with $H_a: \mu > 16500$, the critical value is $z_{0.05} = 1.645$.
Since $z < 1.645$, we accept H_0 at $\alpha = 0.05$.
(d) p-value$= P(Z > z) = P(Z > -1.92) = 1 - P(Z \leq -1.92) = 1 - 0.02743 = 0.97257$ (using Table T3).
See TI-84 Plus ***normalcdf*()** function below for computing p-value:

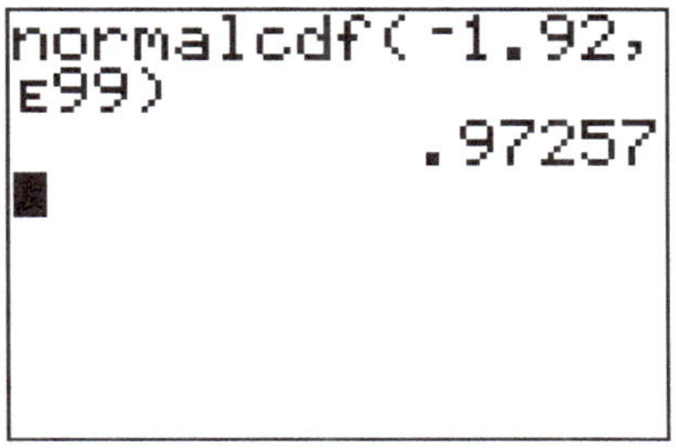

Conclusion: Since $p\text{-}value > 0.05$, we accept the null hypothesis at $\alpha = 0.05$.

(e) The conclusion in (c) made by the critical value approach and the one in (d) made by the p-value approach are identical.

Minitab Instruction for one-sample Z-test (summary data)

Ref: Example 10.7

(1) Select Stat>Basic Statistics>1- Sample Z…

(2) Choose **Summarized data** option and type Sample size: 45, Sample mean: 28, and Known standard deviation: 4, Check mark **Perform hypothesis test** and type Hypothesized mean: 30.

(3) Use **Options**… to specify ***Confidence level***: 95, Specify **Alternative hypothesis**: Mean>Hypothesized mean.

(4) Click [OK] two times.

Note that if a **sample** is given instead of summarized data, in step (2) choose **one or more samples, each in a column** option, specify **Known standard deviation**: 4, Check mark **Perform hypothesis test** and type Hypothesized mean: 30.

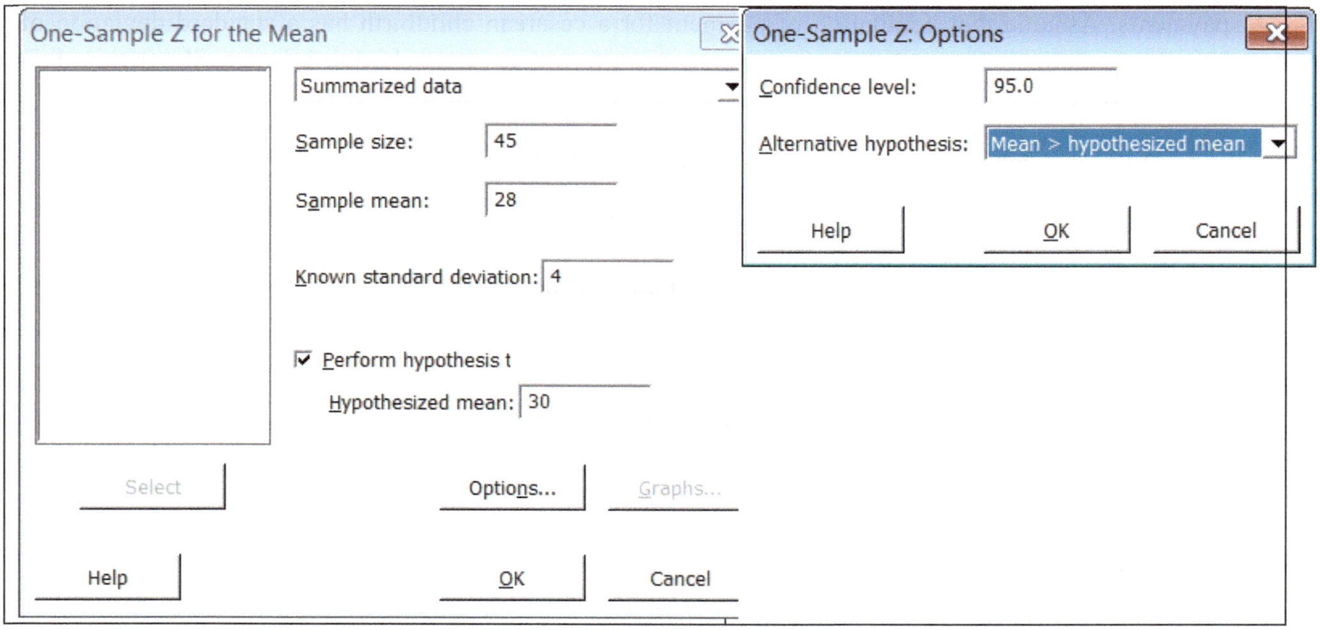

See the output below for one-sample Z-test:

One-Sample Z

Test of μ = 30 vs > 30

The assumed standard deviation = 4

N	Mean	SE Mean	95% Lower Bound	Z	P
45	28.000	0.596	27.019	-3.35	1.000

TI 84-Plus Calculator for one-sample Z-Test

One-sample Z-test with summary statistics (Stats)

Ref: Example 10.9

(1) Press STAT, select TESTS, choose **1: Z-Test...**and press **ENTER.**
(2) Move the cursor to **Stats** and press **ENTER**.
(3) Specify as appropriate: μ_0: 30, σ: 4 , $\bar{x}$: 28, n: 45.
(4) For μ: choose $\neq \mu_0$ for a two-tailed test.
(5) Move the cursor to **Calculate** and press **ENTER**

See the option selections and results in figures below:

(a) (b) (c) (d)

In figure (c) above, the observed value of $z = -3.3541$ and p-value= $7.9634 \times 10^{-4} \cong 0.0008$, what was found using the standard normal table.

Note that if **Data** are given instead of summary statistics (**Stats**), in step (2) above we move the cursor to **Data** and follow with appropriate options; see the figure (d) above. Use **List**: L1 (if data is entered in L1, or as appropriate) and **Freq:1**.

Section 10.3: T-test for a Mean when σ is unknown

Recall that the test statistic to test $H_0: \mu = \mu_0$ when σ is known is the Z-Test statistic given by

$$Z = \frac{\bar{x} - \mu_0}{\sigma/\sqrt{n}}$$

What if the population standard deviation σ is unknown? Of course, the standard deviation σ will be replaced by its estimate s by the sample.

Indeed, to test $H_0: \mu = \mu_0$ when σ is unknown, we use the T-test statistic given by

$$T = \frac{\bar{x} - \mu_0}{s/\sqrt{n}}$$

Under the null hypothesis, the above test statistic follows a t-distribution with $(n-1)$ degrees of freedom (df).

Assumption for T-test

The following assumptions are required for the validity of the T-test:

(1) Have an SRS.
(2) Either sample size is large ($n \geq 30$) or the population is normal.
(3) The population standard deviation σ is unknown.

Computation of p-value for T-test

Following the similar computational procedure as in section 10.2, the computation of p-value for the test $H_0: \mu = \mu_0$ when σ is unknown is given by

p-value$= P(T > t)$ if $H_a: \mu > \mu_0$ (right-tailed test)

p-value$= P(T < t)$ if $H_a: \mu < \mu_0$ (left-tailed test)

p-value$= P(T < -t) + P(T > t) = 2 \times P(T > |t|)$ if $H_a: \mu \neq \mu_0$ (two-tailed test)

where the test statistic T follows a t-distribution with $(n-1)\ df$ and t is the observed value of the test statistic.

Decision rule

(1) Reject H_0 if p-value $\leq \alpha$.
(2) Accept H_0 if we fail to reject H_0 (i.e., p-value$> \alpha$).

However, unlike the area under the standard normal curve, there is no unique t-table to find the area under the t-curve because different degrees of freedom correspond to different t-curves. Therefore, we cannot use any table to compute p-value; we can find p-value using technology. For making decision about acceptance or rejection of the hypothesis, we generally use t-table with selected critical values for varying values of the degrees of freedom. Technology generally gives a p-value.

Definition: Critical value for a T-test

For testing $H_0: \mu = \mu_0$ (when σ is unknown), the critical value of a T-test at level α is

(a) t_α if $H_a: \mu > \mu_0$
(b) $-t_\alpha$ if $H_a: \mu < \mu_0$
(c) $\pm t_{\alpha/2}$ if $H_a: \mu \neq \mu_0$

where t_α and $-t_\alpha$ are values under the curve of a t-distribution with $df = n - 1$ having area α to its right and left, respectively. Note that, by the symmetry of the t-curve, $-t_\alpha = t_{1-\alpha}$.

The table below provides critical value of t-distribution for selected values of degrees of freedom and α.

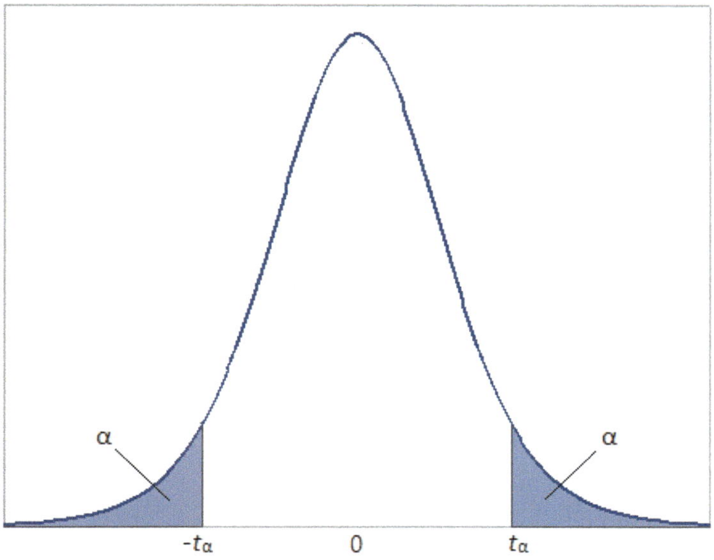

df	$t_{0.1}$	$t_{.05}$	$t_{0.01}$
3	1.638	2.353	4.541
4	1.533	2.132	3.747
22	1.321	1.717	2.508
23	1.319	1.714	2.500
24	1.318	1.711	2.492
25	1.316	1.708	2.485
26	1.315	1.706	2.479
100	1.29	1.66	2.364
200	1.286	1.653	2.345
Inf	1.282	1.645	2.326

Decision rule using critical value

At a significance level α, we reject $H_0: \mu = \mu_0$ (when σ is unknown) if

(1) $t > t_\alpha$ when $H_a: \mu > \mu_0$ (i.e., right-tailed test)

(2) $t < -t_\alpha$ when $H_a: \mu < \mu_0$ (i.e., left-tailed test)

(3) $|t| > t_{\alpha/2}$ when $H_a: \mu \neq \mu_0$ (i.e., two-tailed test)

where t_α and $-t_\alpha$ are as defined above.

Example 10.11

Suppose we wish to test $H_0: \mu = 10.5$ vs $H_a: \mu > 10.5$ on the basis of a sample of 25 observations from a normal population. The sample results in $\bar{x} = 11.3$ and $s = 2.17$.

(a) Write out an appropriate test statistic for the test and find the value of the test statistic.

(b) Find the degrees of freedom for the test.

(c) Find critical value of the test at $\alpha = 0.05$ and make a conclusion regarding the test.

(d) Find critical value of the test at $\alpha = 0.01$ and make a conclusion regarding the test.

(e) What is your comment on conclusions from (c) and (d)?

(f) Find p-value of the test using any technology.

(g) Make a conclusion about the test on the basis of the p-value at $\alpha = 0.05$.

(h) Make a conclusion about the test on the basis of the p-value at $\alpha = 0.01$.

(i) What is your comment on conclusions from (g) and (h)?

(j) Given $\alpha = 0.05$, is there any difference in the conclusions using critical and p-value approaches?

Solution

(a) Test statistic is $T = \frac{\bar{x}-\mu_0}{s/\sqrt{n}}$.

Here, we have $\bar{x} = 11.3$, $\mu_0 = 10.5$, $s = 2.17$ and $n = 25$ and hence the observed value of the test statistic is $t = \frac{11.3-10.5}{2.17/\sqrt{25}} = 1.84$

(b) $df = n - 1 = 25 - 1 = 24$.

(c) With $df = 24$ and $\alpha = 0.05$, the critical value of the test is $t_\alpha = t_{0.05} = 1.711$.

Conclusion: Since $t > t_\alpha$, we reject $H_0: \mu = 10.5$ vs $H_a: \mu > 10.5$ at $\alpha = 0.05$.

df	$t_{0.1}$	$t_{0.05}$	$t_{0.025}$	$t_{0.01}$	$t_{0.005}$	$t_{0.001}$	$t_{0.0005}$
23	1.319	1.714	2.069	2.500	2.807	3.485	3.768
24	1.318	1.711	2.064	2.492	2.797	3.467	3.745
25	1.316	1.708	2.060	2.485	2.787	3.450	3.725

(d) With $df = 24$ and $\alpha = 0.01$, the critical value of the test is $t_\alpha = t_{0.01} = 2.492$.

Conclusion: Since $t < t_\alpha$, we fail to reject $H_0: \mu = 10.5$ vs $H_a: \mu > 10.5$ at $\alpha = 0.01$.

(e) We reject H_0 at higher value of α (i.e., $\alpha = 0.05$) and failed to reject H_0 at lower of α (i.e., $\alpha = 0.01$).

(f) Since $H_a: \mu > 10.5$, p-value$= P(T > 1.84) = 0.039$.

To get this result, we used TI-84 Plus: function $\boldsymbol{tcdf(lv, uv, df)}$, where lv=lower value, uv=upper value. Here, lv=1.84, $uv = 1E99$ for ∞ and $df = 24$. See the output using the function:

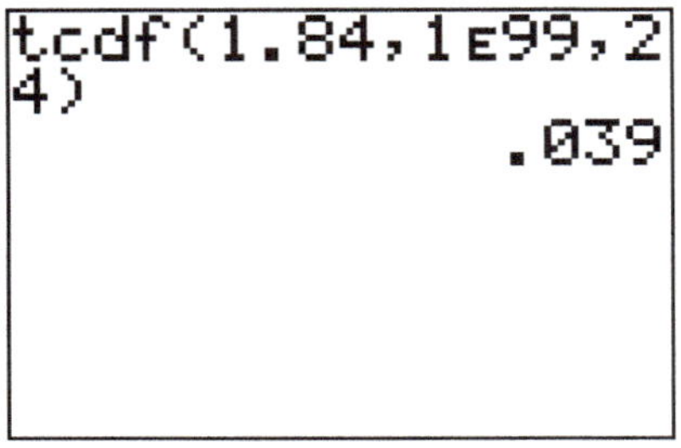

(g) Since p-value$<\alpha = 0.05$, reject H_0 at $\alpha = 0.05$.
(h) Since p-value$>\alpha = 0.01$, we fail to reject H_0 at $\alpha = 0.01$.
(i) We reject H_0 at higher value of α (i.e., $\alpha = 0.05$) and failed to reject H_0 at lower α (i.e., $\alpha = 0.01$), an identical conclusion using critical value approach.
(j) Given $\alpha = 0.05$, there is no difference in the conclusions using critical and p-value approaches.

TI 84-Plus Calculator for one-sample T-Test

One-sample T-Test with summary statistics (Stats)

Ref: Example 10.11

(1) Press STAT, select TESTS, choose **2: T-Test…**and press **ENTER.**
(2) Move the cursor to **Stats** and press **ENTER**.
(3) Specify as appropriate: μ_0: 10.5, $\bar{x}$: 11.3, Sx: 2.17 , n: 25.
(4) For μ: choose $>\mu_0$ (because we have a right-tailed test).
(5) Move the cursor to **Calculate** and press **ENTER**

See the option selections and results in figures below:

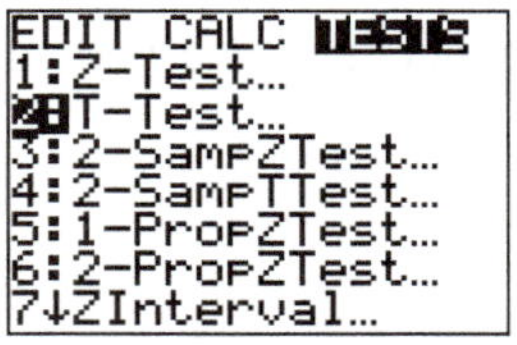

(a)

(b)

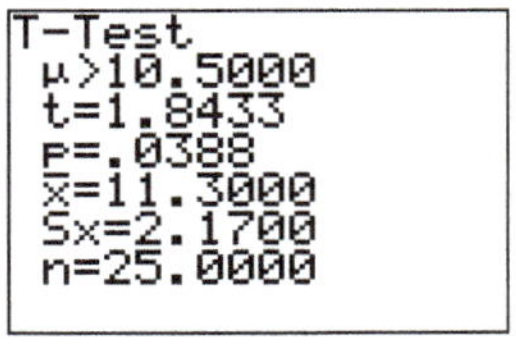

(c)

Note that if Data are given instead of summary statistics (**Stats**), in step (2) above we move the cursor to **Data** and follow with appropriate options.

Section 10.4: Test for two population means

Let us have two independent SRSs, one of size n_1 from a population with mean μ_1 and standard deviation σ_1, and the other of size n_2 from a population with mean μ_2 and standard deviation σ_2. Let $\bar{x}_1$ and s_1^2 be the mean and variance of the first sample, and $\bar{x}_2$ and s_2^2 be the mean and variance of the second sample. We want to test $H_0: \mu_1 = \mu_2$ against the following alternatives:

$$H_a: \mu_1 > \mu_2$$
$$H_a: \mu_1 < \mu_2$$
$$H_a: \mu_1 \neq \mu_2$$

Assumptions required for testing $H_0: \mu_1 = \mu_2$

The following three sets of assumptions are required for testing $H_0: \mu_1 = \mu_2$:

(1) Two SRSs are independent,
(2) Either samples are large or populations are normal, and
(3) Any of the followings:
(a) Population variances σ_1^2 and σ_2^2 are known (Case I: Two-sample Z-Test).
(b) Population variances σ_1^2 and σ_2^2 are equal but unknown (Case II: Two-sample pooled T-Test).
(c) Population variances σ_1^2 and σ_2^2 are unequal and unknown (Case III: Two-sample non-pooled T-Test).

Case I: Two-sample Z-Test

The test of $H_0: \mu_1 = \mu_2$, when σ_1^2 and σ_2^2 are known, is called the ***two-sample Z-test***. The test statistic for the test is given by

$$Z = \frac{(\bar{x}_1 - \bar{x}_2) - (\mu_1 - \mu_2)}{\sqrt{\frac{\sigma_1^2}{n_1} + \frac{\sigma_2^2}{n_2}}}$$

where Z follows a N(0,1) distribution under the null hypothesis. As usual, an observed value of this test statistic is denoted by z.
The decision about acceptance or rejection of the null hypothesis can be made using critical value or p-value as described in section 10.2.

Decision rule using critical value approach

Given α, we reject $H_0: \mu_1 = \mu_2$ if

(1) $z > z_\alpha$ when $H_a: \mu_1 > \mu_2$ (i.e., right-tailed test)
(2) $z < -z_\alpha$ when $H_a: \mu_1 < \mu_2$ (i.e., left-tailed test)
(3) $|z| > z_{\alpha/2}$ when $H_a: \mu_1 \neq \mu_2$ (i.e., two-tailed test)

where z_α and $-z_\alpha$ are critical values for right- and left-tailed tests described in section 10.2. Note that z_α and $-z_\alpha$ are values under the standard normal curve with area α to its right and left, respectively. Also, by the symmetry of normal curve, $z_{1-\alpha} = -z_\alpha$.

Example 10.12

Suppose that for a sample of 10 payments of Verizon customers, the mean of monthly bills, including taxes and fees, is \$145, and from a sample of 12 payments of Sprint customers, the mean of monthly bills, including taxes and fees, is \$143. Suppose that the payments of bills of Verizon and Sprint customers approximately follow normal distributions with unknown means and known standard deviations $\sigma_1 =$ \$10 and $\sigma_2 =$ \$12, respectivly.
We wish to test $H_0: \mu_1 = \mu_2$ against $H_a: \mu_1 > \mu_2$ at $\alpha = 0.05$. Answer the following questions:

(a) What test should you use to solve the problem?
(b) Write out the appropriate formula for the test statistic.
(c) Find an observed value of the test statistic.
(d) Find critical value for the test and make a conclusion about the test at $\alpha = 0.05$.
(e) Compute the p-value and make a conclusion of the test.

(f) What is the difference in your conclusions from (d) and (e)?

Solution

(a) We should use a two-sample Z-test.

(b) The test statistic is given by

$$Z = \frac{(\bar{x}_1 - \bar{x}_2) - (\mu_1 - \mu_2)}{\sqrt{\frac{\sigma_1^2}{n_1} + \frac{\sigma_2^2}{n_2}}}$$

(c) We have $\bar{x}_1 = 145$, $\bar{x}_2 = 143$, $n_1 = 10$, $n_2 = 12$, $\sigma_1 = \$10$ and $\sigma_2 = \$12$. Then, the observed value of the test statistic under $H_0: \mu_1 = \mu_2$ is

$$z = \frac{(145 - 143) - (0)}{\sqrt{\frac{10^2}{10} + \frac{12^2}{12}}} = \frac{2}{4.69} = 0.43$$

(d) Since $H_a: \mu_1 > \mu_2$, the critical value at $\alpha = 0.05$ is $z_\alpha = z_{0.05} = 1.645$. Since $z < z_\alpha$, we fail to reject H_0 at $\alpha = 0.05$.

TI-84 Plus Calculator function ***invNorm***() can be used to find critical value by providing the left area of 0.95 as argument of the function.

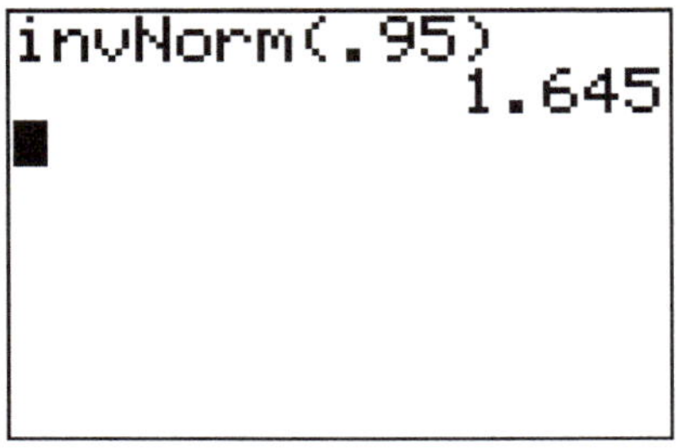

(e) Since $H_a: \mu_1 > \mu_2$, p-value= $P(Z > z) = P(Z > 0.43) = 1 - P(z \leq 0.43) = 1 - 0.6664 = 0.3336$ (Using Table T3).

Computing p-value in TI – 84 Plus calculator:

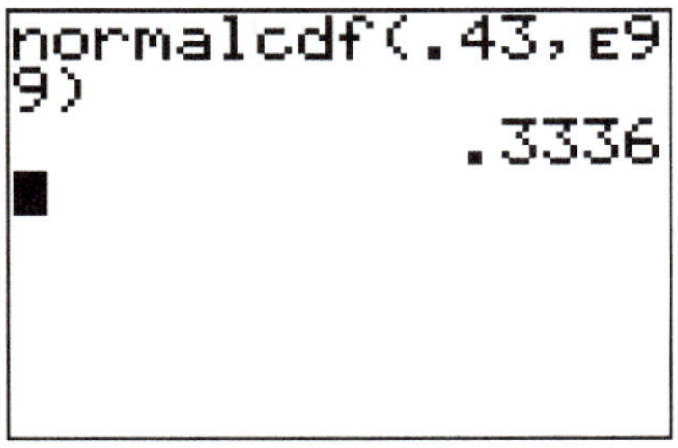

Since p-value> α, we fail to reject H_0 at $\alpha = 0.05$.

(f) There is no difference in conclusions from (d) and (e).

Example 10.13

Revisit Example 10.12 and test $H_0: \mu_1 = \mu_2$ against $H_a: \mu_1 < \mu_2$ at $\alpha = 0.05$. Make a conclusion of the test using a

(a) critical value.

(b) p-value.

Solution

We have $z = 0.43$.

(a) At $\alpha = 0.05$, the test $H_0\colon \mu_1 = \mu_2$ against $H_a\colon \mu_1 < \mu_2$ has the critical value $-z_\alpha = -1.645$ (using Table T3).
TI-84 Plus Calculator function $\boldsymbol{invNorm}()$ can be used to find critical value by providing the left area of 0.05 as argument of the function.

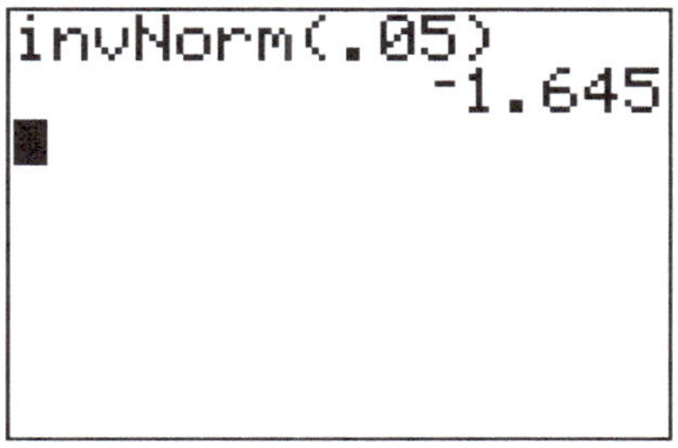

Conclusion: Since $z > -z_\alpha$, we fail to reject H_0 at $\alpha = 0.05$.

(b) Since $H_a\colon \mu_1 < \mu_2$, the p-value$= P(Z < z) = P(Z < 0.43) = 0.6664$ (using the Table T3).
We get identical p-value in the TI – 84 Plus calculator $\boldsymbol{normalcdf}$ () function:

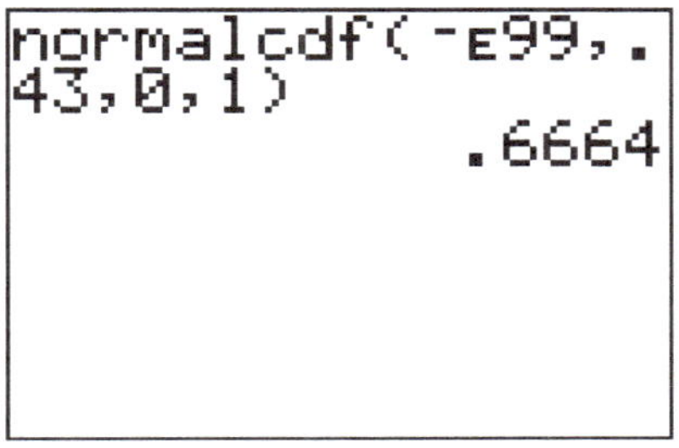

Conclusion: Since p-value$> \alpha$, we fail to reject H_0 at $\alpha = 0.05$.

Example 10.14

Revisit Example 10.12 and test $H_0\colon \mu_1 = \mu_2$ against $H_a\colon \mu_1 \neq \mu_2$, at $\alpha = 0.05$.
Make a conclusion of the test using a

(a) critical value.
(b) p-value.

Solution

We have $z = 0.43$.

(a) Since $H_a\colon \mu_1 \neq \mu_2$, at $\alpha = 0.05$, the right critical value is $z_{\alpha/2} = 1.96$ (using Table T3).
TI-84 Plus Calculator function $\boldsymbol{invNorm}()$ can be used to find critical value by providing the left area of 0.975 as argument of the function.

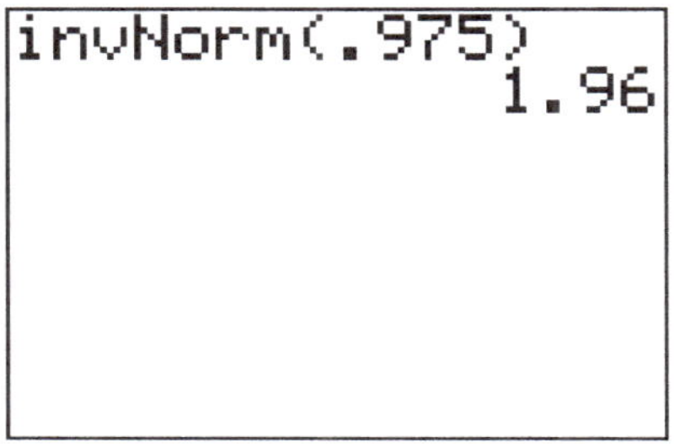

Conclusion: Since $|z| < z_{\alpha/2}$, we fail to reject H_0 at $\alpha = 0.05$.

(b) Since $H_a\colon \mu_1 \neq \mu_2$, the p-value$= 2 \times P(Z > z) = 2 \times P(Z > 0.43) = 2 \times [1 - P(Z \leq 0.43)] = 2 \times (1 - 0.6664) = 2 \times 0.3336 = 0.6672$ (Using Table T3).
We get identical p-value in the TI – 84 Plus calculator $\boldsymbol{normalcdf}$ () function:

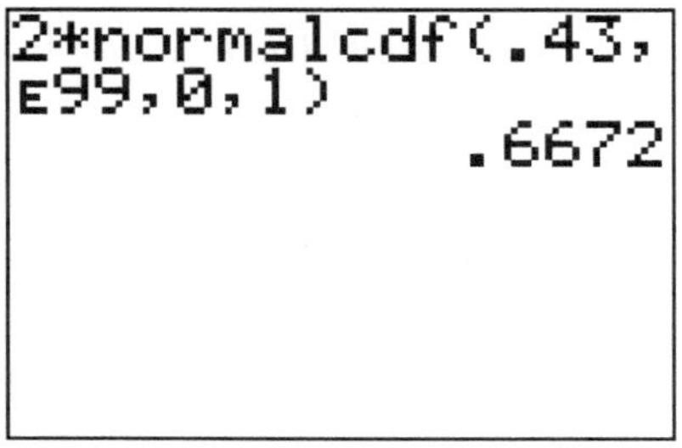

Conclusion: Since p-value> α, we fail to reject H_0 at $\alpha = 0.05$.

TI-84 Plus Calculator Instruction for a Two-sample Z-test

Two-sample Z-test with summary statistics

Ref: Example 10.12

(1) Press STAT, select TESTS, choose **3: 2-SampZTest…**and press **ENTER.**
(2) Move the cursor to **Stats** and press **ENTER**.
(3) Specify as appropriate: $\bar{x}_1$: 145, σ_1: 10 , n_1: 10, $\bar{x}_2$: 143, σ_2: 12 , n_2: 12.
(4) For μ_1: Choose > μ_2 for a right-tailed test.
(5) Move the cursor to Calculate and press ENTER

See the option selections and results in figures below:

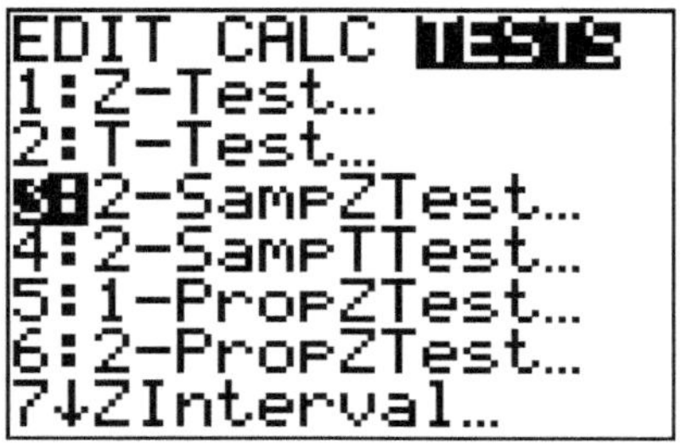

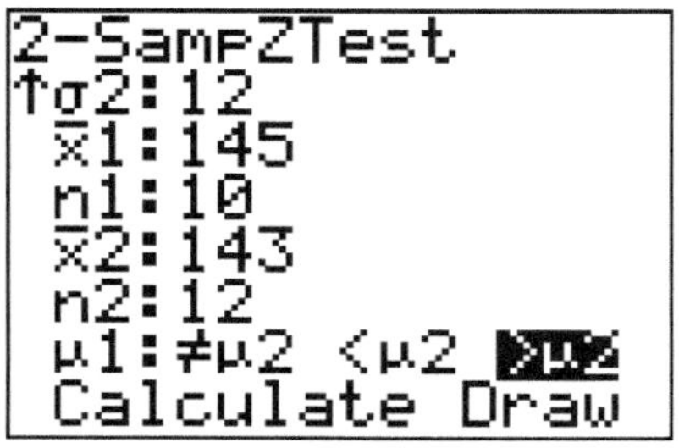

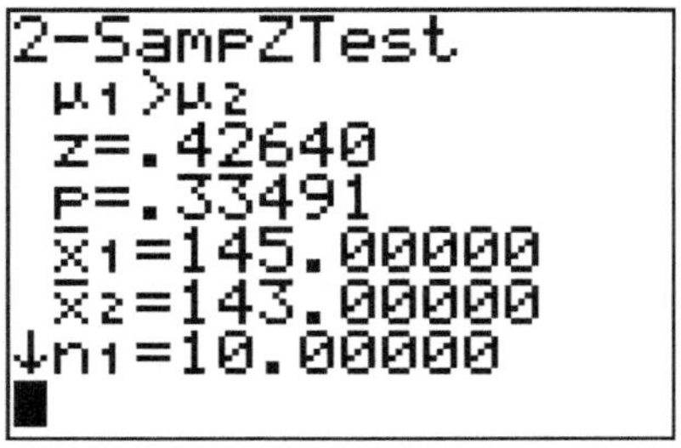

Case II: Two-sample pooled T-Test

The test statistic of test $H_0: \mu_1 = \mu_2$, when σ_1^2 and σ_2^2 are unknown but equal, is called the ***two-sample pooled T-test***. The test statistic for the test is given by

$$T = \frac{(\bar{x}_1 - \bar{x}_2) - (\mu_1 - \mu_2)}{s_p \times \sqrt{\frac{1}{n_1} + \frac{1}{n_2}}}$$

where $s_p^2 = \frac{(n_1-1)s_1^2+(n_2-1)s_2^2}{n_1+n_2-2}$ is called the pooled estimate of the common variance $\sigma^2 (= \sigma_1^2 = \sigma_2^2)$.

The statistic T follows a t-distribution with df =$(n_1 + n_2 - 2)$. An observed value of the test statistic T is denoted by t.

Decision rule using critical value approach

The observed value t beyond the critical value in the direction of the alternative provides evidence against the null hypothesis. Therefore, at significance level α, we reject $H_0: \mu_1 = \mu_2$ if

(1) $t > t_\alpha$ when $H_a: \mu_1 > \mu_2$ (i.e., right-tailed test)
(2) $t < -t_\alpha$ when $H_a: \mu_1 < \mu_2$ (i.e., left-tailed test)

(3) $|t| > t_{\alpha/2}$ when $H_a: \mu_1 \neq \mu_2$ (i.e., two-tailed test)

where t_α and $-t_\alpha$ are values under the curve of a t-distribution with $df = (n_1 + n_2 - 2)$ having area α to its right and left, respectively.

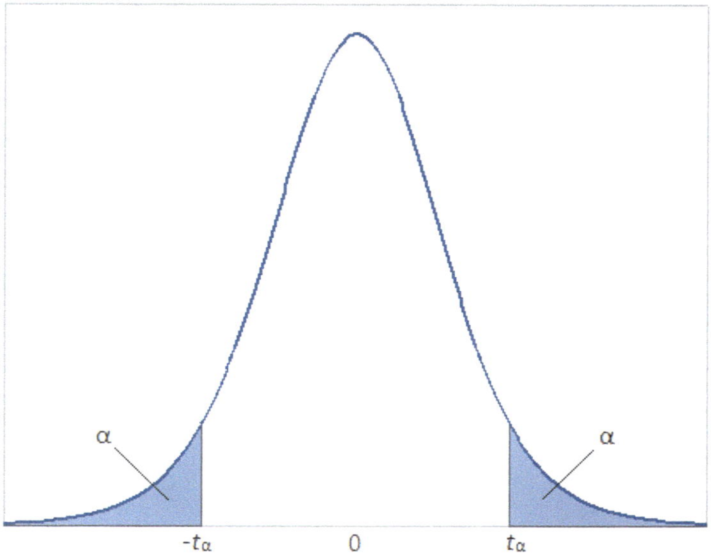

Computation of p-value for two-sample T-test

Following the similar computational procedure as in section 10.2, the computation of p-value for the test $H_0: \mu_1 = \mu_2$ when σ_1 and σ_2 are unknown is given by

p-value$= P(T > t)$ if $H_a: \mu_1 > \mu_2$ (right-tailed test)

p-value$= P(T < t)$ if $H_a: \mu_1 < \mu_2$ (left-tailed test)

p-value$= P(T < -t) + P(T > t) = 2 \times P(T > |t|)$ if when $H_a: \mu_1 \neq \mu_2$ (i.e., two-tailed test)

where the test statistic T follows a t-distribution with $df = (n_1 + n_2 - 2)$ df and t is the observed value of the test statistic.

Decision rule

(1) Reject H_0 if p-value $\leq \alpha$.

(2) Accept H_0 if we fail to reject H_0 (i.e., p-value$> \alpha$).

Example 10.15

Suppose two independent samples, one of size 22 males, and the other of size 20 females, reveal that the mean age of diagnosis of diabetes for male and female are 44 and 41 years and standard deviations are 9.15 and 8.2 years, respectively. Assume that the populations of age of diagnosis of diabetes among males and females approximately follow normal distributions with unknown means and **unknown but equal standard deviations**.

Test $H_0: \mu_1 = \mu_2$ against $H_a: \mu_1 > \mu_2$ at $\alpha = 0.05$ by answering the following questions:

(a) What test should you use?

(b) Write out the appropriate formula for the test statistic.

(c) Find an observed value of the test statistic.

(d) Compute df for the test.

(e) Find critical value for the test and make a conclusion of the test.

(f) Compute p-value using any technology and make a conclusion of the test.

Solution

(a) We have two populations with unknown and equal standard deviations. Therefore, we should use the two-sample pooled T-test.

(b) The test statistic is given by

$$T = \frac{(\bar{x}_1 - \bar{x}_2) - (\mu_1 - \mu_2)}{s_p \times \sqrt{\frac{1}{n_1} + \frac{1}{n_2}}}$$

(c) Here, $\bar{x}_1 = 44,\ \bar{x}_2 = 41;\ s_1 = 9.15;\ s_2 = 8.2;\ n_1 = 22;\ n_2 = 20$. Then, we have

$$s_p^2 = \frac{(n_1 - 1)s_1^2 + (n_2 - 1)s_2^2}{n_1 + n_2 - 2} = \frac{(22 - 1)(9.15)^2 + (20 - 1)(8.2)^2}{22 + 20 - 2} = 75.98$$

$$s_p = \sqrt{75.98} = 8.71$$

The observed value of the test statistic is

$$t = \frac{(44 - 41) - 0}{8.71 \times \sqrt{\frac{1}{22} + \frac{1}{20}}} = \frac{3}{2.691} = 1.115$$

(d) The degrees of freedom is $df = 22 + 20 - 2 = 40$.

df	$t_{0.1}$	$t_{0.05}$	$t_{0.025}$	$t_{0.01}$	$t_{0.005}$	$t_{0.001}$	$t_{0.0005}$
39	*1.304*	*1.685*	*2.023*	*2.426*	*2.708*	*3.313*	*3.558*
40	*1.303*	*1.684*	*2.021*	*2.423*	*2.704*	*3.307*	*3.551*

(e) Since $H_a\!:\mu_1 > \mu_2$, with $df = 40$, the critical value at $\alpha = 0.05$ is $t_\alpha = t_{0.05} = 1.684$.
Conclusion: Since $t < 1.684$, we accept H_0 at 5% level of significance.

(f) Since $H_a\!:\mu_1 > \mu_2$, p-value$= P(T > t) = P(T > 1.115) = 0.136$.

In order to compute the p-value, we use TI-84 Plus: function $tcdf(lv, uv, df)$, where lv=lower value, uv=upper value. Here, lv=1.115, $uv = 1E99$ for ∞ and $df = 40$. See the output using the function:

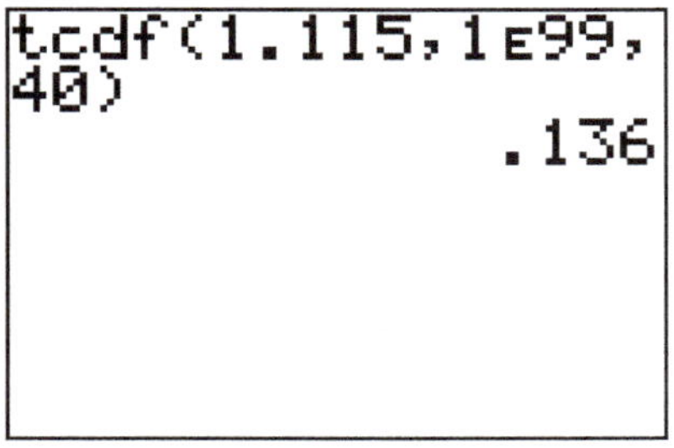

Conclusion: Since p-value > 0.05, we accept H_0 at 5% level of significance.

Example 10.16

In reference to Example 10.15, test $H_0\!:\mu_1 = \mu_2$ against $H_a\!:\mu_1 < \mu_2$ at $\alpha = 0.05$ using

(a) critical value approach.

(b) p-value approach.

Use the fact that we have $t = 1.115$ and $df = 40$ from Example 10.15.

Solution

(a) The observed value of the test statistic T is $t = 1.115$ and $df = 40$. Since $H_a: \mu_1 < \mu_2$, with $df = 40$, the critical value at $\alpha = 0.05$ is $-t_\alpha = -t_{0.05} = -1.684$. Since $t > -t_\alpha$, we accept H_0 at 5% level of significance.

(b) Since $H_a: \mu_1 < \mu_2$, p-value= $P(T < t) = P(T < 1.115) = 0.864$.
Conclusion: Since p-value > 0.05, we accept H_0 at 5% level of significance.

In order to compute the p-value in (b), we use TI-84 Plus: function $tcdf(lv, uv, df)$, where lv=lower value, uv=upper value. Here, lv=$-1E99$ for $-\infty$, $uv = 1.115$ and $df = 40$. See the output using the function:

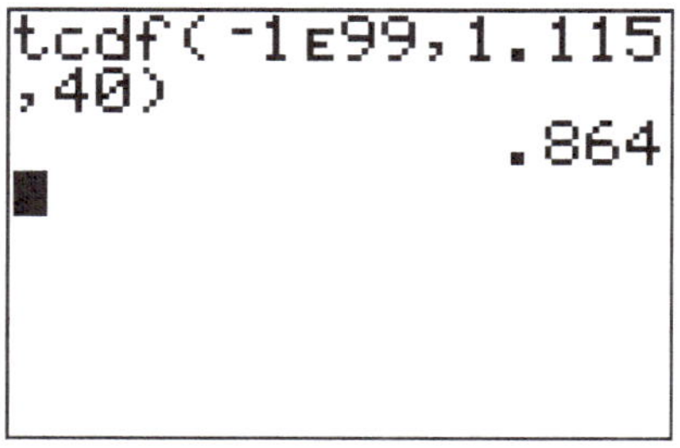

Example 10.17

In reference to Example 10.15, test $H_0: \mu_1 = \mu_2$ against $H_a: \mu_1 \neq \mu_2$ at $\alpha = 0.05$ using

(a) critical value approach.

(b) p-value approach.

Use the fact that we have $t = 1.115$ and $df = 40$ from Example 10.15.

Solution

(a) We have $t = 1.115$ and $df = 40$. Since $H_a: \mu_1 \neq \mu_2$, with $df = 40$, the critical value at $\alpha = 0.05$ is $\pm t_{\alpha/2} = \pm t_{0.025} = \pm 2.021$.

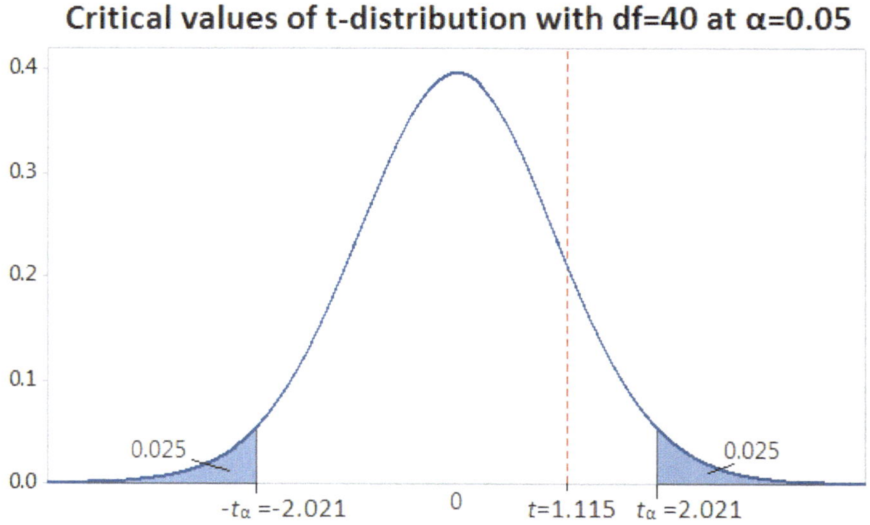

Conclusion: Since $|t| < t_{\alpha/2}$, we accept H_0 at 5% level of significance.

(b) Since $H_a: \mu_1 \neq \mu_2$, p-value= $2 \times P(T > |t|) = 2 \times P(T > 1.115) = 2 \times .136 = .272$.
In order to compute p-value we used the TI-84 Calculator function **tcdf**() as shown below:

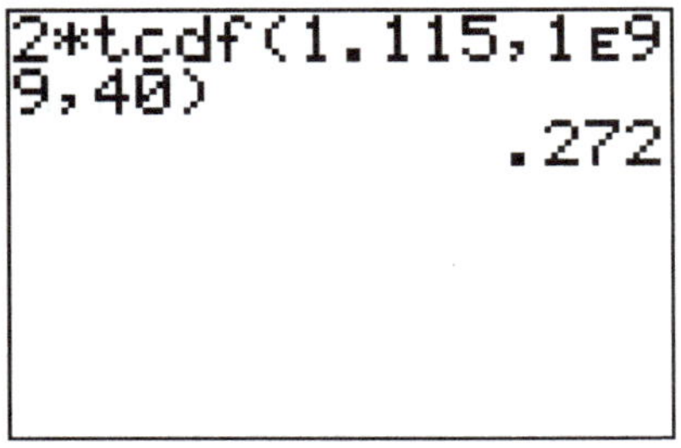

Conclusion: Since p-value> 0.05, we accept H_0 at 5% level of significance.

Case III: Two-sample non-pooled T-test

The test of $H_0: \mu_1 = \mu_2$, when σ_1^2 and σ_2^2 are unknown and unequal, is called the ***two-sample non-pooled T-test***. The test statistic for the test is given by

$$T = \frac{(\bar{x}_1 - \bar{x}_2) - (\mu_1 - \mu_2)}{\sqrt{\frac{s_1^2}{n_1} + \frac{s_2^2}{n_2}}}$$

where T follows an approximate t-distribution with an approximate degrees of freedom (df) calculated using the formula:

$$df = \frac{\left(\frac{s_1^2}{n_1} + \frac{s_2^2}{n_2}\right)^2}{\left(\frac{1}{n_1 - 1}\right) \times \left(\frac{s_1^2}{n_1}\right)^2 + \left(\frac{1}{n_2 - 1}\right) \times \left(\frac{s_2^2}{n_2}\right)^2}$$

The df is rounded down to the nearest integer. Once the df and the observed value of T, denoted by t, are obtained, the decision about acceptance or rejection of the null hypothesis can be made using the same rule described in Case II.

Example 10.18

Reference to Example 10.15, assume that the population variances are ***unknown and unequal***. Test $H_0: \mu_1 = \mu_2$ against $H_a: \mu_1 \neq \mu_2$ at $\boldsymbol{\alpha} = \mathbf{0.1}$ by answering the following questions:

(a) What test should you use for the test?
(b) Write out the appropriate formula for the test statistic.
(c) Find an observed value of the test statistic.
(d) Compute df for the test.
(e) Find critical value for the test and make a conclusion of the test.

Solution

(a) We should use the two-sample non-pooled T-test.
(b) The formula of the test statistic is given by

$$T = \frac{(\bar{x}_1 - \bar{x}_2) - (\mu_1 - \mu_2)}{\sqrt{\left(\frac{s_1^2}{n_1} + \frac{s_2^2}{n_2}\right)}}$$

(c) We have, $\bar{x}_1 - \bar{x}_2 = 44 - 41 = 3$; $s_1 = 9.15$; $s_2 = 8.2$; $n_1 = 22$; $n_2 = 20$. Therefore, an observed value of the test statistic is

$$t = \frac{3-0}{\sqrt{\left(\frac{9.15^2}{22} + \frac{8.2^2}{20}\right)}} = \frac{3}{\sqrt{7.17}} = 1.12$$

(d) The test statistic T follows an approximate t-distribution with approximate degrees of freedom (df) given by

$$df = \frac{\left(\frac{s_1^2}{n_1} + \frac{s_2^2}{n_2}\right)^2}{\left(\frac{1}{n_1 - 1}\right) \times \left(\frac{s_1^2}{n_1}\right)^2 + \left(\frac{1}{n_2 - 1}\right) \times \left(\frac{s_2^2}{n_2}\right)^2}$$

$$df = \frac{\left(\frac{s_1^2}{n_1} + \frac{s_2^2}{n_2}\right)^2}{\left(\frac{1}{n_1 - 1}\right) \times \left(\frac{s_1^2}{n_1}\right)^2 + \left(\frac{1}{n_2 - 1}\right) \times \left(\frac{s_2^2}{n_2}\right)^2} = \frac{\left(\frac{9.15^2}{22} + \frac{8.2^2}{20}\right)^2}{\left(\frac{1}{21}\right) \times \left(\frac{9.15^2}{22}\right)^2 + \left(\frac{1}{19}\right) \times \left(\frac{8.2^2}{20}\right)^2} = \frac{51.37}{0.69 + 0.59}$$

$$= 40.13$$

Rounding down to the nearest integer, $df = 40$.

(e) Since $H_a: \mu_1 \neq \mu_2$, the critical value at $\alpha = 0.1$ is $t_{\alpha/2} = t_{0.05} = 1.648$. Since $|t| = 1.12 < 1.648$, we accept the null hypothesis at $\alpha = 0.1$.

TI 84 Plus Calculator Instruction for a Two-sample T-test

Two-sample pooled T-test with summary statistics

Ref: Example 10.15

(1) Press STAT, select TESTS, choose **4: 2-SampTTest…**and press **ENTER.**

(2) Move the cursor to **Stats** and press **ENTER**.

(3) Specify as appropriate: $\bar{x}_1$: 44, S_{x1}: 9.15 , n_1: 22, $\bar{x}_2$: 41, S_{x2}: 8.2 , n_2: 20.

(4) For μ_1: Choose $> \mu_2$ [for Example 10.16: $< \mu_2$; for Example 10.17: $\neq \mu_2$].

(5) For Pooled: **Yes**, ENTER.

(6) Move the cursor to Calculate and press ENTER

See the option selections and results in figures below:

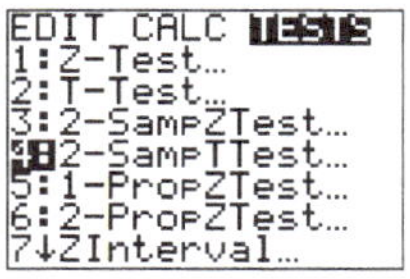

(A)

(B)

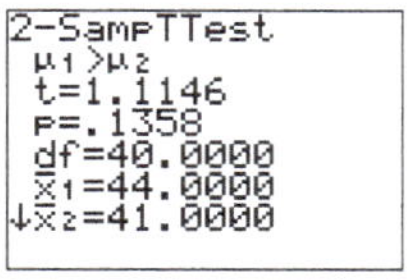

(a)

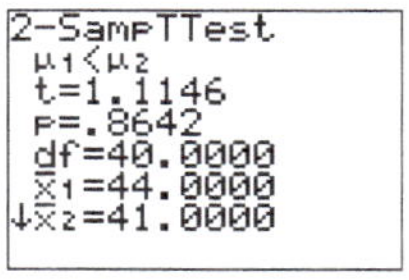

(b)

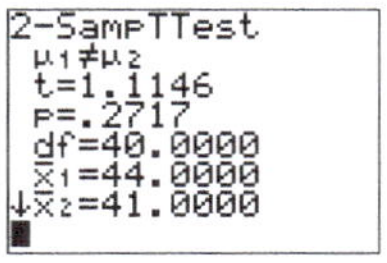

(c)

Observed values of t and p-values for Examples 10.15, 10.16, 10.17 appear in figures (a)-(c) above, respectively.

Two-sample non-pooled T-Test

Ref: Example 10.18

(1) Press STAT, select TESTS, choose **4: 2-SampTTest…**and press **ENTER.**

(2) Move the cursor to **Stats** and press **ENTER**.

(3) Specify as appropriate: $\bar{x}_1$: 44, S_{x1}: 9.15 , n_1: 22, $\bar{x}_2$: 41, S_{x2}: 8.2 , n_2: 20.

(4) For μ_1: Choose $\neq \mu_2$.

(5) For Pooled: **No**, ENTER.
(6) Move the cursor to Calculate and press ENTER

See the option selections and result in figures below:

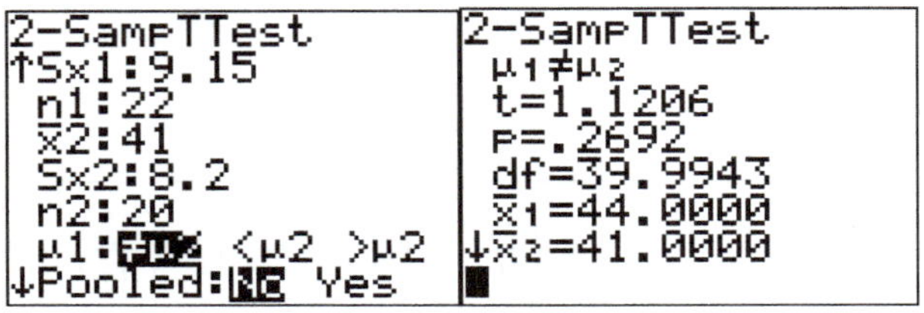

Note that the observed value of $t = 1.121$ and p-value=0.2692, which means that we accept the null hypothesis at $\alpha = 0.1$, an identical conclusion found in Example 10.18 using the critical value approach.

Minitab Instruction for a Two-sample T-test

For summary data we follow the steps below:

(1) Select Stat>Basic Statistics>2- Sample t…
(2) Choose **Summarized data** option and type Sample size, Sample mean, and Sample standard deviation for Sample 1 and Sample 2.
(3) Use option to specify ***Confidence level***: 95, Specify Hypothesized difference:0, Alternative hypothesis: Difference ≠ hypothesized difference.
(4) For **pooled T-test**, check mark **Assume equal variances** option, and
(5) For **non-pooled T-test** uncheck **Assume equal variances** option.
(6) Click [OK] two times.

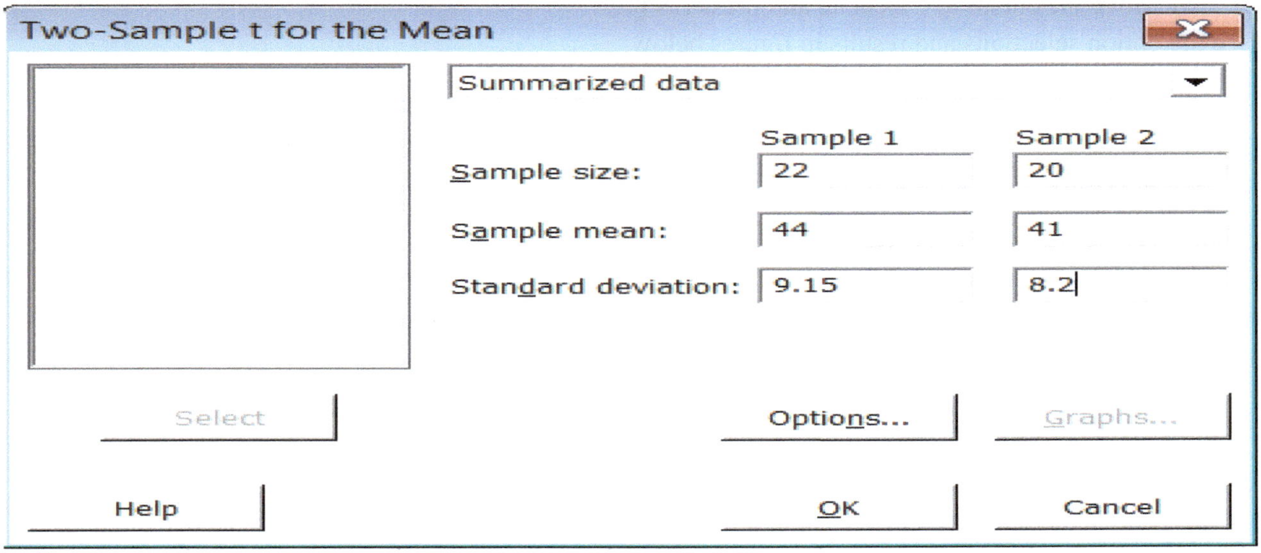

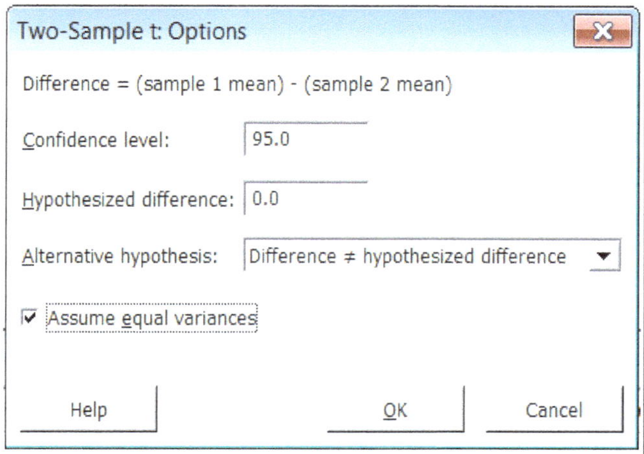

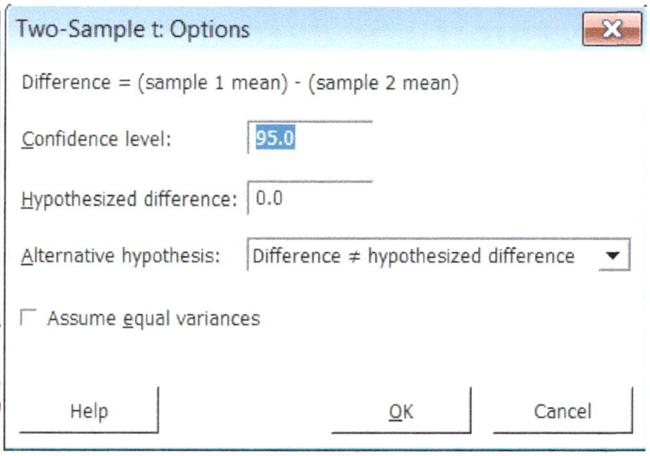

Minitab output

Ref: Example 10.15

Two-Sample T-Test and CI (Pooled T-test)

Sample	N	Mean	StDev	SE Mean
1	22	44.00	9.15	2.0
2	20	41.00	8.20	1.8

Difference = μ (1) - μ (2)
Estimate for difference: 3.00
95% CI for difference: (-2.44, 8.44)
T-Test of difference = 0 (vs ≠): T-Value = 1.11 **P-Value = 0.272** DF = 40
Both use Pooled StDev = 8.7117

Ref: Example 10.18

Two-Sample T-Test and CI (Non-pooled T-test)

Sample	N	Mean	StDev	SE Mean
1	22	44.00	9.15	2.0
2	20	41.00	8.20	1.8

Difference = μ (1) - μ (2)
Estimate for difference: 3.00
95% CI for difference: (-2.42, 8.42)
T-Test of difference = 0 (vs ≠): T-Value = 1.12 **P-Value = 0.269** DF = 39

Section 10.5: Paired T-test

Often, we are interested in two populations whose measurements are paired. For example, let μ_1 and μ_2 be the means of two populations that represent before and after measurements or measurements for two different treatments, etc. We wish to test $H_0: \mu_1 - \mu_2 = 0$ against $H_a: \mu_1 - \mu_2 \neq 0$, or equivalently, by letting $\mu_d = \mu_1 - \mu_2$, test $H_0: \mu_d = 0$ against $H_a: \mu_d \neq 0$. This test is called a ***paired T-test***.

In order to carry out a paired T-test, we need a sample of n paired measurements: $(x_1, y_1), (x_2, y_2), \dots, (x_n, y_n)$.

Let $d_i = x_i - y_i$; $i = 1,2, \dots, n$; $\bar{d} = \bar{x} - \bar{y}$ and $s_d^2 = \frac{\sum(d_i - \bar{d})^2}{n-1}$.

Using above notations, the test statistic to test $H_0: \mu_d = 0$ against $H_a: \mu_d \neq 0$ is given by

$$T = \frac{\bar{d} - \mu_d}{s_d/\sqrt{n}}$$

Under null hypothesis, the statistic T follows a t-distribution with $(n-1)$ degrees of freedom.

Assumptions for paired T-test

The paired T-test requires the following assumptions:

(1) The observations of paired differences represent an SRS,

(2) Either the sample is large ($n \geq 30$) or the population of paired differences follows a normal distribution,

(3) Standard deviation of the population of paired differences is unknown.

Decision rule for paired T-test using critical value approach

Reject $H_0: \mu_d = 0$ at significance level α, if

(1) $t > t_\alpha$ when $H_a: \mu_d > 0$ (i.e., right-tailed test)

(2) $t < -t_\alpha$ when $H_a: \mu_d < 0$ (i.e., left-tailed test)

(3) $|t| > t_{\alpha/2}$ when $H_a: \mu_d \neq 0$ (i.e., two-tailed test)

where t_α and $-t_\alpha$ are values under the curve of a t-distribution with $df = n - 1$ having area α to its right and left, respectively.

Decision rule for paired T-test using p value approach

Reject $H_0: \mu_d = 0$ at significance level α if p-value$\leq \alpha$, where p-value is computed as follows

p-value$= P(T > t)$ if $H_a: \mu_d > 0$ (right-tailed test)

p-value$= P(T < t)$ if $H_a: \mu_d < 0$ (left-tailed test)

p-value$= 2 \times P(T > |t|)$ if when $H_a: \mu_d \neq 0$ (i.e., two-tailed test)

where the test statistic T follows a t-distribution with $df = n - 1$ and t is the observed value of T.

Example 10.19

We have a sample of 5 ordered pairs of domestic (x) and international (y) airline fuel cost (in dollars/gallon) from the U.S. scheduled carriers reported by months of 2014. We wish to test if there is any difference in the average fuel cost using a paired T-test.

x	y	$d = x - y$
2.98	2.93	0.05
2.89	2.81	0.08
2.98	2.84	0.14
3.09	2.95	0.14
2.99	2.93	0.06

Source: *Bureau of Transportation Statistics F41 Schedule P12A as of 6/25/2015. For more detail about the source visit http://www.transtats.bts.gov/fuel.asp. Sample is generated in Minitab.*
Assume that the population of differences in domestic and international fuel cost follows a normal distribution.
Under usual notations, we have $\bar{d} = 0.094$ and $s_d = .0434$.

(a) Set up null and alternative hypotheses for the test.
(b) Write out the appropriate test statistic.
(c) Find the value of the test statistic and degrees of freedom (df).
(d) Find the critical value for the test at 5% level of significance and make a conclusion of the test.
(e) Find p-value for the test using any **technology** and make a conclusion of the test at $\alpha = 0.05$.
(f) Do the conclusion in (d) and (e) differ?

Solution

(a) The null hypothesis is $H_0: \mu_d = 0$ and alternative is $H_a: \mu_d \neq 0$.
(b) The test statistic is $T = \frac{\bar{d} - \mu_d}{s_d/\sqrt{n}}$
(c) The value of the test statistic is $t = \frac{0.094 - 0}{0.0434/\sqrt{5}} = 4.843$ and $df = n - 1 = 5 - 1 = 4$.
(d) Since $H_a: \mu_d \neq 0$, at $\alpha = 0.05$ and $df = 4$, the critical value is $t_{\alpha/2} = t_{0.025} = 2.776$.

df	$t_{0.1}$	$t_{0.05}$	$t_{0.025}$	$t_{0.01}$	$t_{0.005}$	$t_{0.001}$	$t_{0.0005}$
3	*1.638*	*2.353*	*3.182*	*4.541*	*5.841*	*10.215*	*12.924*
4	*1.533*	*2.132*	*2.776*	*3.747*	*4.604*	*7.173*	*8.610*

Conclusion: Since $|t| > t_{0.025} = 2.776$, we reject $H_0: \mu_d = 0$ at $\alpha = 0.05$ against $H_a: \mu_d \neq 0$.

(e) Since $H_a: \mu_d \neq 0$, p-value$= 2 \times P(T > 4.843) = 0.008$ (see the TI-84 Plus snapshot below for the computation of the p-value).

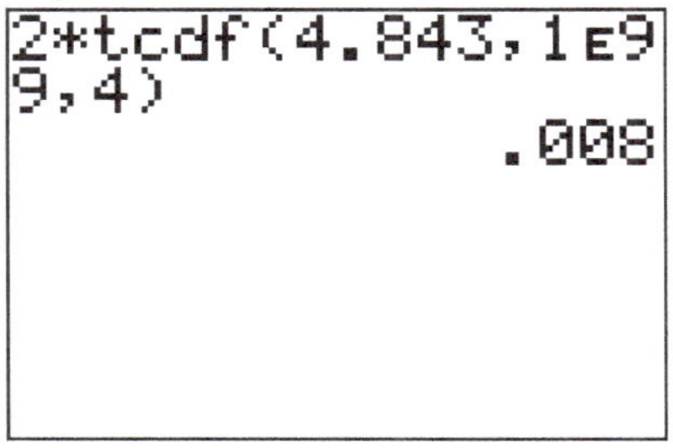

Conclusion: Since p-value $= 0.008 < 0.05$, we reject the null hypothesis at 5% level of significance.

(f) Conclusions from (d) using critical value and (e) using p-value approaches are identical.

Example 10.20

A paired sample of 10 domestic itinerary fares (in dollars) is considered from the top 100 ranked US Airports for Quarter 4 of 2012 (x) and 2013 (y). The sample data appears in table below:

Rank	Airport	Q42012 (x)	Q42013 (y)	d=x-y
83	Knoxville, TN	470.19	402.89	67.30
84	White Plains, NY	396.17	414.21	-18.04
31	Kansas City, MO	356.73	363.36	-6.62
55	Palm Beach, FL	316.14	337.82	-21.68
34	Oakland, CA	302.99	304.06	-1.08
46	Hartford, CT	408.09	381.76	26.33
41	Cleveland, OH	461.46	458.53	2.93
33	Raleigh/Durham, NC	354.24	372.33	-18.09
17	Baltimore, MD	323.29	328.25	-4.96
58	Oklahoma City, OK	411.52	400.42	11.10
Average		380.08	376.36	3.72

Source: Bureau of Transportation Statistics, Origin and Destination Survey: DB1BTicket
http://www.rita.dot.gov/bts/airfares/programs/economics_and_finance/air_travel_price_index/html/table_09.html

Suppose that the population of differences of itinerary fares in Quarter 4 of 2012 and 2013 for the top 100 Airports follows a normal distribution with unknown mean and standard deviation.
Given the sample, it follows that $\bar{d} = 3.72$ and $s_d = 26.7$.
We wish to test $H_0: \mu_d = 0$ against $H_a: \mu_d \neq 0$ at 5% level of significance.

(a) What test should you use?
(b) Write out the test statistic and find the observed value of the test statistic.
(c) Find the degrees of freedom for the test.
(d) Find the critical value for the test and make a conclusion about the test at 5% level of significance.
(e) Find p-value of the test and make a conclusion about the test at $\alpha = 0.05$.

Solution

(a) We should use a paired T-test for the test.
(b) The test statistic is $T = \frac{\bar{d}-\mu_d}{s_d/\sqrt{n}}$ and the value of the test statistic is $t = \frac{3.72-0}{26.7/\sqrt{10}}$=0.44.
(c) The $df = n - 1 = 10 - 1 = 9$.
(d) Since $H_a: \mu_d \neq 0$, at 5% level of significance, the critical value is $t_{\alpha/2} = t_{0.025} = 2.262$.

df	$t_{0.1}$	$t_{0.05}$	$t_{0.025}$	$t_{0.01}$	$t_{0.005}$	$t_{0.001}$	$t_{0.0005}$
8	*1.397*	*1.860*	2.306	2.896	3.355	4.501	5.041
9	*1.383*	*1.833*	2.262	2.821	3.250	4.297	4.781
10	*1.372*	*1.812*	2.228	2.764	3.169	4.144	4.587

Conclusion: Since $|t| < t_{0.025}$, we fail to reject the null hypothesis H_0.

(e) Since $H_a: \mu_d \neq 0$, p-value= $2 \times P(T > 0.44) = 0.6703$ (see the TI-84 Plus snapshot below for the computation of the p-value).

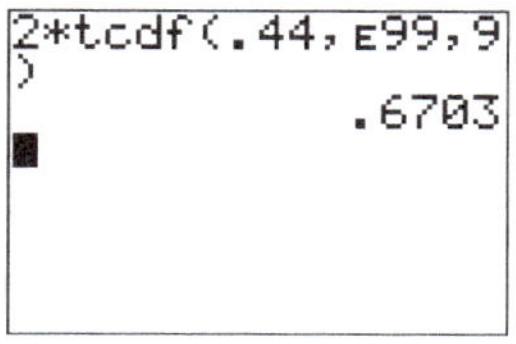

Conclusion: Since p-value = 0.6703 > 0.05, we fail to reject the null hypothesis at 5% level of significance.

Minitab steps for paired T-Test

Case I: Each sample is a column

Ref: Example 10.19

(1) In Minitab worksheet, type x and y in columns C1 and C2, respectively.
(2) Select Stat>Basic Statistics>Paired t…
(3) Choose **Each sample is in a column** option and Sample 1: select x for and for Sample 2: select y.
(4) Use option to specify ***Confidence level***: 95, Specify Hypothesized difference:0, Alternative hypothesis: Difference ≠ hypothesized difference
(5) Click [OK] two times.

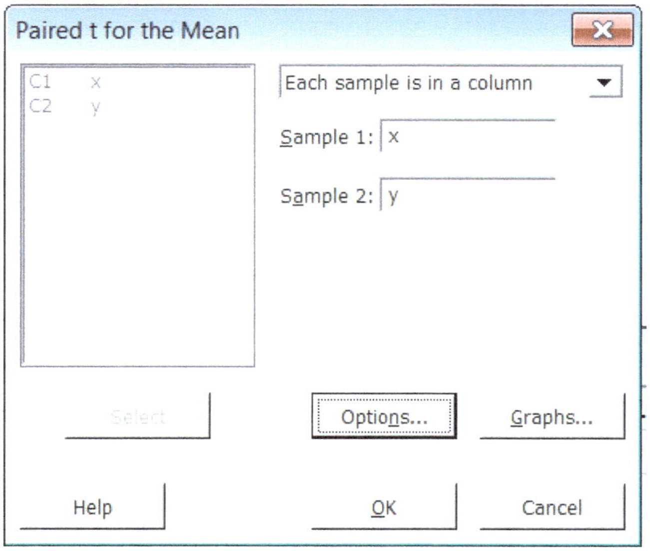

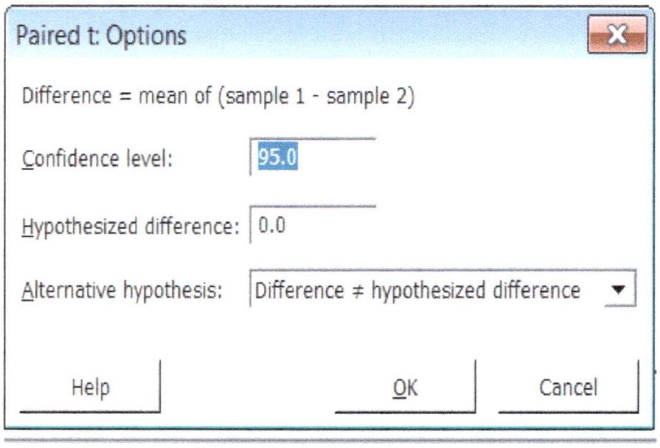

See the output below:

Paired T for x - y

	N	Mean	StDev	SE Mean
x	5	2.9860	0.0709	0.0317
y	5	2.8920	0.0626	0.0280
Difference	5	0.0940	0.0434	0.0194

95% CI for mean difference: (0.0402, 0.1478)
T-Test of mean difference = 0 (vs ≠ 0): **T-Value = 4.85 P-Value = 0.008**

Case II: When summarized data are given

Ref: Example 10.20

(1) Select Stat>Basic Statistics>Paired t…
(2) Choose option **Summarized data**, and specify **sample size:10**, **sample mean:** 3.72, **Standard deviation:** 26.7,
(3) Use option to specify ***Confidence level***: 95, and Specify Hypothesized difference:0, Alternative hypothesis: Difference ≠ hypothesized difference.
(4) Click [OK] two times.

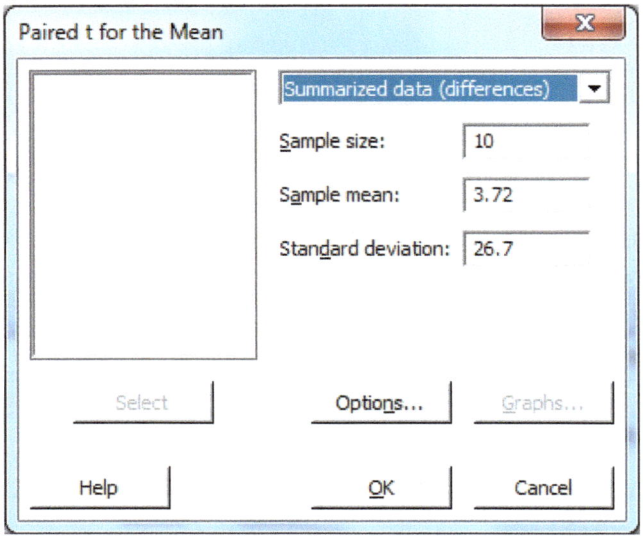

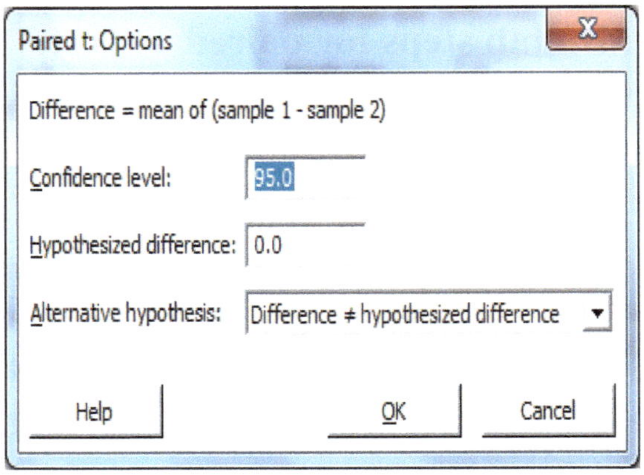

Paired T-Test and CI

	N	Mean	StDev	SE Mean
Difference	10	3.72	26.70	8.44

95% CI for mean difference: (-15.38, 22.82)
T-Test of mean difference = 0 (vs ≠ 0):
T-Value = 0.44 P-Value = 0.670

Chapter 10: Practice Problems

Section 10.1

(10.1) How much money did you spend on your favorite non-academic book last year?
On the basis of a sample of 30 students in a large class, the average amount spent on non-academic book last year was estimated to be \$18.47. We want to test the hypothesis that the average amount spent on non-academic book last year by all students in the class was actually \$21.50.

How should you express the null and alternative hypotheses in this scenario?

(10.2) A researcher wants to test if the average prices spent on history and non-history books are different. Let the unknown average price spent on history and non-history books be μ_1 and μ_2. Set up the null and alternative hypotheses for this test. Identify your choice of the test to be one-tailed or two tailed.

(10.3) As reported by the Office of Dietary Supplements (ODS), National Institutes of Health (November 16, 2012), the recommended dietary allowance (RDA) for pregnant women aged 19-50 years is 1,000 mg/day (Source: http://ods.od.nih.gov/factsheets/Calcium-HealthProfessional/).
We might be interest to test if, on average, a pregnant woman with an income below the poverty level gets the RDA of 1,000 mg/day. Specify the followings for the test:

(a) The null hypothesis.
(b) The alternative hypothesis.
(c) Classify the hypothesis test as two-tailed, left-tailed, or right-tailed.
(d) Justify if it is reasonable to set up the alternative to be $H_a: \mu > 1000$ or $H_a: \mu \neq 1000$ in (b).

(10.4) Suppose a student is interested to see if there is any significant gender difference between the average times it takes to finish an exam. In testing a hypothesis, a student specifies the null and the alternative hypotheses as follows:
$H_0: \bar{x}_1 = \bar{x}_2$versus
$H_a: \bar{x}_1 \neq \bar{x}_2$
What is your opinion about the set up of the null and alternative hypotheses above?

Section 10.2

(10.5) We wish to test a hypothesis $H_0: \mu = 10$ against $H_a: \mu \neq 10$, where $\sigma = 2.5$. On the basis of a sample of size 30, the sample mean is found to be 11.5.

(a) The test statistic for the test is $Z = \frac{\bar{x}-\mu_0}{\sigma/\sqrt{n}}$. Find the value of the test statistic.
(b) Which of the followings will be used to compute p-value of the test?
  (i) p-value= $P(Z > z)$
  (ii) p-value= $P(Z < z)$
  (iii) p-value= $2P(Z \geq |z|)$
  (iv) All of the above
(c) Compute the p-value for the test.
(d) Make a conclusion about the test at 5% level of significance.

(10.6) For testing $H_0: \mu = 5$ against $H_a: \mu > 5$, where σ is a known value, the value of the test statistic is found to be $z = 2.5$.

(a) At $\alpha = 0.05$, what is the conclusion about the test using a critical value approach?
(b) What is the conclusion about the test using a p-value approach?

(10.7) For testing $H_0: \mu = 2$ against $H_a: \mu < 2$, where σ is a known value, the test statistic has the value $z = 1.3$.

(a) Find the critical value for making a conclusion about the test at $\alpha = 0.05$.
(b) What conclusion can you make by comparing $z = 1.3$ with the critical value?
(c) Compute a p-value and make a conclusion about the test at $\alpha = 0.05$.
(d) How does the conclusion in (c) differ from the conclusion in (b)?

(10.8) Let μ be the unknown mean of number of credit cards that students in a large class have. For a test $H_0: \mu = 3$ against $H_a: \mu \neq 3$, where σ is a known value, the test statistic has the value $z = 2.01$.
(a) Make a conclusion about the test at $\alpha = 0.05$ using a critical value approach.
(b) Make a conclusion about the test at $\alpha = 0.05$ using a p-value approach.
(c) Do you find any difference in the conclusions of (a) and (b)?

(10.9) The posted speed limit of vehicles in a neighborhood is 25 miles per hour. It is believed that the average speed at which cars pass by is higher than the actual limit. In order to test the belief, a sample of 30 cars is considered and the mean speed is found to be 27.5 miles/hour. Suppose that speed of all cars passing through the neighborhood follows a normal distribution with standard deviation $\sigma = 4$ miles/hour.
(a) Write out the null and alternative hypotheses for the test.
(b) What test is appropriate in this scenario?
(c) Find the observed value of the test statistic.
(d) What is the conclusion about your test at 5% level of significance?
(e) Find the p-value for the test and make a conclusion at $\alpha = 0.05$ and $\alpha = 0.001$.

(10.10) We are testing $H_0: \mu = 30$ versus $H_a: \mu > 30$, where σ is equal to 15. Suppose that a sample of $n = 45$ measurements reveals a sample mean of $\bar{x} = 35$. The test statistic for this test is $Z = \frac{\bar{x}-\mu_0}{\sigma/\sqrt{n}}$.
(a) Find the observed value of the test statistic.
(b) Find the critical value of the test at $\alpha = 0.05$ and make a conclusion of the test.
(c) Calculate the p-value of the test and make a conclusion of the test at $\alpha = 0.05$.

(10.11) For a test $H_0: \mu = 10\ versus\ H_a: \mu \neq 10$, a sample of size $n = 36$ is considered and the sample mean turns out to be 9.23. Suppose that the population standard deviation is 3.5.
(a) What test should you consider for the test?
(b) Compute the value of the test statistic.
(c) Find the critical value for the test at 5% level of significance and comment on the test.
(d) Find p-value for the test and make a conclusion about the test at 5% level of significance.

(10.12) To test $H_0: \mu = 15\ versus\ H_a: \mu > 15$, a sample of size 45 is drawn and the mean is found to be 16.5. Suppose that the population standard deviation is 4.
(a) What test should you consider for the test?
(b) Compute the value of the test statistic.
(c) Find the critical value for the test at 5% level of significance and comment on the test.
(d) Find p-value for the test and make a conclusion about the test at 5% level of significance.

(10.13) Suppose that on the basis of a sample of 20 bill payments of T-Mobile customers, the mean of bills is \$118. It is also known that the bills of T-Mobile customers approximately follow a normal distribution with standard deviation \$10. Test $H_0: \mu = 120$ versus $H_a: \mu > 120$ at $\alpha = 0.1$ by answering the following questions:

(a) What test should you use?
(b) Write out the formula of the test statistic and compute the value of the test statistic.
(c) Find the critical value of the test and make a conclusion about the test.
(d) Compute p-value for the test and make a conclusion about the test on the basis of the p-value.

(10.14) Reference to Exercise 10.13, test $H_0: \mu = 120$ versus $H_a: \mu < 120$ at $\alpha = 0.1$ by answering questions (a)-(d).

(10.15) Reference to Exercise 10.13, test $H_0: \mu = 120$ versus $H_a: \mu \neq 120$ at $\alpha = 0.1$ by answering questions (a)-(d).

Section 10.3

(10.16) For a test $H_0: \mu = 10$ against $H_a: \mu \neq 10$, where σ is unknown, an observed value of the test statistic is found to be $t = 2.1$ for a sample of size 35. Answer the following questions regarding the test:

(a) What do you call this test?
(b) What are the degrees of freedom for the test statistic?
(c) Find the critical value at 5% level of significance. What is the conclusion about the test?
(d) Use any technology to compute the p-value for the test.
(e) Make a conclusion about the test on the basis of p-value you obtained in (d) at 5% level of significance.

(10.17) For testing $H_0: \mu = 5$ against $H_a: \mu > 5$, where σ is unknown, the value of the test statistic is found to be $t = 1.95$ using a sample of size 45. Answer the following questions regarding the test:

(a) What do you call this test?
(b) What are the degrees of freedom for the test statistic?
(c) Find the critical value at 5% level of significance. What is the conclusion about the test?
(d) Use any technology to compute the p-value for the test.
(e) Make a conclusion about the test on the basis of p-value you obtained in (d) at 5% level of significance.

(10.18) For testing $H_0: \mu = 2$ against $H_a: \mu < 2$, where σ is unknown, the test statistic has the value $t = 1.3$ on the basis of a sample of size $n = 49$.

(a) Find the critical value for making a conclusion about the test at $\alpha = 0.05$.
(b) What conclusion can you make by comparing $t = 1.3$ with the critical value?
(c) Compute the p-value and make a conclusion about the test at $\alpha = 0.05$.
(d) How does the conclusion in (c) differ from the conclusion in (b)?

(10.19) Let μ be the unknown mean of number of credit cards that students in a large class have. For a test $H_0: \mu = 3$ against $H_a: \mu \neq 3$, where σ is unknown, the test statistic has the value $t = 3.21$ using a sample of size 30.

(a) Make a conclusion about the test at $\alpha = 0.05$ using a critical value approach.

(b) Make a conclusion about the test at $\alpha = 0.05$ using a p-value approach.
(c) Do you find any difference in the conclusions of (a) and (b)?

(10.20) The speed limit of vehicles in a neighborhood is 25 miles/hour. We wish to test $H_0: \mu = 25$ vs $H_a: \mu > 25$, i.e., the average speed at which cars are driven is higher than the actual limit. To test the hypothesis, a sample of 30 cars is considered and the mean speed is found to be 27.5 miles/hour and $s = 7.3$.

(a) What test should you use for the test?
(b) Explain how the assumptions of the test are satisfied.
(c) Find the observed value of the test statistic.
(d) What are the degrees of freedom for the test?
(e) Find the critical value of the test at $\alpha = 0.05$ and make a conclusion about the test.
(f) Find the p-value for the test and make a conclusion at $\alpha = 0.05$ and $\alpha = 0.001$.

(10.21) Reference to the Exercise 10.20, suppose we wish to test $H_0: \mu = 25$ vs $H_a: \mu \neq 25$.

(a) Does the test statistic change with the change in the alternative hypothesis?
(b) Find the observed value of the test statistic and make a conclusion about your test at 5% level of significance using critical value approach.
(c) What is the p-value for the test?
(d) Make a conclusion about the test on the basis of the p-value.

(10.22) We wish to test $H_0: \mu = 15\ versus\ H_a: \mu > 15$ for a population mean. On the basis of a sample of size 45, it turns out that the sample mean is 16.5 and sample standard deviation is 3.5.

(a) Name the test you will be using here.
(b) Compute the value of the test statistic.
(c) Find the critical value for the test at 5% level of significance and comment on the test.
(d) What is the p-value for the test? Make a conclusion about the test using the p-value at 5% level of significance.

(10.23) Suppose we wish to test $H_0: \mu = 30$ versus $H_a: \mu > 30$. Suppose that a sample of $n = 45$ measurements reveals a sample mean of $\bar{x} = 35$ and $s = 12$.

(a) Write out the test statistic.
(b) Compute the observed value of the test statistic.
(c) Find the critical value of the test at $\alpha = 0.05$ and make a conclusion of the test.
(d) Calculate the p-value of the test and make a conclusion of the test at $\alpha = 0.05$ and $\alpha = 0.01$.

(10.24) On the basis of a sample of 25 payments for Sprint customers, the average and standard deviation of monthly bills are found to be \$143 and \$10.9, respectively, including taxes and fees. We want to test $H_0: \mu = \$140$ versus $H_a: \mu > \$140$.

Assume that the population of monthly payments of Sprint customers follows approximately a normal distribution.

(a) What test should you use? Write out the test statistic.
(b) Compute the observed value of the test statistic. What is your conclusion about the test at $\alpha = 0.05$ and $\alpha = 0.01$?

(c) What is the p-value of the test? What is your conclusion about the test on the basis of the p-value at $\alpha = 0.05$ and $\alpha = 0.01$?

(10.25) Suppose that on the basis of a sample of 20 monthly bill payments of T-Mobile customers, the mean and standard deviation of bills are found to be $\bar{x} = \$118$ and $s = \$11.5$. It is also known that the monthly bills of T-Mobile customers approximately follow a normal distribution with an unknown mean and standard deviation. Test $H_0: \mu = 120$ versus $H_a: \mu > 120$ at $\alpha = 0.05$ by answering the following questions:

(a) What test should you use for the test?
(b) Find the observed value of the test statistic.
(c) What are the degrees of freedom for the test?
(d) Find the critical value of the test and make a conclusion about your test at 5% level of significance.
(e) Find the p-value for the test and make a conclusion at $\alpha = 0.05$.

(10.26) In reference to Exercise 10.25, test $H_0: \mu = 120$ versus $H_a: \mu < 120$ at $\alpha = 0.05$ by answering questions (a)-(e).

(10.27) In reference to Exercise 10.25, test $H_0: \mu = 120$ versus $H_a: \mu \neq 120$ at $\alpha = 0.05$ by answering questions (a)-(e).

Section 10.4

(10.28) Suppose that on the basis of two independent samples, we have $\bar{x}_1 = 44$, $\bar{x}_2 = 41$, $s_1 = 9.15$, $s_2 = 8.2$, $n_1 = 22$ and $n_2 = 20$. Suppose that the two populations are approximately normal with a common unknown variance. We wish to test $H_0: \mu_1 = \mu_2$ versus $H_a: \mu_1 > \mu_2$. Answer the following questions:

(a) What test should you use for the test?
(b) Are the assumptions of the test satisfied in this problem?
(c) Find the value of the test statistic.
(d) What are the degrees of freedom for the test?
(e) Make a conclusion about the test by setting α to 0.1 and 0.05 using the critical value approach.
(f) Find p-value of the test using any technology. Make conclusions about the test by setting α to 0.1 and 0.05.

(10.29) Reference to Exercise 10.28, test $H_0: \mu_1 = \mu_2$ versus $H_a: \mu_1 < \mu_2$ and answer questions (a)-(f).

(10.30) Reference to Exercise 10.28, test $H_0: \mu_1 = \mu_2$ versus $H_a: \mu_1 \neq \mu_2$ by answering questions (a)-(f).

(10.31) Suppose that on the basis of two independent samples, we have $\bar{x}_1 = 44$, $\bar{x}_2 = 41$, $n_1 = 22$ and $n_2 = 20$. Suppose that the two populations are approximately normal and population variances are known to be $\sigma_1^2 = 100$ and $\sigma_2^2 = 64$. Test $H_0: \mu_1 = \mu_2$ versus $H_a: \mu_1 > \mu_2$ by answering the following questions:

(a) What is the name of the test?

(b) Write out the formula of the test statistic.
(c) Compute the value of the test statistic.
(d) Make conclusions about the test by setting α to 0.1, 0.05 and 0.01 using critical value approach.
(e) Find p-value for the test and make conclusion about the test using p-value by setting α to 0.1, 0.05 and 0.01.

(10.32) Reference to Exercise 10.31, test $H_0: \mu_1 = \mu_2$ versus $H_a: \mu_1 < \mu_2$ and answer questions (a)-(e).

(10.33) Reference to Exercise 10.31, test $H_0: \mu_1 = \mu_2$ versus $H_a: \mu_1 \neq \mu_2$ and answer questions (a)-(e).

(10.34) Suppose that on the basis of a sample of 15 monthly payments of Verizon customers, the average and standard deviation of bills are \$147 and \$9.5, respectively, including taxes and fees. On the basis of a sample of 20 monthly payments for Sprint customers, the average and standard deviation of bills are found to be \$143 and \$10.9, respectively, including taxes and fees. Suppose the payments of monthly bills for Verizon and Sprint customers approximately follow normal distributions with unknown means and unknown but equal standard deviations. Test $H_0: \mu_1 = \mu_2$ versus $H_a: \mu_1 > \mu_2$ by answering the following questions:
(a) Name the test to be used here.
(b) Write out the appropriate test statistic.
(c) Compute the value of the test statistic.
(d) What are the degrees of freedom for the statistic?
(e) Find critical values for the test at $\alpha = 0.1$, 0.05 and 0.01 and make a conclusion about the test.
(f) Compute p-value of the test and make a conclusion about the test at $\alpha = 0.1$, 0.05 and 0.01.

(10.35) Reference to Exercise 10.34, test $H_0: \mu_1 = \mu_2$ versus $H_a: \mu_1 < \mu_2$ by answering questions (a)-(f).

(10.36) Reference to Exercise 10.34, test $H_0: \mu_1 = \mu_2$ versus $H_a: \mu_1 \neq \mu_2$ by answering questions (a)-(f).

(10.37) Use the fact of Exercise 10.34 that $\bar{x}_1 = \$147$, $\bar{x}_2 = \$143$, $n_1 = 15$ and $n_2 = 20$. It is also known that the standard deviations of payments of monthly bills of all Verizon and Sprint customers are known to be \$11 and \$9.5, respectively. Given the new scenario, test $H_0: \mu_1 = \mu_2$ versus $H_a: \mu_1 > \mu_2$ by answering the following questions:
(a) What do you call the test?
(b) Write out the appropriate test statistic.
(c) Compute the value of the test statistic.
(d) Find critical values for the test by setting α to 0.1 and 0.05.
(e) Make a conclusion about the test for each α in (d).

(10.38) For Exercise 10.37, test $H_0: \mu_1 = \mu_2$ versus $H_a: \mu_1 < \mu_2$ by answering questions in (a)-(e).

(10.39) For Exercise 10.37, test $H_0: \mu_1 = \mu_2$ versus $H_a: \mu_1 \neq \mu_2$ by answering questions in (a)-(e).

Section 10.5

(10.40) Distinguish between a paired t-test and two-sample t-test. Give some examples where each of them can be applied.

(10.41) Suppose that for a sample of 36 paired differences, we have $\bar{d} = 1.54$ and $s_d = 2.21$. Test $H_0: \mu_d = 0$ against $H_a: \mu_d > 0$ at 5% level of significance. Test the hypothesis by answering the following questions:

(a) What test should you use?
(b) What assumptions are required for the validity of the test? Are these assumptions satisfied by the problem?
(c) Write out the test statistic and find the observed value of the test statistic.
(d) What are the degrees of freedom for the test?
(e) Find the critical value for the test and make a conclusion about the test at 5% level of significance.
(f) Find p-value of the test and make a conclusion about the test at $\alpha = 0.05$.

(10.42) In reference to Exercise 10.41, answer questions (a)-(f) for testing $H_0: \mu_d = 0$ against $H_a: \mu_d < 0$ at $\alpha = 0.05$.

(10.43) In reference to Exercise 10.41, answer questions (a)-(f) for testing $H_0: \mu_d = 0$ against $H_a: \mu_d \neq 0$ at $\alpha = 0.05$.

(10.44) Suppose that a sample of 9 video game players rated a game on a 10-point scale after playing the game first and second time. The scale consists of numbers 1 through 10, with a lower number indicating the lower satisfaction rate. The 9 ordered pair data appears in the table below, where x and y refer to the rating after playing the game first and second time, respectively.

x	8	7	10	6	8	7	8	7	4
y	7	7	8	6	7	7	9	6	5
$d = x - y$	1	0	2	0	1	0	-1	1	-1

One can verify that $\bar{d} = 0.33$ and $s_d = 1$. Test $H_0: \mu_d = 0$ against $H_a: \mu_d > 0$ at 5% level of significance by answering the following questions:

(a) Write out the formula of the test statistic and compute its value.
(b) What are the degrees of freedom?
(c) Find the critical value or p-value for the test.
(d) Make a conclusion about the test.

(10.45) In reference to Exercise 10.44, answer questions (a)-(d) for testing $H_0: \mu_d = 0$ against $H_a: \mu_d < 0$ at 5% level of significance.

(10.46) In reference to Exercise 10.44, answer questions (a)-(d) for testing $H_0: \mu_d = 0$ against $H_a: \mu_d \neq 0$ at 5% level of significance.

Chapter 10: Answer to selected problems

(10.1) (a) The null hypothesis is $H_0: \mu = \$21.50$ (b) The alternative hypothesis should be $H_a: \mu \neq 21.5$ (two-tailed).

(10.2) (a) The null hypothesis is $H_0: \mu_1 = \mu_2$ (b) The alternative hypothesis is $H_a: \mu_1 \neq \mu_2$ (two-tailed).

(10.3) (a) $H_0: \mu = 1000$ (b) $H_a: \mu < 1000$ (c) Left-tailed test (d) Specifying the alternative to be $H_a: \mu > 1000$ or $H_a: \mu \neq 1000$ is not interesting to us because it is of great concern to us to know if $\mu < 1000$.

(10.4) The null and alternative hypotheses are always statements about population parameters, not about statistics.

(10.5) (a) Z=3.29 (b) (iii) (c) 0.0010 (d) Reject $H_0: \mu = 10$ against $H_a: \mu \neq 10$

(10.6)

(a) At $\alpha = 0.05$, $z_\alpha = z_{0.05} = 1.645$. Since $z = 2.5 > z_\alpha$, reject $H_0: \mu = 5$.
(b) p-value=$P(z > 2.5) = 0.0062 < \alpha = 0.05$. So, reject $H_0: \mu = 5$.

(10.7)

(a) At $\alpha = 0.05$, $-z_\alpha = -z_{0.05} = -1.645$.
(b) Since $z = 1.3 > -z_\alpha$, fail to reject $H_0: \mu = 2$.
(c) p-value=$P(z < 1.3) = 0.9032$. So, fail to reject $H_0: \mu = 2$.
(d) The conclusion in (c) and (d) are the same.

(10.8)

(a) At $\alpha = 0.05$, $z_{\alpha/2} = z_{0.025} = 1.96$. Since $z = 2.01 > 1.96$, reject $H_0: \mu = 3$.
(b) p-value=$2 \times P(z > |z|) = 2 \times P(z > 2.01) = 2 \times 0.02222 = 0.04444 < \alpha = 0.05$. So, reject $H_0: \mu = 3$.
(c) There is no difference in the conclusion of (b) and (c).

(10.9) (a) $H_0: \mu = 25$ vs $H_a: \mu > 25$ (b) Z-test (c) $z = 3.42$ (d) At $\alpha = 0.05$, $z_\alpha = z_{0.05} = 1.645$. Since $z = 3.42 > 1.96$, reject $H_0: \mu = 25$ (e) p-value= $P(z > 4.32) = 1 - p(z < 4.32) = 0.0003$; Since p-value $= 0.0003 < \alpha = 0.05$, reject H_0 at $\alpha = 0.05$. Also, p-value $= 0.0003 < \alpha = 0.001$, reject H_0 at $\alpha = 0.001$.

(10.10) (a) $z = 2.24$ (b) At $\alpha = 0.05$, $z_\alpha = z_{0.05} = 1.645$. Since $z = 2.24 > 1.645$, reject $H_0: \mu = 30$ (c) p-value= $P(z > 2.24) = 1 - p(z < 2.24) = 0.01255$; Since p-value $< \alpha = 0.05$, reject H_0 at $\alpha = 0.05$.

(10.11) (a) Z-Test (b) $z = -1.32$ (c) At $\alpha = 0.05$, $z_{\alpha/2} = z_{0.025} = 1.96$. Since $|z| = 1.32 < 1.96$, fail to reject $H_0: \mu = 10$ (d) p-value= $2 \times P(z > |z|) = 2 \times P(z > 1.32) = 2 \times \{1 - p(z < 1.32)\} = 0.18684$; Since p-value$> \alpha = 0.05$, fail to reject H_0 at $\alpha = 0.05$.

(10.12) (a) Z-Test (b) $z = 2.52$ (c) At $\alpha = 0.05$, the critical value is $z_{\alpha/2} = z_{0.05} = 1.645$. Since $z = 2.52 > 1.645$, reject $H_0: \mu = 15$ (d) p-value$= 0.0059$; Since p-value$< \alpha = 0.05$, reject H_0 at $\alpha = 0.05$.

(10.16) (a) T-test. (b) $df = 34$. (c) At $\alpha = 0.05$, the critical value is $t_{\alpha/2} = t_{0.025} = 2.032$. Since $|t| = 2.1 > 2.032$, reject H_0. (d) p-value$= 2 \times P(t > |t|) = 2 \times P(t > 2.1) = 0.0432$, using TI-84 calculator. (e) Since p-value$< \alpha = 0.05$, reject H_0.

(10.17) (a) T-test. (b) $df = 44$. (c) At $\alpha = 0.05$, the critical value is $t_\alpha = t_{0.05} = 1.68$. Since $|t| = 1.95 > 1.68$, reject H_0. (d) p-value$= P(t > 1.95) = 0.0288$, using TI-84 calculator. (e) Since p-value$< \alpha = 0.05$, reject H_0.

(10.18)

(a) For a T-test with $df = 48$ and $\alpha = 0.05$, the critical value for a left tailed test is $-t_\alpha = -t_{0.05} = -1.677$.
(b) Since $t = 1.3 > -t_{0.05}$, fail to reject H_0.
(c) p-value$= P(t < 1.3) = 0.9001$, using TI-84 calculator.
(d) Since p-value$> \alpha = 0.05$, fail to reject H_0.

(10.20)

(a) T-test.
(b) A sample is considered, the sample is large ($n = 30$) and population standard deviation is unknown. Thus, the assumptions of T-test are met.
(c) $t = 1.88$
(d) $df = 29$.
(e) At $\alpha = 0.05$, the critical value is $t_\alpha = t_{0.05} = 1.699$. Since $t = 1.88 > 1.699$, reject H_0.
(f) p-value$= P(t > 1.88) = 0.0351$, using TI-84 calculator.
Since p-value$< \alpha = 0.05$, reject H_0 at $\alpha = 0.05$. On the other hand, since p-value$> \alpha = 0.001$, fail to reject H_0 at $\alpha = 0.001$.

(10.21)

(a) No, we still use the T-test.
(b) $t = 1.88$
(c) with $df = 29$ and $\alpha = 0.05$, the critical value is $t_{\alpha/2} = t_{0.025} = 2.045$. Since $t = 1.88 < 2.045$, we fail to reject H_0.
(d) p-value$= 2 \times P(t > |t|) = 2 \times P(t > 1.88) = 2 \times 0.0351 = 0.0702$.
(e) Since p-value$> \alpha = 0.05$, we fail to reject H_0 at $\alpha = 0.05$.

(10.23)

(a) $T = \frac{\bar{x} - \mu_0}{s/\sqrt{n}}$.
(b) $t = 2.80$
(c) With $df = 44$ and $\alpha = 0.05$, the critical value for a right tailed test is $t_\alpha = t_{0.05} = 1.68$. Since $t = 2.80 > 1.68$, reject H_0.
(d) p-value$= P(t > 2.80) = 0.0038$, using TI-84 calculator.
Since p-value$< \alpha = 0.05$, reject H_0 at $\alpha = 0.05$. On the other hand, since p-value$> \alpha = 0.001$, fail to reject H_0 at $\alpha = 0.001$.

(10.25)

(a) T-test.

(b) $t = -0.78$

(c) $df = 19$

(d) With $df = 19$ and $\alpha = 0.05$, the critical value for a right tailed test is $t_\alpha = t_{0.05} = 1.729$. Since $t = -0.78 < 1.68$, we fail to reject H_0.

(e) p-value= $P(t > -0.78) = 0.7775$, using TI-84 calculator.
Since p-value> $\alpha = 0.05$, we fail to reject H_0 at $\alpha = 0.05$.

(10.28)

(a) Two-sample pooled T-test.

(b) Yes (Two sample are given, two populations are normal and unknown common population standard).

(c) $t = 1.115$.

(d) $df = n_1 + n_2 - 2 = 22 + 20 - 2 = 40$.

(e) At given $\alpha = 0.1$ and $df = 40$, the critical value of a right tailed t-test is $t_\alpha = t_{0.1} = 1.303$. Since $t < t_{0.1}$, fail to reject H_0 at $\alpha = 0.1$.
In a similar manner, $\alpha = 0.05$ and $df = 40$, $t_\alpha = t_{0.05} = 1.684$. Since $t < t_{0.1}$, fail to reject H_0at $\alpha = 0.05$.

(f) p-value= $P(t > 1.115) = 0.1358$, using TI-84 calculator. Since p-value> $\alpha = 0.1$, we fail to reject H_0 at $\alpha = 0.1$. Also, p-value> $\alpha = 0.05$, we fail to reject H_0 at $\alpha = 0.05$

(10.29)

(a), (b), (c) and (d) are same as in (10.28).

(e) At $\alpha = 0.1, 0.05$, critical values of left tail test are $-t_{0.1} = -1.303$, $-t_{0.05} = -1.684$.
Since $t = 1.115 > -t_\alpha$ for $\alpha = 0.1$ and 0.05, we fail to reject H_0 at $\alpha = 0.1$ and 0.05.

(f) p-value= $P(t < 1.115) = 1 - 0.1358 = 0.8642$, using TI-84 calculator. Since p-value> α for $\alpha = 0.1$ and 0.05, we fail to reject H_0 at $\alpha = 0.1$ and 0.05.

(10.31)

(a) Two-sample Z-test.

(b) The test statistic is given by $Z = \frac{(\bar{x}_1 - \bar{x}_2)}{\sqrt{\frac{\sigma_1^2}{n_1} + \frac{\sigma_2^2}{n_2}}}$ under H_0.

(c) Observed value of the test statistic is $z = 1.08$.

(d) At $\alpha = 0.1, 0.05$ and 0.01, the critical value of a right tailed Z-test (z_α) are 1.28, 1.645 and 2.33, respectively. Since $z < z_\alpha$ for $\alpha = 0.1, 0.05$ and 0.01, we fail to reject H_0 at $\alpha = 0.1, 0.05$ and 0.01.

(e) The p-value is $P(z > 1.08) = 0.1405$. Since p-value> $\alpha = 0.1, 0.05, 0.01$, we fail to reject H_0 at $\alpha = 0.1, 0.05$ and 0.01.

(10.32)

(a), (b) and (c) are same as in (10.31).

(d) Do yourself.

(e) p-value $= P(z < 1.08) = 0.8595$. Since p-value> $\alpha = 0.1, 0.05, 0.01$, we fail to reject H_0 at $\alpha = 0.1, 0.05$ and 0.01.

(10.33)

(a), (b) and (c) are same as in (10.31).

(d) At $\alpha = 0.1, 0.05$ and 0.01, the critical value for a two tailed Z-test ($z_{\alpha/2}$) are 1.645, 1.96 and 2.58, respectively. Since $|z| < z_{\alpha/2}$ for $\alpha = 0.1, 0.05$ and 0.01, we fail to reject H_0 at $\alpha = 0.1, 0.05$ and 0.01.

(e) The p-value=$2 \times P(z > |z|) = 2 \times P(z > 1.08) = 0.281$. Since p-value$> \alpha = 0.1, 0.05, 0.01$, we fail to reject H_0 at $\alpha = 0.1, 0.05$ and 0.01.

(10.34)

(a) Two-sample pooled T-test.

(b) The test statistic is $T = \frac{(\bar{x}_1 - \bar{x}_2)}{s_p \times \sqrt{\frac{1}{n_1} + \frac{1}{n_2}}}$ under H_0.

(c) $t = 1.134$.

(d) $df = 33$.

(e) With $df = 33$, $t_{0.1} = 1.308$, $t_{0.05} = 1.692$, $t_{0.01} = 2.445$. Since $t < t_{0.1}$, $t_{0.1}$ and $t_{0.1}$, we fail to reject H_0 at $\alpha = 0.1, 0.05$ and 0.01.

(f) p-value= 0.1325. Since p-value$> \alpha = 0.1, 0.05$ and 0.01, we reject H_0 at $\alpha = 0.1, 0.05$ and 0.01.

(10.35) (a), (b), (c) and (d) are same as in (10.34) (e) Try yourself (f) p-value $= P(t < 1.134) = 0.8675$. Since p-value$> \alpha = 0.1, 0.05, 0.01$, we fail to reject H_0 at $\alpha = 0.1, 0.05$ and 0.01.

(10.36) (a), (b), (c) and (d) are same as in (10.34) (e) Try yourself (f) p-value $= 2 \times P(t > |t|) = 2 \times P(t > 1.134) = 0.265$. Since p-value$> \alpha = 0.1, 0.05, 0.01$, we fail to reject H_0 at $\alpha = 0.1, 0.05$ and 0.01.

(10.41) (a) Paired t-test (b) Try yourself (c) $t = 4.18$ (d) $df = 35$ (e) With $df = 35$ and $\alpha = 0.05$, $t_{0.05} = 1.69$. Since $t > t_{0.05}$, we reject H_0 at $\alpha = 0.05$ (f) p-value=0.000. Since p-value$< \alpha = 0.05$, we reject H_0 at $\alpha = 0.05$.

(10.42) (a), (b), (c) and (d) are same as in (10.41) (e) With $df = 35$ and $\alpha = 0.05$, $-t_{0.05} = -1.69$. Since $t > -t_{0.05}$, we fail to reject H_0 at $\alpha = 0.05$ (f) p-value $= P(t < 4.18) = 0.9999$. Since p-value$>$ α, we fail to reject H_0 at $\alpha = 0.05$.

(10.43) (a), (b), (c) and (d) are same as in (10.41) (e) With $df = 35$ and $\alpha = 0.05$, $t_{\alpha/2} = t_{0.025} = 2.030$. Since$|t| > t_{0.025}$, reject H_0 at $\alpha = 0.05$ (f) p-value $= 0.000$. Since p-value$< \alpha$, we reject H_0 at $\alpha = 0.05$.

(10.44)

(a) $t = 0.99$.

(b) Try yourself.

(c) With $df = 8$ and $\alpha = 0.05$, $t_{0.05} = 1.86$. p-value=0.176.

(d) Since $t < t_{0.05}$, we fail to reject H_0 at $\alpha = 0.05$ using the critical value approach. Using the p-value approach, since p-value$> \alpha = 0.05$, we fail to reject H_0 at $\alpha = 0.05$.

CHAPTER 11
INFERENCE ABOUT PROPORTIONS

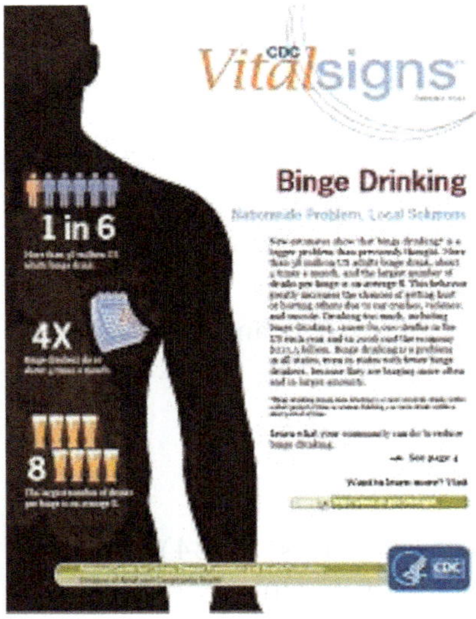

Source: CDC website at http://www.cdc.gov/alcohol/fact-sheets/binge-drinking.htm

Introduction

In this chapter, we address the one and two sample inference procedures for proportions via estimation and test of hypothesis. The specific objectives include the followings:

- Confidence interval estimate of a single population proportion
- Margin of error and sample size for estimating proportion
- Test of hypothesis for a single population proportion
- Confidence interval estimate of difference of two population proportions
- Test of hypothesis for one and two population proportions

Section 11.1: Estimation of a single proportion

As we have discussed in Chapter 8, given a population, the proportion of subjects with a particular characteristic is called the **population proportion**. For example, population proportions could be the proportion of people with salmonella infection in the USA, proportion of people in the USA buying dietary supplements, proportion of smokers in the USA, proportion of students in a university with blood type A, proportion of freshman students in a university already have a driver license, proportion of students in the USA whose parents are making any charitable donation, etc.

We can define population proportion by the following equation

$$p = X/N$$

where,
p is population proportion, or simply proportion,
X is the number of subjects in the population possessing a particular characteristic and
N is the population size (i.e., the number of subjects in the population).

Example 11.1

As per Centers for Disease Control and Prevention, CDC, binge drinking is the most common pattern of excessive alcohol use (i.e., men consume 5 or more drinks, and women consume 4 or more drinks, in about 2 hours) in the United States. According to national surveys, as reported in the CDC website, 1 in 6 U.S. adults binge drinks four times a month, consuming eight drinks per binge. Then, the proportion of U.S. adults who binge drink four times a month is

$$p = 1/6 = 0.17$$

This means that about 17% of U.S. adults binge drink four times a month.
Source: http://www.cdc.gov/alcohol/fact-sheets/binge-drinking.htm

In general, the population proportion p is unknown, and given a sample, we wish to estimate p. In order to estimate p, we need a sample of subjects and count how many of the subjects in the sample possess the characteristic.

Definition: Point estimate of proportion *p*

A **point estimate** of unknown population proportion p is the sample proportion

$$\hat{p} = \frac{x}{n}$$

where,
x is the number of individuals in the sample possessing the particular characteristic,
n is the sample size, and
$\hat{p}$ is the sample proportion, called a point estimate of p.

Example 11.2

Suppose, in a sample of 255 people, 125 people are found to be smokers. Then, the sample proportion of smokers is

$$\hat{p} = \frac{x}{n} = \frac{125}{255} = 0.49$$

This means 49% of subjects in the sample are smokers. A point estimate of unknown proportion of smokers is 0.49.

Example 11.3

In a sample of 185 people from a big city, 80 people buy dietary supplements containing calcium. Find the point estimate of the unknown proportion of people buying the dietary supplements containing calcium in the city.

Solution

Let p be the unknown proportion of people buying the dietary supplements containing calcium in the city. Then, the point estimate of p is the sample proportion

$$\hat{p} = \frac{x}{n} = \frac{80}{185} = 0.43$$

Therefore, the point estimate of p is 0.43.

Properties of $\hat{p}$

As discussed in Chapter 8, we have following properties of $\hat{p}$:

(1) The mean of all possible sample proportions $\hat{p}$ is $\mu(\hat{p}) = p$.

(2) The standard deviation of $\hat{p}$ is $(\hat{p}) = \sqrt{\frac{p(1-p)}{n}}$. The standard deviation of $\hat{p}$ is also called the standard error of $\hat{p}$.

(3) An estimate of the standard deviation of $\hat{p}$ is $\hat{\sigma}(\hat{p}) = \sqrt{\frac{\hat{p}(1-\hat{p})}{n}}$.

(4) When n is large, $\hat{p}$ has approximately a normal distribution $N\left(p, \sqrt{\frac{p(1-p)}{n}}\right)$.

(5) It is recommended that n should be considered large if both np and $n(1-p)$ are at least 5 to hold the property (4).

Confidence interval estimate of p

When n is large, a $100(1-\alpha)\%$ confidence interval for p is $[\hat{p} - m,\ \hat{p} + m]$, where $m = z_{\alpha/2}\sqrt{\frac{\hat{p}(1-\hat{p})}{n}}$ is the margin of error of the desired CI of p and $z_{\alpha/2}$ is the z-value with an area $\alpha/2$ to its right of the standard normal curve.

It is recommended that n should be considered large enough so that both $n\hat{p}$ and $n(1-\hat{p})$ are at least 5.

Example 11.4

In a sample of 255 people, 125 people are found to be smokers.

(a) Find a point estimate of the proportion of smokers in the population.

(b) Find an estimate of the standard error of estimate in (a).

(c) Find the margin of error for a 95% CI estimate of p.

(d) Compute a 95% CI estimate of p.

Solution

(a) The point estimate of p is $\hat{p} = \frac{125}{255} = 0.49$.

(b) The estimate of the standard error of $\hat{p}$ is $\hat{\sigma}(\hat{p}) = \sqrt{\frac{\hat{p}(1-\hat{p})}{n}} = \sqrt{\frac{0.49(1-0.49)}{255}} = 0.031$.

(c) The margin of error for the 95% CI is $m = z_{\alpha/2}\sqrt{\frac{\hat{p}(1-\hat{p})}{n}} = 1.96 \times 0.031 = 0.061$.

(d) The 95% CI of the proportion of smokers in the population is

$$[\hat{p} - m, \hat{p} + m] = [0.49 - 0.061, 0.49 + 0.061] = [0.429, 0.551].$$

Example 11.5

Suppose that a researcher takes a sample of 185 people from a big city and 80 people in the sample buy dietary supplements containing calcium.

(a) Find a point estimate of the proportion of people in the city buying the dietary supplements containing calcium.
(b) Find an estimate of the standard error of estimate in (a).
(c) Find the margin of error for a 90% CI estimate of p.
(d) Find the 90% CI estimate for the proportion of people in the city buying the dietary supplements containing calcium.

Solution

(a) The point estimate of p is $\hat{p} = \frac{80}{185} = 0.43$.

(b) An estimate of the standard error of $\hat{p}$ is $\hat{\sigma}(\hat{p}) = \sqrt{\frac{\hat{p}(1-\hat{p})}{n}} = \sqrt{\frac{0.43(1-0.43)}{185}} = 0.036$.

(c) The margin of error for the 90% CI is $m = z_{\alpha/2}\sqrt{\frac{\hat{p}(1-\hat{p})}{n}} = 1.645 \times 0.036 = 0.06$.

(d) The 90% CI of the proportion of people buying the dietary supplements containing calcium in the population is

$$[\hat{p} - m, \hat{p} + m] = [0.43 - 0.06, 0.43 + 0.06] = [0.37, 0.49].$$

Sample size determination for proportion

Often, researchers wish to determine the sample size for a specified margin of error for estimating confidence interval of the true population proportion. In order to determine the sample size, we need to solve the equation $m = z_{\alpha/2}\sqrt{\frac{\hat{p}(1-\hat{p})}{n}}$ for n, which yields

$$n = \hat{p}(1 - \hat{p})\left(\frac{z_{\alpha/2}}{m}\right)^2$$

What value of $\hat{p}$ is to be considered in above formula? Well, use an estimate of p from a previous study or use $\hat{p} = 0.5$ if no prior knowledge of $\hat{p}$ is available. The choice of $\hat{p} = 0.5$ maximizes the value of $\hat{p}(1 - \hat{p})$, and hence guarantees the largest possible sample size for a desired margin of error and confidence interval. See the table and graph below to see how $\hat{p}(1 - \hat{p})$ gets maximized at $\hat{p} = 0.5$.

$\hat{p}$	$\hat{p}\,(1-\hat{p})$
0.1	0.09
0.2	0.16
0.3	0.21
0.4	0.24
0.5	**0.25**
0.6	0.24
0.7	0.21
0.8	0.16
0.9	0.09

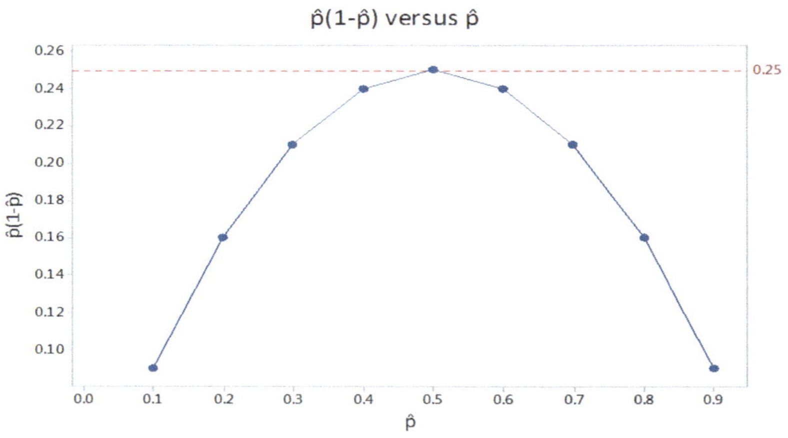

Example 11.6

In reference to the Example 11.5, suppose that the researcher wants a 95% CI for the proportion of people in the city buying dietary supplements containing calcium with a 2% margin of error. Determine the sample size in the following two situations and comment on comparing the two results:

(a) $\hat{p} = 0.43$ as is obtained in the sample.

(b) No prior knowledge of $\hat{p}$ exists.

Solution

(a) For a 95% CI, $z_{\alpha/2} = z_{0.025} = 1.96$. Here, $m = 2\% = 0.02$. Therefore,

$$n = \hat{p}(1-\hat{p})\left(\frac{z_{\alpha/2}}{m}\right)^2 = 0.43(1-0.43)\left(\frac{1.96}{0.02}\right)^2 = 2353.94$$

Rounding up to the nearest integer, we would recommend a sample of size 2354 for estimating a 95% CI of the true proportion.

(b) When no prior knowledge of $\hat{p}$ exists, we consider $\hat{p} = 0.5$. Then,

$$n = \hat{p}(1-\hat{p})\left(\frac{z_{\alpha/2}}{m}\right)^2 = 0.50(1-0.50)\left(\frac{1.96}{0.02}\right)^2 = 2401$$

By comparing the two results, larger sample size is recommended in (b) as compared with (a).

Minitab Instruction for CI for a Single Proportion

Ref: Example 11.5

We follow the steps below to carry out a CI for a proportion

(1) Select **Stat>Basic Statistics> 1 Proportion**…

(2) Choose Summarized Data and enter values for **Number of events:**80, **Number of trials:** 185.

(3) Use Options to decide **Confidence level**: 90 and **Method:** Normal approximation

(4) Click OK two times.

Minitab dialog box with selection options and results from aforementioned steps are given below:

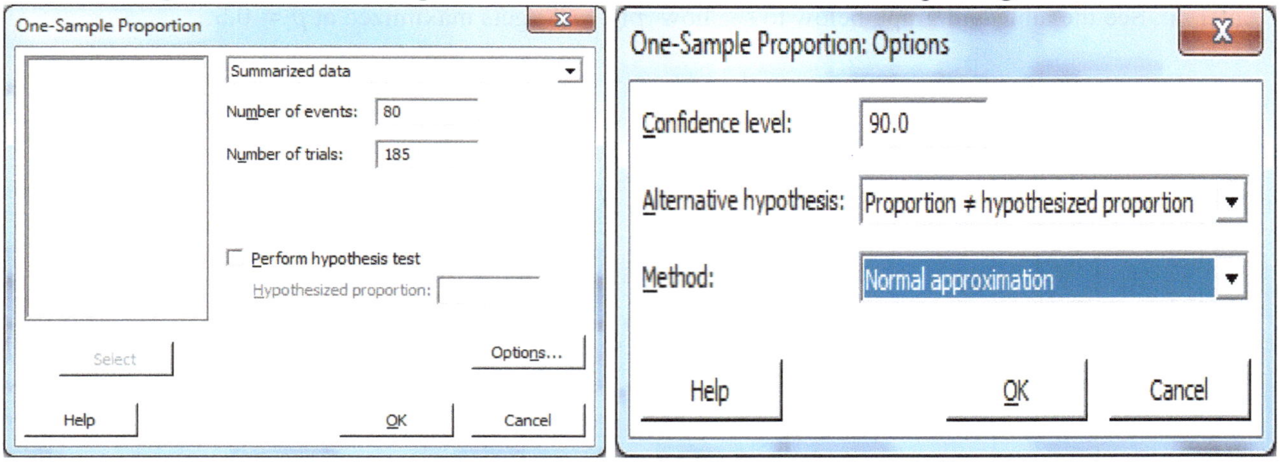

Minitab Output

Test and CI for One Proportion

```
Sample   X    N   Sample p          90% CI
1       80  185   0.432432  (0.372521, 0.492344)
```

TI-84 Plus Calculator for 1-PorpZInt

Ref: Example 11.5

Follow the steps below to carry out a **1-PropZInt** in TI 84 Plus calculator:

(1) Press STAT, move cursor to **TESTS** and then down to select **A: 1-PropZInt ...**and press ENTER.

(2) Specify appropriate values for X:80, n:185, and C-level:.9.

(3) Move the cursor to Calculate and press ENTER.

See figures below for selection options and the output:

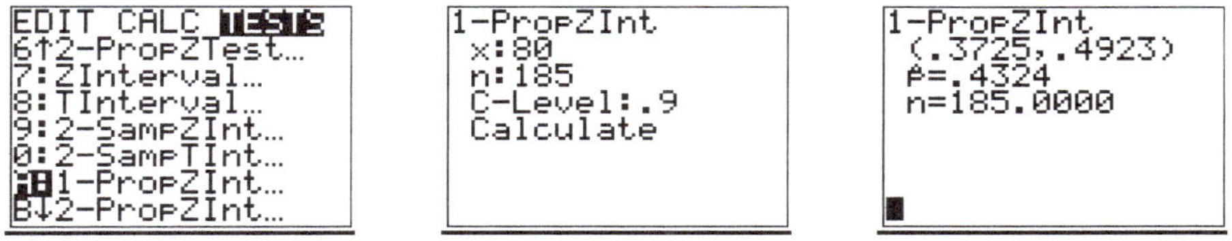

Section 11.2: Test of hypothesis about a single proportion

In this section, we want to test hypothesis about a single proportion using a Z-Test. We will test the null hypothesis $H_0: p = p_0$ against either of the following alternative hypotheses:

$$H_a: p > p_0 \ (right\ tailed\ alternative)$$
$$H_a: p < p_0 (left\ tailed\ alternative)$$
$$H_a: p \neq p_0 \ (two\ tailed\ alternative)$$

Test statistic

The test statistic for this test under H_0 is given by

$$Z = \frac{\hat{p} - p_0}{\sqrt{\frac{p_0(1-p_0)}{n}}}$$

which has a standard normal distribution for a large sample size n.

p-value

The p-value for testing $H_0: p = p_0$ is computed by:

$$p\text{-value} = P(Z > z) \ if \ H_a: p > p_0$$
$$p\text{-value} = P(Z < z) \ if \ H_a: p < p_0$$
$$p\text{-value} = 2P(Z > |z|) \ if \ H_a: p \neq p_0$$

where Z is the test statistic distributed as $N(0,1)$ and z is the observed value of the test statistic given the sample.

Decision rule

Reject the null hypothesis at α, if the p-value $\leq \alpha$.

Example 11.7

According to CDC, about 19% of all adults aged 18 years or older smoke cigarettes in the USA. A sample from a large community is intended to test if the similar claim exists in the community. On the basis of a sample of size 35, the proportion of smokers among adults aged 18 years or older is found to be 0.17. Test $H_0: p = 0.19$ against $H_a: p \neq 0.19$ and make a conclusion to see if the proportion of smokers in the community is 0.19 or not at 5% level of significance.

(a) Calculate the value of the test statistic z.
(b) Calculate the p-value to test H_0 versus H_a at $\alpha = 0.05$ and conclude about the test.
(c) Use p-value to test H_0 versus H_a by setting α to 0.10, 0.01 and 0.001.

Solution

We have $\hat{p} = 0.17, p_0 = 0.19, n = 35, \alpha = 0.05$

Test statistic: $Z = \frac{\hat{p}-p_0}{\sqrt{\frac{p_0(1-p_0)}{n}}}$

(a) $z = \frac{0.17-0.19}{\sqrt{\frac{0.19(1-0.19)}{35}}} = -0.30$

(b) p-value$= 2P(Z > |-0.30|) = 2P(Z > 0.30) = 2 \times (1 - 0.6179) = 2 \times 0.3821 = 0.7642$.
Since p-value is greater than 0.05, we fail to reject the null hypothesis at 5% level of significance.

(c) Accept H_0 when α is set to 0.10, 0.01 and 0.001 as well, since p-value calculated in part (b) is greater than $\alpha = 0.10, 0.01$ and 0.001.

Example 11.8

In a sample of 185 people in a big city, 80 people buy dietary supplements containing calcium. A researcher wants to test $H_0: p = 0.47$ against $H_a: p > 0.47$.

(a) Write out the test statistic for the test.
(b) Find the observed value of the test statistic.
(c) Compute the p-value for testing H_0 against H_a.
(d) What conclusion can you make about the test at 5% level of significance?

Solution

We have $\hat{p} = \frac{80}{185} = 0.43, p_0 = 0.47, n = 185, \alpha = 0.05$

(a) Test statistic is $Z = \frac{\hat{p}-p_0}{\sqrt{\frac{p_0(1-p_0)}{n}}}$.

(b) The observed value of the test statistic is $z = \frac{0.43-0.47}{\sqrt{\frac{0.47(1-0.47)}{185}}} = \frac{-0.04}{0.0367} = -1.09$.

(c) p-value$= P(Z > -1.09) = 1 - 0.1379 = 0.8621$

(d) Accept the null hypothesis at 5% level of significance since $p\ value > 0.05$.

Minitab Instruction for a Test of Hypothesis about a Single Proportion

Ref: Example 11.8

We follow the steps below to perform a test of hypothesis for a proportion.

(1) Select **Stat>Basic Statistics> 1 Proportion**...

(2) Choose Summarized Data and enter values for **Number of events:**80, **Number of trials:** 185. Check mark **Perform hypothesis test** and enter **Hypothesized proportion:** 0.47.

(3) Use **Options…** to decide α. To specify $\alpha = 0.05$, for a greater than alternative or a less than alternative specify **Confidence level**: 95. Choose **Method:** Normal approximation

(4) Click OK two times.

Minitab dialog box with selection options and results from aforementioned steps are given below:

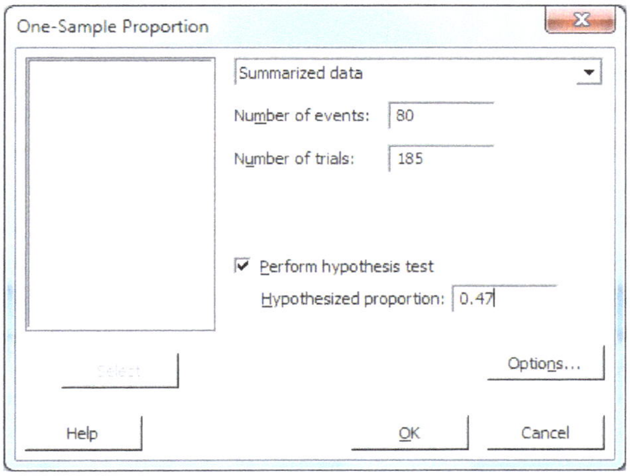

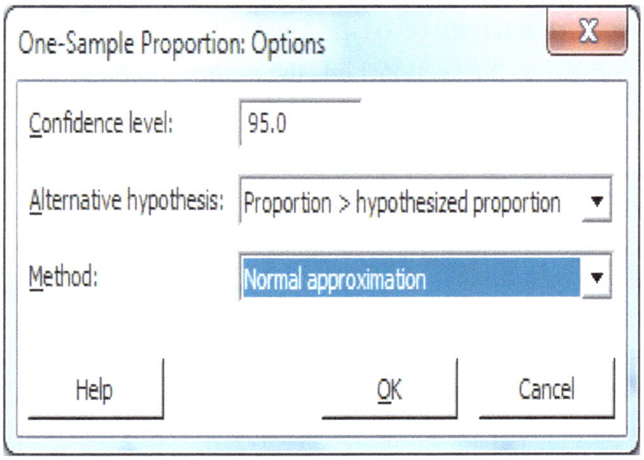

Test and CI for One Proportion

```
Test of p = 0.47 vs p > 0.47

Sample   X    N  Sample p  95% Lower Bound  Z-Value  P-Value
1       80  185  0.432432         0.372521    -1.02    0.847

Using the normal approximation.
```

TI-84 Plus Instruction for Test of Hypothesis of a Single Proportion

Ref: Example 11.8

Follow the steps below to carry out a **1-PropZTest** in TI 84 Plus calculator:

(1) Press **STAT**, move cursor to **TESTS**, and select **5:1-PropZTest**…and press ENTER.

(2) Specify appropriate values for P0:0.47, X:80, n:185,Prop: >P0

(3) Move cursor to Calculate and press ENTER.

See figures below for selection options and corresponding output:

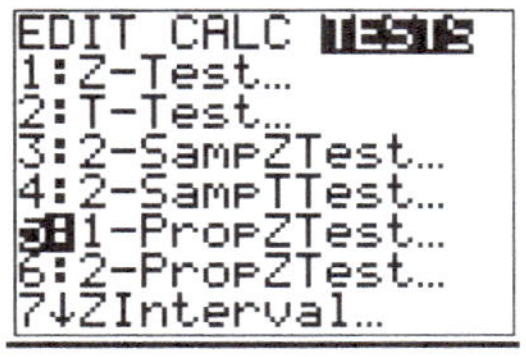

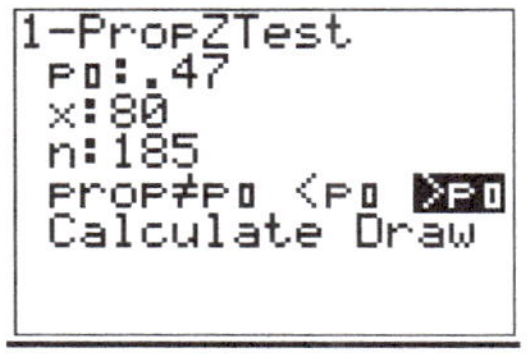

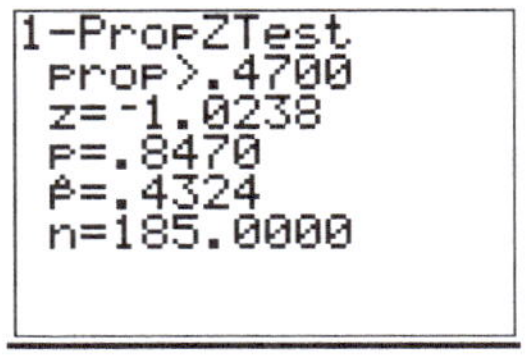

Section 11.3: Estimating difference of two proportions

In this section, we will estimate difference of two population proportions using two independent samples. Given two independent samples of size n_1 and n_2 from two populations with unknown proportions p_1 and p_2, respectively, we wish to estimate the difference $p_1 - p_2$.

Point estimate of $p_1 - p_2$

Similar to the section 11.1, the point estimates of p_1 and p_2 are given by

$$\hat{p}_1 = \frac{x_1}{n_1}$$
$$\hat{p}_2 = \frac{x_2}{n_2}$$

where x_1 and x_2 are number of subjects with a given characteristic in two samples of sizes n_1 and n_2, respectively.
Then, a point estimate of $p_1 - p_2$ is $\hat{p}_1 - \hat{p}_2$.

The following formulas apply in reference to the estimation of $p_1 - p_2$:

(1) The mean of $\hat{p}_1 - \hat{p}_2$ is $(\hat{p}_1 - \hat{p}_2) = p_1 - p_2$.
(2) The standard deviation (also called the standard error) of $\hat{p}_1 - \hat{p}_2$ is $\sigma(\hat{p}_1 - \hat{p}_2) = \sqrt{\frac{p_1(1-p_1)}{n_1} + \frac{p_2(1-p_2)}{n_2}}$.
(3) An estimate of the standard error of $\hat{p}_1 - \hat{p}_2$ is $\hat{\sigma}(\hat{p}_1 - \hat{p}_2) = \sqrt{\frac{\hat{p}_1(1-\hat{p}_1)}{n_1} + \frac{\hat{p}_2(1-\hat{p}_2)}{n_2}}$.
(4) When the samples are large, $\hat{p}_1 - \hat{p}_2$ has approximately a normal distribution given by $N\left(p_1 - p_2, \sqrt{\frac{p_1(1-p_1)}{n_1} + \frac{p_2(1-p_2)}{n_2}}\right)$.
(5) It is recommended that n_1 and n_2 should be considered large if n_1p_1, $n_1(1-p_1)$, n_2p_2, and $n_2(1-p_2)$ are all at least 5 so as to hold the property (4).

Confidence interval estimate of $p_1 - p_2$

When the samples are large, a $100(1-\alpha)\%$ CI of $p_1 - p_2$ is given by

$$\hat{p}_1 - \hat{p}_2 \pm z_{\alpha/2}\sqrt{\frac{\hat{p}_1(1-\hat{p}_1)}{n_1} + \frac{\hat{p}_2(1-\hat{p}_2)}{n_2}}$$

where $z_{\alpha/2}$ is the z-value with an area $\alpha/2$ to its right of the standard normal curve. The quantity $z_{\alpha/2}\sqrt{\frac{\hat{p}_1(1-\hat{p}_1)}{n_1} + \frac{\hat{p}_2(1-\hat{p}_2)}{n_2}}$ is called the margin of error, m, of the CI. Then, a $100(1-\alpha)\%$ CI of $p_1 - p_2$ is given by

$$[(\hat{p}_1 - \hat{p}_2) - m, (\hat{p}_1 - \hat{p}_2) + m]$$

where $m = z_{\alpha/2}\sqrt{\frac{\hat{p}_1(1-\hat{p}_1)}{n_1} + \frac{\hat{p}_2(1-\hat{p}_2)}{n_2}}$ is the margin of error.

Example 11.9

It is believed that 43% Americans buy dietary supplements with calcium. Suppose, a researcher claims that the percent of Americans who buy dietary supplements in two large cities differ. In one city, 63 Americans out of 150 randomly selected Americans buy dietary supplements with calcium, whereas in the other city, 67 out of 170 randomly selected Americans buy dietary supplements with calcium.

(a) Find a point estimate of the difference in the proportions of Americans buying dietary supplements with calcium in two cities.
(b) Find the estimate of the standard deviation of your estimate in (a).
(c) Compute the margin of error for a 90% CI of $p_1 - p_2$.
(d) Find a 90% CI estimate of the difference in the proportions of Americans buying dietary supplements with calcium in two cities.

Solution

(a) $\hat{p}_1 - \hat{p}_2 = \frac{63}{150} - \frac{67}{170} = 0.42 - 0.39 = 0.03.$

(b) $\hat{\sigma}(\hat{p}_1 - \hat{p}_2) = \sqrt{\frac{\hat{p}_1(1-\hat{p}_1)}{n_1} + \frac{\hat{p}_2(1-\hat{p}_2)}{n_2}} = \sqrt{\frac{0.42(1-0.42)}{150} + \frac{0.39(1-0.39)}{170}} = 0.055.$

(c) Margin of error for a 90% CI is $m = z_{0.05}\sqrt{\frac{\hat{p}_1(1-\hat{p}_1)}{n_1} + \frac{\hat{p}_2(1-\hat{p}_2)}{n_2}} = 1.645 \times 0.055 = 0.09.$

(d) A 90% CI for the difference in the proportion is [0.03-0.09, 0.03+0.09]=[-0.06,0.12].

Technology for CI Estimate of Difference of Two Proportions

Minitab Instruction for CI of Difference of Two Proportions

Ref: Example 11.9

We follow the steps below to find out a CI of difference of two proportions

(1) Select **Stat>Basic Statistics> 2 Proportions**...
(2) Choose Summarized Data and enter values for **Number of events:** Sample-1: 63 and Sample-2: 67, **Number of trials:** Sample-1: 150 and Sample-2: 170.
(3) Use Options to decide **Confidence level**: 90 and **Test Method:** Estimate the proportions separately.
(4) Click OK two times.

Minitab output and the dialog box selection options are given below:

Test and CI for Two Proportions

```
Sample   X    N   Sample p
1       63  150   0.420000
2       67  170   0.394118

Difference = p (1) - p (2)
Estimate for difference:  0.0258824
90% CI for difference:  (-0.0646390, 0.116404)
Test for difference = 0 (vs ≠ 0):  Z = 0.47  P-Value = 0.638
```

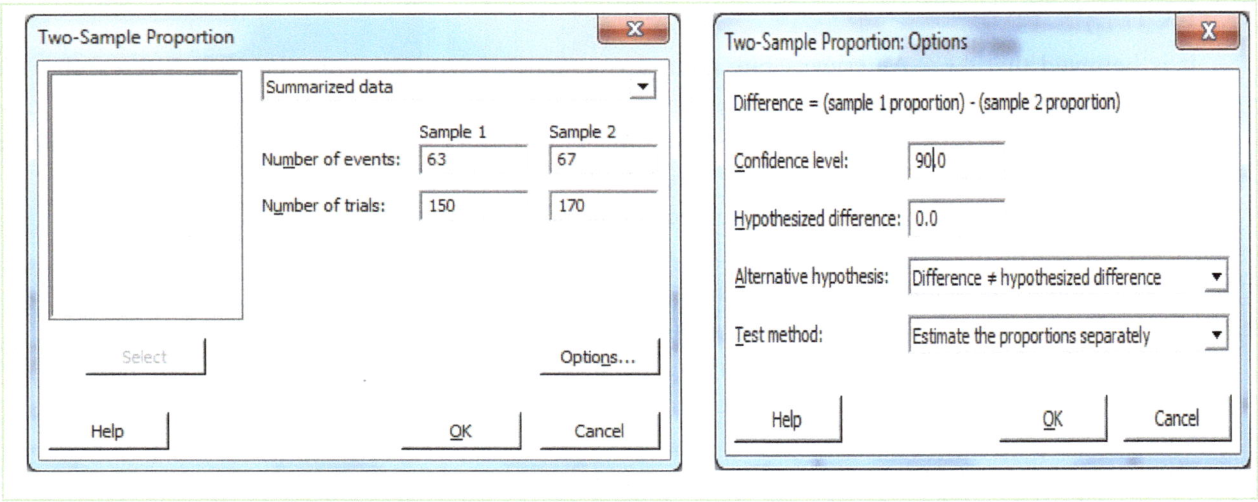

TI-84 Plus Instruction for CI of Difference of Two Proportions

Ref: Example 11.9

Follow the steps below to carry out a **2-PropZInt** in TI 84 Plus calculator:

(1) Press **STAT**, move cursor to **TESTS**, and then down to **B: 2-PropZInt**...and press ENTER.
(2) Specify appropriate values for X1:63, n1:150, X2:67 and n2:170 and C-level:.9.
(3) Move the cursor to Calculate and press ENTER.

See figures below for selection options and the output:

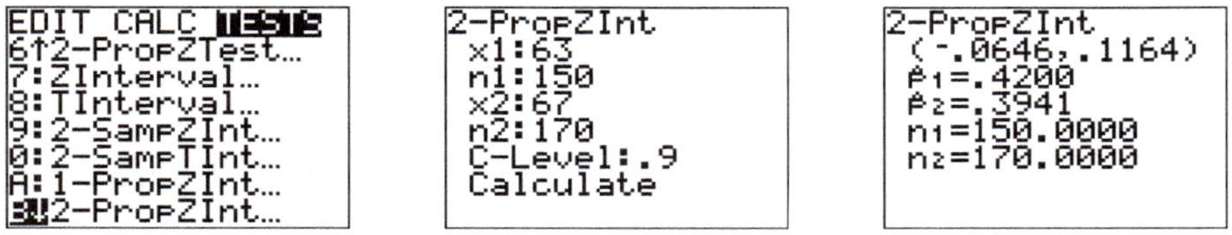

Section 11.4: Test of hypothesis for difference of two proportions

In this section, we will test hypothesis regarding the two population proportions using two independent samples. We want to test the null hypothesis $H_0: p_1 = p_2$ against any of the following alternative hypotheses:

$$H_a: p_1 > p_2 \ (right\ tailed\ alternative)$$
$$H_a: p_1 < p_2 \ (left\ tailed\ alternative)$$
$$H_a: p_1 \neq p_2 \ (two\ tailed\ alternative)$$

Test statistic

The test statistic for this test under H_0 is given by

$$Z = \frac{\hat{p}_1 - \hat{p}_2}{\sqrt{\hat{p}(1-\hat{p})\left(\frac{1}{n_1} + \frac{1}{n_2}\right)}}$$

where $\hat{p} = \frac{X_1+X_2}{n_1+n_2}$ and Z follows a standard normal distribution for large sample sizes.

p-value

The p-value for testing $H_0: p_1 = p_2$ is computed by:

$$p\text{-value} = P(Z > z) \; if \; H_a: p_1 > p_2$$
$$p\text{-value} = P(Z < z) \; if \; H_a: p_1 < p_2$$
$$p\text{-value} = 2P(Z > |z|) \; if \; H_a: p_1 \neq p_2$$

where Z is the test statistic distributed as $N(0,1)$ and z is the observed value of the test statistic given the sample.

Decision rule

Reject the null hypothesis at α, if the p-value $\leq \alpha$.

Example 11.10

Reference to Example 11.9, suppose the researcher wants to test $H_0: p_1 = p_2$ against the following alternative hypotheses:

(a) $H_a: p_1 > p_2$
(b) $H_a: p_1 < p_2$
(c) $H_a: p_1 \neq p_2$
at 5% level of significance.

Make use of the fact that 63 out of 150 randomly selected Americans in one city buy dietary supplements with calcium, whereas in the other city, 67 out of 170 randomly selected Americans buy dietary supplements with calcium.

Write out the formula of the test statistic and the value of the test statistic for these tests. Compute the p-value for each of the alternatives in (a)-(c) and conclude about the test at 5% level of significance.

Solution

The test statistic is

$$Z = \frac{\hat{p}_1 - \hat{p}_2}{\sqrt{\hat{p}(1-\hat{p})\left(\frac{1}{n_1} + \frac{1}{n_2}\right)}}$$

Here, $\hat{p}_1 - \hat{p}_2 = \frac{63}{150} - \frac{67}{170} = 0.42 - 0.39 = 0.03$; $\hat{p} = \frac{X_1+X_2}{n_1+n_2} = \frac{63+67}{150+170} = 0.41$.

Then, the observed value of the test statistic is

$$z = \frac{0.03}{\sqrt{0.41 \times (1-0.41) \times \left(\frac{1}{150} + \frac{1}{170}\right)}} = \frac{0.03}{0.055} = 0.55$$

(a) p-value $= P(Z > 0.55) = 1 - 0.7088 = 0.2912$.
Since p-value>0.05, we accept $H_0: p_1 = p_2$ at 5% level of significance.
(b) p-value$= P(Z < 0.55) = 0.7088$.
Since p-value>0.05, we accept $H_0: p_1 = p_2$ at 5% level of significance.

(c) p-value$= 2P(Z > 0.55) = 2 \times (1 - 0.7088) = 2 \times 0.2912 = 0.5824$.

Since p-value>0.05, we accept $H_0\colon p_1 = p_2$ at 5% level of significance.

Technology for Test of Hypothesis of Difference of Two Proportions

Minitab Instruction for test of difference of two proportions

Ref: Example 11.10 (a), (b) and (c)

We follow the steps below to carry out a test of difference of two proportions.

(1) Select **Stat>Basic Statistics> 2 Proportions**...

(2) Choose Summarized Data and enter values for **Number of events:** Sample-1: 63 and Sample-2: 67, **Number of trials:** Sample-1: 150 and Sample-2: 170.

(3) Use Options to decide α and Alternative hypothesis. Specify $\alpha = 0.05$ by specifying **Confidence level**: 95

(a) For a greater than alternative, as in Example 11.10(a), choose **Alternative hypothesis**: Difference >hypothesized difference.

(b) For a less than alternative, as in Example 11.10(b), choose **Alternative hypothesis**: Difference <hypothesized difference.

(c) For a two sided alternative, as in Example 11.10(c), choose **Alternative hypothesis**: Difference ≠hypothesized difference.

(4) For **Test Method:** Use the pooled estimate of the proportion.

(5) Click OK two times.

Minitab dialog box with selection options and results from aforementioned steps are given below:

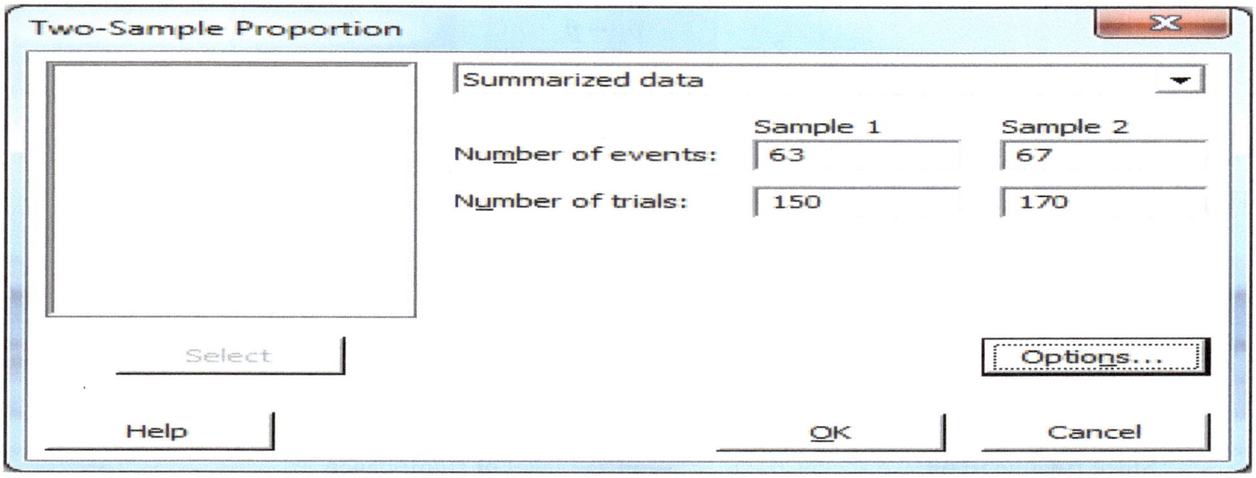

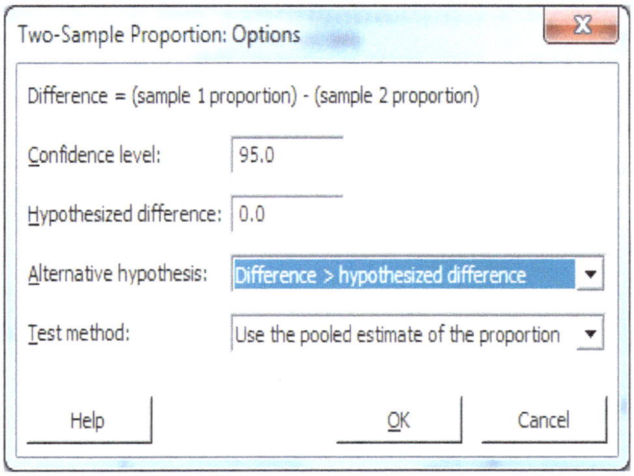

Test and CI for Two Proportions

```
Sample   X    N   Sample p
1       63  150   0.420000
2       67  170   0.394118

Difference = p (1) - p (2)
Estimate for difference:
0.0258824
95% lower bound for difference:
-0.0646390

Test for difference = 0 (vs > 0):
Z = 0.47  P-Value = 0.319
```

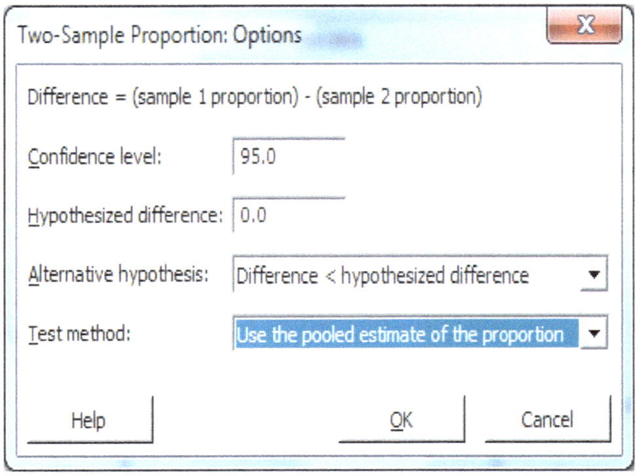

Test and CI for Two Proportions

```
Sample   X    N   Sample p
1       63  150   0.420000
2       67  170   0.394118

Difference = p (1) - p (2)
Estimate for difference:
0.0258824
95% upper bound for difference:
0.116404

Test for difference = 0 (vs < 0):
Z = 0.47  P-Value = 0.681
```

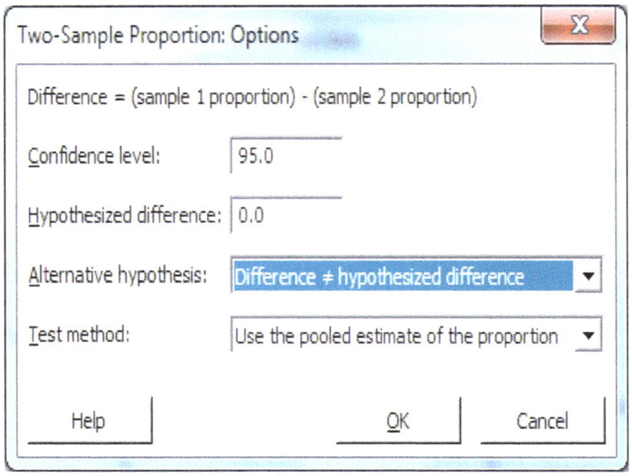

Test and CI for Two Proportions

```
Sample   X    N   Sample p
1       63  150   0.420000
2       67  170   0.394118


Difference = p (1) - p (2)
Estimate for difference:
0.0258824
95% CI for difference:  (-
0.0819805, 0.133745)

Test for difference = 0 (vs ≠ 0):
Z = 0.47  P-Value = 0.638
```

TI-84 Plus Instruction for CI of Difference of Two Proportions

Ref: Example 11.10

Follow the steps below to carry out a **2-PropZTest** in TI 84 Plus calculator:

(1) Press **STAT**, move cursor to **TESTS**, and select **6:2-PropZTest**…
(2) Specify appropriate values for X1:63, n1:150, X2:67 and n2:170.
(3) For p-values in Example 11.10 (a), (b) and (c) choose >P2, <P2 and ≠P2, respectively, as shown in figures (a), (b) and (c).

For selection options see figures (a), (b) and (c), and for results see figures (A), (B) and (C), respectively.

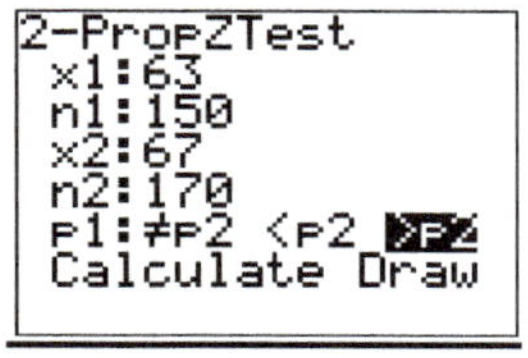

(a)

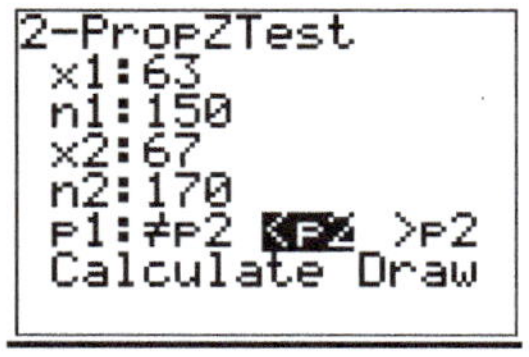

(b)

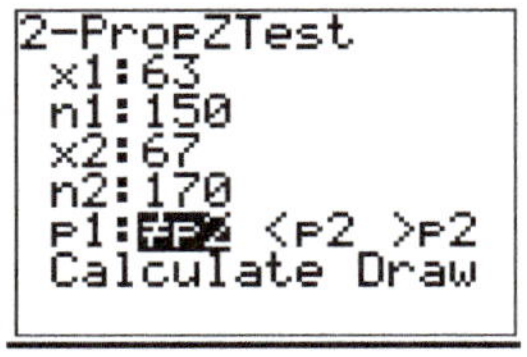

(c)

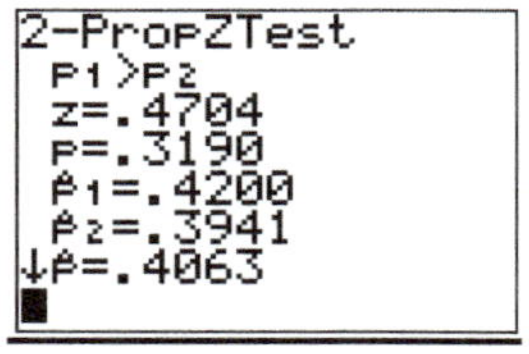

(A)

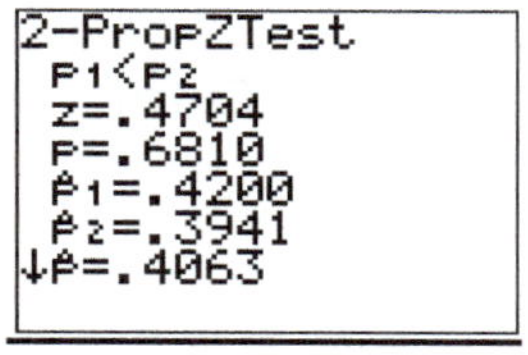

(B)

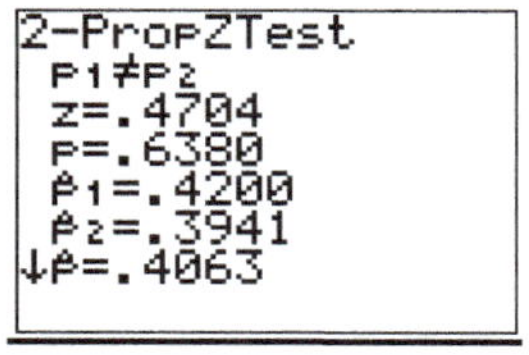

(C)

Chapter 11: Practice problems

(11.1) In a sample of 45 students, 23 students are PC users. A researcher wants to estimate the unknown proportion of students who use PC. Answer the following questions:

(a) What is the point estimate of the proportion of PC users?
(b) Find an estimate of the standard deviation of your estimate in (a).
(c) What is the 95% CI of the proportion of PC users?
(d) If a 95% CI is sought with a 2% margin of error, what sample size would you recommend?

(11.2) A recent poll suggests that 56% Americans own a Smartphone. An instructor wants to know if the percent of Smartphone owners in a large class agree with the poll result. A sample of 30 students reveals that 19 of them own a Smartphone. Answer the following questions:

(a) Give a point estimate of unknown proportion of students own a Smartphone.
(b) What would be the standard deviation of the estimate in (a)?
(c) Find 90% and 95% CIs of the true proportion.
(d) What sample size will be reasonable for a 95% CI of the true proportion with 1.5% margin of error?

(11.3) In reference to the problem 11.2, we want to test the null hypothesis $H_0: p = 0.56$ against $H_a: p > 0.56$.

(a) Find the value of the test statistic.
(b) Find the p-value for the test.
(c) What is your conclusion at 5% level of significance using the p value approach?
(d) Find the critical value at 5% level of significance and make a conclusion about the test.

(11.4) As per CDC, 19% of all adults aged 18 years or older in the USA smoke cigarettes. A researcher is interested to test if the percent of smokers in a large community is the same as that of the national percent. A sample of 100 adults from the community reveals that sample proportion of smokers is 0.17.
(a) How should you set up the null and alternative hypotheses for the test?
(b) Is the proposed test in (a) one or two-tailed?
(c) Write out the test statistic for the test in (a).
(d) What is the value of the test statistic?
(e) Make a conclusion of the test in (a) at α =0.05.

(11.5) On the basis of two samples of adult males and females, each of size 100, the proportion of smokers among adult males and females were found to be 0.22 and 0.17, respectively. Let the unknown proportions of smoker among males and females be denoted by p_1 and p_2, respectively. Answer the following questions:
(a) Find out a point estimate of $p_1 - p_2$.
(b) Find a 95% confidence interval of $p_1 - p_2$.
(c) Suppose we wish to test if the proportion of smoker among males is higher than the proportion of smoker among females. Specify null and alternative hypotheses for the test and specify if the test is one or two-tailed.
(d) Compute the value of the test statistic for the specified test in (c).
(e) Make a conclusion of your test in (c) at significance level 0.05.

(11.6) Among the included sources, shell eggs are responsible for the highest proportion of foodborne Salmonella serotype Enteritidis (SE) infections during the baseline period (2007-2009). In a sample of 83 SE infections, 34 SE infections are due to the shell eggs.
(a) Find an estimate of the proportion of illnesses attributable to shell eggs in the population.
(b) Find an estimate of the standard error of estimate in (a).
(c) Find a 95% CI of the proportion of foodborne SE infections.
Source: Result of CDC study at http://www.cdc.gov/foodsafety/ifsac/projects/completed.html

(11.7) It is believed that 43% Americans buy dietary supplements with calcium. Suppose, a researcher claims that the percent of Americans who buy dietary supplements in two large cities differ. In one city, 63 out of 150 randomly selected Americans buy dietary supplements with calcium, whereas in the other city, 67 out of 170 randomly selected Americans buy dietary supplements with calcium.
(a) Find a point estimate of the difference in the proportions of Americans buying dietary supplements with calcium in two cities.
(b) Find the estimate of the standard error of estimate in (a).
(c) Compute the margin of error for a 95% CI of $p_1 - p_2$.
(d) Find a 95% CI estimate of the difference in the proportions of Americans buying dietary supplements with calcium in two cities.

(11.8) Suppose in a sample of 100 male Facebook users, 39 say that they use Facebook for seeing photos and videos. On the other hand, in a sample of 100 female Facebook users, 54 say that they use Facebook for seeing photos and videos.

(a) Find a point estimate of the difference in the proportions of male and female Facebook users using Facebook for seeing photos and videos in the population of male and female Facebook users.
(b) Find the estimate of standard error of estimate in (a).
(c) Compute the margin of error for a 95% CI of $p_1 - p_2$.
(d) Find a 95% CI estimate of the difference in the proportions of male and female Facebook users using Facebook for seeing photos and videos in the population of male and female Facebook users.

Source: Derived from http://www.pewresearch.org/fact-tank/2014/02/03/6-new-facts-about-facebook/

(11.9) For problem (11.8), we wish to test $H_0: p_1 = p_2$ against the alternative hypothesis $H_a: p_1 < p_2$.

(a) Write out the test statistic for the test.
(b) Compute the value of the test statistic.
(c) Report p value or critical value for the test and make a conclusion about the test at 5% level of significance.
(d) Make a conclusion about the test at 10% level of significance.
(e) If we wish to test $H_0: p_1 = p_2$ against the alternative hypothesis $H_a: p_1 \neq p_2$ at 5% level of significance, report p value or critical value for the test and make a conclusion about the test.

(11.10) Suppose in a sample of 100 female Facebook users, 43 say that they use Facebook for seeing entertaining and funny posts. On the other hand, in a sample of 100 male Facebook users, 35 say that they use Facebook for seeing entertaining and funny posts.

(a) Find a point estimate of difference in the proportions of female and male Facebook users using Facebook for seeing entertaining and funny posts in the population of female and male Facebook users.
(b) Find the estimate of standard error of estimate in (a).
(c) Compute the margin of error for a 95% CI of $p_1 - p_2$.
(d) Find a 95% CI estimate of the difference in the proportions of female and male Facebook users using Facebook for seeing entertaining and funny posts in the population of male and female Facebook users.

Source: Derived from http://www.pewresearch.org/fact-tank/2014/02/03/6-new-facts-about-facebook/

(11.11) For problem (11.10), we wish to test $H_0: p_1 = p_2$ against the alternative hypothesis $H_a: p_1 > p_2$.

(a) Write out the test statistic for the test.
(b) Compute the value of the test statistic.
(c) Report p value or critical value for the test and make a conclusion about the test at 5% level of significance.
(d) Make a conclusion about the test at 10% level of significance.
(e) If we wish to test $H_0: p_1 = p_2$ against the alternative hypothesis $H_a: p_1 \neq p_2$ at 5% level of significance, report p value or critical value for the test and make a conclusion about the test.

Chapter 11: Answer to selected problems

(11.1) (a) A point estimate, $\hat{p} = 0.51$. (b) An estimate of standard deviation is 0.0745. (c) 95% CI of the proportion of PC users is [0.365, 0.657]. (d) $n = 2401$.

(11.2) (a) A point estimate, $\hat{p} = 0.63$. (b) An estimate of standard deviation is 0.088. (c) 90% CI is [0.489,0.778]; 95% CI is [0.461, 0.806]. (d) $n = 3980$.

(11.3)

(a) The value of the test statistic is z=0.81.

(b) p-value= 0.209.

(c) Since p-value> $\alpha = 0.05$, we fail to reject the null hypothesis at 5% level of significance.

(d) $z < z_{0.05} = 1.645$, and therefore, we fail to reject the null hypothesis at 5% level of significance, an identical conclusion obtained in (c).

(11.4) (a) $H_0: p = 0.19$ versus $H_a: p \neq 0.19$. (b) Two-tailed test. (c) Try yourself, (d) $z = -0.5098$ (e) p-value=0.6102, fail to reject null hypothesis at 5% level of significance.

(11.5)

(a) A point estimate of $p_1 - p_2$ is 0.05.

(b) 95% confidence interval of $p_1 - p_2$ is [-0.0596, 0.1596].

(c) $H_0: p_1 = p_2$ versus $H_a: p_1 > p_2$, right tailed test.

(d) $z = 0.89$

(e) p-value = 0.186>0.05; fail to reject the null hypothesis at 5% level of significance.

(11.6) 0.41, (b) 0.054, (c) A 90% CI of p is $[0.30 - 0.51]$.

(11.7) (a) 0.026, (b) $\sqrt{\frac{0.42*(1-0.42)}{150} + \frac{0.39*(1-0.39)}{170}} = 0.055$ (c) $m = 0.108$ (d) $[-0.082, 0.134]$.

(11.8) (a) -0.15, (b) $\sqrt{\frac{0.39*(1-0.39)}{100} + \frac{0.54*(1-0.54)}{100}} = 0.0697$ (c) $m = 0.1366$ (d) $[-0.2867, -0.0133]$.

(11.9) (a) Try yourself (b) $z = -2.127$, (c) p-value=0.017, reject H_0 at $\alpha = 0.05$. (d) p-value=0.017, reject H_0 at $\alpha = 0.10$ (e) p-value=0.033, reject H_0 at $\alpha = 0.05$.

(11.10) (a) 0.08 (b) 0.0687, (c) $m = 0.1347$ (d) $[-0.0547, 0.2147]$.

(11.11) (a) Try yourself (b) $z = 1.1597$, (c) p-value=0.1230>0.05, fail to reject H_0 at $\alpha = 0.05$. (d) p-value=0.1230>0.10, so fail to reject H_0 at $\alpha = 0.10$ (e) p-value=0.2461>0.05, so fail to reject H_0 at $\alpha = 0.05$.

CHAPTER 12
INFERENCE ABOUT VARIANCES

Introduction

In this chapter, we will introduce two popular sampling distributions, namely, Chi-square and F distributions. The Chi-square distribution is useful for (1) estimating a population variance or standard deviation via confidence interval estimate, (2) testing hypothesis regarding a specified value of the single variance, (3) testing independence of two categorical variables, (4) testing homogeneity of proportions, and (5) testing goodness of fit of a distribution. Chapter 13 is devoted to address applications of a Chi-square distribution as mentioned in (3)-(5) above. An F distribution is useful for testing the equality of two population variances and testing the homogeneity of several treatment means in the analysis of variance.

The specific objectives of this chapter, however, include the followings:

- Estimate an unknown population variance and standard deviation via confidence interval by incorporating a Chi-square distribution.
- Test of hypothesis about a single population variance using a Chi-square distribution.
- Test of hypothesis about two population variances using an F distribution.

Section 12.1: Chi-square distribution basics

In this section, we address a new statistical distribution, called the ***chi-square distribution***. This distribution is a basis of the confidence interval estimate of the population variance σ^2 and hypothesis testing about it.
We would like to address the chi-square distribution as a definition.

Definition: Chi-square distribution

If s^2 is the sample variance of a sample of size n taken from a normal distribution with variance σ^2, then the sampling distribution of the statistic

$$\frac{(n-1)s^2}{\sigma^2}$$

is a ***chi-square distribution*** with $(n-1)$ degrees of freedom.

A chi-square distribution is denoted by the square of a Greek letter χ, χ^2 (pronounced kigh-square). The shape of a chi-square distribution is right skewed. As the degrees of freedom increases the shape gradually appears to be symmetric and bell-shaped. In Figure 12.1, we have several chi-square distributions with varying degrees of freedom.

Figure 12.1: Chi-Square Distributions with 5, 10, 15, 20 and 25 degrees of freedom

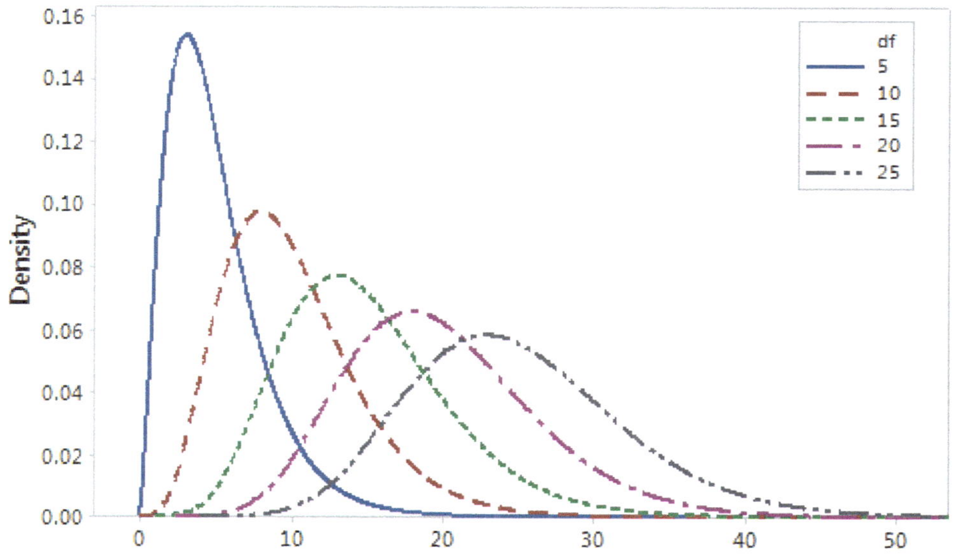

As we see in Figure 12.1, the curve of a chi-square distribution approaches a symmetric curve similar to the normal curve as degrees of freedom increases.

Characteristics of a Chi-square distribution

Some of the important characteristics of a chi-square distribution can be summarized as follows:

- As follows from the definition, values of a chi-square variable are always positive.
- A chi-square distribution is determined by its degrees of freedom.
- Area under the curve of a chi-square distribution is 1.
- Shape of a chi-square distribution is right skewed.
- As the degrees of freedom increases the shape appears to be symmetric and bell-shaped, gradually.

Notation χ^2_α and $\chi^2_{1-\alpha}$

χ^2_α refers to the point on the horizontal axis under the curve of a chi-square distribution with area α to its right. On the other hand, $\chi^2_{1-\alpha}$ refers to the point on the horizontal axis under the curve of a chi-square distribution with area α to its left. See the figure below for the points χ^2_α and $\chi^2_{1-\alpha}$ under the curve of a chi-square distribution.

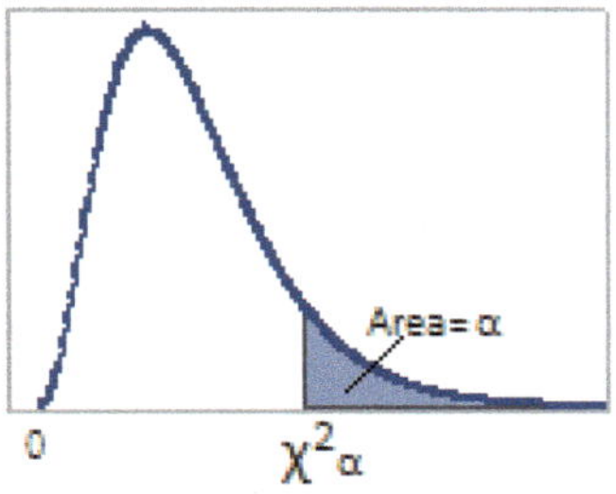

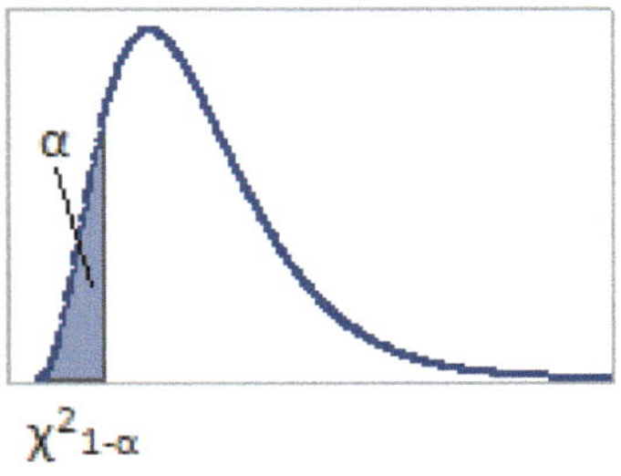

Various χ^2_α values are tabulated in **Table T5** in the Appendix for selected values of degrees of freedom and α. Below is a part of the Table T5 appears in the Appendix.

df	$\chi^2_{.995}$	$\chi^2_{.99}$	$\chi^2_{.975}$	$\chi^2_{.95}$	$\chi^2_{.90}$	$\chi^2_{.10}$	$\chi^2_{.05}$	$\chi^2_{.025}$	$\chi^2_{.01}$	$\chi^2_{.005}$
1	*0.000*	*0.000*	*0.001*	*0.004*	*0.016*	*2.706*	*3.841*	*5.024*	*6.635*	*7.879*
2	*0.010*	*0.020*	*0.051*	*0.103*	*0.211*	*4.605*	*5.991*	*7.378*	*9.210*	*10.597*
3	*0.072*	*0.115*	*0.216*	*0.352*	*0.584*	*6.251*	*7.815*	*9.348*	*11.345*	*12.838*
4	*0.207*	*0.297*	*0.484*	*0.711*	*1.064*	*7.779*	*9.488*	*11.143*	*13.277*	*14.860*
5	*0.412*	*0.554*	*0.831*	*1.145*	*1.610*	*9.236*	*11.070*	*12.833*	*15.086*	*16.750*
6	*0.676*	*0.872*	*1.237*	*1.635*	*2.204*	*10.645*	*12.592*	*14.449*	*16.812*	*18.548*
7	*0.989*	*1.239*	*1.690*	*2.167*	*2.833*	*12.017*	*14.067*	*16.013*	*18.475*	*20.278*
8	*1.344*	*1.646*	*2.180*	*2.733*	*3.490*	*13.362*	*15.507*	*17.535*	*20.090*	*21.955*

In above table, when $\alpha = 0.05$, with df=5, $\chi^2_{.05} = 11.070$. This means that area to the right of the chi-square point 11.07 with the df=5 is 0.05, which has been shown in Figure 12.2.

Figure 12.2: Area under the curve of a Chi-square distribution with df=5 to the right of 11.07.

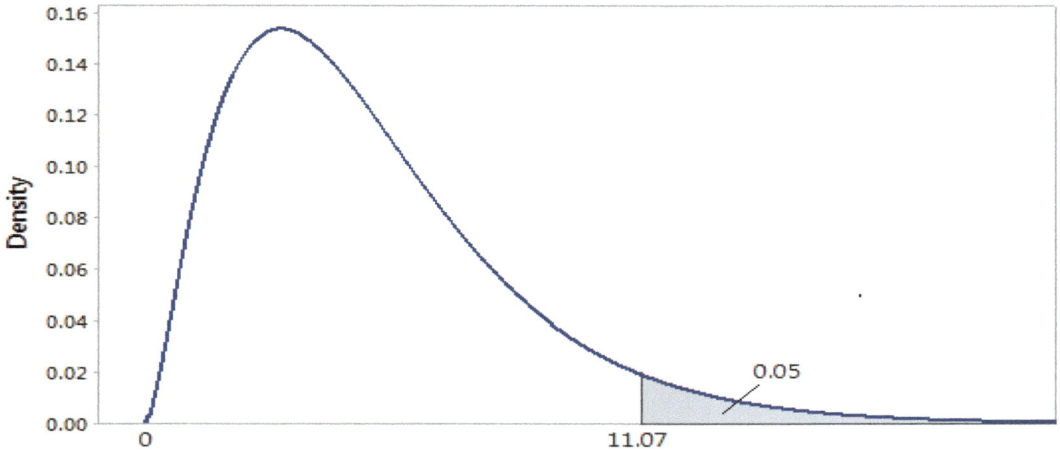

With df=5, two chi-square points with area 0.95 in the middle are shown in Figure 12.3.

Figure 12.3: Area under the curve of a Chi-square distribution with df=5 between 0.831 and 12.833.

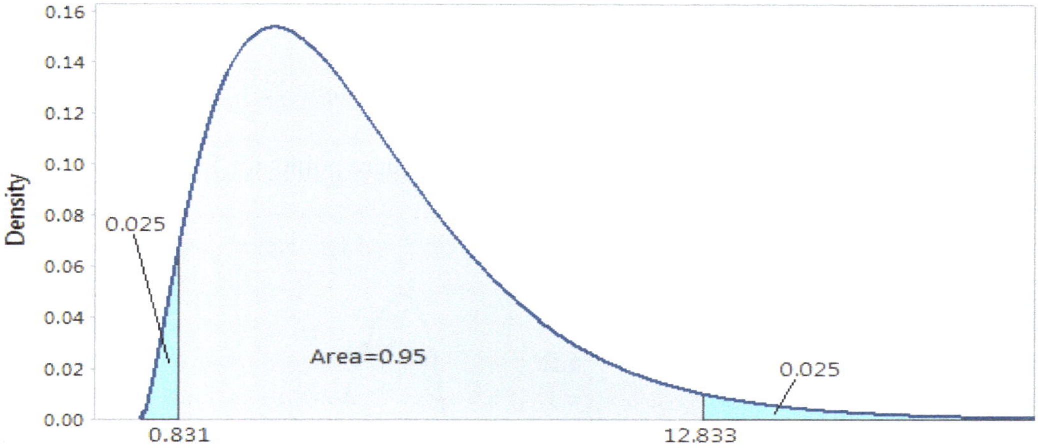

In terms of the notation of the Table 5, with df=5, $\chi^2_{.025} = 12.833$ and $\chi^2_{.975} = 0.831$. In other words, the area to the left of 0.831 is 1-0.975=0.025 and to the left of 12.833 is 1-0.025=0.975. The area between chi-square points 0.831 and 12.833 is 0.95, which has been shown in Figure 12.3.

df	$\chi^2_{.995}$	$\chi^2_{.99}$	$\chi^2_{.975}$	$\chi^2_{.95}$	$\chi^2_{.90}$	$\chi^2_{.10}$	$\chi^2_{.05}$	$\chi^2_{.025}$	$\chi^2_{.01}$	$\chi^2_{.005}$
1	0.000	0.000	0.001	0.004	0.016	2.706	3.841	5.024	6.635	7.879
2	0.010	0.020	0.051	0.103	0.211	4.605	5.991	7.378	9.210	10.597
3	0.072	0.115	0.216	0.352	0.584	6.251	7.815	9.348	11.345	12.838
4	0.207	0.297	0.484	0.711	1.064	7.779	9.488	11.143	13.277	14.860
5	0.412	0.554	0.831	1.145	1.610	9.236	11.070	12.833	15.086	16.750
6	0.676	0.872	1.237	1.635	2.204	10.645	12.592	14.449	16.812	18.548
7	0.989	1.239	1.690	2.167	2.833	12.017	14.067	16.013	18.475	20.278
8	1.344	1.646	2.180	2.733	3.490	13.362	15.507	17.535	20.090	21.955
9	1.735	2.088	2.700	3.325	4.168	14.684	16.919	19.023	21.666	23.589

Example 12.1

Find the values of the following chi-square points with degrees of freedom df=7 and interpret the results.

(a) $\chi^2_{.995}$
(b) $\chi^2_{.95}$
(c) $\chi^2_{.10}$
(d) $\chi^2_{.01}$

Solution

With df=7

(a) $\chi^2_{.995} = 0.989$. This means that the area to the right of the chi-square point 0.989 with df=7 is 0.995.

(b) $\chi^2_{.95} = 2.167$. This means that the area to the right of the chi-square point 2.167 with df=7 is 0.95.

(c) $\chi^2_{.10} = 12.017$. This means that the area to the right of the chi-square point 12.017 with df=7 is 0.10.

(d) $\chi^2_{.01} = 18.475$. This means that the area to the right of the chi-square point 18.475 with df=7 is 0.01.

See Figure 12.4 (a)-(d) for area to the right of the chi-square points for Example 12.1.

Figure 12.4: Area under the chi-square curves to the right of the chi-square points for Example 12.1 (a)-(d).

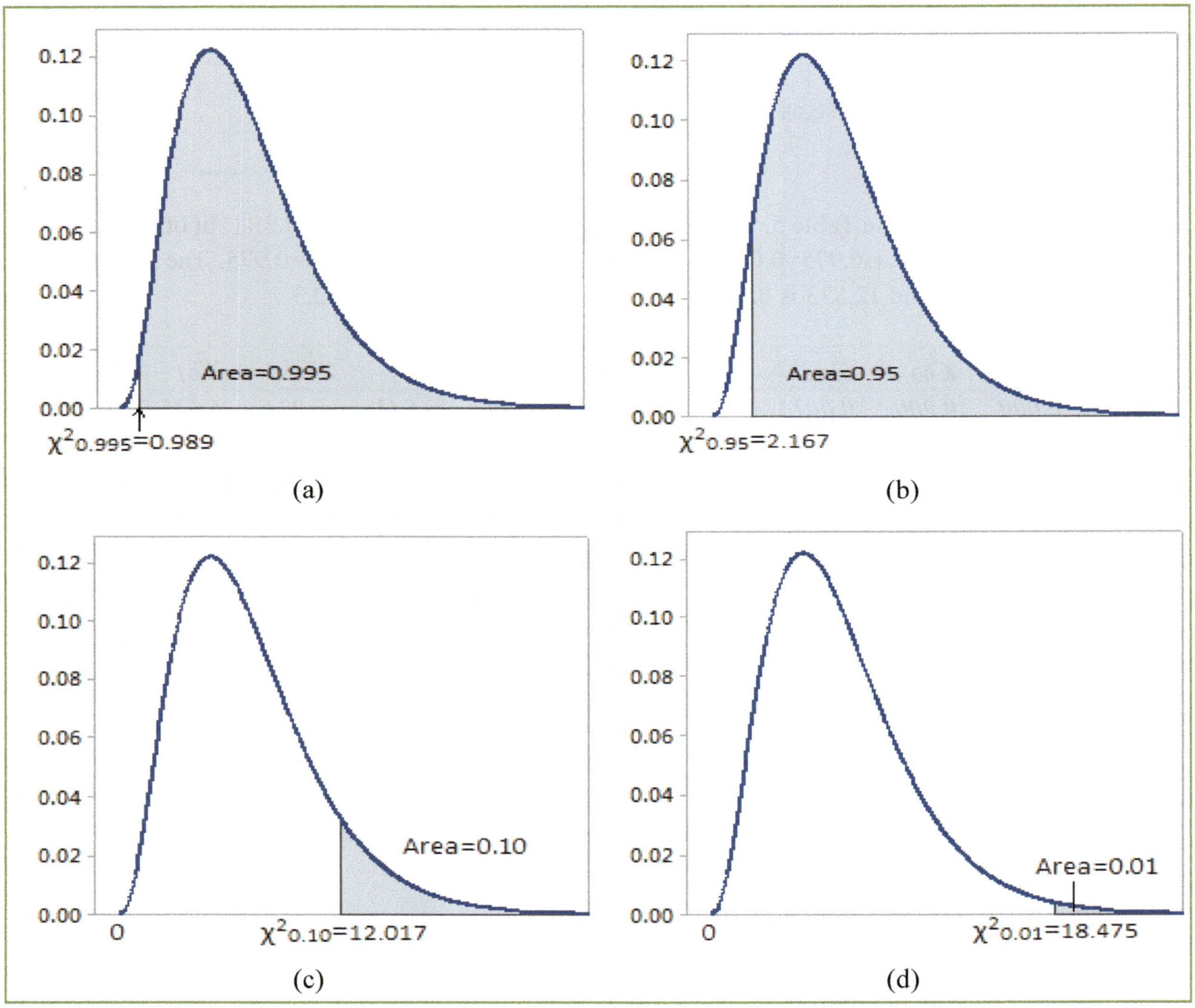

Section 12.2: Estimation of a population variance σ^2

In this section, we want to find confidence interval estimate of an unknown population variance σ^2 and test hypothesis regarding σ^2. The point estimate of σ^2 is the sample variance s^2 given by $s^2 = \frac{\Sigma(X-\bar{X})^2}{n-1}$.

Confidence interval estimation of σ^2

As mentioned, if s^2 is the sample variance of a sample from a normal distribution with a variance σ^2, then the sampling distribution of the statistic

$$\chi^2 = \frac{(n-1)s^2}{\sigma^2}$$

is a chi-squared distribution with $(n-1)$ degrees of freedom. Therefore, for a $100(1-\alpha)\%$ confidence interval estimate of σ^2 we use the fact that

$$P\left(\chi^2_{1-(\alpha/2)} \le \frac{(n-1)s^2}{\sigma^2} \le \chi^2_{\alpha/2}\right) = 1-\alpha$$

where α is a number between 0 and 1.
After some algebraic manipulations of above equation, we have

$$P\left(\frac{(n-1)s^2}{\chi^2_{\alpha/2}} \le \sigma^2 \le \frac{(n-1)s^2}{\chi^2_{1-(\alpha/2)}}\right) = 1-\alpha$$

Then, the interval $\left[\frac{(n-1)s^2}{\chi^2_{\alpha/2}}, \frac{(n-1)s^2}{\chi^2_{1-(\alpha/2)}}\right]$ contains the variance σ^2 with confidence level $C = 1-\alpha$. For example, a 95% confidence interval (i.e., $\alpha = 0.05$) estimate of σ^2 is constructed by

$$\left[\frac{(n-1)s^2}{\chi^2_{0.05/2}}, \frac{(n-1)s^2}{\chi^2_{1-(0.05/2)}}\right] = \left[\frac{(n-1)s^2}{\chi^2_{0.025}}, \frac{(n-1)s^2}{\chi^2_{0.975}}\right].$$

If the sample size $n = 6$, i.e., $df = 5$, then the 95% confidence interval estimate of σ^2 is constructed by $\left[\frac{5\times s^2}{\chi^2_{0.025}}, \frac{5\times s^2}{\chi^2_{0.975}}\right] = \left[\frac{5\times s^2}{12.833}, \frac{5\times s^2}{0.833}\right]$.

Once we know the CI estimate of σ^2, we can find the CI estimate of σ, by taking the positive square root of the CI estimate of σ^2.

Confidence interval estimation of σ

A $100(1-\alpha)\%$ confidence interval estimate of σ follows from the positive square root of confidence interval estimate of σ^2:

$$P\left(\sqrt{\frac{(n-1)s^2}{\chi^2_{\alpha/2}}} \le \sigma \le \sqrt{\frac{(n-1)s^2}{\chi^2_{1-(\alpha/2)}}}\right) = 1-\alpha$$

Then, the interval $\left[\sqrt{\frac{(n-1)s^2}{\chi^2_{\alpha/2}}}, \sqrt{\frac{(n-1)s^2}{\chi^2_{1-(\alpha/2)}}}\right]$ contains the population standard deviation σ with confidence level $C = 1-\alpha$. For example, a 95% confidence interval (i.e., $\alpha = 0.05$) estimate of σ is constructed by

$$\left[\sqrt{\frac{(n-1)s^2}{\chi^2_{0.05/2}}}, \sqrt{\frac{(n-1)s^2}{\chi^2_{1-(0.05/2)}}}\right] = \left[\sqrt{\frac{(n-1)s^2}{\chi^2_{0.025}}}, \sqrt{\frac{(n-1)s^2}{\chi^2_{0.975}}}\right].$$

If the sample size $n = 6$, i.e., $df = 5$, then the 95% confidence interval estimate of σ is constructed by $\left[\sqrt{\frac{5\times s^2}{\chi^2_{0.025}}}, \sqrt{\frac{5\times s^2}{\chi^2_{0.975}}}\right] = \left[\sqrt{\frac{5\times s^2}{12.833}}, \sqrt{\frac{5\times s^2}{0.833}}\right]$.

Example 12.2

Below is a sample of size 6 from a normal population generated using Minitab. For the computational convenience, the sample data points are rounded up to the one decimal point. You can verify that the sample variance for the sample data points is $s^2 = 4.102$.

13.9 9.9 11.8 10.8 14.4 14.7.

Answer the following questions:

(a) What is the point estimate of σ^2?
(b) Find a 90% confidence interval estimate of σ^2.
(c) Find a 90% confidence interval estimate of σ.

Solution

(a) The point estimate of σ^2 is $s^2 = 4.102$.

(b) For a 90% confidence interval estimates of σ^2, we need to evaluate $\left[\frac{(n-1)s^2}{\chi^2_{\alpha/2}}, \frac{(n-1)s^2}{\chi^2_{1-(\alpha/2)}}\right]$. Here, $n = 6$, $s^2 = 4.102$ and $\alpha = 0.10$. Therefore, $\chi^2_{\alpha/2} = \chi^2_{0.05}$ and $\chi^2_{1-(\alpha/2)} = \chi^2_{0.95}$. With $df = n - 1 = 6 - 1 = 5$, $\chi^2_{0.05} = 11.07$ and $\chi^2_{0.95} = 1.145$ (Using the Appendix Table T5). Then, $\left[\frac{(n-1)s^2}{\chi^2_{\alpha/2}}, \frac{(n-1)s^2}{\chi^2_{1-(\alpha/2)}}\right] = \left[\frac{5\times 4.102}{11.070}, \frac{5\times 4.102}{1.145}\right] = [1.85, 17.91]$, which is the desired 90% CI estimates of σ^2.

(c) A 90% CI for σ is $\left[\sqrt{1.85}, \sqrt{17.91}\right] = [1.36, 4.23]$

Minitab Instruction for CI for Variance or Standard deviation

Ref: Example 12.2

We follow the steps below to carry out a CI for a proportion

(1) Select **Stat>Basic Statistics> 1 Variance**...
(2) Choose Sample variance or Sample standard deviation and enter corresponding value and sample size: **Sample size**: 6 **Sample variance**: 4.102. Choose one or more samples, each in one column option to input sample data points.
(3) Use Options to decide **Confidence level**: 90, (default is **Confidence level**: 95).
(4) Click OK two times.

Minitab dialog box with selection options and results from aforementioned steps are given below:

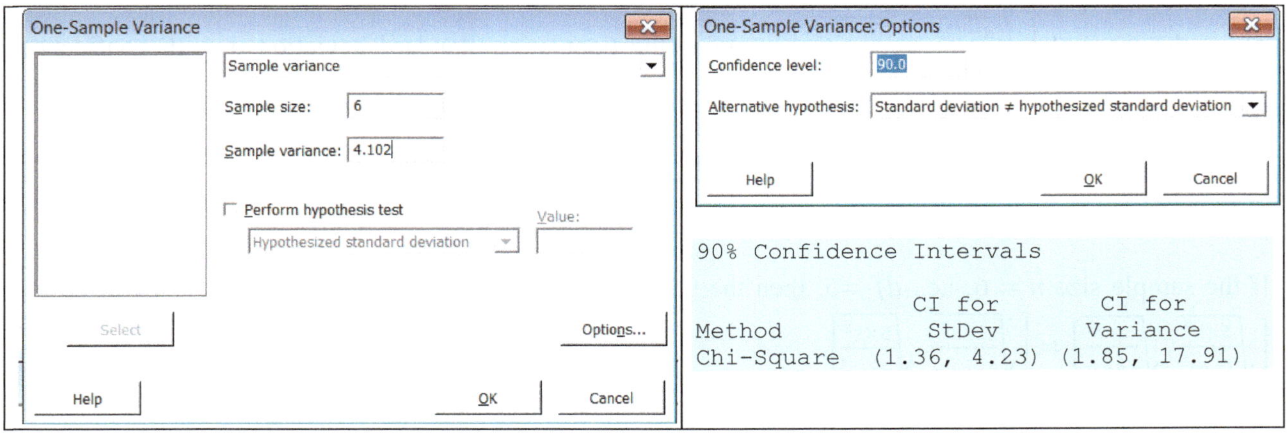

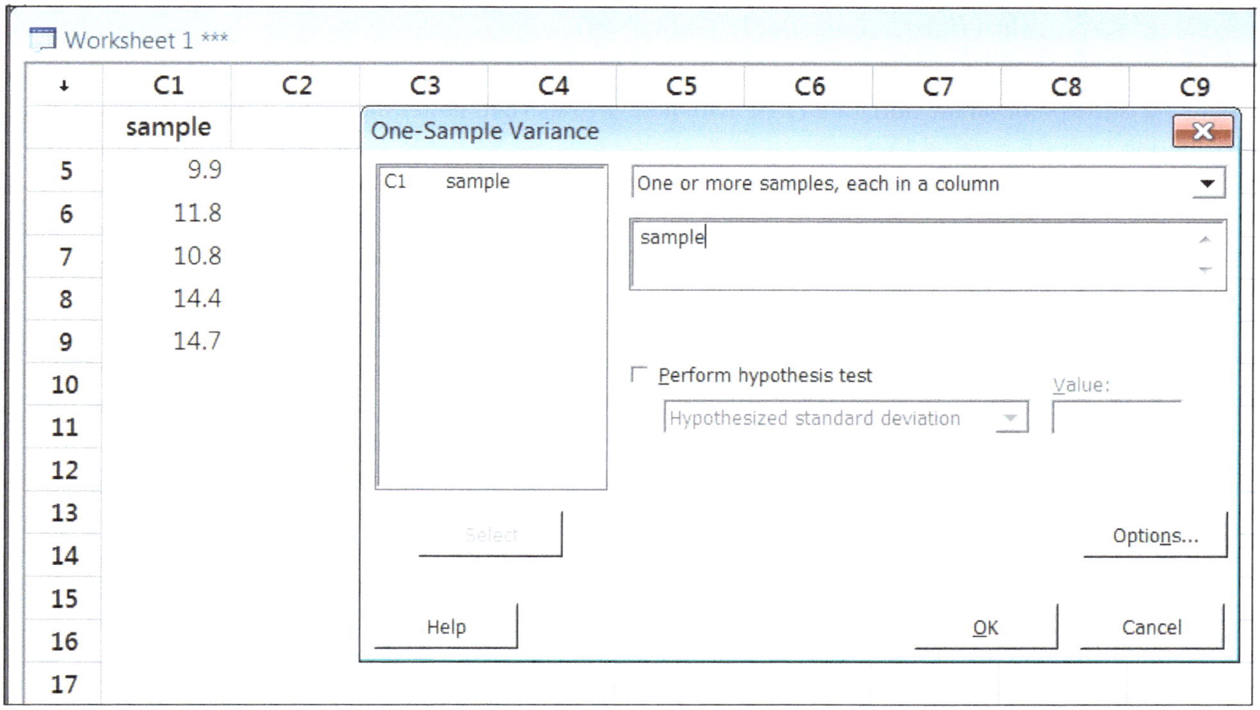

As we see Minitab output displays 90% confidence interval for standard deviation and variance. The Minitab output is identical what we found in Example 12.2.

Section 12.3: Test of hypothesis about σ^2

We want to test the null hypothesis $H_0\colon \sigma^2 = \sigma_0^2$ against the following alternative hypotheses:

$$H_a\colon \sigma^2 > \sigma_0^2$$
$$H_a\colon \sigma^2 < \sigma_0^2$$
$$H_a\colon \sigma^2 \neq \sigma_0^2$$

Test statistic

The test statistic for this test under H_0 is given by

$$\chi^2 = \frac{(n-1)s^2}{\sigma_0^2}$$

which has a chi-squared distribution with $df = (n-1)$. Compare the observed value of the χ^2, χ_{obs}^2, with the critical value to decide about rejection of the null hypothesis.

Decision rule

Using the critical value approach, we reject the null hypothesis at significance level α if

(1) $\chi^2 > \chi^2_\alpha$ when $H_a: \sigma^2 > \sigma_0^2$ (i.e., right-tailed test)
(2) $\chi^2 < \chi^2_{1-\alpha}$ when $H_a: \sigma^2 < \sigma_0^2$ (i.e., left-tailed test)
(3) $\chi^2 > \chi^2_{\alpha/2}$ or $\chi^2 < \chi^2_{1-(\alpha/2)}$ when $H_a: \sigma^2 \neq \sigma_0^2$ (i.e., two-tailed test)

where χ^2_α is the point on the horizontal axis with area α to its right under the curve of a chi-square distribution with $df = (n-1)$.n.

The points χ^2_α and $\chi^2_{1-\alpha}$ are critical values for right and left tailed tests, respectively. The points $\chi^2_{\alpha/2}$ and $\chi^2_{1-(\alpha/2)}$ are the critical values for a two tailed test.

***p*-value**

Using p-value approach reject null hypothesis at significance level α if

$$p - value \leq \alpha$$

The p-value for testing $H_0: \sigma^2 = \sigma_0^2$ is computed by:

$$p\text{-value} = P(\chi^2 > \chi^2_{obs}) \ if \ H_a: \sigma^2 > \sigma_0^2$$

$$p\text{-value} = P(\chi^2 < \chi^2_{obs}) \ if \ H_a: \sigma^2 < \sigma_0^2$$

$$p\text{-value} = 2 \times min\{P(\chi^2 > \chi^2_{obs}), P(\chi^2 < \chi^2_{obs})\} \ if \ H_a: \sigma^2 \neq \sigma_0^2$$

where χ^2 is the Chi-square distribution with degrees of freedom $(n-1)$.

Example 12.3

In reference to the Example 12.2, we wish to test the null hypothesis $H_0: \sigma^2 = 4$ against the following alternative hypotheses at $\alpha = 0.05$:

(a) $H_a: \sigma^2 > 4$
(b) $H_a: \sigma^2 < 4$
(c) $H_a: \sigma^2 \neq 4$

We have for Example 12.2, $n = 6$ and $s^2 = 4.102$. Make a conclusion about the test for each of alternatives (a)-(c) using critical value approach.

Solution

We have:

$n = 6$, $s^2 = 4.102$ and $\sigma_0^2 = 4$

$$\chi^2 = \frac{(n-1)s^2}{\sigma_0^2} = \frac{5 \times s^2}{4} = 5.13$$

With $df = n - 1 = 6 - 1 = 5$ and $\alpha = 0.05$, $\chi^2_\alpha = \chi^2_{0.05} = 11.07$ and $\chi^2_{1-\alpha} = \chi^2_{0.95} = 1.145$, $\chi^2_{\alpha/2} = \chi^2_{0.025} = 12.833$ and $\chi^2_{1-\alpha/2} = \chi^2_{0.975} = 0.931$ (using Table T5).

(a) Since $\chi^2 = 5.13 < \chi^2_{0.05} = 11.07$, we fail to reject the null hypothesis $H_0: \sigma^2 = 4$ against $H_a: \sigma^2 > 4$.
(b) Since $\chi^2 = 5.13 > \chi^2_{0.95} = 1.145$, we fail to reject the null hypothesis $H_0: \sigma^2 = 4$ against $H_a: \sigma^2 < 4$.
(c) Since $\chi^2 = 5.13 < \chi^2_{0.025} = 12.833$ and $\chi^2 = 5.13 > \chi^2_{0.975} = 0.931$, we fail to reject $H_0: \sigma^2 = 4$ against $H_a: \sigma^2 \neq 4$.

Example 12.4

Reference to the Example 12.3, find p-value for null hypothesis $H_0: \sigma^2 = 4$ against each of the following alternative hypotheses:

(a) $H_a: \sigma^2 > 4$
(b) $H_a: \sigma^2 < 4$
(c) $H_a: \sigma^2 \neq 4$

at $\alpha = 0.05$ using any technology.

Solution

In Example 12.3, we have $\chi^2 = \frac{(n-1)s^2}{\sigma_0^2} = \frac{5 \times s^2}{4} = 5.13$ and $df = n - 1 = 6 - 1 = 5$. Below we use TI-84 Plus calculator function $\chi^2 cdf()$ for computing p-value.

(a) For $H_a: \sigma^2 > 4$, p-value= $P(\chi^2 > 5.13) = 0.40$. Since $p - value > \alpha = 0.05$, we fail to reject the null hypothesis $H_0: \sigma^2 = 4$ against $H_a: \sigma^2 > 4$.
(b) For $H_a: \sigma^2 < 4$, p-value= $P(\chi^2 < 5.13) = 0.60$. Since $p - value > \alpha = 0.05$, we fail to reject $H_0: \sigma^2 = 4$ against $H_a: \sigma^2 < 4$.
(c) For $H_a: \sigma^2 \neq 4$, p-value= $2 \times min\{P(\chi^2 > 5.13), P(\chi^2 < 5.13)\} = 2 \times 0.40 = 0.80$. Since $p - value > \alpha = 0.05$, we fail to reject $H_0: \sigma^2 = 4$ against $H_a: \sigma^2 \neq 4$

One can verify above computations by executing TI-84 Plus calculator $\chi^2 cdf()$ function as follows:

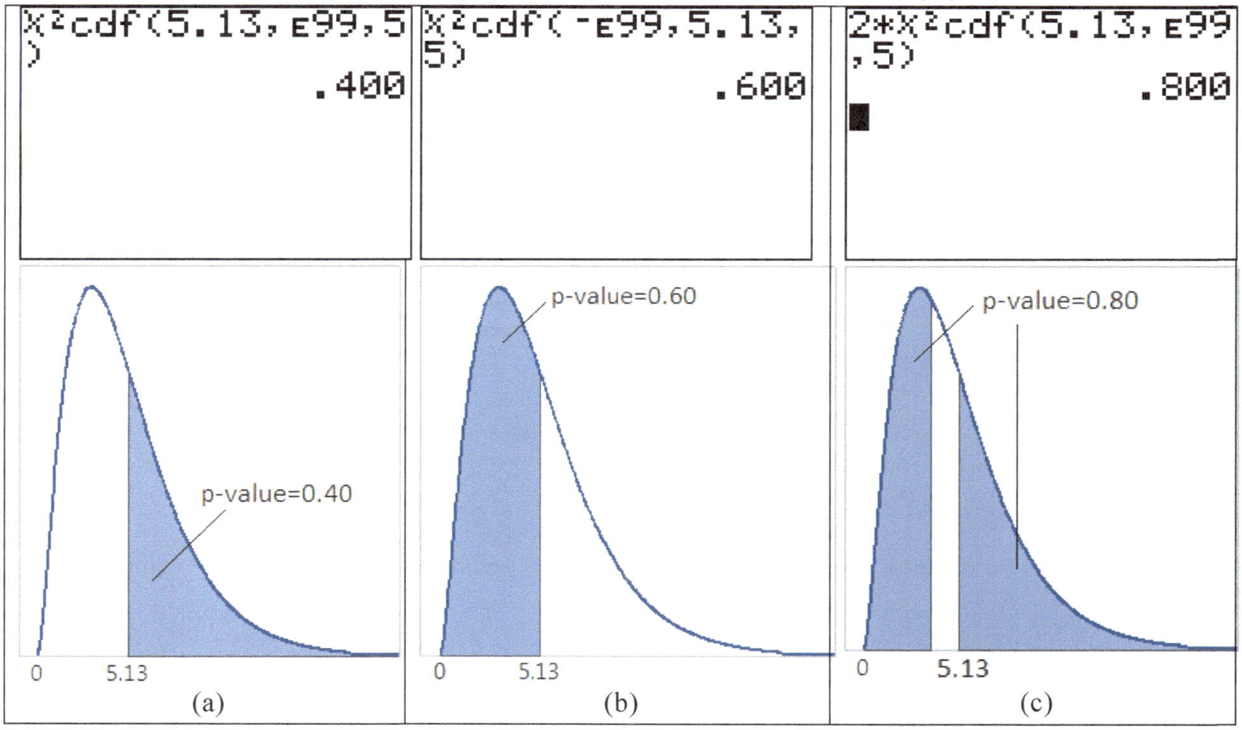

Implementing Minitab for testing hypothesis about σ^2

Ref: Example 12.4

We follow the steps below to carry out the test for population variance σ^2:

(1) Select **Stat>Basic Statistics> 1 Variance**…
(2) Choose Sample variance or Sample standard deviation and enter corresponding value and sample size: **Sample size**: 6 **Sample variance**: 4.102. Choose one or more samples, each in one column option to input sample data points.

(3) Check the box saying **Perform hypothesis test** and specify **hypothesized variance** or **hypothesized standard deviation** value: 4.

(4) Choose Options to decide **Confidence level**: 95 (default) for a 95% CI or as desired such as 90 for a 90% CI. Also, specify Alternative hypothesis: Variance>hypothesized variance, Variance<hypothesized variance or Variance≠hypothesized variance for (a), (b) and (c) respectively.

(5) Click OK two times.

Minitab dialog box with selection options and results from aforementioned steps are given below:

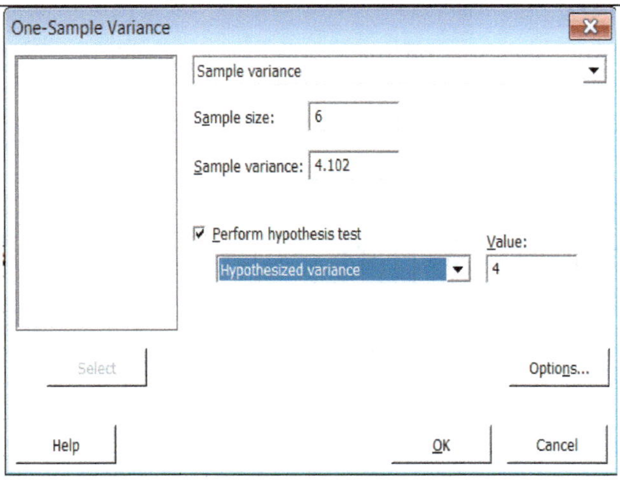

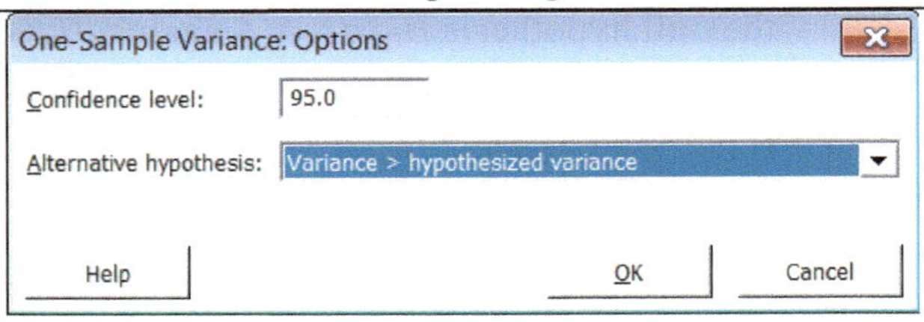

(a)

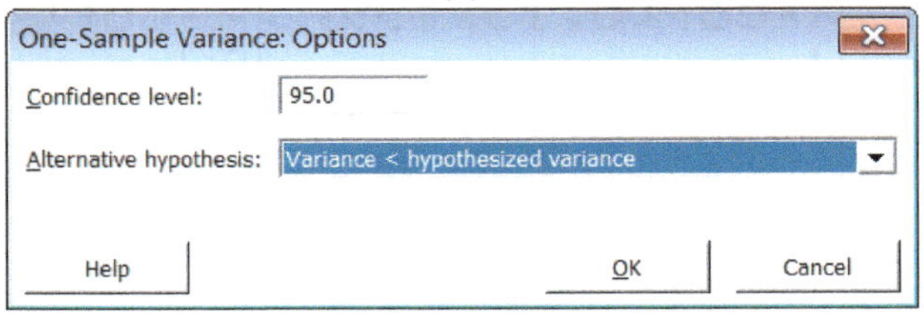

(b)

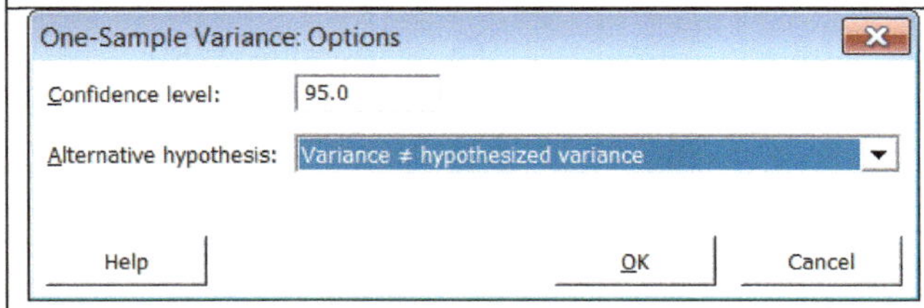

(c)

```
Tests
Alternative hypothesis: σ-squared > 4

                 Test
Method       Statistic  DF  P-Value
Chi-Square        5.13   5    0.401
```

(A)

```
Tests
Alternative hypothesis: σ-squared < 4

                 Test
Method       Statistic  DF  P-Value
Chi-Square        5.13   5    0.599
```

(B)

```
Tests
Alternative hypothesis: σ-squared ≠ 4

                 Test
Method       Statistic  DF  P-Value
Chi-Square        5.13   5    0.801
```

(C)

Example 12.5

A random sample of sugar content (in grams) from various General Mills' cereals is considered from a population of list of 78 General Mills' cereal reported in http://cereal.findthebest.com/. The 10 sample observations are reported below:

8	7	16	9	1	10	9	9	10	1

As can easily be verified that the sample standard deviation $s = 4.397$ or sample variance $s^2 = 19.334$. Assume that sugar content in the population of General Mills' cereals follows a normal distribution. Answer the followings:

(a) Find a point estimate of population variance σ^2.
(b) Find a 95% CI estimate of σ^2.
(c) Find a 95% CI estimate of σ.
(d) Test the null hypothesis $H_0: \sigma^2 = 16$ against $H_a: \sigma^2 > 16$ at $\alpha = 0.05$.

Solution

(a) The point estimate of σ^2 is $s^2 = 19.334$.

(b) For a 95% CI estimate of σ^2, we need to evaluate $\left[\frac{(n-1)s^2}{\chi^2_{\alpha/2}}, \frac{(n-1)s^2}{\chi^2_{1-(\alpha/2)}}\right]$, where $n = 10$, $s^2 = 19.334$, $\alpha = 0.05$ and $df = n - 1 = 10 - 1 = 9$. For $df = 9$, $\chi^2_{\alpha/2} = \chi^2_{0.025} = 19.023$ and $\chi^2_{1-(\alpha/2)} = \chi^2_{0.975} = 2.7$ (using Table T5).
Then, $\left[\frac{(n-1)s^2}{\chi^2_{\alpha/2}}, \frac{(n-1)s^2}{\chi^2_{1-(\alpha/2)}}\right] = \left[\frac{9\times19.334}{19.023}, \frac{9\times19.334}{2.7}\right] = [9.147, 64.447]$, which is the 95% CI estimate of σ^2.

(c) A 95% CI for σ is $\left[\sqrt{9.147}, \sqrt{64.447}\right] = [3.024, 8.028]$.

(d) We have $\sigma_0^2 = 16$ and $\chi^2 = \frac{(n-1)s^2}{\sigma_0^2} = \frac{9\times19.334}{16} = 10.875$. For $df = 9$ and $\alpha = 0.05$, $\chi^2_{0.05}$ =16.919 (using Table T5). Since $\chi^2 = 10.875 < \chi^2_{0.05} = 16.919$, we fail to reject the null hypothesis $H_0: \sigma^2 = 16$ against $H_0: \sigma^2 > 16$.

Section 12.4: F distribution basics

In this section, we introduce a new statistical distribution, called an ***F distribution***. We will use this distribution to test the equality of two population variances, i.e., to test the null hypothesis $H_0: \sigma_1^2 = \sigma_2^2$. Below is the definition of the F distribution.

Definition: F distribution

Suppose s_1^2 and s_2^2 are two sample variances from independent samples of respective sizes n_1 and n_2 from two normal distributions having variances σ_1^2 and σ_2^2, respectively. Then, under $H_0: \sigma_1^2 = \sigma_2^2$ or equivalently, $H_0: \frac{\sigma_1^2}{\sigma_2^2} = 1$, the sampling distribution of the statistic

$$F = \frac{s_1^2}{s_2^2}$$

is an F distribution with numerator degrees of freedom $df_1 = (n_1 - 1)$ and denominator degrees of freedom $df_2 = (n_2 - 1)$.

The shape of F distribution depends on two parameters, called numerator degrees of freedom $df_1 = (n_1 - 1)$ and denominator degrees of freedom $df_2 = (n_2 - 1)$. Generally, the shape of the F distribution is right skewed. In Figure 12.5, we have three curves of F distributions with numerator degrees of freedom $df_1 = 5$ and denominator degrees of freedom $df_2 = 5, 10$ and 15.

Figure 12.5: F curves with df_1= 5 and df_2= 5, 10 and 15.

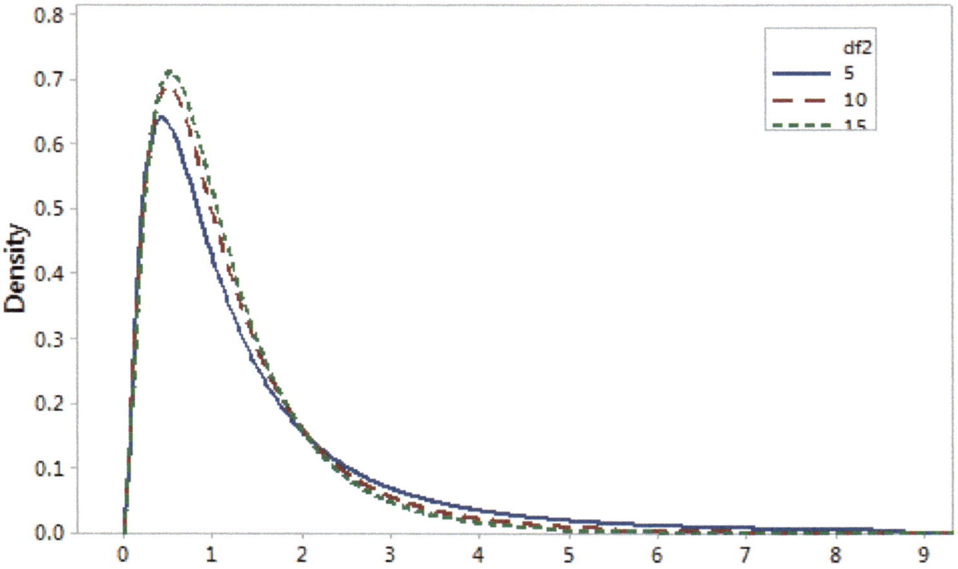

As we see from Figure 12.5, as denominator degrees of freedom df_2 increases, thickness of the right tail decreases.

Characteristics of an F distribution

Some of the important characteristics of an F distribution can be summarized as follows:

- Values of an F variable are always positive.
- An F distribution is determined by two parameters, the numerator degrees of freedom df_1= $(n_1 - 1)$ and the denominator degrees of freedom df_2= $(n_2 - 1)$.
- Area under the curve of an F distribution is 1.
- Shape of an F distribution is right skewed.

Notation F_α

F_α refers to the point on the horizontal axis under the curve of an F-distribution with numerator degrees of freedom df_1 and denominator degrees of freedom df_2 having an area α to its right.

Various values of F_α are tabulated in Tables T6-T9 in the Appendix for selected values of numerator and denominator degrees of freedom and α. $F_{0.05}$ is the value of F point with area 0.05 to its right for specified values of df_1 and df_2.

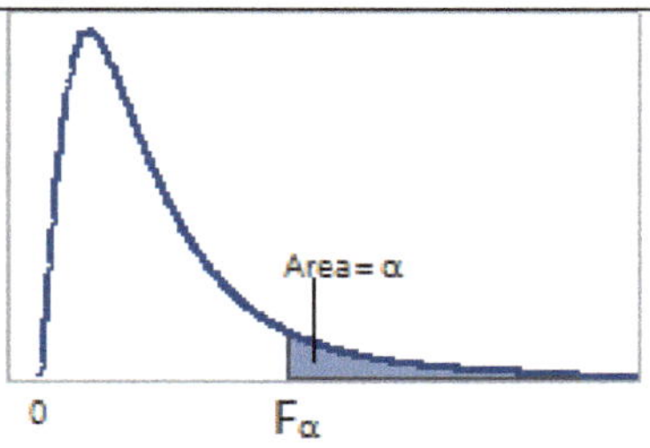

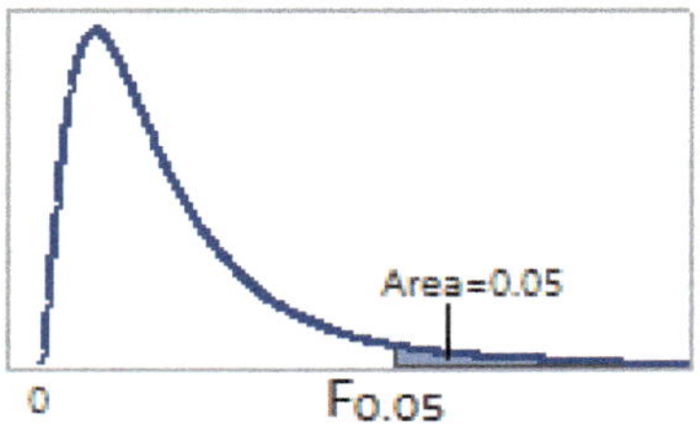

Below is a part of the Table T7 for $\alpha = 0.05$ showing an F point for which area to its right is 0.05 for selected values of df_1 and df_2.

	Numerator degrees of freedom df_1													
df_2	4	5	6	7	8	9	10	11	12	13	14	15	16	17
4	6.39	6.26	6.16	6.09	6.04	6.00	5.96	5.94	5.91	5.89	5.87	5.86	5.84	5.83
5	5.19	5.05	4.95	4.88	4.82	4.77	4.74	4.70	4.68	4.66	4.64	4.62	4.60	4.59
6	4.53	4.39	4.28	4.21	4.15	4.10	4.06	4.03	4.00	3.98	3.96	3.94	3.92	3.91
7	4.12	3.97	3.87	3.79	3.73	3.68	3.64	3.60	3.57	3.55	3.53	3.51	3.49	3.48
8	3.84	3.69	3.58	3.50	3.44	3.39	3.35	3.31	3.28	3.26	3.24	3.22	3.20	3.19
9	3.63	3.48	3.37	3.29	3.23	3.18	3.14	3.10	3.07	3.05	3.03	3.01	2.99	2.97
10	3.48	3.33	3.22	3.14	3.07	3.02	2.98	2.94	2.91	2.89	2.86	2.85	2.83	2.81
11	3.36	3.20	3.09	3.01	2.95	2.90	2.85	2.82	2.79	2.76	2.74	2.72	2.70	2.69
12	3.26	3.11	3.00	2.91	2.85	2.80	2.75	2.72	2.69	2.66	2.64	2.62	2.60	2.58

In above table ($\alpha = 0.05$), for example, with $df_1 = 5$ and $df_2 = 10$, $F_{0.05} = 3.33$. This means that area to the right of the F point 3.33 with numerator degrees of freedom $df_1 = 5$ and denominator degrees of freedom $df_2 = 10$ is 0.05, which has been shown in Figure 12.6.

Figure 12.6: Area under the F curve with $df_1 = 5$ and $df_2 = 10$ to the right of 3.33

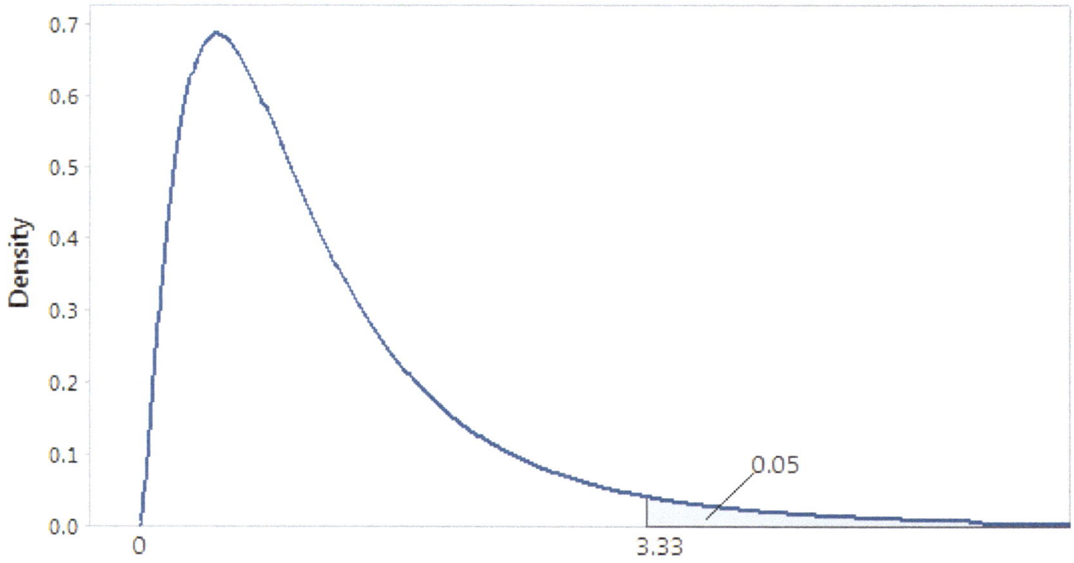

Using technology, we can find that for $df_1 = 5$ and $df_2 = 10$, with area 0.95 in the middle, the two F points are 0.151 and 4.236, shown in Figure 12.7.

Figure 12.7: Area under the curve of an F distribution with $df_1 = 5$ and $df_2 = 10$ between F points 0.151 and 4.236.

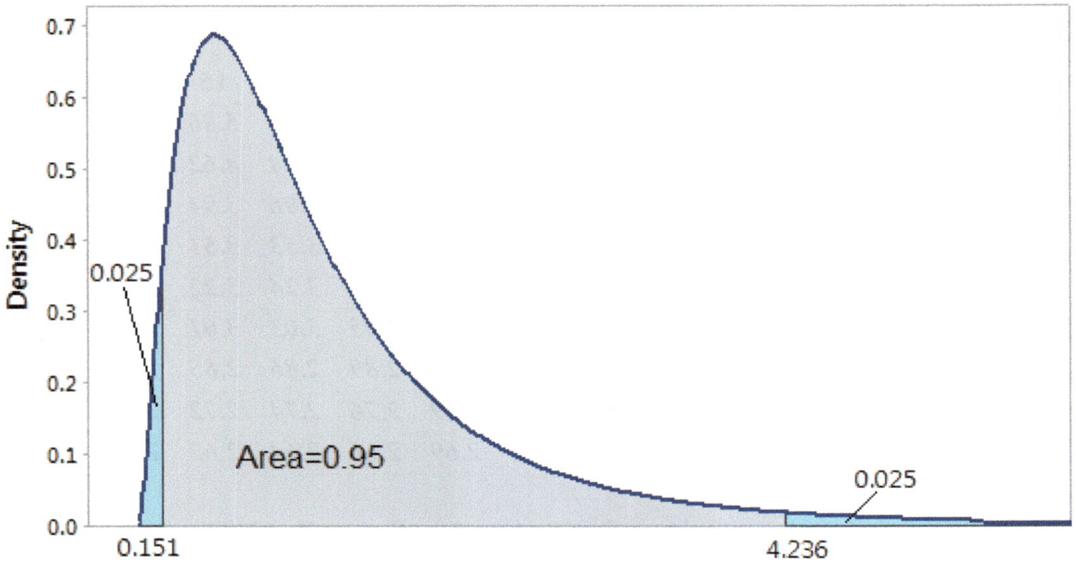

With different choices of degrees of freedoms df_1 and df_2, , the two F points with area 0.95 in the middle will be different.

Section 12.5: F test for equality of two population variances

In this section, we want to test the null hypothesis $H_0: \sigma_1^2 = \sigma_2^2$ against any of the following alternative hypotheses:

i. $H_a: \sigma_1^2 > \sigma_2^2$
ii. $H_a: \sigma_1^2 < \sigma_2^2$
iii. $H_a: \sigma_1^2 \neq \sigma_2^2$

An equivalent set of hypotheses can be derived by dividing both sides by σ_2^2:

$$H_0: \frac{\sigma_1^2}{\sigma_2^2} = 1$$

i. $H_a: \frac{\sigma_1^2}{\sigma_2^2} > 1$
ii. $H_a: \frac{\sigma_1^2}{\sigma_2^2} < 1$
iii. $H_a: \frac{\sigma_1^2}{\sigma_2^2} \neq 1$

The test statistic for testing $H_0: \sigma_1^2 = \sigma_2^2$ or equivalently, $H_0: \frac{\sigma_1^2}{\sigma_2^2} = 1$, is called an F statistic.

Test statistic

The test statistic to test $H_0: \sigma_1^2 = \sigma_2^2$ or equivalently $H_0: \frac{\sigma_1^2}{\sigma_2^2} = 1$ is given by

$$F = \frac{s_1^2}{s_2^2}$$

which follows an F distribution with the numerator degrees of freedom $df_1 = (n_1 - 1)$ and the denominator degrees of freedom $df_2 = (n_2 - 1)$, under the null hypothesis.

An important note about F test

By writing the higher sample variance in the numerator of F statistic, we can use the inequality > for the first two alternatives $H_a: \frac{\sigma_1^2}{\sigma_2^2} > 1$ and $H_a: \frac{\sigma_1^2}{\sigma_2^2} < 1$. Therefore, we can compare the observed value of F with the right tailed critical value F_α for one tailed test or critical value $F_{\alpha/2}$ for two tailed test. In such cases, numerator degrees of freedom df_1 will refer to the degrees of freedom of the higher sample variance appearing in the numerator. Given these circumstances, we assume that n_1 is the sample size from which the larger sample variance is obtained and n_2 is the sample size from which the smaller sample variance is obtained.

Decision using critical value approach

Reject $H_0: \sigma_1^2 = \sigma_2^2$ (or equivalently, $H_0: \frac{\sigma_1^2}{\sigma_2^2} = 1$) at significance level α if

(1) $F > F_\alpha$ when the test is one tailed, i.e., $H_a: \sigma_1^2 > \sigma_2^2$ or $H_a: \sigma_1^2 < \sigma_2^2$

(2) $F > F_{\alpha/2}$ when the test is two tailed, i.e., $H_a: \sigma_1^2 \neq \sigma_2^2$

where F_α and $F_{\alpha/2}$ refer to F values with area α and $\alpha/2$ to its right, respectively, under F curve with $(n_1 - 1)$ and $(n_2 - 1)$ degrees of freedom. Here, n_1 is the sample size for larger sample variance placed in the numerator and n_2 is the sample size for smaller sample variance placed in the denominator of the F statistic.

Decision using p-value approach

Using p-value approach, reject $H_0: \sigma_1^2 = \sigma_2^2$ (or equivalently, $H_0: \frac{\sigma_1^2}{\sigma_2^2} = 1$) at significance level α if

$$p\text{-value} \leq \alpha$$

The p-value for testing $H_0: \sigma_1^2 = \sigma_2^2$ is computed by:

$$p\text{-value} = P(F > F_{obs}) \ if \ H_a: \sigma_1^2 > \sigma_2^2$$

$$p\text{-value} = P(F < F_{obs}) \ if \ H_a: \sigma_1^2 < \sigma_2^2$$

$$p\text{-value} = 2 \times min\{P(F > F_{obs}), P(F < F_{obs})\} \ if \ H_a: \sigma_1^2 \neq \sigma_2^2$$

where F is the F-distribution with numerator degrees of freedom $(n_1 - 1)$ for larger sample placed in the numerator and denominator degrees of freedom $(n_2 - 1)$ for smaller sample variance placed in the denominator of the F statistic.

Assumptions for the validity of an F test

An F test requires the following two assumptions:

(1) The two samples are independent.

(2) The two populations samples come from are normal.

Example 12.6

Suppose two independent samples of sizes $n_1 = 10$ and $n_2 = 7$ are taken from two normal populations having variances σ_1^2 and σ_2^2 whose estimates from the samples are $s_1^2 = 50$ and $s_2^2 = 25$. Test

(a) $H_0: \sigma_1^2 = \sigma_2^2$ against $H_a: \sigma_1^2 > \sigma_2^2$ at $\alpha = 0.05$ and make your conclusion about the test.
(b) $H_0: \sigma_1^2 = \sigma_2^2$ against $H_a: \sigma_1^2 \neq \sigma_2^2$ at $\alpha = 0.05$ and make your conclusion about the test.

Solution

Note that, $s_1^2 = 50 > s_2^2 = 25$. Then,

$$F = \frac{s_1^2}{s_2^2} = \frac{50}{25} = 2$$

(a) Here $df_1 = 10 - 1 = 9$ and $df_2 = 7 - 1 = 6$. With degrees of freedom 9 and 6 and $\alpha = 0.05$, $F_\alpha = F_{0.05} = 4.10$. Since $F = 2 < F_{0.05} = 4.10$, we fail to reject H_0 at $\alpha = 0.05$.
(b) With degrees of freedom 9 and 6 and $\alpha = 0.05$, $F_{\alpha/2} = F_{0.025} = 5.52$. Since $F < F_{0.025}$, we fail to reject H_0 at $\alpha = 0.05$.

Example 12.7

Suppose two independent samples of sizes $n_1 = 7$ and $n_2 = 10$ are taken from two normal populations having variances σ_1^2 and σ_2^2 whose estimates from the samples are $s_1^2 = 10$ and $s_2^2 = 50$. Test

(a) $H_0: \sigma_1^2 = \sigma_2^2$ against $H_a: \sigma_1^2 < \sigma_2^2$ at $\alpha = 0.05$ and make your conclusion about the test.
(b) $H_0: \sigma_1^2 = \sigma_2^2$ against $H_a: \sigma_1^2 \neq \sigma_2^2$ at $\alpha = 0.05$ and make your conclusion about the test.

Solution

Note that, $s_2^2 = 50 > s_1^2 = 10$. Then,

$$F = \frac{s_2^2}{s_1^2} = \frac{50}{10} = 5$$

(a) With degrees of freedom 9 and 6 and $\alpha = 0.05$, $F_\alpha = F_{0.05} = 4.10$. Since $F > F_{0.05}$, we reject H_0 at $\alpha = 0.05$.
(b) With degrees of freedom 9 and 6 and $\alpha = 0.05$, $F_{\alpha/2} = F_{0.025} = 5.52$. Since $F < F_{0.025}$, we fail to reject H_0 at $\alpha = 0.05$.

Example 12.8

Suppose we have two samples, one of size 7 and the other of size 9, of calorie count from different serving size of fast food. Two samples are generated using data of serving size and calorie available at http://www.acaloriecounter.com/fast-food.php. Two samples are as follows:

x	785	550	710	415	600	500	720		
y	780	610	690	300	500	450	560	460	437

Two sample variances turn out to be $s_1^2 = 17756$ and $s_2^2 = 21140$. Assume that the two populations are approximately normally distributed.

Answer the followings:

(a) Test $H_0: \sigma_1^2 = \sigma_2^2$ against $H_a: \sigma_1^2 < \sigma_2^2$ at $\alpha = 0.05$.
(b) Test $H_0: \sigma_1^2 = \sigma_2^2$ against $H_a: \sigma_1^2 \neq \sigma_2^2$ at $\alpha = 0.05$.

(c) How does the test in (a) differ from the test $H_0: \frac{\sigma_1^2}{\sigma_2^2} = 1$ against $H_a: \frac{\sigma_1^2}{\sigma_2^2} < 1$ at $\alpha = 0.05$?

Solution

Note that, $s_2^2 = 21140 > s_1^2 = 17756$. Then,

$$F = \frac{s_2^2}{s_1^2} = \frac{21140}{17756} = 1.19$$

With numerator degrees of freedom $9 - 1 = 8$, denominator degrees of freedom $7 - 1 = 6$ and $\alpha = 0.05$, $F_\alpha = F_{0.05} = 4.15$ and $F_{\alpha/2} = F_{0.025} = 5.60$ (see T7 in the Appendix).

(a) Since $F = 1.19 < F_{0.05} = 4.15$, we fail to reject $H_0: \sigma_1^2 = \sigma_2^2$ in favor of $H_a: \sigma_1^2 < \sigma_2^2$ at $\alpha = 0.05$.

(b) Since $F = 1.19 < F_{0.025} = 5.60$, we fail to reject $H_0: \sigma_1^2 = \sigma_2^2$ in favor of $H_a: \sigma_1^2 \neq \sigma_2^2$ at $\alpha = 0.05$.

(c) The test in (a) and the one specified in (c) are equivalent and hence the conclusion are identical.

Example 12.9

Reference to Example 12.8, use a p-value approach to conclude about

(a) Test $H_0: \sigma_1^2 = \sigma_2^2$ against $H_a: \sigma_1^2 < \sigma_2^2$ at $\alpha = 0.05$.

(b) Test $H_0: \sigma_1^2 = \sigma_2^2$ against $H_a: \sigma_1^2 > \sigma_2^2$ at $\alpha = 0.05$.

(c) Test $H_0: \sigma_1^2 = \sigma_2^2$ against $H_a: \sigma_1^2 \neq \sigma_2^2$ at $\alpha = 0.05$.

Note that using the fact $s_2^2 = 21140 > s_1^2 = 17756$, we have $F = \frac{s_2^2}{s_1^2} = \frac{21140}{17756} = 1.19$.

Solution

(a) For $H_a: \sigma_1^2 < \sigma_2^2$, p-value= $P(F < 1.19) = 0.5721$, which implies that we fail to reject the null hypothesis at $\alpha = 0.05$. We computed this p-value using TI-84 Plus calculator function $\boldsymbol{Fcdf}()$ with arguments $numerator\ df$, $denominator\ df$ and $observed\ F\ value$:

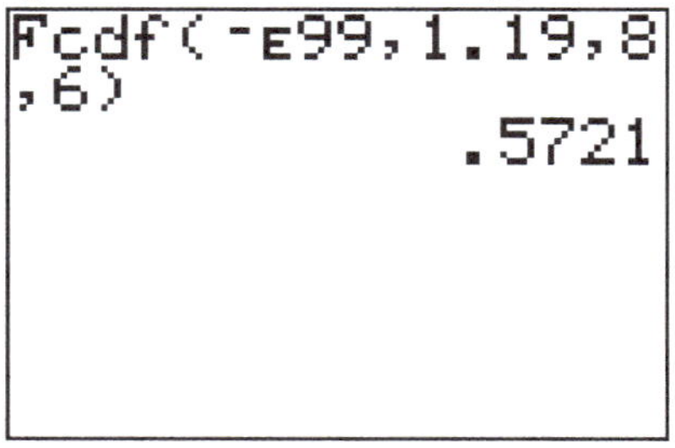

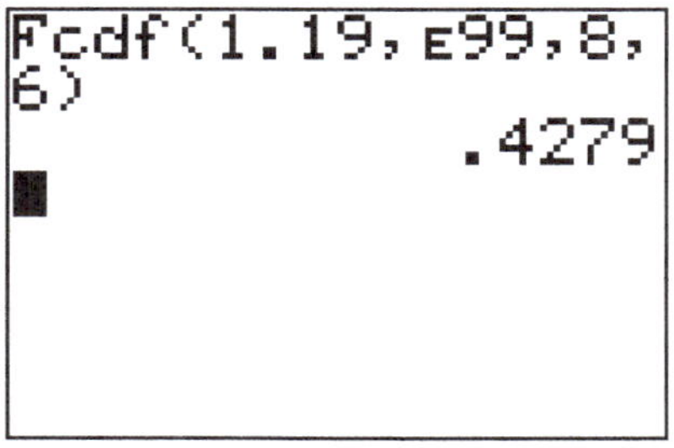

(b) For $H_a: \sigma_1^2 > \sigma_2^2$, p-value= $P(F > 1.19) = 0.4279$, which implies that we fail to reject the null hypothesis at $\alpha = 0.05$.

(c) For $H_a: \sigma_1^2 \neq \sigma_2^2$, p-value= $2 \times min\{P(F > F_{obs}), P(F < F_{obs})\} = 2 \times 0.4279 = 0.8558$, which implies that we reject the null hypothesis at $\alpha = 0.05$.

Implementing Minitab F test for equality of two variances

Ref: Example 12.9

We will use two samples data points to perform equality of two population variances in Minitab using the following steps.

(1) In Minitab worksheet, we type x in C1 and y in C2.

(2) Select **Stat>Basic Statistics> 2 Variance…**
(3) Choose each sample is in its own column and specify sample names: Sample 1: y (with higher sample variance in the numerator), Sample 2: x (with smaller sample variance, in the denominator).
(4) Use Options to decide **Confidence level**: 95 (default), Hypothesized ratio: 1, Alternative hypothesis: ratio>hypothesized ratio.
(5) Check mark **Use test and confidence intervals based on normal distribution**.
(6) Click OK two times.

Minitab dialog box with selection options and results from aforementioned steps are given below:

↓	C1	C2	C3
	x	y	
1	785	780	
2	550	610	
3	710	690	
4	415	300	
5	600	500	
6	500	450	
7	720	560	
8		460	
9		437	

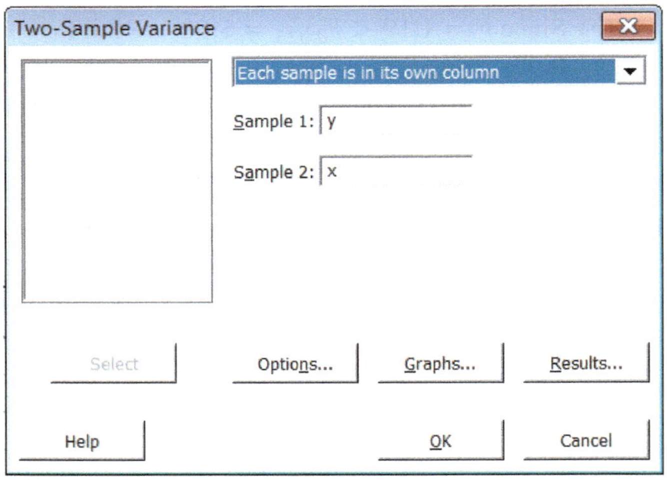

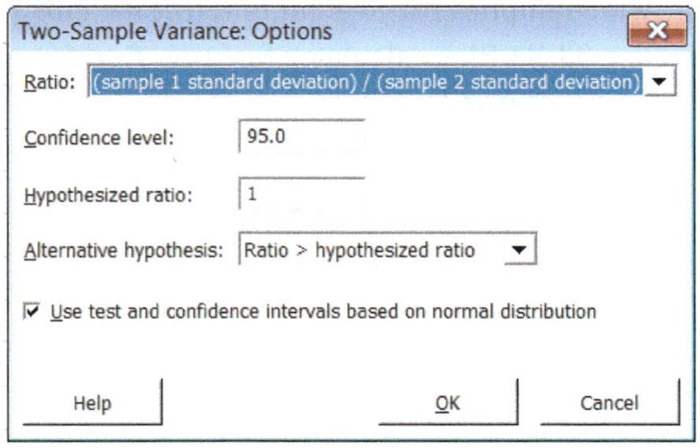


```
Variable  N    StDev   Variance
Y         9  145.395   21139.61
X         7  133.251   17755.95

                          Test
 Method  DF1  DF2  Statistic  P-Value
F          8    6       1.19    0.428

Fail to reject the null hypothesis
since p-value>0.05.
```

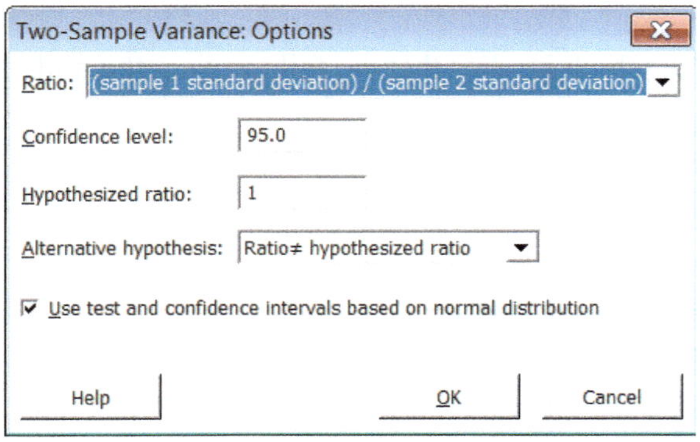


```
Variable  N    StDev   Variance
Y         9  145.395   21139.61
X         7  133.251   17755.95

                          Test
Method  DF1  DF2  Statistic  P-Value
F         8    6       1.19    0.855

Fail to reject the null hypothesis
since p-value>0.05.
```

Chapter 12: Practice problems

(12.1) Explain the meaning of (a) χ^2_α (b) $\chi^2_{1-\alpha}$.
(12.2) With degrees of freedom $df = 8$ and $\alpha = 0.05$, find the value of (a) χ^2_α (b) $\chi^2_{1-\alpha}$.
(12.3) With numerator degrees of freedom $df_1 = 5$ and denominator degrees of freedom $df_2 = 10$, find the value of each of the followings for specified α:

(a) F_α at $\alpha = 0.05$
(b) F_α at $\alpha = 0.10$
(c) F_α at $\alpha = 0.025$
(d) F_α at $\alpha = 0.01$

(12.4) We want to test $H_0: \sigma^2 = \sigma_0^2$ against $H_a: \sigma^2 > \sigma_0^2$.

(a) What is the meaning of σ_0^2 for the test above?
(b) What is the distributional assumption to test $H_0: \sigma^2 = \sigma_0^2$ against $H_a: \sigma^2 > \sigma_0^2$?

(12.5) We want to test $H_0: \sigma^2 = \sigma_0^2$ at significance level α against

(a) $H_a: \sigma^2 > \sigma_0^2$
(b) $H_a: \sigma^2 < \sigma_0^2$
(c) $H_a: \sigma^2 \neq \sigma_0^2$

The test statistic for this test under H_0 is given by $\chi^2 = \frac{(n-1)s^2}{\sigma_0^2}$ which has a chi-squared distribution with $df = (n-1)$. Under the usual notations, chi-square points χ^2_α, $\chi^2_{1-\alpha}$, $\chi^2_{\alpha/2}$ and $\chi^2_{1-(\alpha/2)}$ are critical values for testing $H_0: \sigma^2 = \sigma_0^2$ depending on the type of alternative hypothesis. Using the critical value approach, at significance level α

(a) Reject $H_0: \sigma^2 = \sigma_0^2$ against $H_a: \sigma^2 > \sigma_0^2$ if ______________________.
(b) Reject $H_0: \sigma^2 = \sigma_0^2$ against $H_a: \sigma^2 < \sigma_0^2$ if ______________________.
(c) Reject $H_0: \sigma^2 = \sigma_0^2$ against $H_a: \sigma^2 \neq \sigma_0^2$ if ______________________.

(12.6) Let df_1 and df_2 be the numerator and denominator degrees of freedom, respectively, for an F distribution. Use Tables T6-T9 in the Appendix, as appropriate, to find $F_{0.10}$ using:

(a) $df_1 = 5$ and $df_2 = 10$.
(b) $df_1 = 10$ and $df_2 = 5$.
(c) $df_1 = 5$ and $df_2 = 7$.
(d) $df_1 = 15$ and $df_2 = 20$.

(12.7) Let df_1 and df_2 be the numerator and denominator degrees of freedom, respectively, for an F distribution. Use Tables T6-T9 in the Appendix, as appropriate, to find $F_{0.05}$ using:

(a) $df_1 = 5$ and $df_2 = 10$.
(b) $df_1 = 10$ and $df_2 = 5$.
(c) $df_1 = 5$ and $df_2 = 7$.
(d) $df_1 = 15$ and $df_2 = 20$.

(12.8) Let df_1 and df_2 be the numerator and denominator degrees of freedom, respectively, for an F distribution. Use Tables T6-T9, as appropriate, to find the followings:

(a) $F_{0.10}$ with $df_1 = 5$ and $df_2 = 10$.
(b) $F_{0.10}$ with $df_1 = 10$ and $df_2 = 5$.
(c) $F_{0.05}$ with $df_1 = 5$ and $df_2 = 7$.

(d) $F_{0.01}$ with $df_1 = 15$ and $df_2 = 20$.
(e) $F_{0.025}$ with $df_1 = 15$ and $df_2 = 20$.

(12.9) A sample of size $n = 8$ of sugar content (in grams) per serving size of various General Mills' cereals are considered from a population of list of 78 General Mills' cereals reported in http://cereal.findthebest.com/. The sample observations are as follows:

8 7 16 9 1 10 9 9

It reveals that the sample variance $s^2 = 16.84$. Assume that sugar content in the population of General Mills' cereals follows a normal distribution.
Answer the followings:

(a) What is the point estimate of population variance σ^2?
(b) Find a 95% confidence interval estimate of σ^2.
(c) Find a 95% confidence interval estimate of σ.

(12.10) Reference to the Exercise (12.9), we wish to test $H_0\colon \sigma^2 = 15$ against $H_a\colon \sigma^2 > 15$ at $\alpha = 0.05$. To this end, answer the following questions:

(a) Compute the value of the test statistic.
(b) Find critical value of the test at $\alpha = 0.05$ and make a conclusion about the test.
(c) Use any technology to compute the p-value and make a conclusion about the test using the p - value approach.

(12.11) Reference to the Exercise (12.9), we wish to test $H_0\colon \sigma^2 = 20$ against $H_a\colon \sigma^2 < 20$ at $\alpha = 0.05$. To this end, answer the following questions:

(a) Compute the value of the test statistic.
(b) Find critical value of the test at $\alpha = 0.05$ and make a conclusion about the test.
(c) Use any technology to compute the p-value and make a conclusion about the test using the p - value approach.

(12.12) Reference to the Exercise (12.9), we wish to test $H_0\colon \sigma^2 = 16$ against $H_a\colon \sigma^2 \neq 16$ at $\alpha = 0.05$. To this end, answer the following questions:

(a) Compute the value of the test statistic.
(b) Find critical value of the test at $\alpha = 0.05$and make a conclusion about the test.
(c) Use any technology to compute the p-value and make a conclusion about the test using the p - value approach.

(12.13) The amount of carbohydrate (in grams) per serving size of various General Mills' cereals are considered from a list of 78 General Mills' cereals (see http://cereal.findthebest.com/). The amounts of carbohydrate (in grams) from a sample of 10 General Mills' cereals are as follows:

23 28 41 22 20 23 22 27 23 20

It reveals that the sample variance $s^2 = 38.77$. Assume that the amount of carbohydrate in the population of General Mills' cereals follows an approximate normal distribution.
Answer the followings:

(a) What is the value of the point estimate of population variance σ^2.
(b) Find a 90% confidence interval estimate of σ^2.
(c) Find a 90% confidence interval estimate of σ.

(12.14) Reference to the Exercise (12.13), we wish to test $H_0\colon \sigma^2 = 35$ against $H_a\colon \sigma^2 > 35$ at $\alpha = 0.10$. To this end, answer the following questions:

(d) Compute the value of the test statistic.
(e) Find critical value of the test at $\alpha = 0.10$ and make a conclusion about the test.
(f) Use any technology to compute the p-value and make a conclusion about the test using the p-value approach.

(12.15) Reference to the Exercise (12.13), we wish to test $H_0: \sigma^2 = 40$ against $H_a: \sigma^2 < 40$ at $\alpha = 0.05$. To this end, answer the following questions:
(d) Compute the value of the test statistic.
(e) Find critical value of the test at $\alpha = 0.10$ and make a conclusion about the test.
(f) Use any technology to compute the p-value and make a conclusion about the test using the p-value approach.

(12.16) Reference to the Exercise (12.13), we wish to test $H_0: \sigma^2 = 38$ against $H_a: \sigma^2 \neq 38$ at $\alpha = 0.05$. To this end, answer the following questions:
(d) Compute the value of the test statistic.
(e) Find critical value of the test at $\alpha = 0.10$ and make a conclusion about the test.
(f) Use any technology to compute the p-value and make a conclusion about the test using the p-value approach.

(12.17) Suppose we have two samples, one of size 10 and the other of size 12, of calorie count from different serving size of fast food. Two samples are generated using data of serving size and calorie available at http://www.acaloriecounter.com/fast-food.php. Two samples are as follows:

x	550	560	550	490	490	500	610	500	610	450		
y	780	610	690	300	500	450	560	460	437	557	550	490

Two sample variances turn out to be $s_1^2 = 2876.7$ and $s_2^2 = 15620.9$. Assume that the two Answer the followings:
(a) Test $H_0: \sigma_1^2 = \sigma_2^2$ against $H_a: \sigma_1^2 < \sigma_2^2$ at $\alpha = 0.10$.
(b) Test $H_0: \sigma_1^2 = \sigma_2^2$ against $H_a: \sigma_1^2 \neq \sigma_2^2$ at $\alpha = 0.10$.
(c) How does the test in (b) differ from the test $H_0: \frac{\sigma_1^2}{\sigma_2^2} = 1$ against $H_a: \frac{\sigma_1^2}{\sigma_2^2} \neq 1$?

(12.18) Suppose, given two samples, we have the following information:

$$n_1 = 15,\ s_1^2 = 64,\ n_2 = 10 \text{ and } s_2^2 = 49.$$

Assume that the two populations are approximately normal. Answer the followings:
(a) Test $H_0: \sigma_1^2 = \sigma_2^2$ against $H_a: \sigma_1^2 < \sigma_2^2$ at $\alpha = 0.05$.
(b) Test $H_0: \sigma_1^2 = \sigma_2^2$ against $H_a: \sigma_1^2 \neq \sigma_2^2$ at $\alpha = 0.05$.
(c) How does the test in (b) differ from the test $H_0: \frac{\sigma_1^2}{\sigma_2^2} = 1$ against $H_a: \frac{\sigma_1^2}{\sigma_2^2} \neq 1$?

Chapter 12: Answer to selected problems

(12.2) (a) 15.507 (b) 2.733; **(12.3)** (a) 3.33 (b) 2.52 (c) 4.24 (d) 5.64

(12.5) (a) $\chi^2 > \chi^2_{\alpha}$ (b) $\chi^2 < \chi^2_{1-\alpha}$ (c) $\chi^2 > \chi^2_{\alpha/2}$ or $\chi^2 < \chi^2_{1-\alpha/2}$.

(12.7) (a) 3.33 (b) 4.74 (c) 5.05 (d) 2.20; **(12.8)** (a) 2.52 (b) 3.30 (c) 3.97 (d) 3.09 (e) 2.57

(12.9) (a) 16.84 (b) The 95% CI of σ^2 is $\left[\frac{7\times s^2}{\chi^2_{0.025}}, \frac{7\times s^2}{\chi^2_{0.975}}\right] = [7.4, 89.8]$ (c) The 95% CI of σ is $\left[\sqrt{7.4}, \sqrt{89.8}\right] = [2.72, 8.4]$

(12.10) We have: $n = 8$, $s^2 = 16.84$ and $\sigma_0^2 = 15$.

(a) Then, $\chi^2 = \frac{(n-1)s^2}{\sigma_0^2} = \frac{(8-1)\times 16.84}{15} = 7.9$

(b) With $df = n - 1 = 8 - 1 = 7$ and $\alpha = 0.05$, $\chi^2_{\alpha} = \chi^2_{0.05}$ =14.067. Since $\chi^2 = 7.9 < \chi^2_{0.05} =$ 14.067, we fail to reject H_0 at $\alpha = 0.05$.

(c) p-value= $P(\chi^2 > 7.9)$ =0.443. Since p-value> α, we fail to reject H_0 at $\alpha = 0.05$.

(12.11) We have: $n = 8$, $s^2 = 16.84$ and $\sigma_0^2 = 20$.

(a) Then, $\chi^2 = \frac{(n-1)s^2}{\sigma_0^2} = \frac{(8-1)\times 16.84}{20} = 5.9$

(b) With $df = 7$ and $\alpha = 0.05$, $\chi^2_{1-\alpha} = \chi^2_{0.95}$ =2.167. Since $\chi^2 = 7.9 > \chi^2_{0.95} = 2.167$, we fail to reject H_0 at $\alpha = 0.05$.

(c) p-value= $P(\chi^2 < 5.9)$ =0.3416. Since p-value> α, we fail to reject H_0 at $\alpha = 0.05$.

(12.12) We have: $n = 8$, $s^2 = 16.84$ and $\sigma_0^2 = 16$.

(a) Then, $\chi^2 = \frac{(n-1)s^2}{\sigma_0^2} = \frac{(8-1)\times 16.84}{16} = 7.4$

(b) With $df = 7$ and $\alpha = 0.05$, $\chi^2_{\alpha/2} = \chi^2_{0.025} = 16.013$. Since $\chi^2 = 7.4 < \chi^2_{0.025} = 16.013$, we fail to eject H_0 at $\alpha = 0.05$.

(c) p-value= $2 \times min\{P(\chi^2 > 7.4), P(\chi^2 < 7.4)\} = 2 \times min\{0.4942, 0.5058\} = 0.9884$. Since p-value> α, we fail to reject H_0 at $\alpha = 0.05$.

(12.13)-(12.16) See steps similar to (12.9)-(12.12), respectively.

(12.17) $n_1 = 10$, $s_1^2 = 2876.7$, $n_2 = 12$ and $s_2^2 = 15620.9$. $F = \frac{15620.9}{2876.7} = 5.4$. df_1=11, df_2= 9.

(a) At $\alpha = 0.10$, the critical value is $F_{0.10} = 2.40$. Reject H_0 at $\alpha = 0.10$.

(b) At $\alpha = 0.10$, the critical value for two-tailed test is $F_{0.05} = 3.10$. Reject H_0 at $\alpha = 0.10$.

CHAPTER 13
CHI-SQUARE TESTS

Introduction

In this chapter, we devote to Chi-square tests for

(a) Independence of two categorical variables,
(b) Goodness-of-fit of an observed frequency distribution to a probability specification, and
(c) Homogeneity of k distinct proportions of outcome.

We will provide applications and technology implementation using Minitab and TI 84 plus calculator.

Section 13.1: A Chi-Square Test for Independence

Two categorical variables are often cross-classified by different categories in the form of a two-way table, where rows represent categories of one variable and columns represent categories of the other variable. Such a table is called a ***contingency table***. If the table contains R rows and C columns, it is called an $R \times C$ (read as R by C) contingency table or simply an $R \times C$ table.

Example 13.1

Suppose, on the basis of a sample, we have the following summary table of tax reform opinion by gender.

	Tax reform opinion			
Gender	Yes	No	Undecided	Total
Male	25	14	5	44
Female	27	13	6	46
Total	52	27	11	90

In the table above, two rows for gender and three columns for tax reform opinion are forming a 2×3 contingency table, with $R = 2$ and $C = 3$.

Given a contingency table, our objective is to test if one categorical variable is independent of the other variable. For contingency table in Example 13.1, we wish to test if gender and tax reform opinion are independent of each other. In other words, the tax reform opinion is gender dependent. Formally, we specify null and alternative hypotheses for this example as follows:

$$H_0: Tax\ reform\ opinion\ is\ independent\ of\ gender$$

versus

$$H_a: Tax\ reform\ opinion\ is\ not\ independent\ of\ gender.$$

In general, we want to test

$$H_0\text{: Row and column variables are independent}$$

versus

$$H_a\text{: Row and column variables are not independent.}$$

In the process of the test, we basically compare the observed cell frequencies with the expected cell frequencies using a Chi-square test statistic. Before we proceed any further, let us define some terminologies that we will use for carrying out the test.

Definitions

An $R \times C$ (read as R by C) ***contingency table*** is a summary of two categorical variables in a table with R rows representing categories of one variable and C columns representing categories of the other variable.
The ***observed frequency*** of a cell is the number of responses in the cell from a sample of subjects.
The ***expected frequency*** of a cell is the predicted frequency computed using observed frequencies under the assumption of independence of two categorical variables.

For computational simplicities, a 2×3 contingency table such as in Example 13.1, can be presented in the following form:

Variable 1	Variable 2			Total
	Label 1	Label 2	Label 3	
Label 1	O_{11}	O_{12}	O_{13}	R_1
Label 2	O_{21}	O_{22}	O_{23}	R_2
Total	C_1	C_2	C_3	N

Let i be the row suffix and j be the column suffix of the table. Then, in the table above
$O_{i,j}$=Observed frequency in (i, j)th cell.
R_i= ith row total
C_j= jth column total
N= Grand total

In reference to Example 13.1, we have:

$$O_{1,1} = 25, O_{1,2} = 14, O_{1,3} = 5;\ O_{2,1} = 27, O_{2,2} = 13, O_{2,3} = 6$$
$$R_1 = 44, R_2 = 46;\ C_1 = 52, C_2 = 27, C_3 = 11; N = 90$$

Now, we need to compute the expected frequency of (i, j)th cell, $E_{i,j}$, so as to compare it with $O_{i,j}$. How to compute the expected cell frequency $E_{i,j}$?

By the independence of two categorical variables, the expected frequency of (i, j)th cell is computed using the formula

$$E_{i,j} = \frac{R_i \times C_j}{N},$$

$$i.e., \textit{expected frequency of } (i,j)\textit{th cell} = \frac{i\textit{th row total} \times j\textit{th column total}}{\textit{Grand total}}$$

Using this formula, we obtain:

$$E_{1,1} = \frac{R_1 \times C_1}{N} = \frac{44 \times 52}{90} = 25.42$$

$$E_{1,2} = \frac{R_1 \times C_2}{N} = \frac{44 \times 27}{90} = 13.20$$

$$E_{1,3} = \frac{R_1 \times C_3}{N} = \frac{44 \times 11}{90} = 5.38$$

Similarly, other expected frequencies can easily be computed.

Chi-square Test Statistic and Decision Rule

Now that we are familiar with the observed and expected cell frequencies, we wish to compare them to decide about the rejection or acceptance of the null hypothesis of independence of two categorical variables. As usual, we need a test statistic, and then compare the value of the test statistic with a known standard.

Test Statistic

Under the null hypothesis of independence of two categorical variables, the test statistic for the test is given by

$$\chi^2_{obs} = \sum_{Sum:\, all\ i,j} \frac{\left(O_{i,j} - E_{i,j}\right)^2}{E_{i,j}}$$

The test statistic χ^2_{obs} follows a chi-square distribution with $(R-1) \times (C-1)$ degrees of freedom, where R is the total number of rows and C is the total number of columns of the contingency table.

Decision rule using a critical value approach

The larger difference between observed and expected frequencies provides the evidence in the favor of the alternative hypothesis. Therefore, we reject the null hypothesis if the value of the χ^2_{obs} test statistic is large. This is tantamount to saying that reject the null hypothesis if $\chi^2_{obs} > \chi^2_{\alpha}$ at level of significance α. Here, χ^2_{obs} is the observed value of the Chi-square test statistic for a given contingency table and χ^2_{α} is the value of a Chi-square distribution with $df = (R-1)(C-1)$ with area α to its right. χ^2_{α} is also known as critical value of the test.

Example 13.2

In reference to Example 13.1, we want to test if tax reform opinion is independent of gender. Answer the following questions about the test:

(a) Set up the null and alternative hypothesis about the test.
(b) Find the expected frequency for each cell.
(c) Find the value of the Chi-square test statistic χ^2_{obs}.
(d) Determine the degrees of freedom of the Chi-square test statistic.
(e) Make a conclusion about the test at 5% level of significance.

Solution

Gender	Tax reform opinion Yes	No	Undecided	Total
Male	25 (**25.42**)	14 (**13.20**)	5 (**5.38**)	44
Female	27 (**26.58**)	13 (**13.80**)	6 (**5.62**)	46
Total	52	27	11	90

(a) H_0: *Tax reform opinion is independent of gender* versus H_a: *Tax reform opinion is not independent of gender.*

(b) Using the formula $E_{i,j} = \frac{R_i \times C_j}{N}$, we get $E_{1,1} = \frac{R_1 \times C_1}{N} = \frac{44 \times 52}{90} = 25.42$, $E_{1,2} = \frac{R_1 \times C_2}{N} = \frac{44 \times 27}{90} = 13.20$, and $E_{1,3} = \frac{R_1 \times C_3}{N} = \frac{44 \times 11}{90} = 5.38$. The other expected frequencies are calculated in a similar way and are written in bold in parenthesis in the table.

(c) The observed value of the test statistic reference to the given contingency table is

$$\chi^2_{obs} = \frac{(25-25.42)^2}{25.42} + \frac{(14-13.20)^2}{13.20} + \frac{(5-5.38)^2}{5.38} + \frac{(27-26.58)^2}{26.58} + \frac{(13-13.80)^2}{13.80} + \frac{(6-5.62)^2}{5.62} = 0.161$$

(d) The degrees of freedom of the Chi-square test statistic, $df = (2-1) \times (3-1) = 2$.

(e) With $df = 2$ and $\alpha = 0.05$, $\chi^2_\alpha = \chi^2_{0.05} = 5.991$. Since $\chi^2_{obs} = 0.161 < \chi^2_{0.05} = 5.991$, we failed to reject the null hypothesis at 5% level of significance. This means, at 5% level of significance, the tax reform opinion is independent of gender.

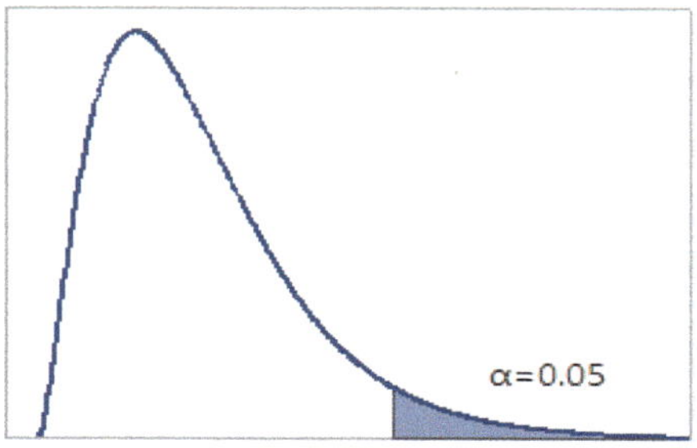

$\chi^2_\alpha = 5.991$

df	$\chi^2_{.10}$	$\chi^2_{.05}$	$\chi^2_{.01}$	$\chi^2_{.005}$
1	*2.706*	*3.841*	*6.635*	*7.879*
2	*4.605*	*5.991*	*9.210*	*10.597*
3	*6.251*	*7.815*	*11.345*	*12.838*
4	*7.779*	*9.488*	*13.277*	*14.860*
5	*9.236*	*11.070*	*15.086*	*16.750*
6	*10.645*	*12.592*	*16.812*	*18.548*

Decision rule using a *p*-value approach

Note that, the way the Chi-square test statistic is defined, its value will always be positive. We reject the null hypothesis if the value of the χ^2_{obs} is large. In other words, reject the null hypothesis if the value of the χ^2_{obs} is far from zero in the positive direction (i.e., far right of zero), leaving a smaller area to the right of χ^2_{obs}. For a Chi-square test of independence, p-value is the area to the right of χ^2_{obs} under a chi-square distribution. Therefore, smaller p-value is the indication of the larger χ^2_{obs} value, supporting the alternative hypothesis.

In summary, using a p-value approach, reject the null hypothesis if the p-value defined by

$$p - value = P(\chi^2 > \chi^2_{obs}) < \alpha,$$

where α is the desired level of significance.

Following the definition, p-value for Example 13.2 is computed by

$$p - value = P(\chi^2 > 0.161) = 0.923$$

One can verify above computation in TI-84 graphing calculator using 8: $\chi^2 cdf(\chi^2_{obs}, E99, df)$:

$$\chi^2 cdf(0.161, E99, 2) = 0.923$$

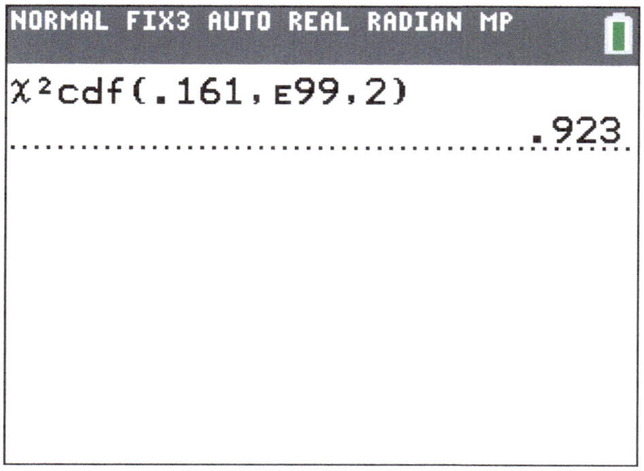

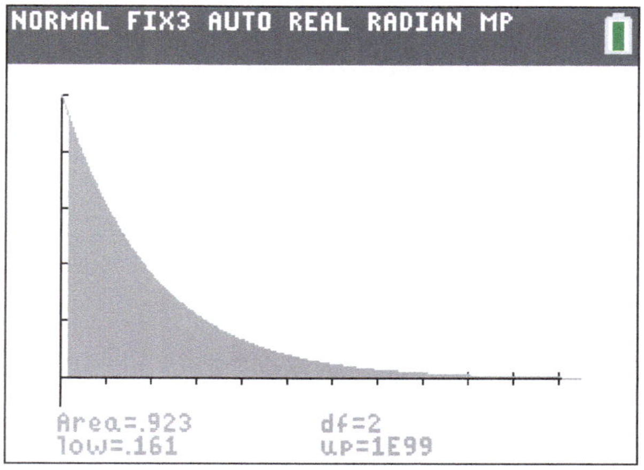

Note that p-value> $\alpha = 0.05$. Therefore, we fail to reject the null hypothesis at 5% level of significance, an identical conclusion we have made using the critical value approach.

Assumptions for the validity of the Chi-square test of independence

(1) The contingency table is formed from a sample of observed data values, and
(2) The expected frequency of each cell is at least 5; otherwise, combine two adjacent rows or columns so the expected frequency of each cell becomes at least 5.

What follows from assumption (2) is that after computing expected cell frequencies, it is important that we verify whether the expected cell frequency of each cell is at least five. If expected frequency of any cell is less than 5, we need to combine two adjacent rows or columns so as to get the expected frequency of 5 or higher and then adjust the degrees of freedom for the table accordingly.

Implementing Technology for a Chi-square Test of Independence

Below we provide instruction for performing a Chi-square test of independence using Minitab and TI 84 plus Calculator.

Minitab Instruction for a Chi-square Test of Independence

Below are Minitab steps to carry out a Chi-square test of independence (***ref: Example 13.2***).

(1) Type Gender in C1, and Tax reform opinion: Yes, No and Undecided in C2, C3 and C4, respectively.
(2) Select **Stats>Tables>Chi-Squared Test for Association**…
(3) Choose **Summarized data in a two-way table option**.
(4) For **Columns containing the table**: select C2, C3 and C4 (Yes-Undecided).
(5) For **Row Labels:** type Gender and **Column Labels:** type Tax reform opinion.
(6) Use **Statistics…** option to check mark desired options and click [OK] two times.

See Minitab dialog box option selections and output from the Minitab session window in Figure 13.1 (a)-(d).

TI 84 Plus Instruction for a Chi-square Test of Independence

(1) Press $2^{ND}[x^{-1}]$ to select MATRIX and move the cursor to EDIT and press ENTER to select 1: [A]
(2) Specify the matrix dimension to be 2×3 and press ENTER.
(3) Enter each observed frequency in the matrix as appear in the contingency table.
(4) Press STAT, move the cursor to TESTS, and then select $C{:}\chi^2 - Test \ldots$ (using ALPHA PRGM [C]) and Press ENTER.
(5) Make sure to select **Observed: [A]** and **Expected: [B]**, and then move the cursor to Calculate and press ENTER.

Below are TI 84 Plus steps to perform a Chi-square test of independence (***ref: Example 13.2***).

See the option selections and the output for above steps in Figure 13.2 (a)-(e).

Figure 13.1: Minitab dialog box option selections and output from the Minitab session window.

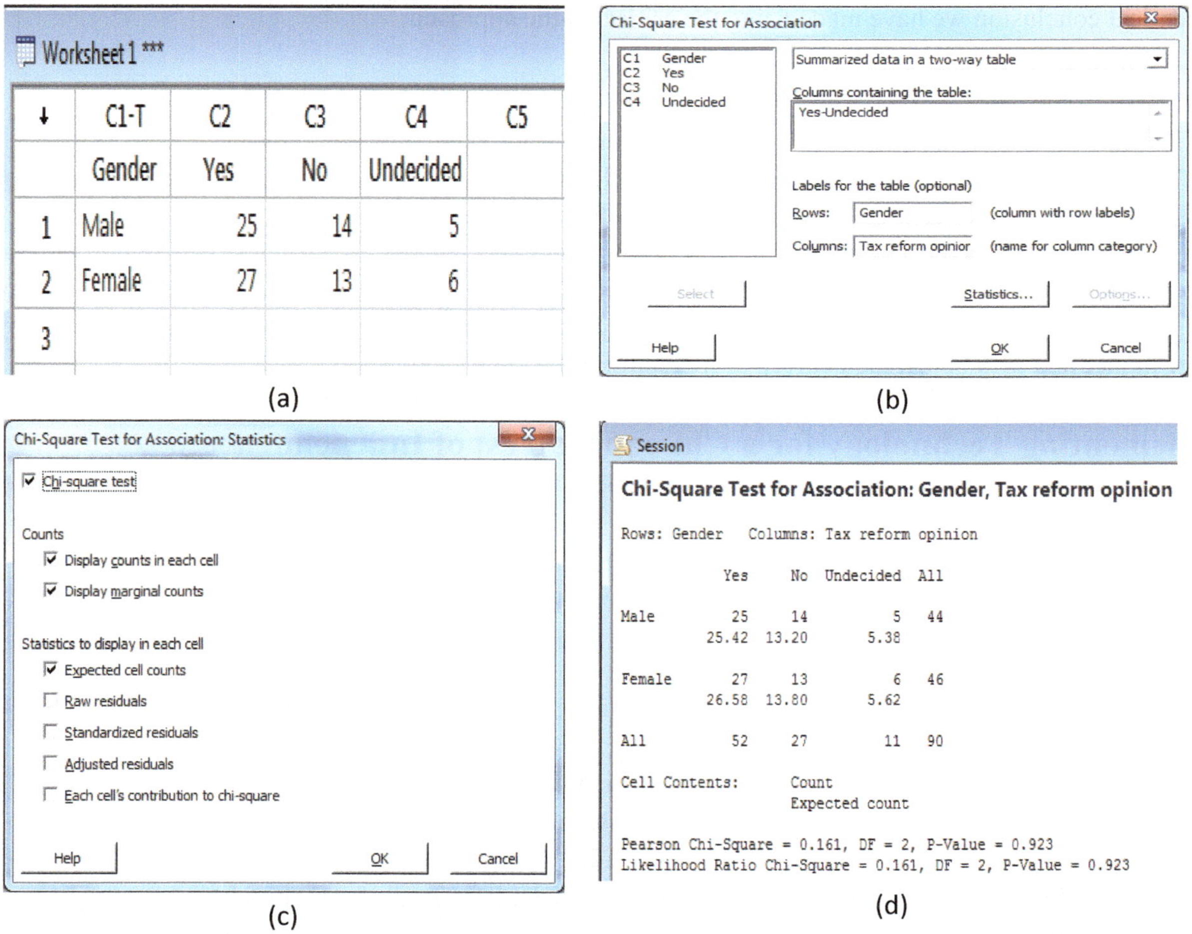

Figure 13.2: TI 84 plus option selections and output for Example 13.2.

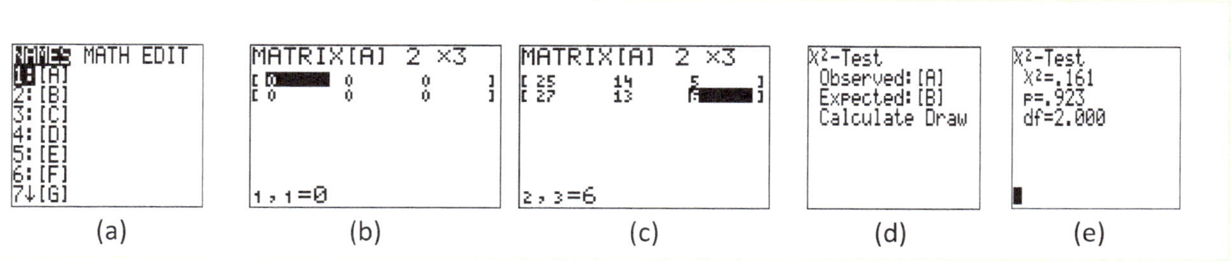

Note that the p-value for the test is $0.923 > \alpha = 0.05$. Therefore, we fail to reject the null hypothesis at 5% level of significance, the same conclusion we have made using the critical value approach.

Section 13.2: A Chi-Square Goodness of Fit Test

Often, researchers wish to test if the observed frequency distribution of a categorical variable is subject to a given probability specification. For example, suppose an item may fall into any of 5 possible categories and we want to test if five categories are subject to different probability specifications.

Example 13.3

Suppose an item is produced by any of the 5 different machines. Given a defective item, the producer believes that it is produced by machines 1, 2, 3, 4 and 5 with respective probabilities 0.25, 0.25, 0.20, 0.20 and 0.10.

The producer wants to test the hypothesis

$$H_0: p_1 = 0.25, \quad p_2 = 0.25, \quad p_3 = 0.20, \quad p_4 = 0.20, \quad p_5 = 0.10 \; versus$$
$$H_a: at\ least\ one\ of\ the\ above\ specifications\ is\ untrue$$

This is an example of a **goodness-of-fit test** because the producer wants to test if the proportions of defective from different machines are subject to different probability specifications.

In order to perform the test, we compare observed frequency from a sample to the expected frequency under the assumption of the null hypothesis by means of a Chi-square test statistic.

Definition: Goodness-of-fit test

The ***goodness-of-fit test*** (GOF) refers to a Chi-square test to verify if an observed frequency distribution conforms to a specified probability distribution.

Null and alternative hypotheses for this test are specified as

$$H_0: c\ distinct\ categories\ have\ probabilities\ p_1, p_2, \dots, p_c$$
$$H_a: at\ least\ one\ probability\ as\ specified\ in\ H_0\ is\ untrue$$

Expected frequency

Given a sample of size n, and c probabilities $p_1, p_2, \dots, p_c$ specified in the null hypothesis, the ***expected frequencies*** for c categories are computed by

$$E_1 = n \times p_1, E_2 = n \times p_2, \ldots, E_c = n \times p_c$$

Test statistic for GOF

The test statistic for a GOF test is the **Chi-square statistic** defined by

$$\chi^2 = \sum_{i=1}^{c} \frac{(O_i - E_i)^2}{E_i}$$

which follows a Chi-square distribution with $(c-1)$ degrees of freedom. Given a category i, $i = 1,2,\ldots,c$, O_i and E_i are observed and expected frequency of category i, respectively. Given a sample, an observed value of the χ^2 is denoted by χ^2_{obs}.

Decision rule

Reject the null hypothesis at α if $\chi^2_{obs} > \chi^2_{\alpha}$, and otherwise, fail to reject the null hypothesis. Here, χ^2_{α} is the critical value of a Chi-square distribution with $df = (c-1)$ with area α to its right.

Example 13.4

In reference to Example 13.3, the producer wants to test the hypothesis

$$H_0: p_1 = 0.25,\ p_2 = 0.25,\ p_3 = 0.20,\ p_4 = 0.20,\ p_5 = 0.10 \text{ versus}$$

$$H_a: \textit{at least one of the above specifications is untrue.}$$

Suppose, on the basis of a sample of 50 defective items produced during a specified time, the following frequency distribution is observed.

Machine	1	2	3	4	5
Observed frequency	9	10	8	11	12

Answer the following questions:

(a) Write observed frequency using the O_i notation.
(b) Find the expected frequency using the E_i notation.
(c) Find the value of the Chi-square test statistic and the degrees of freedom to test the null hypothesis.
(d) Make your conclusion about the test at 5% level of significance.

Solution

(a) Let O_i, $i = 1,2,\ldots,5$, be the observed frequency of defectives from machine i. Then, $O_1 = 9$, $O_2 = 10$, $O_3 = 8$, $O_4 = 11$ and $O_5 = 12$.

(b) Under the null hypothesis, the expected frequency of defectives from machine i is

$$E_i = np_i = 50 \times p_i$$

Then, $E_1 = 50 \times 0.25 = 12.5$, $E_2 = np_2 = 50 \times 0.25 = 12.5$, and similarly, $E_3 = 10, E_4 = 10$ and $E_5 = 5$.

(c) The test statistic $\chi^2 = \sum_{i=1}^{5} \frac{(O_i - E_i)^2}{E_i}$ follows a Chi-square distribution with $df = 5 - 1 = 4$. The observed value of the test statistic is $\chi^2_{obs} = \frac{(9-12.5)^2}{12.5} + \frac{(10-12.5)^2}{12.5} + \frac{(8-10)^2}{10} + \frac{(11-10)^2}{10} + \frac{(12-5)^2}{5} = 11.78$.

(d) With $df = 4$ and $\alpha = 0.05$, $\chi^2_\alpha = \chi^2_{0.05}$=9.488. Since $\chi^2_{obs} = 11.78 > \chi^2_{0.05} = 9.488$, we reject the null hypothesis at 5% level of significance.

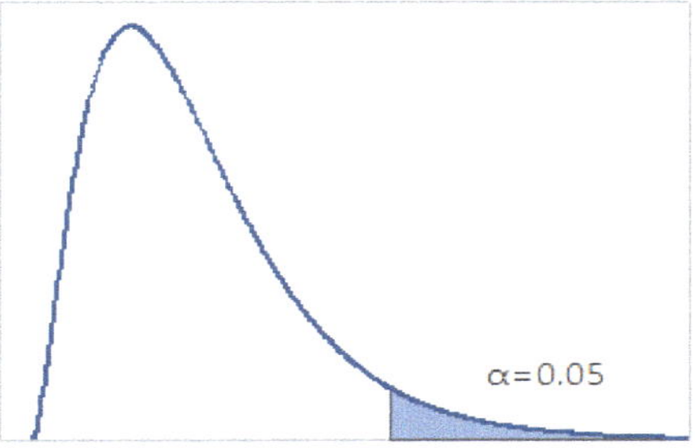

$\chi^2_\alpha = 9.488$

df	$\chi^2_{.10}$	$\chi^2_{.05}$	$\chi^2_{.01}$	$\chi^2_{.005}$
1	*2.706*	*3.841*	*6.635*	*7.879*
2	*4.605*	*5.991*	*9.210*	*10.597*
3	*6.251*	*7.815*	*11.345*	*12.838*
4	*7.779*	*9.488*	*13.277*	*14.860*
5	*9.236*	*11.070*	*15.086*	*16.750*
6	*10.645*	*12.592*	*16.812*	*18.548*

Implementing Technology for a Chi-square Goodness-of-fit

Below we perform a Chi-square goodness-of-fit using Minitab and TI 84 plus Calculator.

Minitab Instruction for a Chi-square Test of Goodness-of-fit

In reference to **Example 13.4**, we can implement a Chi-square test of goodness-of-fit in Minitab using the following steps.

(1) Type Machine type in C1, Observed frequency in C2 and specified proportions in C3. We label them as Machine, Count and p, respectively.
(2) Select **Stats>Tables>Chi-Square Goodness-of-fit Test (One Variable)...**
(3) For **Observed counts:** select Count, click **Specific proportions** and select p.

See the dialog box with the option selections and output from the Minitab session window in Figure 13.3.

Figure 13.3: Minitab dialog option selections and output in the Minitab session window for Example 13.4

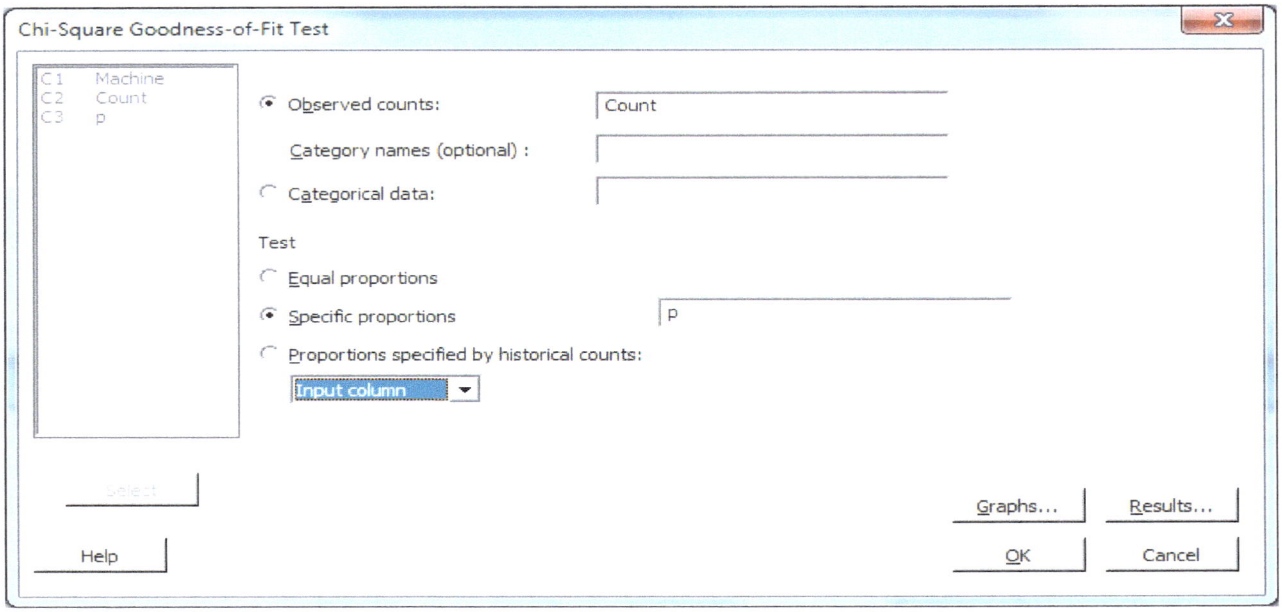

Chi-Square Goodness-of-Fit Test for Observed Counts in Variable: Count

Category	Observed	Test Proportion	Expected	Contribution to Chi-Sq
1	9	0.25	12.5	0.98
2	10	0.25	12.5	0.50
3	8	0.20	10.0	0.40
4	11	0.20	10.0	0.10
5	12	0.10	5.0	9.80

N	DF	Chi-Sq	P-Value
50	4	11.78	0.019

Note that the p-value for the test is $0.019 < \alpha = 0.05$. Therefore, we fail to reject the null hypothesis at 5% level of significance. We have reached the same conclusion using the critical value approach.

TI 84 Plus Instruction for a Chi-square Goodness-of-fit Test

In reference to **Example 13.4**, we can implement the Chi-square test of goodness-of-fit in TI 84 plus calculator by undertaking the following steps.

(1) Enter Observed count in L1 and Expected count in L2.

(2) Press STAT, move the cursor to TESTS and select $D\colon \chi^2 GOF - Test$ ... (by using ALPHA x^{-1} [D]) and Press ENTER.

(3) Be sure to select **Observed: L1**, **Expected: L2**, and df: 4.

(4) Move cursor to **Calculate** and press ENTER.

See the option selections and output from the steps above in Figure 13.4 (a)-(d).

Figure 13.4: TI 84 plus option selections and output for Example 13.4.

(a) (b) (c) (d)

Section 13.3: A Chi-Square Test for Homogeneity of Proportions

In section 13.2, we test whether c categories conform to desired specifications. Often, we want to test if all c possible categories are equally likely. In other words, we test the equality of all c proportions. For example, in reference to example 13.3, suppose we want to test

$$H_0\colon p_1 = p_2 = p_3 = p_4 = p_5 = 0.20 \; versus$$

$$H_a: at\ least\ one\ of\ the\ equalities\ does\ not\ hold$$

This is an example of a test of homogeneity of proportions, and is a special form of goodness-of-fit test.

Definition: Test of homogeneity proportions

A ***test of homogeneity of proportions*** refers to the test that c proportions are identical via a Chi-square test. The test can be carried out under the following three cases:

(1) To test equality of c proportions of a distribution
(2) To test equality of c binomial proportions
(3) To test equality of c proportions for each of r rows in an $r \times c$ table with predetermined row or column totals.

Below, we wish to briefly discuss all three cases, along with corresponding hypothesis and test:

Case I: Test of equality of c proportions of a distribution

In a multinomial distribution, we wish to test if c distinct categories are equally likely, i.e., test

$$H_0: p_1 = p_2 = \cdots = p_c\ versus$$
$$H_a: at\ least\ one\ of\ the\ equalities\ does\ not\ hold$$

Test statistic

The test statistic for the test is given by

$$\chi^2 = \sum_{i=1}^{c} \frac{(O_i - E_i)^2}{E_i}$$

where,
$i = 1,2, \dots, c$
c is the number of categories,
O_i is the observed frequency of ith category,
E_i is the expected frequency of ith category computed by

$$E_i = n \times p_i$$

The statistic χ^2 follows a Chi-square distribution with $df = c - 1$.

Case II: Test of equality of c binomial proportions

We wish to test if c independent binomial proportions are identical, i.e., test

$$H_0: p_1 = p_2 = \cdots = p_c\ versus$$
$$H_a: c\ binomial\ proportions\ are\ not\ all\ equal$$

To carry out the test, we require c samples of sizes $n_1, n_2, \dots, n_c$ from c binomial populations arranged in a $2 \times c$ contingency table as follows:

Samples	1	2	...	c	Total
Successes	x_1	x_2	...	x_c	r_1
Failures	$n_1 - x_1$	$n_2 - x_2$	...	$n_c - x_c$	r_2
Total	n_1	n_2	...	n_c	N

In the table above,
$i = 1, 2; j = 1, 2, \dots, c$.
x_j and $n_j - x_j$ are the number of successes and failures, respectively, in sample j of size n_j.

r_1 and r_2 are total number of successes and failures in c samples, respectively.
N is the grand total in c samples.

Test statistic

The test statistic for the test

$$H_0: p_1 = p_2 = \cdots = p_c \ versus$$
$$H_a: c \ binomial \ proportions \ are \ not \ all \ equal$$

is given by

$$\chi^2 = \sum\nolimits_{all\ i,j} \frac{\left(O_{i,j} - E_{i,j}\right)^2}{E_{i,j}}$$

where
$O_{i,j}$ is the observed frequency in (i,j)th cell,
$E_{i,j}$ is the expected frequency in (i,j)th cell, computed by $E_{i,j} = \frac{r_i \times n_j}{N}$
The statistic χ^2 follows a Chi-square distribution with $df = (2-1) \times (c-1) = (c-1)$.

Case III: Test of equality of c proportions in $r \times c$ table

In this case, we wish to test if c categories of a categorical variable conform in proportion, for each of r categories of another categorical variable with predetermined row and column totals. In other words, we wish to test

$$H_0: for\ each\ row\ category\ c\ column\ proportions\ are\ all\ equal$$
$$H_a: for\ at\ least\ one\ row\ category\ c\ column\ proportions\ are\ not\ all\ equal$$

To carry out the test, we require an $r \times c$ contingency table as follows:

$i \mid j$	1	2	...	c	Total
1	$O_{1,1}$	$O_{1,2}$	...	$O_{1,c}$	R_1
2	$O_{2,1}$	$O_{2,2}$	...	$O_{2,c}$	R_2
$\vdots$	$\vdots$	$\vdots$		$\vdots$	$\vdots$
r	$O_{r,1}$	$O_{r,2}$	...	$O_{r,c}$	R_r
Total	C_1	C_2	...	C_c	N

In the table above, for $i = 1, 2, \cdots, r; j = 1, 2, \cdots, c;$
$O_{i,j}$ is the observed frequency of (i,j)th cell,
R_i is the ith row total,
C_j is the jth column total and
N is the total count of observed frequency in the contingency table.

Test statistic

The test statistic for this test is defined by

$$\chi^2 = \sum\nolimits_{all\ i,j} \frac{\left(O_{i,j} - E_{i,j}\right)^2}{E_{i,j}}$$

where, $E_{i,j} = \frac{R_i \times C_j}{N}$ is the expected frequency of (i,j)th cell. The test statistic χ^2 follows a Chi-square distribution with $df = (r-1) \times (c-1)$.

Decision Rule

For each of the three cases, let the observed value of the test statistic χ^2 be χ^2_{obs}. Then, we reject H_0 at α if $\chi^2_{obs} > \chi^2_{\alpha}$, and otherwise, fail to reject the null hypothesis, where χ^2_{α} is the critical value of the Chi-square distribution.

Using p-value approach reject the null hypothesis if p-value$=P(\chi^2 > \chi^2_{obs}) < \alpha$.

Example 13.5

In reference to Example 13.4, suppose we wish to test

$$H_0: p_1 = p_2 = p_3 = p_4 = p_5 = 0.20 \ versus$$
$$H_a: at\ least\ one\ of\ the\ equalities\ does\ not\ hold$$

Perform the test above, at significance level $\alpha = 0.05$, using a sample of 50 defective items produced during a certain period of time with the following observed frequency distribution:

Machine	1	2	3	4	5
Observed frequency	9	10	8	11	12

Solution

We have, $O_1 = 9$, $O_2 = 10$, $O_3 = 8$, $O_4 = 11$ and $O_5 = 12$.

Under the null hypothesis, the expected frequencies of defective from all machines are the same, that is, $E_i = 50 \times 0.20 = 10$, for all $i = 1,2,\ldots,5$. Then, the observed value of the Chi-square statistic is

$$\chi^2_{obs} = \frac{(9-10)^2}{10} + \frac{(10-10)^2}{10} + \frac{(8-10)^2}{10} + \frac{(11-10)^2}{10} + \frac{(12-10)^2}{10} = 1$$

We have, $df = 5 - 1 = 4$; $\chi^2_{0.05}$=9.488. Since $\chi^2_{obs} = 1 < \chi^2_{0.05} = 9.488$, we fail to reject the null hypothesis at 5% level of significance.

Example 13.6

Three independent samples of 277 Non-Hispanic white (NH White), 176 Non-Hispanic black (NH Black) and 219 Hispanic people aged 12–15 years are considered.

We wish to test if the proportion of people with 2 hours or less of TV viewing and computer use, per day, in three populations are the same. The table below provides the frequency distribution of three samples with the status of TV viewing and computer use per day.

2 hours or less of TV and computer use per day	NH White	NH black	Hispanic	Total
Yes	81	35	58	174
No	196	141	161	498
Total	277	176	219	672

Sources: CDC/NCHS, National Health and Nutrition Examination Survey (NHANES) and NHANES National Youth Fitness Survey, 2012.
http://www.cdc.gov/nchs/data/databriefs/db157_table.pdf#3

Let p_1, p_2 and p_3 be the true proportions of Non-Hispanic white, Non-Hispanic black and Hispanic people aged 12–15 years with 2 hours or less of TV viewing and computer use per day, respectively. Perform a test of equality of proportions at 5% level of significance by answering the following questions:

(a) What do you call this test?
(b) Set up the null and alternative hypotheses for the test.
(c) Find the expected frequencies for all observed cell frequencies.
(d) Find the observed value of the test statistic.
(e) Find the df and the critical value.
(f) What is your conclusion about the test at 5% level of significance?

Solution

2 hours or less of TV and computer use per day	NH White	NH black	Hispanic	Total
Yes	81 (**71.72**)	35 (**45.57**)	58 (**56.71**)	174
No	196 (**205.28**)	141(**130.43**)	161 (**162.29**)	498
Total	277	176	219	672

(a) This test is referred to as the homogeneity of independent binomial proportions.
(b) $H_0: p_1 = p_2 = p_3 \ versus \ H_a: p_1, p_2 \ and \ p_3 \ are \ not \ all \ equal.$
(c) We compute the expected frequency using the formula $E_{i,j} = \frac{r_i \times n_j}{N}$. We have $E_{1,1} = \frac{174*277}{672} = 71.72$, $E_{1,2} = \frac{174*176}{672} = 45.57$, $E_{1,3} = \frac{174*219}{672} = 56.71$ and $E_{2,1} = \frac{498*277}{672} = 205.28$. Other expected frequencies are computed in a similar manner. All expected frequencies are written in bold in parenthesis in the table.
(d) The observed value of the test statistic is

$$\chi^2_{obs} = \frac{(81-71.72)^2}{71.72} + \frac{(35-45.57)^2}{45.57} \cdots + \frac{(161-162.29)^2}{162.29} = 4.97$$

(e) $df = (2-1) \times (3-1) = 2$, with $\alpha = 0.05$, the critical value is $\chi^2_{0.05} = 5.991$.
(f) Since $\chi^2_{obs} = 4.97 < \chi^2_{0.05} = 5.991$, we fail to reject the null hypothesis at 5% level of significance.

Implementing Technology for a Chi-square Test of Homogeneity

Let us now consider implementing a Chi-square test of homogeneity of proportions using Minitab and TI-84 Plus calculator.

Minitab Instruction for Chi-square Test of Homogeneity

Follow the steps below to implement a Chi-square test of homogeneity of proportions in Minitab (**ref: Example 13.5**):

(1) Type Machine in C1 and Count in C2 columns as before in section 13.2.
(2) Select **Stats>Tables>Chi-Square Goodness-of-fit Test (One Variable)…**
(3) For Observed counts: type **Count** and Check mark **Equal proportions** and click OK.

See the dialog box option selections and output from the Minitab session window in Figure 13.5.

TI 84 Plus Instruction for Chi-squared Test of Homogeneity

Below are TI 84 Plus steps to perform a Chi-squared test of homogeneity (**ref: Example 13.5**).

(1) Enter Observed count in L1 and Expected count in L2.
(2) Press STAT, move the cursor to TESTS and select $D{:}\chi^2 GOF - Test$ … (using ALPHA x^{-1} [D]) and Press ENTER.
(3) Make sure to have Observed: L1, Expected: L2, and df: 4.
(4) Move the cursor to Calculate and press ENTER.

See the option selections and the output for above steps in Figure 13.6.

Figure 13.5: Minitab dialog box option selections and output for the Chi-square test of homogeneity for Example 13.5

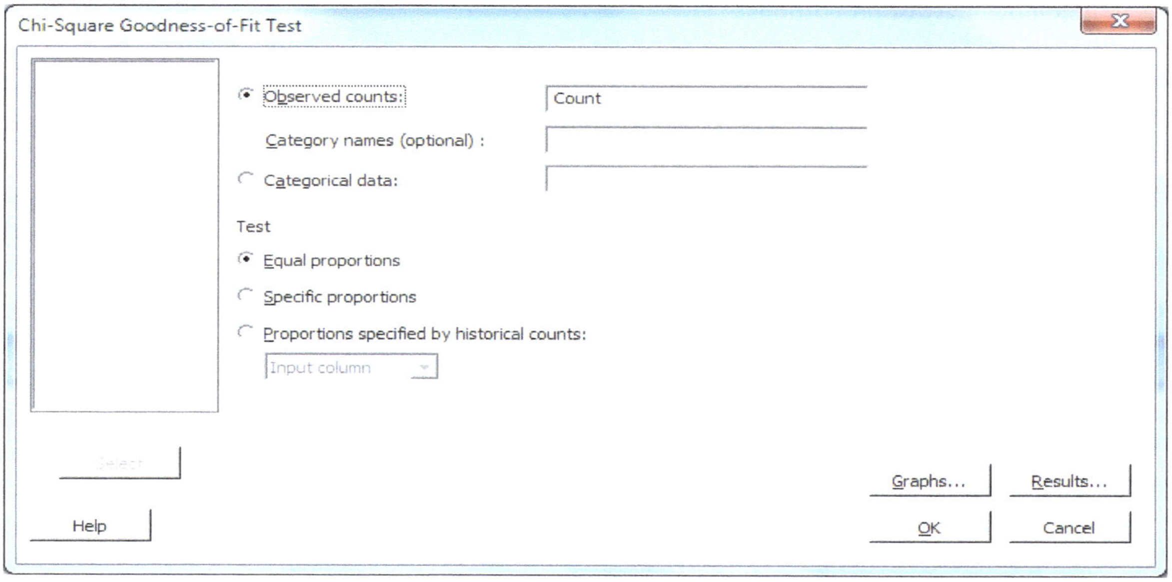

Chi-Square Goodness-of-Fit Test for Observed Counts in Variable: Count

Category	Observed	Test Proportion	Expected	Contribution to Chi-Sq
1	9	0.2	10	0.1
2	10	0.2	10	0.0
3	8	0.2	10	0.4
4	11	0.2	10	0.1
5	12	0.2	10	0.4

N	DF	Chi-Sq	P-Value
50	4	1	0.910

Figure 13.6: TI 84 plus option selections and output for the Chi-square test of homogeneity for Example 13.5.

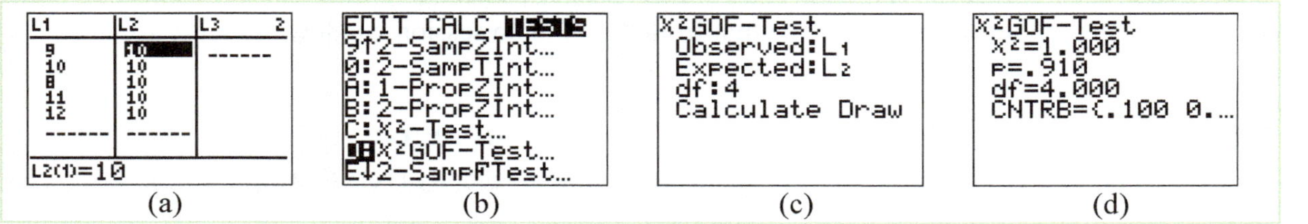

From the results of Minitab and TI 84 Plus calculator output, it has found that the p-value for the test is $0.910 > \alpha = 0.05$. Therefore, we to fail to reject the null hypothesis at 5% level of significance, the same conclusion that was made using the critical value approach.

Chapter 13: Practice problems

(13.1) Explain the purpose of the following tests:
- (a) test of independence
- (b) test of goodness of fit
- (c) test of homogeneity of proportions

(13.2) Given a contingency table, find degrees of freedom, critical value and decide whether the null hypothesis of independence of two categorical variables should be rejected or not for each of the following $R \times C$ contingency tables at $\alpha = 0.05$:
- (a) R=2, C=2, $\chi^2_{obs} = 4.33$
- (b) R=2, C=2, $\chi^2_{obs} = 3.21$
- (c) R=2, C=3, $\chi^2_{obs} = 5.01$
- (d) R=3, C=2, $\chi^2_{obs} = 7.24$
- (e) R=3, C=3, $\chi^2_{obs} = 5.27$
- (f) R=3, C=3, $\chi^2_{obs} = 11.21$
- (g) R=3, C=2, $\chi^2_{obs} = 2.51$
- (h) R=2, C=3, $\chi^2_{obs} = 6.25$

(13.3) Given a contingency table, find degrees of freedom, critical value and decide whether the null hypothesis of independence of two categorical variables should be rejected or not for each of the following $R \times C$ contingency tables at $\alpha = 0.1$:
- (a) R=2, C=2, $\chi^2_{obs} = 2.53$
- (b) R=2, C=2, $\chi^2_{obs} = 7.99$
- (c) R=2, C=3, $\chi^2_{obs} = 10.02$
- (d) R=3, C=2, $\chi^2_{obs} = 3.92$
- (e) R=3, C=3, $\chi^2_{obs} = 5.27$
- (f) R=3, C=3, $\chi^2_{obs} = 14.85$
- (g) R=3, C=2, $\chi^2_{obs} = 2.53$

(h) R=2, C=3, $\chi^2_{obs} = 12.25$

(13.4) For Exercise (13.3) (a)-(h), find degrees of freedom, critical value and decide whether the null hypothesis of independence of two categorical variables should be rejected or not at $\alpha = 0.01$.

(13.5) Find p-value for each of Exercises (13.2) (a)-(h) using TI 84 plus calculator and decide whether the null hypothesis of independence of two categorical variables should be rejected at $\alpha = 0.05$.

(13.6) Find p-value for each of Exercises (13.3) (a)-(h) using TI 84 plus calculator and decide whether the null hypothesis of independence of two categorical variables should be rejected at $\alpha = 0.1$ and 0.01.

(13.7) On the basis of a sample of 150 infant deaths occurred under age 1 year, the following 2 by 2 contingency table has been observed classified by gender and type of infant death, neonatal (death under 28 days) and post-neonatal (28 days- 11 months) .

	Infant death		
Gender	Neonatal	Post-neonatal	Total
Male	57	28	85
Female	49	16	65
Total	106	44	150

Source: A multinomial sample using the data of Table D, National Vital Statistics Reports; Vol. 61, No. 4, May 8, 2013.

To test if the levels of gender and infant death are independent,

(a) Specify null and alternative hypotheses for the test.
(b) Find expected frequency of each cell.
(c) Specify test statistic and find the observed value of the test statistic.
(d) Find the degrees of freedom of the test statistic.
(e) Conclude about the test by the critical value approach at 5% level of significance.

(13.8) As of January, 2011, the percent distribution of Facebook users by age group is given in the following table.

	Facebook users	
age group	Percent (2011)	Count (2014)
13-17	9%	6
18-24	31%	27
25-34	22%	30
35-54	27%	38
55+	11%	19

Denoting five age groups by 1, 2, 3, 4 and 5, respectively, we want to test if the percentage distribution of the face book users in January 2014 is the same as that of 2011. In other words, among January 2014 users of Facebook, we want to test

$$H_0: p_1 = 0.09, p_2 = 0.31, p_3 = 0.22, p_4 = 0.27, p_5 = 0.11 \; versus$$

$$H_a\text{: } at\ least\ one\ of\ the\ equalities\ does\ not\ hold$$

On the basis of a sample of 120 facebook users, the distribution of the number of users in January 2014 appears in the last column in the table.

Source: Count (2014) is a multinomial sample of size 120 using population percentage as specified in http://istrategylabs.com/2014/01/3-million-teens-leave-facebook-in-3-years-the-2014-facebook-demographic-report/.

(a) What is the test is about?
(b) Write out the appropriate test statistic for the test.
(c) Find the observed value of the test statistic and df.
(d) Make a conclusion of the test at $\alpha = 0.05$.

(13.9) In reference to the problem in (13.8), the researcher wants to test

$$H_0: p_1 = p_2 = p_3 = p_4 = p_5 = 0.20\ versus$$
$$H_a\text{: } at\ least\ one\ of\ the\ equalities\ does\ not\ hold$$

at $\alpha = 0.05$ for January 2014 Facebook users.

(a) What is the name of the test?
(b) Specify the test statistic for the test.
(c) Find the observed value of the test statistic.
(d) Find out the df of the test statistic.
(e) Make a conclusion about the test at 5% level of significance.

(13.10) According to the 2014 American Red Cross, the percentage distribution of blood type among Caucasians in the USA is as follows:

Blood Type:	O	A	B	AB
Percentage	45%	40%	11%	4%

Source: http://www.redcrossblood.org/learn-about-blood/blood-types; 2014 American Red Cross, Blood Types and the Populations

A certain blood donation center wants to test if the Caucasian people who recently donated blood have the specified proportions of US people. Suppose on the basis of a sample of 120 Caucasians who donated blood recently, the following frequency distribution is observed:

Blood Type:	O	A	B	AB
Frequency:	50	49	14	7

(a) Specify the null and alternative hypothesis for the test.
(b) What do you call the test?
(c) Specify the test statistic and find the observed value of the test statistic.
(d) Find out the df of the test statistic and make a conclusion about the test at 5% level of significance.

(13.11) A recent study reports distribution of women with invasive breast cancer by ethnicity as follows:

Ethnicity	Proportion
Non-Hispanic white	0.72
Hispanic white	0.09
Black	0.11
Others	0.08

Source: *JAMA.* 2015;313(2):165-173. doi:10.1001/jama.2014.17322

A group of physicians are interested to see if the result is consistent with the facility they are associated with. Suppose, in a sample of 125 women with invasive breast cancer they found the following distribution:

Ethnicity	Frequency
Non-Hispanic white	85
Hispanic white	12
Black	15
Others	13

How will you test if these data are consistent with the study result? Perform the test at 5% level of significance.

(13.12) Following a recent study, the table below provides the proportion of people with Alzheimer's disease in the United States by age.

Age of diagnosis	Proportion
85+ years	0.38
75-84 years	0.44
65-74 years	0.15
<65 years	0.04

The proportions may not add to exactly 1 due to the rounding.
Source: Alzheimer's Association, 2014 Alzheimer's Disease Facts and Figures, *Alzheimer's & Dementia,* Volume 10, Issue 2. doi:10.1016/j.jalz.2014.02.001

Suppose, in a sample of 145 patients diagnosed with Alzheimer's disease in a treatment facility, we observed the following distribution:

Age of diagnosis	Frequency
85+ years	50
75-84 years	60
65-74 years	20
<65 years	15

How will you test if these data are consistent with the study result? Perform the test at 5% level of significance.

(13.13) A recent study found the following proportional distribution of patients with small cell lung cancer (SCLC) who have at least five years of survival by their stage of diagnosis:

Stage	Proportion
I	0.52
II	0.32
III	0.13
IV	0.03

Source: Derived from SEER information at http://www.cancer.org/cancer/lungcancer-smallcell/detailedguide/small-cell-lung-cancer-survival-rates

A researcher wants to see if those who have been treated in a given facility are consistent with the study result. Suppose the researcher observed the following distribution of SCLC patients with at least 5 years of survival by their stage of diagnosis:

Stage	Frequency
I	90
II	58
III	25
IV	7

How will you test if the data is consistent with the study result? Perform the test at 5% level of significance.

(13.14) Native to tropical America, *Mirabilis jalapa* plants are well-known as four-o'clocks because their flowers open in the late afternoon. Controlled by a single gene locus with two alleles expressing incomplete dominance, individual four-o'clock plants can have flowers in red, white or pink: homozygotes (RR) with one allele are red flowers, and homozygotes with other allele (rr) are white flowers and heterozygotes (Rr) are pink flowers. By the Mendelian genetic principles, self-pollinated pink-flowered plants should produce progeny with red, white and pink flowers in a 1:1:2 ratio.

A researcher observed 25 red flowers, 17 white flowers and 48 pink flowers among 90 progeny of self-pollinated pink-flowered plants. The researcher is interested to know if these data are conforms to the Mendelian model.

Answer the following questions to justify the consistency of the data with Mendelian genetic model.

(a) What kind of test is this problem about?
(b) Specify null and alternative hypotheses for this test.
(c) Write out the observed frequency using the O_i notation.
(d) Find the expected frequency using E_i notation.
(e) Write out the Chi-square test statistic and find the value of the statistic.
(f) Find the degrees of freedom and critical value at $\alpha = 0.05$ for the test.

(g) Make your conclusion about the test at $\alpha = 0.05$ (i.e., 5% level of significance).

(13.15) Three independent samples of 277 Non-Hispanic white (NH White), 176 Non-Hispanic black (NH Black) and 219 Hispanic people aged 12–15 years are considered. The table below provides the frequency distribution of three samples with the status of TV viewing per day.

2 hours or less of viewing	NH White	NH black	Hispanic	Total
Yes	182	94	150	174
No	95	82	69	498
Total	277	176	219	672

Sources: CDC/NCHS, National Health and Nutrition Examination Survey (NHANES) and NHANES National Youth Fitness Survey, 2012.
http://www.cdc.gov/nchs/data/databriefs/db157_table.pdf#3

We wish to test if the proportion of people with 2 hours or less of TV viewing, per day, are the same. Let p_1, p_2 and p_3 be true proportions of NH White, NH Black and Hispanic people aged 12–15 years with 2 hours or less of TV viewing per day, respectively. Test

$$H_0: p_1 = p_2 = p_3 \; versus$$
$$H_a: p_1, p_2 \; and \; p_3 \; are \; not \; all \; equal$$

at 5% level of significance.

(a) What is the name of the test or what do we call this test?
(b) Find expected frequencies for all observed cell frequencies.
(c) Write out the test statistic for the test.
(d) Find the observed value of the test statistic.
(e) Find the df and the critical value at 5% level of significance.
(f) What is your conclusion about the test at 5% level of significance?

Chapter 13: Answer to selected problems

(13.2)

(a) $df = 1$, $\chi^2_{0.05} = 3.841$. Since $\chi^2_{obs} = 4.33 > \chi^2_{0.05}$, we reject the null hypothesis of independence at $\alpha = 0.05$.

(b) $df = 1$, $\chi^2_{0.05} = 3.841$. Since $\chi^2_{obs} = 3.21 < \chi^2_{0.05}$, we fail to reject the null hypothesis of independence at $\alpha = 0.05$.

(c) $df = 2$, $\chi^2_{0.05} = 5.991$. Since $\chi^2_{obs} = 5.01 < \chi^2_{0.05}$, we fail to reject the null hypothesis of independence at $\alpha = 0.05$.

(d) $df = 2$, $\chi^2_{0.05} = 5.991$. Since $\chi^2_{obs} = 7.24 > \chi^2_{0.05}$, we reject the null hypothesis of independence at $\alpha = 0.05$.

(e) $df = 4$, $\chi^2_{0.05} = 9.488$. Since $\chi^2_{obs} = 5.27 < \chi^2_{0.05}$, we fail to reject the null hypothesis of independence at $\alpha = 0.05$.

(f) $df = 4$, $\chi^2_{0.05} = 9.488$. Since $\chi^2_{obs} = 11.21 > \chi^2_{0.05}$, we reject the null hypothesis of independence at $\alpha = 0.05$.

(g) $df = 2$, $\chi^2_{0.05} = 5.991$. Since $\chi^2_{obs} = 2.51 < \chi^2_{0.05}$, we fail to reject the null hypothesis of independence at $\alpha = 0.05$.

(h) $df = 2$, $\chi^2_{0.05} = 5.991$. Since $\chi^2_{obs} = 6.25 > \chi^2_{0.05}$, we reject the null hypothesis of independence at $\alpha = 0.05$.

(13.5)

(a) $df = 1$ $\chi^2_{obs} = 4.33$, p-value=$P(\chi^2 > 4.33) = 0.0374$. Since p-value$< \alpha = 0.05$, we reject the null hypothesis of independence at $\alpha = 0.05$.

(b) $df = 1$, $\chi^2_{obs} = 3.21$, p-value=$P(\chi^2 > 3.21) = 0.0732$. Since p-value$> \alpha = 0.05$, we fail to reject the null hypothesis of independence at $\alpha = 0.05$.

(c) $df = 2$, $\chi^2_{obs} = 5.01$, p-value=$P(\chi^2 > 5.01) = 0.0817$. Since p-value$> \alpha = 0.05$, we fail to reject the null hypothesis of independence at $\alpha = 0.05$.

(d) $df = 2$, $\chi^2_{obs} = 7.24$, p-value=$P(\chi^2 > 7.24) = 0.0268$. Since p-value$<\alpha = 0.05$, we reject the null hypothesis of independence at $\alpha = 0.05$.

(e) $df = 4$, $\chi^2_{obs} = 5.27$, p-value=$P(\chi^2 > 5.01) = 0.2607$. Since p-value$> \alpha = 0.05$, we fail to reject the null hypothesis of independence at $\alpha = 0.05$.

(f) $df = 4$, $\chi^2_{obs} = 11.21$, p-value=$P(\chi^2 > 11.21) = 0.0243$. Since p-value$< \alpha = 0.05$, we reject the null hypothesis of independence at $\alpha = 0.05$.

(g) $df = 2$, $\chi^2_{obs} = 2.51$, p-value=$P(\chi^2 > 2.51) = 0.2851$. Since p-value$> \alpha = 0.05$, we fail to reject the null hypothesis of independence at $\alpha = 0.05$.

(h) $df = 2$, $\chi^2_{obs} = 6.25$, p-value=$P(\chi^2 > 6.25) = 0.0439$. Since p-value$< \alpha = 0.05$, we reject the null hypothesis of independence at $\alpha = 0.05$.

(13.7) (a) H_0: Gender and infant death are independent versus H_a: Gender and infant death are not independent (b) See the expected frequency in the table below of the Minitab output (c)Try yourself, $\chi^2_{obs} = 1.232$ (d) $df = 1$ (e) Fail to reject H_0 at $\alpha = 0.05$.

```
Rows: Gender   Columns: Infant death
                  Neonatal  Post-neonatal

Male          60.07          24.93
Female        45.93          19.07

Cell Contents:       Expected count

Pearson Chi-Square = 1.232, DF = 1, P-Value = 0.267
Likelihood Ratio Chi-Square = 1.244, DF = 1, P-Value = 0.265
```

(13.8) (a) Goodness of fit test (b) Try yourself (c) $\chi^2_{obs} = 8.9374$, $df = 4$ (d) and p-value=0.063. So, fail to reject H_0 at $\alpha = 0.05$. See the Minitab output in table below:

```
                    Test                        Contribution
Category  Observed  Proportion  Expected     to Chi-Sq
1                6        0.09      10.8       2.13333
2               27        0.31      37.2       2.79677
3               30        0.22      26.4       0.49091
4               38        0.27      32.4       0.96790
5               19        0.11      13.2       2.54848

  N  DF    Chi-Sq  P-Value
120   4  8.93740    0.063
```

(13.9) (a) Test of homogeneity of proportions (b) Try yourself (c) $\chi^2_{obs} = 24.5833$ (d) $df = 4$ (e) p-value=0.000. So, reject H_0 at $\alpha = 0.05$. See the Minitab output in table below:

```
                    Test                Contribution
Category  Observed  Proportion  Expected     to Chi-Sq
1                6         0.2        24       13.5000
2               27         0.2        24        0.3750
3               30         0.2        24        1.5000
4               38         0.2        24        8.1667
5               19         0.2        24        1.0417
  N  DF    Chi-Sq  P-Value
120   4  24.5833    0.000
```

(13.10)-(13.14) are same as (13.8).

APPENDIX FOR STATISTICAL TABLES

In this appendix, we add some statistical tables that are popularly used when technology is not available or want to explore fun using tables.

Below is the list of tables:

Table T1: Binomial Probability Table for Selected Values of n and p

Table T2: Poisson Probability Table for Selected Values of Mean

Table T3: Cumulative Area under a Standard Normal Curve

Table T4: t-Table with upper αth percentile t_α

Table T5: χ^2-Table with upper αth percentile χ^2_α

Table T6: F-Table with upper percentile $F_{0.10}$

Table T7: F Table with upper percentile $F_{0.05}$

Table T8: F Table with upper percentile $F_{0.025}$

Table T9: F Table with upper percentile $F_{0.01}$

Table T1: Binomial Probability Table for Selected Values of n and p

$$p(x) = \binom{n}{x} p^x (1-p)^{n-x}; \quad x = 0, 1, 2, \ldots, n$$

$n = 3$	p										
x	0.05	0.10	0.20	0.30	0.40	0.50	0.60	0.70	0.80	0.90	0.95
0	0.8574	0.7290	0.5120	0.3430	0.2160	0.1250	0.0640	0.0270	0.0080	0.0010	0.0001
1	0.1354	0.2430	0.3840	0.4410	0.4320	0.3750	0.2880	0.1890	0.0960	0.0270	0.0071
2	0.0071	0.0270	0.0960	0.1890	0.2880	0.3750	0.4320	0.4410	0.3840	0.2430	0.1354
3	0.0001	0.0010	0.0080	0.0270	0.0640	0.1250	0.2160	0.3430	0.5120	0.7290	0.8574
$n = 4$	p										
x	0.05	0.10	0.20	0.30	0.40	0.50	0.60	0.70	0.80	0.90	0.95
0	0.8145	0.6561	0.4096	0.2401	0.1296	0.0625	0.0256	0.0081	0.0016	0.0001	0
1	0.1715	0.2916	0.4096	0.4116	0.3456	0.2500	0.1536	0.0756	0.0256	0.0036	0.0005
2	0.0135	0.0486	0.1536	0.2646	0.3456	0.3750	0.3456	0.2646	0.1536	0.0486	0.0135
3	0.0005	0.0036	0.0256	0.0756	0.1536	0.2500	0.3456	0.4116	0.4096	0.2916	0.1715
4	0	0.0001	0.0016	0.0081	0.0256	0.06250	0.1296	0.2401	0.4096	0.6561	0.8145
$n = 5$	p										
x	0.05	0.10	0.20	0.30	0.40	0.50	0.60	0.70	0.80	0.90	0.95
0	0.7738	0.5905	0.3277	0.1681	0.0778	0.0312	0.0102	0.0024	0.0003	0	0
1	0.2036	0.3280	0.4096	0.3602	0.2592	0.1562	0.0768	0.0283	0.0064	0.0004	0
2	0.0214	0.0729	0.2048	0.3087	0.3456	0.3125	0.2304	0.1323	0.0512	0.0081	0.0011
3	0.0011	0.0081	0.0512	0.1323	0.2304	0.3125	0.3456	0.3087	0.2048	0.0729	0.0214
4	0	0.0005	0.0064	0.0284	0.0768	0.1562	0.2592	0.3602	0.4096	0.3280	0.2036
5	0	0	0.0003	0.0024	0.0102	0.0312	0.0778	0.1681	0.3277	0.5905	0.7738
$n = 6$	p										
x	0.05	0.10	0.20	0.30	0.40	0.50	0.60	0.70	0.80	0.90	0.95
0	0.7351	0.5314	0.2621	0.1176	0.0467	0.0156	0.0041	0.0007	0.0001	0	0
1	0.2321	0.3543	0.3932	0.3025	0.1866	0.0937	0.0369	0.0102	0.0015	0.0001	0
2	0.0305	0.0984	0.2458	0.3241	0.3110	0.2344	0.1382	0.0595	0.0154	0.0012	0.0001
3	0.0021	0.0146	0.0819	0.1852	0.2765	0.3125	0.2765	0.1852	0.0819	0.0146	0.0021
4	0.0001	0.0012	0.0154	0.0595	0.1382	0.2344	0.3110	0.3241	0.2458	0.0984	0.0305
5	0	0.0001	0.0015	0.0102	0.0369	0.0938	0.1866	0.3025	0.3932	0.3543	0.2321
6	0	0	0.0001	0.0007	0.0041	0.0156	0.0467	0.1176	0.2621	0.5314	0.7351
$n = 7$	p										
x	0.05	0.10	0.20	0.30	0.40	0.50	0.60	0.70	0.80	0.90	0.95
0	0.6983	0.4783	0.2097	0.0824	0.0280	0.0078	0.0016	0.0002	0	0	0
1	0.2573	0.3720	0.3670	0.2471	0.1306	0.0547	0.0172	0.0036	0.0004	0	0
2	0.0406	0.1240	0.2753	0.3177	0.2613	0.1641	0.0774	0.0250	0.0043	0.0002	0
3	0.0036	0.0230	0.1147	0.2269	0.2903	0.2734	0.1935	0.0972	0.0287	0.0026	0.0002
4	0.0002	0.0026	0.0287	0.0972	0.1935	0.2734	0.2903	0.2269	0.1147	0.0230	0.0036
5	0	0.0002	0.0043	0.0250	0.0774	0.1641	0.2613	0.3177	0.2753	0.1240	0.0406
6	0	0	0.0004	0.0036	0.0172	0.0547	0.1306	0.2471	0.3670	0.3720	0.2573
7	0	0	0	0.0002	0.0016	0.0078	0.0280	0.0824	0.2097	0.4783	0.6983

Table T1: Binomial Probability Table for Selected Values of n and p

$$p(x) = \binom{n}{x} p^x (1-p)^{n-x};\ \ x = 0, 1, 2, \ldots, n$$

$n = 8$					p						
x	0.05	0.10	0.20	0.30	0.40	0.50	0.60	0.70	0.80	0.90	0.95
0	0.6634	0.4305	0.1678	0.0576	0.0168	0.0039	0.0007	0.0001	0	0	0
1	0.2793	0.3826	0.3355	0.1977	0.0896	0.0313	0.0079	0.0012	0.0001	0	0
2	0.0515	0.1488	0.2936	0.2965	0.209	0.1094	0.0413	0.0100	0.0011	0	0
3	0.0054	0.0331	0.1468	0.2541	0.2787	0.2187	0.1239	0.0467	0.0092	0.0004	0
4	0.0004	0.0046	0.0459	0.1361	0.2322	0.2734	0.2322	0.1361	0.0459	0.0046	0.0004
5	0	0.0004	0.0092	0.0467	0.1239	0.2187	0.2787	0.2541	0.1468	0.0331	0.0054
6	0	0	0.0011	0.0100	0.0413	0.1094	0.2090	0.2965	0.2936	0.1488	0.0515
7	0	0	0.0001	0.0012	0.0079	0.0313	0.0896	0.1977	0.3355	0.3826	0.2793
8	0	0	0	0.0001	0.0007	0.0039	0.0168	0.0576	0.1678	0.4305	0.6634

$n = 9$					p						
x	0.05	0.10	0.20	0.30	0.40	0.50	0.60	0.70	0.80	0.90	0.95
0	0.6302	0.3874	0.1342	0.0404	0.0101	0.002	0.0003	0	0	0	0
1	0.2985	0.3874	0.3020	0.1556	0.0605	0.0176	0.0035	0.0004	0	0	0
2	0.0629	0.1722	0.3020	0.2668	0.1612	0.0703	0.0212	0.0039	0.0003	0	0
3	0.0077	0.0446	0.1762	0.2668	0.2508	0.1641	0.0743	0.0210	0.0028	0.0001	0
4	0.0006	0.0074	0.0661	0.1715	0.2508	0.2461	0.1672	0.0735	0.0165	0.0008	0
5	0	0.0008	0.0165	0.0735	0.1672	0.2461	0.2508	0.1715	0.0661	0.0074	0.0006
6	0	0.0001	0.0028	0.0210	0.0743	0.1641	0.2508	0.2668	0.1762	0.0446	0.0077
7	0	0	0.0003	0.0039	0.0212	0.0703	0.1612	0.2668	0.3020	0.1722	0.0629
8	0	0	0	0.0004	0.0035	0.0176	0.0605	0.1556	0.3020	0.3874	0.2985
9	0	0	0	0	0.0003	0.0020	0.0101	0.0404	0.1342	0.3874	0.6302

n =10					p						
x	0.05	0.10	0.20	0.30	0.40	0.50	0.60	0.70	0.80	0.90	0.95
0	0.5987	0.3487	0.1074	0.0282	0.0060	0.0010	0.0001	0	0	0	0
1	0.3151	0.3874	0.2684	0.1211	0.0403	0.0098	0.0016	0.0001	0	0	0
2	0.0746	0.1937	0.3020	0.2335	0.1209	0.0439	0.0106	0.0014	0.0001	0	0
3	0.0105	0.0574	0.2013	0.2668	0.2150	0.1172	0.0425	0.0090	0.0008	0	0
4	0.0010	0.0112	0.0881	0.2001	0.2508	0.2051	0.1115	0.0368	0.0055	0.0001	0
5	0.0001	0.0015	0.0264	0.1029	0.2007	0.2461	0.2007	0.1029	0.0264	0.0015	0.0001
6	0	0.0001	0.0055	0.0368	0.1115	0.2051	0.2508	0.2001	0.0881	0.0112	0.0010
7	0	0	0.0008	0.0090	0.0425	0.1172	0.2150	0.2668	0.2013	0.0574	0.0105
8	0	0	0.0001	0.0014	0.0106	0.0439	0.1209	0.2335	0.3020	0.1937	0.0746
9	0	0	0	0.0001	0.0016	0.0098	0.0403	0.1211	0.2684	0.3874	0.3151
10	0	0	0	0	0.0001	0.0010	0.0060	0.0282	0.1074	0.3487	0.5987

Table T1: Binomial Probability Table for Selected Values of n and p

$$p(x) = \binom{n}{x} p^x (1-p)^{n-x}; \quad x = 0, 1, 2, \ldots, n$$

n = 15	p										
x	0.05	0.10	0.20	0.30	0.40	0.50	0.60	0.70	0.80	0.90	0.95
0	0.4633	0.2059	0.0352	0.0047	0.0005	0	0	0	0	0	0
1	0.3658	0.3432	0.1319	0.0305	0.0047	0.0005	0	0	0	0	0
2	0.1348	0.2669	0.2309	0.0916	0.0219	0.0032	0.0003	0	0	0	0
3	0.0307	0.1285	0.2501	0.1700	0.0634	0.0139	0.0016	0.0001	0	0	0
4	0.0049	0.0428	0.1876	0.2186	0.1268	0.0417	0.0074	0.0006	0	0	0
5	0.0006	0.0105	0.1032	0.2061	0.1859	0.0916	0.0245	0.0030	0.0001	0	0
6	0	0.0019	0.0430	0.1472	0.2066	0.1527	0.0612	0.0116	0.0007	0	0
7	0	0.0003	0.0138	0.0811	0.1771	0.1964	0.1181	0.0348	0.0035	0	0
8	0	0	0.0035	0.0348	0.1181	0.1964	0.1771	0.0811	0.0138	0.0003	0
9	0	0	0.0007	0.0116	0.0612	0.1527	0.2066	0.1472	0.0430	0.0019	0
10	0	0	0.0001	0.0030	0.0245	0.0916	0.1859	0.2061	0.1032	0.0105	0.0006
11	0	0	0	0.0006	0.0074	0.0417	0.1268	0.2186	0.1876	0.0428	0.0049
12	0	0	0	0.0001	0.0016	0.0139	0.0634	0.1700	0.2501	0.1285	0.0307
13	0	0	0	0	0.0003	0.0032	0.0219	0.0916	0.2309	0.2669	0.1348
14	0	0	0	0	0	0.0005	0.0047	0.0305	0.1319	0.3432	0.3658
15	0	0	0	0	0	0	0.0005	0.0047	0.0352	0.2059	0.4633

Table T1: Binomial Probability Table for Selected Values of n and p

$$p(x) = \binom{n}{x} p^x (1-p)^{n-x}; \quad x = 0, 1, 2, \ldots, n$$

n =20	p										
x	0.05	0.10	0.20	0.30	0.40	0.50	0.60	0.70	0.80	0.90	0.95
0	0.3585	0.1216	0.0115	0.0008	0	0	0	0	0	0	0
1	0.3774	0.2702	0.0576	0.0068	0.0005	0	0	0	0	0	0
2	0.1887	0.2852	0.1369	0.0278	0.0031	0.0002	0	0	0	0	0
3	0.0596	0.1901	0.2054	0.0716	0.0123	0.0011	0	0	0	0	0
4	0.0133	0.0898	0.2182	0.1304	0.0350	0.0046	0.0003	0	0	0	0
5	0.0022	0.0319	0.1746	0.1789	0.0746	0.0148	0.0013	0	0	0	0
6	0.0003	0.0089	0.1091	0.1916	0.1244	0.0370	0.0049	0.0002	0	0	0
7	0	0.0020	0.0545	0.1643	0.1659	0.0739	0.0146	0.0010	0	0	0
8	0	0.0004	0.0222	0.1144	0.1797	0.1201	0.0355	0.0039	0.0001	0	0
9	0	0.0001	0.0074	0.0654	0.1597	0.1602	0.0710	0.0120	0.0005	0	0
10	0	0	0.0020	0.0308	0.1171	0.1762	0.1171	0.0308	0.0020	0	0
11	0	0	0.0005	0.0120	0.0710	0.1602	0.1597	0.0654	0.0074	0.0001	0
12	0	0	0.0001	0.0039	0.0355	0.1201	0.1797	0.1144	0.0222	0.0004	0
13	0	0	0	0.0010	0.0146	0.0739	0.1659	0.1643	0.0545	0.0020	0
14	0	0	0	0.0002	0.0049	0.0370	0.1244	0.1916	0.1091	0.0089	0.0003
15	0	0	0	0	0.0013	0.0148	0.0746	0.1789	0.1746	0.0319	0.0022
16	0	0	0	0	0.0003	0.0046	0.0350	0.1304	0.2182	0.0898	0.0133
17	0	0	0	0	0	0.0011	0.0123	0.0716	0.2054	0.1901	0.0596
18	0	0	0	0	0	0.0002	0.0031	0.0278	0.1369	0.2852	0.1887
19	0	0	0	0	0	0	0.0005	0.0068	0.0576	0.2702	0.3774
20	0	0	0	0	0	0	0	0.0008	0.0115	0.1216	0.3585

Table T2: Poisson Probability Table for Selected Values of Mean

$$p(x) = \frac{e^{-\mu}\mu^{x}}{x!};\ x = 0, 1, 2, \ldots$$

	μ									
x	0.2	0.4	0.6	0.8	1	1.2	1.4	1.6	1.8	2
0	0.8187	0.6703	0.5488	0.4493	0.3679	0.3012	0.2466	0.2019	0.1653	0.1353
1	0.1637	0.2681	0.3293	0.3595	0.3679	0.3614	0.3452	0.3230	0.2975	0.2707
2	0.0164	0.0536	0.0988	0.1438	0.1839	0.2169	0.2417	0.2584	0.2678	0.2707
3	0.0011	0.0072	0.0198	0.0383	0.0613	0.0867	0.1128	0.1378	0.1607	0.1804
4	0.0001	0.0007	0.0030	0.0077	0.0153	0.0260	0.0395	0.0551	0.0723	0.0902
5	0	0.0001	0.0004	0.0012	0.0031	0.0062	0.0111	0.0176	0.0260	0.0361
6	0	0	0	0.0002	0.0005	0.0012	0.0026	0.0047	0.0078	0.0120
7	0	0	0	0	0.0001	0.0002	0.0005	0.0011	0.0020	0.0034
8	0	0	0	0	0	0	0.0001	0.0002	0.0005	0.0009

	μ									
x	2.2	2.4	2.6	2.8	3	3.2	3.4	3.6	3.8	4
0	0.1108	0.0907	0.0743	0.0608	0.0498	0.0408	0.0334	0.0273	0.0224	0.0183
1	0.2438	0.2177	0.1931	0.1703	0.1494	0.1304	0.1135	0.0984	0.0850	0.0733
2	0.2681	0.2613	0.2510	0.2384	0.2240	0.2087	0.1929	0.1771	0.1615	0.1465
3	0.1966	0.2090	0.2176	0.2225	0.2240	0.2226	0.2186	0.2125	0.2046	0.1954
4	0.1082	0.1254	0.1414	0.1557	0.1680	0.1781	0.1858	0.1912	0.1944	0.1954
5	0.0476	0.0602	0.0735	0.0872	0.1008	0.1140	0.1264	0.1377	0.1477	0.1563
6	0.0174	0.0241	0.0319	0.0407	0.0504	0.0608	0.0716	0.0826	0.0936	0.1042
7	0.0055	0.0083	0.0118	0.0163	0.0216	0.0278	0.0348	0.0425	0.0508	0.0595
8	0.0015	0.0025	0.0038	0.0057	0.0081	0.0111	0.0148	0.0191	0.0241	0.0298
9	0.0004	0.0007	0.0011	0.0018	0.0027	0.0040	0.0056	0.0076	0.0102	0.0132
10	0.0001	0.0002	0.0003	0.0005	0.0008	0.0013	0.0019	0.0028	0.0039	0.0053

	μ									
x	4.4	4.6	4.8	5	5.2	5.4	5.6	5.8	6	6.2
0	0.0123	0.0101	0.0082	0.0067	0.0055	0.0045	0.0037	0.0030	0.0025	0.0020
1	0.0540	0.0462	0.0395	0.0337	0.0287	0.0244	0.0207	0.0176	0.0149	0.0126
2	0.1188	0.1063	0.0948	0.0842	0.0746	0.0659	0.0580	0.0509	0.0446	0.0390
3	0.1743	0.1631	0.1517	0.1404	0.1293	0.1185	0.1082	0.0985	0.0892	0.0806
4	0.1917	0.1875	0.1820	0.1755	0.1681	0.1600	0.1515	0.1428	0.1339	0.1249
5	0.1687	0.1725	0.1747	0.1755	0.1748	0.1728	0.1697	0.1656	0.1606	0.1549
6	0.1237	0.1323	0.1398	0.1462	0.1515	0.1555	0.1584	0.1601	0.1606	0.1601
7	0.0778	0.0869	0.0959	0.1044	0.1125	0.1200	0.1267	0.1326	0.1377	0.1418
8	0.0428	0.0500	0.0575	0.0653	0.0731	0.0810	0.0887	0.0962	0.1033	0.1099
9	0.0209	0.0255	0.0307	0.0363	0.0423	0.0486	0.0552	0.0620	0.0688	0.0757
10	0.0092	0.0118	0.0147	0.0181	0.0220	0.0262	0.0309	0.0359	0.0413	0.0469

Table T2: Poisson Probability Table for Selected Values of Mean

$$p(x) = \frac{e^{-\mu}\mu^x}{x!};\ x = 0, 1, 2, \ldots$$

	μ									
x	4.4	4.6	4.8	5	5.2	5.4	5.6	5.8	6	6.2
11	0.0037	0.0049	0.0064	0.0082	0.0104	0.0129	0.0157	0.0190	0.0225	0.0265
12	0.0013	0.0019	0.0026	0.0034	0.0045	0.0058	0.0073	0.0092	0.0113	0.0137
13	0.0005	0.0007	0.0009	0.0013	0.0018	0.0024	0.0032	0.0041	0.0052	0.0065
14	0.0001	0.0002	0.0003	0.0005	0.0007	0.0009	0.0013	0.0017	0.0022	0.0029
15	0.0000	0.0001	0.0001	0.0002	0.0002	0.0003	0.0005	0.0007	0.0009	0.0012

	μ									
x	6.6	6.8	7	7.2	7.4	7.6	7.8	8	8.2	8.4
0	0.0014	0.0011	0.0009	0.0007	0.0006	0.0005	0.0004	0.0003	0.0003	0.0002
1	0.0090	0.0076	0.0064	0.0054	0.0045	0.0038	0.0032	0.0027	0.0023	0.0019
2	0.0296	0.0258	0.0223	0.0194	0.0167	0.0145	0.0125	0.0107	0.0092	0.0079
3	0.0652	0.0584	0.0521	0.0464	0.0413	0.0366	0.0324	0.0286	0.0252	0.0222
4	0.1076	0.0992	0.0912	0.0836	0.0764	0.0696	0.0632	0.0573	0.0517	0.0466
5	0.1420	0.1349	0.1277	0.1204	0.1130	0.1057	0.0986	0.0916	0.0849	0.0784
6	0.1562	0.1529	0.1490	0.1445	0.1394	0.1339	0.1282	0.1221	0.1160	0.1097
7	0.1472	0.1486	0.1490	0.1486	0.1474	0.1454	0.1428	0.1396	0.1358	0.1317
8	0.1215	0.1263	0.1304	0.1337	0.1363	0.1381	0.1392	0.1396	0.1392	0.1382
9	0.0891	0.0954	0.1014	0.1070	0.1121	0.1167	0.1207	0.1241	0.1269	0.1290
10	0.0588	0.0649	0.0710	0.0770	0.0829	0.0887	0.0941	0.0993	0.1040	0.1084
11	0.0353	0.0401	0.0452	0.0504	0.0558	0.0613	0.0667	0.0722	0.0776	0.0828
12	0.0194	0.0227	0.0263	0.0303	0.0344	0.0388	0.0434	0.0481	0.0530	0.0579
13	0.0099	0.0119	0.0142	0.0168	0.0196	0.0227	0.0260	0.0296	0.0334	0.0374
14	0.0046	0.0058	0.0071	0.0086	0.0104	0.0123	0.0145	0.0169	0.0196	0.0225
15	0.0020	0.0026	0.0033	0.0041	0.0051	0.0062	0.0075	0.0090	0.0107	0.0126
16	0.0008	0.0011	0.0014	0.0019	0.0024	0.0030	0.0037	0.0045	0.0055	0.0066
17	0.0003	0.0004	0.0006	0.0008	0.0010	0.0013	0.0017	0.0021	0.0026	0.0033
18	0.0001	0.0002	0.0002	0.0003	0.0004	0.0006	0.0007	0.0009	0.0012	0.0015
19	0	0.0001	0.0001	0.0001	0.0002	0.0002	0.0003	0.0004	0.0005	0.0007
20	0	0	0	0	0.0001	0.0001	0.0001	0.0002	0.0002	0.0003

Table T3: Cumulative Area under a Standard Normal Curve

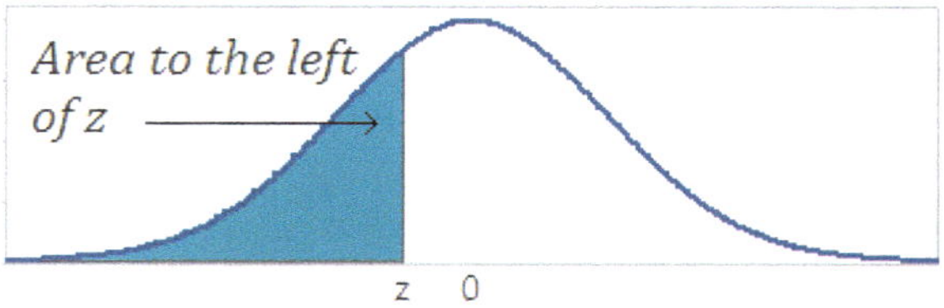

z	.00	.01	.02	.03	.04	.05	.06	.07	.08	.09
-3.3	0.00048	0.00047	0.00045	0.00043	0.00042	0.00040	0.00039	0.00038	0.00036	0.00035
-3.2	0.00069	0.00066	0.00064	0.00062	0.00060	0.00058	0.00056	0.00054	0.00052	0.00050
-3.1	0.00097	0.00094	0.00090	0.00087	0.00084	0.00082	0.00079	0.00076	0.00074	0.00071
-3.0	0.00135	0.00131	0.00126	0.00122	0.00118	0.00114	0.00111	0.00107	0.00104	0.00100
-2.9	0.00187	0.00181	0.00175	0.00169	0.00164	0.00159	0.00154	0.00149	0.00144	0.00139
-2.8	0.00256	0.00248	0.00240	0.00233	0.00226	0.00219	0.00212	0.00205	0.00199	0.00193
-2.7	0.00347	0.00336	0.00326	0.00317	0.00307	0.00298	0.00289	0.00280	0.00272	0.00264
-2.6	0.00466	0.00453	0.00440	0.00427	0.00415	0.00402	0.00391	0.00379	0.00368	0.00357
-2.5	0.00621	0.00604	0.00587	0.00570	0.00554	0.00539	0.00523	0.00508	0.00494	0.00480
-2.4	0.00820	0.00798	0.00776	0.00755	0.00734	0.00714	0.00695	0.00676	0.00657	0.00639
-2.3	0.01072	0.01044	0.01017	0.00990	0.00964	0.00939	0.00914	0.00889	0.00866	0.00842
-2.2	0.01390	0.01355	0.01321	0.01287	0.01255	0.01222	0.01191	0.01160	0.01130	0.01101
-2.1	0.01786	0.01743	0.01700	0.01659	0.01618	0.01578	0.01539	0.01500	0.01463	0.01426
-2.0	0.02275	0.02222	0.02169	0.02118	0.02068	0.02018	0.01970	0.01923	0.01876	0.01831
-1.9	0.02872	0.02807	0.02743	0.02680	0.02619	0.02559	0.02500	0.02442	0.02385	0.02330
-1.8	0.03593	0.03515	0.03438	0.03362	0.03288	0.03216	0.03144	0.03074	0.03005	0.02938
-1.7	0.04457	0.04363	0.04272	0.04182	0.04093	0.04006	0.03920	0.03836	0.03754	0.03673
-1.6	0.05480	0.05370	0.05262	0.05155	0.05050	0.04947	0.04846	0.04746	0.04648	0.04551
-1.5	0.06681	0.06552	0.06426	0.06301	0.06178	0.06057	0.05938	0.05821	0.05705	0.05592
-1.4	0.08076	0.07927	0.07780	0.07636	0.07493	0.07353	0.07215	0.07078	0.06944	0.06811
-1.3	0.09680	0.09510	0.09342	0.09176	0.09012	0.08851	0.08691	0.08534	0.08379	0.08226
-1.2	0.11507	0.11314	0.11123	0.10935	0.10749	0.10565	0.10383	0.10204	0.10027	0.09853
-1.1	0.13567	0.13350	0.13136	0.12924	0.12714	0.12507	0.12302	0.12100	0.11900	0.11702
-1.0	0.15866	0.15625	0.15386	0.15151	0.14917	0.14686	0.14457	0.14231	0.14007	0.13786
-0.9	0.18406	0.18141	0.17879	0.17619	0.17361	0.17106	0.16853	0.16602	0.16354	0.16109
-0.8	0.21186	0.20897	0.20611	0.20327	0.20045	0.19766	0.19489	0.19215	0.18943	0.18673
-0.7	0.24196	0.23885	0.23576	0.23270	0.22965	0.22663	0.22363	0.22065	0.21770	0.21476
-0.6	0.27425	0.27093	0.26763	0.26435	0.26109	0.25785	0.25463	0.25143	0.24825	0.24510
-0.5	0.30854	0.30503	0.30153	0.29806	0.29460	0.29116	0.28774	0.28434	0.28096	0.27760
-0.4	0.34458	0.34090	0.33724	0.33360	0.32997	0.32636	0.32276	0.31918	0.31561	0.31207
-0.3	0.38209	0.37828	0.37448	0.37070	0.36693	0.36317	0.35942	0.35569	0.35197	0.34827
-0.2	0.42074	0.41683	0.41294	0.40905	0.40517	0.40129	0.39743	0.39358	0.38974	0.38591
-0.1	0.46017	0.45620	0.45224	0.44828	0.44433	0.44038	0.43644	0.43251	0.42858	0.42465
-0.0	0.50000	0.49601	0.49202	0.48803	0.48405	0.48006	0.47608	0.47210	0.46812	0.46414

Table T3: Cumulative Area under a Standard Normal Curve

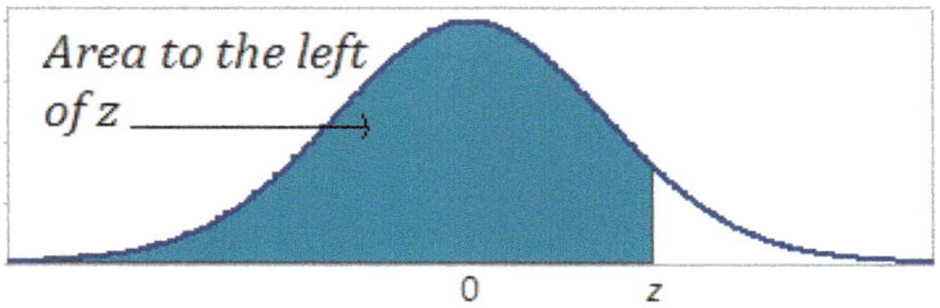

z	.00	.01	.02	.03	.04	.05	.06	.07	.08	.09
0.0	0.50000	0.50399	0.50798	0.51197	0.51595	0.51994	0.52392	0.52790	0.53188	0.53586
0.1	0.53983	0.54380	0.54776	0.55172	0.55567	0.55962	0.56356	0.56749	0.57142	0.57535
0.2	0.57926	0.58317	0.58706	0.59095	0.59483	0.59871	0.60257	0.60642	0.61026	0.61409
0.3	0.61791	0.62172	0.62552	0.62930	0.63307	0.63683	0.64058	0.64431	0.64803	0.65173
0.4	0.65542	0.65910	0.66276	0.66640	0.67003	0.67364	0.67724	0.68082	0.68439	0.68793
0.5	0.69146	0.69497	0.69847	0.70194	0.70540	0.70884	0.71226	0.71566	0.71904	0.72240
0.6	0.72575	0.72907	0.73237	0.73565	0.73891	0.74215	0.74537	0.74857	0.75175	0.75490
0.7	0.75804	0.76115	0.76424	0.76730	0.77035	0.77337	0.77637	0.77935	0.78230	0.78524
0.8	0.78814	0.79103	0.79389	0.79673	0.79955	0.80234	0.80511	0.80785	0.81057	0.81327
0.9	0.81594	0.81859	0.82121	0.82381	0.82639	0.82894	0.83147	0.83398	0.83646	0.83891
1.0	0.84134	0.84375	0.84614	0.84850	0.85083	0.85314	0.85543	0.85769	0.85993	0.86214
1.1	0.86433	0.86650	0.86864	0.87076	0.87286	0.87493	0.87698	0.87900	0.88100	0.88298
1.2	0.88493	0.88686	0.88877	0.89065	0.89251	0.89435	0.89617	0.89796	0.89973	0.90147
1.3	0.90320	0.90490	0.90658	0.90824	0.90988	0.91149	0.91309	0.91466	0.91621	0.91774
1.4	0.91924	0.92073	0.92220	0.92364	0.92507	0.92647	0.92786	0.92922	0.93056	0.93189
1.5	0.93319	0.93448	0.93574	0.93699	0.93822	0.93943	0.94062	0.94179	0.94295	0.94408
1.6	0.94520	0.94630	0.94738	0.94845	0.94950	0.95053	0.95154	0.95254	0.95352	0.95449
1.7	0.95543	0.95637	0.95728	0.95818	0.95907	0.95994	0.96080	0.96164	0.96246	0.96327
1.8	0.96407	0.96485	0.96562	0.96638	0.96712	0.96784	0.96856	0.96926	0.96995	0.97062
1.9	0.97128	0.97193	0.97257	0.97320	0.97381	0.97441	0.97500	0.97558	0.97615	0.97670
2.0	0.97725	0.97778	0.97831	0.97882	0.97932	0.97982	0.98030	0.98077	0.98124	0.98169
2.1	0.98214	0.98257	0.98300	0.98341	0.98382	0.98422	0.98461	0.98500	0.98537	0.98574
2.2	0.98610	0.98645	0.98679	0.98713	0.98745	0.98778	0.98809	0.98840	0.98870	0.98899
2.3	0.98928	0.98956	0.98983	0.99010	0.99036	0.99061	0.99086	0.99111	0.99134	0.99158
2.4	0.99180	0.99202	0.99224	0.99245	0.99266	0.99286	0.99305	0.99324	0.99343	0.99361
2.5	0.99379	0.99396	0.99413	0.99430	0.99446	0.99461	0.99477	0.99492	0.99506	0.99520
2.6	0.99534	0.99547	0.99560	0.99573	0.99585	0.99598	0.99609	0.99621	0.99632	0.99643
2.7	0.99653	0.99664	0.99674	0.99683	0.99693	0.99702	0.99711	0.99720	0.99728	0.99736
2.8	0.99744	0.99752	0.99760	0.99767	0.99774	0.99781	0.99788	0.99795	0.99801	0.99807
2.9	0.99813	0.99819	0.99825	0.99831	0.99836	0.99841	0.99846	0.99851	0.99856	0.99861
3.0	0.99865	0.99869	0.99874	0.99878	0.99882	0.99886	0.99889	0.99893	0.99897	0.99900
3.1	0.99903	0.99906	0.99910	0.99913	0.99916	0.99918	0.99921	0.99924	0.99926	0.99929
3.2	0.99931	0.99934	0.99936	0.99938	0.99940	0.99942	0.99944	0.99946	0.99948	0.99950
3.3	0.99952	0.99953	0.99955	0.99957	0.99958	0.99960	0.99961	0.99962	0.99964	0.99965

Table T4: t-Table with upper αth percentile t_α

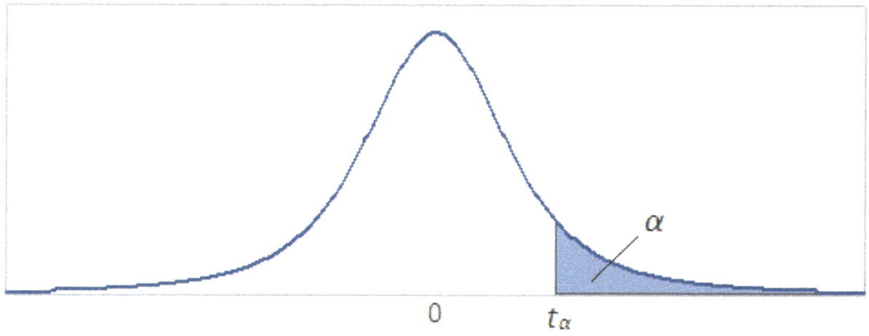

df	$t_{0.1}$	$t_{0.05}$	$t_{0.025}$	$t_{0.01}$	$t_{0.005}$	$t_{0.001}$	$t_{0.0005}$
3	1.638	2.353	3.182	4.541	5.841	10.215	12.924
4	1.533	2.132	2.776	3.747	4.604	7.173	8.610
5	1.476	2.015	2.571	3.365	4.032	5.893	6.869
6	1.440	1.943	2.447	3.143	3.707	5.208	5.959
7	1.415	1.895	2.365	2.998	3.499	4.785	5.408
8	1.397	1.860	2.306	2.896	3.355	4.501	5.041
9	1.383	1.833	2.262	2.821	3.250	4.297	4.781
10	1.372	1.812	2.228	2.764	3.169	4.144	4.587
11	1.363	1.796	2.201	2.718	3.106	4.025	4.437
12	1.356	1.782	2.179	2.681	3.055	3.930	4.318
13	1.350	1.771	2.160	2.650	3.012	3.852	4.221
14	1.345	1.761	2.145	2.624	2.977	3.787	4.140
15	1.341	1.753	2.131	2.602	2.947	3.733	4.073
16	1.337	1.746	2.120	2.583	2.921	3.686	4.015
17	1.333	1.740	2.110	2.567	2.898	3.646	3.965
18	1.330	1.734	2.101	2.552	2.878	3.610	3.922
19	1.328	1.729	2.093	2.539	2.861	3.579	3.883
20	1.325	1.725	2.086	2.528	2.845	3.552	3.850
21	1.323	1.721	2.080	2.518	2.831	3.527	3.819
22	1.321	1.717	2.074	2.508	2.819	3.505	3.792
23	1.319	1.714	2.069	2.500	2.807	3.485	3.768
24	1.318	1.711	2.064	2.492	2.797	3.467	3.745
25	1.316	1.708	2.060	2.485	2.787	3.450	3.725
26	1.315	1.706	2.056	2.479	2.779	3.435	3.707
27	1.314	1.703	2.052	2.473	2.771	3.421	3.690
28	1.313	1.701	2.048	2.467	2.763	3.408	3.674
29	1.311	1.699	2.045	2.462	2.756	3.396	3.659
30	1.310	1.697	2.042	2.457	2.750	3.385	3.646
31	1.309	1.696	2.040	2.453	2.744	3.375	3.633
32	1.309	1.694	2.037	2.449	2.738	3.365	3.622
33	1.308	1.692	2.035	2.445	2.733	3.356	3.611

Table T4: t-Table with upper αth percentile t_α

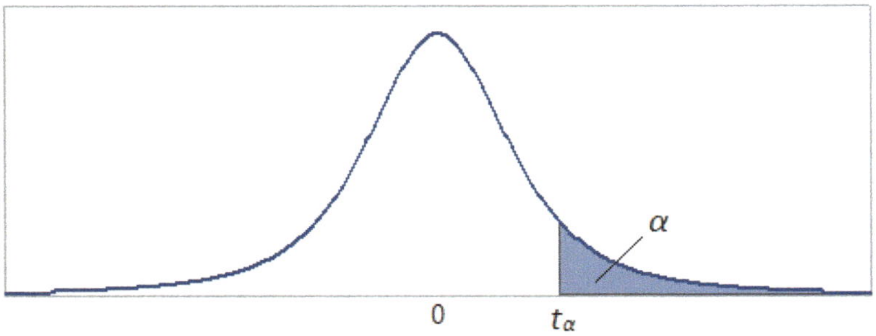

df	$t_{0.1}$	$t_{0.05}$	$t_{0.025}$	$t_{0.01}$	$t_{0.005}$	$t_{0.001}$	$t_{0.0005}$
34	1.307	1.691	2.032	2.441	2.728	3.348	3.601
35	1.306	1.690	2.030	2.438	2.724	3.340	3.591
36	1.306	1.688	2.028	2.434	2.719	3.333	3.582
37	1.305	1.687	2.026	2.431	2.715	3.326	3.574
38	1.304	1.686	2.024	2.429	2.712	3.319	3.566
39	1.304	1.685	2.023	2.426	2.708	3.313	3.558
40	1.303	1.684	2.021	2.423	2.704	3.307	3.551
44	1.301	1.680	2.015	2.414	2.692	3.286	3.526
45	1.301	1.679	2.014	2.412	2.690	3.281	3.520
46	1.300	1.679	2.013	2.410	2.687	3.277	3.515
47	1.300	1.678	2.012	2.408	2.685	3.273	3.510
48	1.299	1.677	2.011	2.407	2.682	3.269	3.505
49	1.299	1.677	2.010	2.405	2.680	3.265	3.500
50	1.299	1.676	2.009	2.403	2.678	3.261	3.496
55	1.297	1.673	2.004	2.396	2.668	3.245	3.476
60	1.296	1.671	2.000	2.390	2.660	3.232	3.460
100	1.290	1.660	1.984	2.364	2.626	3.174	3.390
200	1.286	1.653	1.972	2.345	2.601	3.131	3.340
∞	1.282	1.645	1.960	2.326	2.576	3.090	3.291

Table T5: χ^2-Table with upper αth percentile χ^2_α

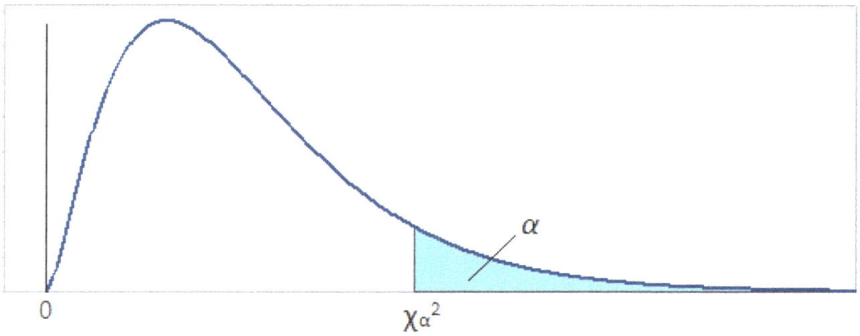

df	$\chi^2_{.995}$	$\chi^2_{.99}$	$\chi^2_{.975}$	$\chi^2_{.95}$	$\chi^2_{.90}$	$\chi^2_{.10}$	$\chi^2_{.05}$	$\chi^2_{.025}$	$\chi^2_{.01}$	$\chi^2_{.005}$
1	0.00	0.00	0.00	0.00	0.02	2.71	3.84	5.02	6.64	7.88
2	0.01	0.02	0.05	0.10	0.21	4.61	5.99	7.38	9.21	10.60
3	0.07	0.12	0.22	0.35	0.58	6.25	7.82	9.35	11.35	12.84
4	0.21	0.30	0.48	0.71	1.06	7.78	9.49	11.14	13.28	14.86
5	0.41	0.55	0.83	1.15	1.61	9.24	11.07	12.83	15.09	16.75
6	0.68	0.87	1.24	1.64	2.20	10.65	12.59	14.45	16.81	18.55
7	0.99	1.24	1.69	2.17	2.83	12.02	14.07	16.01	18.48	20.28
8	1.34	1.65	2.18	2.73	3.49	13.36	15.51	17.54	20.09	21.96
9	1.74	2.09	2.70	3.33	4.17	14.68	16.92	19.02	21.67	23.59
10	2.16	2.56	3.25	3.94	4.87	15.99	18.31	20.48	23.21	25.19
11	2.60	3.05	3.82	4.58	5.58	17.28	19.68	21.92	24.73	26.76
12	3.07	3.57	4.40	5.23	6.30	18.55	21.03	23.34	26.22	28.30
13	3.57	4.11	5.01	5.89	7.04	19.81	22.36	24.74	27.69	29.82
14	4.08	4.66	5.63	6.57	7.79	21.06	23.69	26.12	29.14	31.32
15	4.60	5.23	6.26	7.26	8.55	22.31	25.00	27.49	30.58	32.80
16	5.14	5.81	6.91	7.96	9.31	23.54	26.30	28.85	32.00	34.27
17	5.70	6.41	7.56	8.67	10.09	24.77	27.59	30.19	33.41	35.72
18	6.27	7.02	8.23	9.39	10.87	25.99	28.87	31.53	34.81	37.16
19	6.84	7.63	8.91	10.12	11.65	27.20	30.14	32.85	36.19	38.58
20	7.43	8.26	9.59	10.85	12.44	28.41	31.41	34.17	37.57	40.00
21	8.03	8.90	10.28	11.59	13.24	29.62	32.67	35.48	38.93	41.40
22	8.64	9.54	10.98	12.34	14.04	30.81	33.92	36.78	40.29	42.80
23	9.26	10.20	11.69	13.09	14.85	32.01	35.17	38.08	41.64	44.18
24	9.89	10.86	12.40	13.85	15.66	33.20	36.42	39.36	42.98	45.56
25	10.52	11.52	13.12	14.61	16.47	34.38	37.65	40.65	44.31	46.93
26	11.16	12.20	13.84	15.38	17.29	35.56	38.89	41.92	45.64	48.29
27	11.81	12.88	14.57	16.15	18.11	36.74	40.11	43.20	46.96	49.65
28	12.46	13.57	15.31	16.93	18.94	37.92	41.34	44.46	48.28	50.99
29	13.12	14.26	16.05	17.71	19.77	39.09	42.56	45.72	49.59	52.34
30	13.79	14.95	16.79	18.49	20.60	40.26	43.77	46.98	50.89	53.67

Table T5: χ^2-table with upper αth percentile χ_α^2

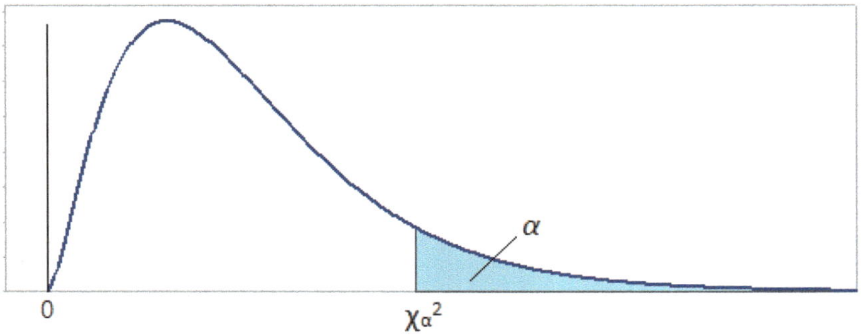

df	$\chi^2_{.995}$	$\chi^2_{.99}$	$\chi^2_{.975}$	$\chi^2_{.95}$	$\chi^2_{.90}$	$\chi^2_{.10}$	$\chi^2_{.05}$	$\chi^2_{.025}$	$\chi^2_{.01}$	$\chi^2_{.005}$
31	14.46	15.66	17.54	19.28	21.43	41.42	44.99	48.23	52.19	55.00
32	15.13	16.36	18.29	20.07	22.27	42.59	46.19	49.48	53.49	56.33
33	15.82	17.07	19.05	20.87	23.11	43.75	47.40	50.73	54.78	57.65
34	16.50	17.79	19.81	21.66	23.95	44.90	48.60	51.97	56.06	58.96
35	17.19	18.51	20.57	22.47	24.80	46.06	49.80	53.20	57.34	60.28
36	17.89	19.23	21.34	23.27	25.64	47.21	51.00	54.44	58.62	61.58
37	18.59	19.96	22.11	24.08	26.49	48.36	52.19	55.67	59.89	62.88
38	19.29	20.69	22.88	24.88	27.34	49.51	53.38	56.90	61.16	64.18
39	20.00	21.43	23.65	25.70	28.20	50.66	54.57	58.12	62.43	65.48
40	20.71	22.16	24.43	26.51	29.05	51.81	55.76	59.34	63.69	66.77
45	24.31	25.90	28.37	30.61	33.35	57.51	61.66	65.41	69.96	73.17
46	25.04	26.66	29.16	31.44	34.22	58.64	62.83	66.62	71.20	74.44
47	25.78	27.42	29.96	32.27	35.08	59.77	64.00	67.82	72.44	75.70
48	26.51	28.18	30.76	33.10	35.95	60.91	65.17	69.02	73.68	76.97
49	27.25	28.94	31.56	33.93	36.82	62.04	66.34	70.22	74.92	78.23
50	27.99	29.71	32.36	34.76	37.69	63.17	67.51	71.42	76.15	79.49
55	31.74	33.57	36.40	38.96	42.06	68.80	73.31	77.38	82.29	85.75
60	35.53	37.49	40.48	43.19	46.46	74.40	79.08	83.30	88.38	91.95
65	39.38	41.44	44.60	47.45	50.88	79.97	84.82	89.18	94.42	98.11
70	43.28	45.44	48.76	51.74	55.33	85.53	90.53	95.02	100.43	104.22
75	47.21	49.48	52.94	56.05	59.80	91.06	96.22	100.84	106.39	110.29
100	67.33	70.07	74.22	77.93	82.36	118.50	124.34	129.56	135.81	140.17
200	152.24	156.43	162.73	168.28	174.84	226.02	233.99	241.06	249.45	255.26

Table T6: F Table with upper αth percentile ($\alpha = 0.10$)

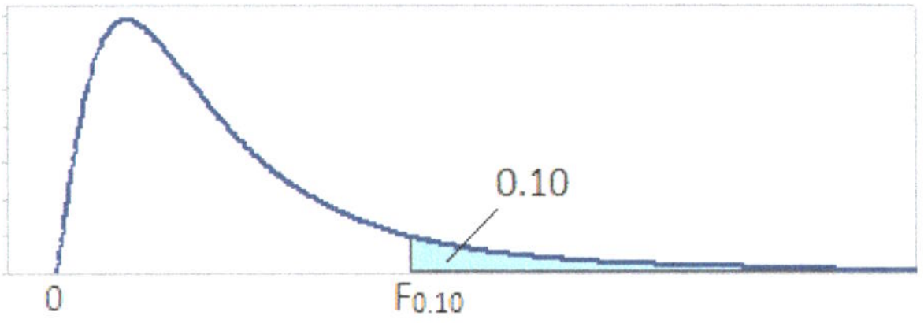

	Numerator degrees of freedom df_1																						
df_2	4	5	6	7	8	9	10	11	12	13	14	15	16	17	18	19	20	21	22	23	24	25	30	40
4	4.11	4.05	4.01	3.98	3.95	3.94	3.92	3.91	3.90	3.89	3.88	3.87	3.86	3.86	3.85	3.85	3.84	3.84	3.84	3.83	3.83	3.83	3.82	3.80
5	3.52	3.45	3.40	3.37	3.34	3.32	3.30	3.28	3.27	3.26	3.25	3.24	3.23	3.22	3.22	3.21	3.21	3.20	3.20	3.19	3.19	3.19	3.17	3.16
6	3.18	3.11	3.05	3.01	2.98	2.96	2.94	2.92	2.90	2.89	2.88	2.87	2.86	2.85	2.85	2.84	2.84	2.83	2.83	2.82	2.82	2.81	2.80	2.78
7	2.96	2.88	2.83	2.78	2.75	2.72	2.70	2.68	2.67	2.65	2.64	2.63	2.62	2.61	2.61	2.60	2.59	2.59	2.58	2.58	2.58	2.57	2.56	2.54
8	2.81	2.73	2.67	2.62	2.59	2.56	2.54	2.52	2.50	2.49	2.48	2.46	2.45	2.45	2.44	2.43	2.42	2.42	2.41	2.41	2.40	2.40	2.38	2.36
9	2.69	2.61	2.55	2.51	2.47	2.44	2.42	2.40	2.38	2.36	2.35	2.34	2.33	2.32	2.31	2.30	2.30	2.29	2.29	2.28	2.28	2.27	2.25	2.23
10	2.61	2.52	2.46	2.41	2.38	2.35	2.32	2.30	2.28	2.27	2.26	2.24	2.23	2.22	2.22	2.21	2.20	2.19	2.19	2.18	2.18	2.17	2.16	2.13
11	2.54	2.45	2.39	2.34	2.30	2.27	2.25	2.23	2.21	2.19	2.18	2.17	2.16	2.15	2.14	2.13	2.12	2.12	2.11	2.11	2.10	2.10	2.08	2.05
12	2.48	2.39	2.33	2.28	2.24	2.21	2.19	2.17	2.15	2.13	2.12	2.10	2.09	2.08	2.08	2.07	2.06	2.05	2.05	2.04	2.04	2.03	2.01	1.99
13	2.43	2.35	2.28	2.23	2.20	2.16	2.14	2.12	2.10	2.08	2.07	2.05	2.04	2.03	2.02	2.01	2.01	2.00	1.99	1.99	1.98	1.98	1.96	1.93
14	2.39	2.31	2.24	2.19	2.15	2.12	2.10	2.07	2.05	2.04	2.02	2.01	2.00	1.99	1.98	1.97	1.96	1.96	1.95	1.94	1.94	1.93	1.91	1.89
15	2.36	2.27	2.21	2.16	2.12	2.09	2.06	2.04	2.02	2.00	1.99	1.97	1.96	1.95	1.94	1.93	1.92	1.92	1.91	1.90	1.90	1.89	1.87	1.85
16	2.33	2.24	2.18	2.13	2.09	2.06	2.03	2.01	1.99	1.97	1.95	1.94	1.93	1.92	1.91	1.90	1.89	1.88	1.88	1.87	1.87	1.86	1.84	1.81
17	2.31	2.22	2.15	2.10	2.06	2.03	2.00	1.98	1.96	1.94	1.93	1.91	1.90	1.89	1.88	1.87	1.86	1.86	1.85	1.84	1.84	1.83	1.81	1.78
18	2.29	2.20	2.13	2.08	2.04	2.00	1.98	1.95	1.93	1.92	1.90	1.89	1.87	1.86	1.85	1.84	1.84	1.83	1.82	1.82	1.81	1.80	1.78	1.75
19	2.27	2.18	2.11	2.06	2.02	1.98	1.96	1.93	1.91	1.89	1.88	1.86	1.85	1.84	1.83	1.82	1.81	1.81	1.80	1.79	1.79	1.78	1.76	1.73
20	2.25	2.16	2.09	2.04	2.00	1.96	1.94	1.91	1.89	1.87	1.86	1.84	1.83	1.82	1.81	1.80	1.79	1.79	1.78	1.77	1.77	1.76	1.74	1.71
21	2.23	2.14	2.08	2.02	1.98	1.95	1.92	1.90	1.87	1.86	1.84	1.83	1.81	1.80	1.79	1.78	1.78	1.77	1.76	1.75	1.75	1.74	1.72	1.69
22	2.22	2.13	2.06	2.01	1.97	1.93	1.90	1.88	1.86	1.84	1.83	1.81	1.80	1.79	1.78	1.77	1.76	1.75	1.74	1.74	1.73	1.73	1.70	1.67
23	2.21	2.11	2.05	1.99	1.95	1.92	1.89	1.87	1.84	1.83	1.81	1.80	1.78	1.77	1.76	1.75	1.74	1.74	1.73	1.72	1.72	1.71	1.69	1.66
24	2.19	2.10	2.04	1.98	1.94	1.91	1.88	1.85	1.83	1.81	1.80	1.78	1.77	1.76	1.75	1.74	1.73	1.72	1.71	1.71	1.70	1.70	1.67	1.64
25	2.18	2.09	2.02	1.97	1.93	1.89	1.87	1.84	1.82	1.80	1.79	1.77	1.76	1.75	1.74	1.73	1.72	1.71	1.70	1.70	1.69	1.68	1.66	1.63
30	2.14	2.05	1.98	1.93	1.88	1.85	1.82	1.79	1.77	1.75	1.74	1.72	1.71	1.70	1.69	1.68	1.67	1.66	1.65	1.64	1.64	1.63	1.61	1.57
40	2.09	2.00	1.93	1.87	1.83	1.79	1.76	1.74	1.71	1.70	1.68	1.66	1.65	1.64	1.62	1.61	1.61	1.60	1.59	1.58	1.57	1.57	1.54	1.51

Table T6: F Table with upper αth percentile ($\alpha = 0.05$)

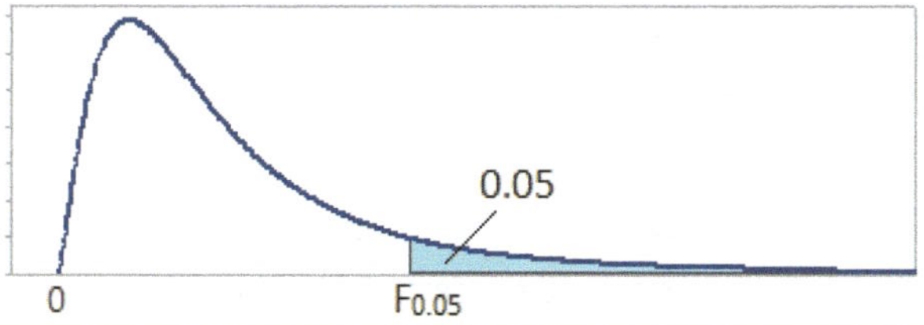

	Numerator degrees of freedom df_1																					
df_2	4	5	6	7	8	9	10	11	12	13	14	15	16	17	18	19	20	21	22	23	24	25	30	40
4	6.39	6.26	6.16	6.09	6.04	6.00	5.96	5.94	5.91	5.89	5.87	5.86	5.84	5.83	5.82	5.81	5.80	5.79	5.79	5.78	5.77	5.77	5.75	5.72
5	5.19	5.05	4.95	4.88	4.82	4.77	4.74	4.70	4.68	4.66	4.64	4.62	4.60	4.59	4.58	4.57	4.56	4.55	4.54	4.53	4.53	4.52	4.50	4.46
6	4.53	4.39	4.28	4.21	4.15	4.10	4.06	4.03	4.00	3.98	3.96	3.94	3.92	3.91	3.90	3.88	3.87	3.86	3.86	3.85	3.84	3.83	3.81	3.77
7	4.12	3.97	3.87	3.79	3.73	3.68	3.64	3.60	3.57	3.55	3.53	3.51	3.49	3.48	3.47	3.46	3.44	3.43	3.43	3.42	3.41	3.40	3.38	3.34
8	3.84	3.69	3.58	3.50	3.44	3.39	3.35	3.31	3.28	3.26	3.24	3.22	3.20	3.19	3.17	3.16	3.15	3.14	3.13	3.12	3.12	3.11	3.08	3.04
9	3.63	3.48	3.37	3.29	3.23	3.18	3.14	3.10	3.07	3.05	3.03	3.01	2.99	2.97	2.96	2.95	2.94	2.93	2.92	2.91	2.90	2.89	2.86	2.83
10	3.48	3.33	3.22	3.14	3.07	3.02	2.98	2.94	2.91	2.89	2.86	2.85	2.83	2.81	2.80	2.79	2.77	2.76	2.75	2.75	2.74	2.73	2.70	2.66
11	3.36	3.20	3.09	3.01	2.95	2.90	2.85	2.82	2.79	2.76	2.74	2.72	2.70	2.69	2.67	2.66	2.65	2.64	2.63	2.62	2.61	2.60	2.57	2.53
12	3.26	3.11	3.00	2.91	2.85	2.80	2.75	2.72	2.69	2.66	2.64	2.62	2.60	2.58	2.57	2.56	2.54	2.53	2.52	2.51	2.51	2.50	2.47	2.43
13	3.18	3.03	2.92	2.83	2.77	2.71	2.67	2.63	2.60	2.58	2.55	2.53	2.51	2.50	2.48	2.47	2.46	2.45	2.44	2.43	2.42	2.41	2.38	2.34
14	3.11	2.96	2.85	2.76	2.70	2.65	2.60	2.57	2.53	2.51	2.48	2.46	2.44	2.43	2.41	2.40	2.39	2.38	2.37	2.36	2.35	2.34	2.31	2.27
15	3.06	2.90	2.79	2.71	2.64	2.59	2.54	2.51	2.48	2.45	2.42	2.40	2.38	2.37	2.35	2.34	2.33	2.32	2.31	2.30	2.29	2.28	2.25	2.20
16	3.01	2.85	2.74	2.66	2.59	2.54	2.49	2.46	2.42	2.40	2.37	2.35	2.33	2.32	2.30	2.29	2.28	2.26	2.25	2.24	2.24	2.23	2.19	2.15
17	2.96	2.81	2.70	2.61	2.55	2.49	2.45	2.41	2.38	2.35	2.33	2.31	2.29	2.27	2.26	2.24	2.23	2.22	2.21	2.20	2.19	2.18	2.15	2.10
18	2.93	2.77	2.66	2.58	2.51	2.46	2.41	2.37	2.34	2.31	2.29	2.27	2.25	2.23	2.22	2.20	2.19	2.18	2.17	2.16	2.15	2.14	2.11	2.06
19	2.90	2.74	2.63	2.54	2.48	2.42	2.38	2.34	2.31	2.28	2.26	2.23	2.21	2.20	2.18	2.17	2.16	2.14	2.13	2.12	2.11	2.11	2.07	2.03
20	2.87	2.71	2.60	2.51	2.45	2.39	2.35	2.31	2.28	2.25	2.22	2.20	2.18	2.17	2.15	2.14	2.12	2.11	2.10	2.09	2.08	2.07	2.04	1.99
21	2.84	2.68	2.57	2.49	2.42	2.37	2.32	2.28	2.25	2.22	2.20	2.18	2.16	2.14	2.12	2.11	2.10	2.08	2.07	2.06	2.05	2.05	2.01	1.96
22	2.82	2.66	2.55	2.46	2.40	2.34	2.30	2.26	2.23	2.20	2.17	2.15	2.13	2.11	2.10	2.08	2.07	2.06	2.05	2.04	2.03	2.02	1.98	1.94
23	2.80	2.64	2.53	2.44	2.37	2.32	2.27	2.24	2.20	2.18	2.15	2.13	2.11	2.09	2.08	2.06	2.05	2.04	2.02	2.01	2.01	2.00	1.96	1.91
24	2.78	2.62	2.51	2.42	2.36	2.30	2.25	2.22	2.18	2.15	2.13	2.11	2.09	2.07	2.05	2.04	2.03	2.01	2.00	1.99	1.98	1.97	1.94	1.89
25	2.76	2.60	2.49	2.40	2.34	2.28	2.24	2.20	2.16	2.14	2.11	2.09	2.07	2.05	2.04	2.02	2.01	2.00	1.98	1.97	1.96	1.96	1.92	1.87
30	2.69	2.53	2.42	2.33	2.27	2.21	2.16	2.13	2.09	2.06	2.04	2.01	1.99	1.98	1.96	1.95	1.93	1.92	1.91	1.90	1.89	1.88	1.84	1.79
40	2.61	2.45	2.34	2.25	2.18	2.12	2.08	2.04	2.00	1.97	1.95	1.92	1.90	1.89	1.87	1.85	1.84	1.83	1.81	1.80	1.79	1.78	1.74	1.69

Table T6: *F* Table with upper αth percentile ($\alpha = 0.025$)

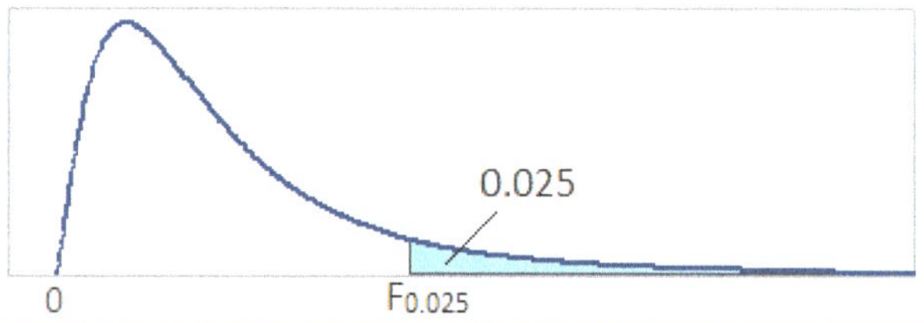

	Numerator degrees of freedom df_1																						
df_2	4	5	6	7	8	9	10	11	12	13	14	15	16	17	18	19	20	21	22	23	24	25	30	40
4	9.60	9.36	9.20	9.07	8.98	8.90	8.84	8.79	8.75	8.71	8.68	8.66	8.63	8.61	8.59	8.58	8.56	8.55	8.53	8.52	8.51	8.50	8.46	8.41
5	7.39	7.15	6.98	6.85	6.76	6.68	6.62	6.57	6.52	6.49	6.46	6.43	6.40	6.38	6.36	6.34	6.33	6.31	6.30	6.29	6.28	6.27	6.23	6.18
6	6.23	5.99	5.82	5.70	5.60	5.52	5.46	5.41	5.37	5.33	5.30	5.27	5.24	5.22	5.20	5.18	5.17	5.15	5.14	5.13	5.12	5.11	5.07	5.01
7	5.52	5.29	5.12	4.99	4.90	4.82	4.76	4.71	4.67	4.63	4.60	4.57	4.54	4.52	4.50	4.48	4.47	4.45	4.44	4.43	4.41	4.40	4.36	4.31
8	5.05	4.82	4.65	4.53	4.43	4.36	4.30	4.24	4.20	4.16	4.13	4.10	4.08	4.05	4.03	4.02	4.00	3.98	3.97	3.96	3.95	3.94	3.89	3.84
9	4.72	4.48	4.32	4.20	4.10	4.03	3.96	3.91	3.87	3.83	3.80	3.77	3.74	3.72	3.70	3.68	3.67	3.65	3.64	3.63	3.61	3.60	3.56	3.51
10	4.47	4.24	4.07	3.95	3.85	3.78	3.72	3.66	3.62	3.58	3.55	3.52	3.50	3.47	3.45	3.44	3.42	3.40	3.39	3.38	3.37	3.35	3.31	3.26
11	4.28	4.04	3.88	3.76	3.66	3.59	3.53	3.47	3.43	3.39	3.36	3.33	3.30	3.28	3.26	3.24	3.23	3.21	3.20	3.18	3.17	3.16	3.12	3.06
12	4.12	3.89	3.73	3.61	3.51	3.44	3.37	3.32	3.28	3.24	3.21	3.18	3.15	3.13	3.11	3.09	3.07	3.06	3.04	3.03	3.02	3.01	2.96	2.91
13	4.00	3.77	3.60	3.48	3.39	3.31	3.25	3.20	3.15	3.12	3.08	3.05	3.03	3.00	2.98	2.96	2.95	2.93	2.92	2.91	2.89	2.88	2.84	2.78
14	3.89	3.66	3.50	3.38	3.29	3.21	3.15	3.09	3.05	3.01	2.98	2.95	2.92	2.90	2.88	2.86	2.84	2.83	2.81	2.80	2.79	2.78	2.73	2.67
15	3.80	3.58	3.41	3.29	3.20	3.12	3.06	3.01	2.96	2.92	2.89	2.86	2.84	2.81	2.79	2.77	2.76	2.74	2.73	2.71	2.70	2.69	2.64	2.59
16	3.73	3.50	3.34	3.22	3.12	3.05	2.99	2.93	2.89	2.85	2.82	2.79	2.76	2.74	2.72	2.70	2.68	2.67	2.65	2.64	2.63	2.61	2.57	2.51
17	3.66	3.44	3.28	3.16	3.06	2.98	2.92	2.87	2.82	2.79	2.75	2.72	2.70	2.67	2.65	2.63	2.62	2.60	2.59	2.57	2.56	2.55	2.50	2.44
18	3.61	3.38	3.22	3.10	3.01	2.93	2.87	2.81	2.77	2.73	2.70	2.67	2.64	2.62	2.60	2.58	2.56	2.54	2.53	2.52	2.50	2.49	2.44	2.38
19	3.56	3.33	3.17	3.05	2.96	2.88	2.82	2.76	2.72	2.68	2.65	2.62	2.59	2.57	2.55	2.53	2.51	2.49	2.48	2.46	2.45	2.44	2.39	2.33
20	3.51	3.29	3.13	3.01	2.91	2.84	2.77	2.72	2.68	2.64	2.60	2.57	2.55	2.52	2.50	2.48	2.46	2.45	2.43	2.42	2.41	2.40	2.35	2.29
21	3.48	3.25	3.09	2.97	2.87	2.80	2.73	2.68	2.64	2.60	2.56	2.53	2.51	2.48	2.46	2.44	2.42	2.41	2.39	2.38	2.37	2.36	2.31	2.25
22	3.44	3.22	3.05	2.93	2.84	2.76	2.70	2.65	2.60	2.56	2.53	2.50	2.47	2.45	2.43	2.41	2.39	2.37	2.36	2.34	2.33	2.32	2.27	2.21
23	3.41	3.18	3.02	2.90	2.81	2.73	2.67	2.62	2.57	2.53	2.50	2.47	2.44	2.42	2.39	2.37	2.36	2.34	2.33	2.31	2.30	2.29	2.24	2.18
24	3.38	3.15	2.99	2.87	2.78	2.70	2.64	2.59	2.54	2.50	2.47	2.44	2.41	2.39	2.36	2.35	2.33	2.31	2.30	2.28	2.27	2.26	2.21	2.15
25	3.35	3.13	2.97	2.85	2.75	2.68	2.61	2.56	2.51	2.48	2.44	2.41	2.38	2.36	2.34	2.32	2.30	2.28	2.27	2.26	2.24	2.23	2.18	2.12
30	3.25	3.03	2.87	2.75	2.65	2.57	2.51	2.46	2.41	2.37	2.34	2.31	2.28	2.26	2.23	2.21	2.20	2.18	2.16	2.15	2.14	2.12	2.07	2.01
40	3.13	2.90	2.74	2.62	2.53	2.45	2.39	2.33	2.29	2.25	2.21	2.18	2.15	2.13	2.11	2.09	2.07	2.05	2.03	2.02	2.01	1.99	1.94	1.88

Table T6: F Table with upper αth percentile ($\alpha = 0.01$)

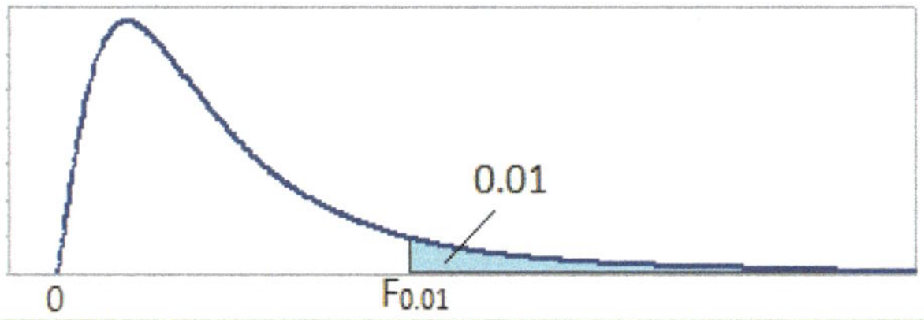

	Numerator degrees of freedom df_1																						
df_2	4	5	6	7	8	9	10	11	12	13	14	15	16	17	18	19	20	21	22	23	24	25	30	40
4	15.98	15.52	15.21	14.98	14.80	14.66	14.55	14.45	14.37	14.31	14.25	14.20	14.15	14.11	14.08	14.05	14.02	13.99	13.97	13.95	13.93	13.91	13.84	13.75
5	11.39	10.97	10.67	10.46	10.29	10.16	10.05	9.96	9.89	9.82	9.77	9.72	9.68	9.64	9.61	9.58	9.55	9.53	9.51	9.49	9.47	9.45	9.38	9.29
6	9.15	8.75	8.47	8.26	8.10	7.98	7.87	7.79	7.72	7.66	7.60	7.56	7.52	7.48	7.45	7.42	7.40	7.37	7.35	7.33	7.31	7.30	7.23	7.14
7	7.85	7.46	7.19	6.99	6.84	6.72	6.62	6.54	6.47	6.41	6.36	6.31	6.28	6.24	6.21	6.18	6.16	6.13	6.11	6.09	6.07	6.06	5.99	5.91
8	7.01	6.63	6.37	6.18	6.03	5.91	5.81	5.73	5.67	5.61	5.56	5.52	5.48	5.44	5.41	5.38	5.36	5.34	5.32	5.30	5.28	5.26	5.20	5.12
9	6.42	6.06	5.80	5.61	5.47	5.35	5.26	5.18	5.11	5.05	5.01	4.96	4.92	4.89	4.86	4.83	4.81	4.79	4.77	4.75	4.73	4.71	4.65	4.57
10	5.99	5.64	5.39	5.20	5.06	4.94	4.85	4.77	4.71	4.65	4.60	4.56	4.52	4.49	4.46	4.43	4.41	4.38	4.36	4.34	4.33	4.31	4.25	4.17
11	5.67	5.32	5.07	4.89	4.74	4.63	4.54	4.46	4.40	4.34	4.29	4.25	4.21	4.18	4.15	4.12	4.10	4.08	4.06	4.04	4.02	4.01	3.94	3.86
12	5.41	5.06	4.82	4.64	4.50	4.39	4.30	4.22	4.16	4.10	4.05	4.01	3.97	3.94	3.91	3.88	3.86	3.84	3.82	3.80	3.78	3.76	3.70	3.62
13	5.21	4.86	4.62	4.44	4.30	4.19	4.10	4.02	3.96	3.91	3.86	3.82	3.78	3.75	3.72	3.69	3.66	3.64	3.62	3.60	3.59	3.57	3.51	3.43
14	5.04	4.69	4.46	4.28	4.14	4.03	3.94	3.86	3.80	3.75	3.70	3.66	3.62	3.59	3.56	3.53	3.51	3.48	3.46	3.44	3.43	3.41	3.35	3.27
15	4.89	4.56	4.32	4.14	4.00	3.89	3.80	3.73	3.67	3.61	3.56	3.52	3.49	3.45	3.42	3.40	3.37	3.35	3.33	3.31	3.29	3.28	3.21	3.13
16	4.77	4.44	4.20	4.03	3.89	3.78	3.69	3.62	3.55	3.50	3.45	3.41	3.37	3.34	3.31	3.28	3.26	3.24	3.22	3.20	3.18	3.16	3.10	3.02
17	4.67	4.34	4.10	3.93	3.79	3.68	3.59	3.52	3.46	3.40	3.35	3.31	3.27	3.24	3.21	3.19	3.16	3.14	3.12	3.10	3.08	3.07	3.00	2.92
18	4.58	4.25	4.01	3.84	3.71	3.60	3.51	3.43	3.37	3.32	3.27	3.23	3.19	3.16	3.13	3.10	3.08	3.05	3.03	3.02	3.00	2.98	2.92	2.84
19	4.50	4.17	3.94	3.77	3.63	3.52	3.43	3.36	3.30	3.24	3.19	3.15	3.12	3.08	3.05	3.03	3.00	2.98	2.96	2.94	2.92	2.91	2.84	2.76
20	4.43	4.10	3.87	3.70	3.56	3.46	3.37	3.29	3.23	3.18	3.13	3.09	3.05	3.02	2.99	2.96	2.94	2.92	2.90	2.88	2.86	2.84	2.78	2.69
21	4.37	4.04	3.81	3.64	3.51	3.40	3.31	3.24	3.17	3.12	3.07	3.03	2.99	2.96	2.93	2.90	2.88	2.86	2.84	2.82	2.80	2.79	2.72	2.64
22	4.31	3.99	3.76	3.59	3.45	3.35	3.26	3.18	3.12	3.07	3.02	2.98	2.94	2.91	2.88	2.85	2.83	2.81	2.78	2.77	2.75	2.73	2.67	2.58
23	4.26	3.94	3.71	3.54	3.41	3.30	3.21	3.14	3.07	3.02	2.97	2.93	2.89	2.86	2.83	2.80	2.78	2.76	2.74	2.72	2.70	2.69	2.62	2.54
24	4.22	3.90	3.67	3.50	3.36	3.26	3.17	3.09	3.03	2.98	2.93	2.89	2.85	2.82	2.79	2.76	2.74	2.72	2.70	2.68	2.66	2.64	2.58	2.49
25	4.18	3.85	3.63	3.46	3.32	3.22	3.13	3.06	2.99	2.94	2.89	2.85	2.81	2.78	2.75	2.72	2.70	2.68	2.66	2.64	2.62	2.60	2.54	2.45
30	4.02	3.70	3.47	3.30	3.17	3.07	2.98	2.91	2.84	2.79	2.74	2.70	2.66	2.63	2.60	2.57	2.55	2.53	2.51	2.49	2.47	2.45	2.39	2.30
40	3.83	3.51	3.29	3.12	2.99	2.89	2.80	2.73	2.66	2.61	2.56	2.52	2.48	2.45	2.42	2.39	2.37	2.35	2.33	2.31	2.29	2.27	2.20	2.11